The Response of Nuclei under Extreme Conditions

ETTORE MAJORANA
INTERNATIONAL SCIENCE SERIES
Series Editor:
Antonino Zichichi
European Physical Society
Geneva, Switzerland

(PHYSICAL SCIENCES)

Recent volumes in the series:

A Continuation Order Plan is available for this series. A continuation order will bring delivery of
each new volume immediately upon publication. Volumes are billed only upon actual shipment.
For further information please contact the publisher.

The Response of Nuclei under Extreme Conditions

Edited by

R. A. Broglia

Università di Milano
Milan, Italy
and
Niels Bohr Institute
Copenhagen, Denmark

and

G. F. Bertsch

Michigan State University
East Lansing, Michigan

Plenum Press • New York and London

Library of Congress Cataloging in Publication Data

International School of Heavy Ion Physics (1986: Erice, Sicily)
    The response of nuclei under extreme conditions.

    (Ettore Majorana international science series. Physical sciences; v. 28)
    "Proceedings of the second course of the International School of Heavy Ion Physics
. . . held October 12–22, 1986, in Erice, Sicily, Italy"—T.p. verso.
    Bibliography: p.
    Includes index.
    1. Heavy ions—Congresses. 2. Heavy ion collisions—Congresses. I. Broglia, R. A.
II. Bertsch, G. III. Title. IV. Series.
QC702.7.H42I63 1986   539.7′2   87-32747
ISBN-13:978-1-4612-8233-4   e-ISBN-13:978-1-4613-0895-9
DOI: 10.1007/978-1-4613-0895-9

Proceedings of the Second Course of the International School of Heavy
Ion Physics on the Response of Nuclei under Extreme Conditions,
held October 12–22, 1986, in Erice, Sicily, Italy

© 1988 Plenum Press, New York
Softcover reprint of the hardcover 1st edition 1988

A Division of Plenum Publishing Corporation
233 Spring Street, New York, N.Y. 10013

DEDICATION

Aage Winther was born sixty years ago in Sao Paolo, Brasil.
He has made his studies at the University of Copenhagen and
from the beginning of his professional career has been associated with the Niels Bohr Institute.

His contributions to the field of Coulomb Excitation and Heavy
Ion Reactions have been seminal, and are well known.

Less known is the central role Aage Winther has played in
bringing clarity and rigor through lectures but specially
through intense, long discussions, on a wide variety of
subjects.

His deep understanding of physics at large and his total com-
mitment to research has made of him an outstanding teacher.
It is my fortune to have been close to him over a long period
of time over which I have profited from his insight, as well
as enjoyed the warm atmosphere that he and Anne Marie have
created for the family and friends.

It is a pleasure to dedicate this Second Course of the International School of Heavy Ion Reactions to celebrate his 60th
birthday.

                                        Ricardo A. Broglia

PARTICIPANTS OF THE SECOND COURSE OF THE INTERNATIONAL SCHOOL OF HEAVY ION PHYSICS, ON THE RESPONSE OF NUCLEI UNDER EXTREME CONDITIONS, HELD OCTOBER 12-22, 1986, IN ERICE, SICILY, ITALY

PREFACE

In recent years, a new field of nuclear research has been opened through the possibility of studying nuclei with very large values of angular momentum, temperature, pressure and number of particles. This development has been closely associated with heavy ion reactions, since collisions between two heavy nuclei are especially effective in producing metastable compound systems with large angular momentum, and in transferring energy which is distributed over the whole nuclear volume.

Under the strain of temperature and of the Coriolis and centrifugal forces, the nucleus displays structural changes which can be interpreted in terms of pairing and shape phase transitions. This was the subject of the lectures of J. D. Garrett, P. J. Twin and S. Levit.

While the rotational motion is, at zero temperature un-damped, the width of giant resonances indicate that the nucleus only oscillates through few periods before the motion is damped by particle decay, and through coupling to the compound nucleus. Temperature and angular momentum influence in an important way the properties of both giant resonances and rotational motion. These subjects were developed by K. Snover, and by P. F. Bortignon and R. A. Broglia, as well as by A. Bracco, A. Dellafiore and F. Matera.

The limits of nuclear stability as a function of mass number at energies not far from the Coulomb barrier, and the interweaving of the variety of degrees of freedom active in the fusion of massive nuclei, were reviewed in the contributions of P. Ambruster, C. Signorini, and A. M. Stefanini.

For those reactions where the two interacting nuclei escape fusion, a variety of regimes have been observed ranging from quasielastic to deep inelastic processes. The interplay between particle transfer and inelastic excitation of collective vibrations, as well as the competition between statistical and memory conserving processes was discussed by M. Baldo, G. Pollarolo and G. Pappalardo.

At energies of the order of the Fermi energy, the interacting ions break in a variety of fragments. The latest results in this field of intermediate energy heavy ion physics were dealt with by G. F. Bertsch, J. Bondorf, J. Knoll and M. Di Toro.

Leaving the Fermi energy regime, relativistic and ultra relativistic heavy ion physics was presented by P. Braun-Munzinger, G. Baur and G. Baym.

The possibilities which will be opened by future heavy ion facilities was discussed by P. Kienle, P. J. Twin, E. Migneco and L. Westerberg.

The overall impression left by the present Course was the richness of the laboratory of many-body physics which heavy ion reactions have opened, and in which phenomena ranging from pairing phase transitions to quark deconfinement can be studied.

It was a privilege to have enjoyed the participation of the outstanding group of lecturers and students which made the second Course of the International School of Heavy Ion Reactions an exciting experience. The efficiency and profesionalism of Dr. A. Gabrielle and his staff contributed in an important way to this result.

We wish especially to thank Prof. A. Zichichi, Director of the Center for Scientific Culture "Ettore Majorana", for the wonderful atmosphere he has created at Erice, and for making this School possible. Last but not least we want to acknowledge the INFN for the financial support provided to the Course.

R. A. Broglia

G. F. Bertsch

CONTENTS

SHAPE AND PAIR CORRELATIONS IN ROTATING NUCLEI

J. D. Garrett

The Niels Bohr Institute
The University of Copenhagen
DK-2100 Copenhagen, Denmark

ABSTRACT

The first lecture considers the difference between the shapes of rotating classical and quantal systems. The experimental determination of nuclear shapes and the variety of shapes encountered in nuclei also is discussed. In the second lecture the bases of pair correlations in spherical and deformed nuclei and its effect on independent particle motion in rotating nuclei is considered. The final lecture explores pair correlations in the limit of rapid rotation, where in the presence of strong Coriolis and centrifugal forces, a transition to an unpaired phase is predicted. In all the lectures a simple heuristic, nonmathematical approach is adhered to, and experimental results are stressed.

PROLOGUE

An advantage of the Sicilian site for this renowned series of schools is the possibility of referring to Greek episodes as well as Roman and Italian sources. Therefore, I would like to start these lectures with the quote that tradition tells us was inscribed on the door to Plato's Academy:

"Let no one enter who is not a mathematician."

Plato, however, was not the only philosopher in ancient Athens. In this series of lectures I would like to adhere to the style of Plato's antithesis. Isocrates, the teacher of many of Greece's great orators, emphasized rhetoric. Indeed ancient and medieval physics was largely narrative. For example, the atomic theory of Epicurus was preserved for modern times by Lucretius' celebrated poem, "On the Nature of Things"[1], and Galileo's fundamental astronomical discoveries were written as dialogues with his contemporary Italians[2]. However from the age of Newton, the second half of the seventeenth century, mathematics became the vocabulary of physics.

In this series of lectures I will adhere to the less modern rhetorical approach trying to convey some exciting recent developments in nuclear spectroscopy in a heuristic manner.

1. THE SHAPE OF RAPIDLY ROTATING OBJECTS - FROM NEWTON TO DARESBURY.

1.1 Classical Objects.

Three hundred years ago Sir Isaac Newton first suggested that a rotating spheroid would become flattened at the poles[3]. This suggestion contradicted the view of the leading contemporary astronomers (e.g. the well-known French astronomer Cassini), who insisted that the best studied macroscopic spheroid, the Earth, was elongated at the poles. They claimed the earth to be a prolate ellipsoid vice Newton's proposed oblate ellipsoid. Fifty years later this controversy was resolved by Maupertuis and Celsius's measurement of the length of one degree of latitude on the surface of the earth in Lapland[4]. This distance compared with previous similar measurements at smaller latitudes fixed the earth to be oblate. Not only is the earth an oblate ellipsoid with an ellipticity of 0.3 percent, but all planets with short rotational periods[5] (Figure 1 and Table 1) and many galaxies[6,7] (Figure 2) are oblate. Indeed, the 10 percent ellipticity of Saturn can be seen with a small telescope - see Figure 1. In fact, if Cassini's observations were sufficient to discover the major gaps in Saturn's rings, why didn't he observe the flattening of Saturn's poles. (This, of course, required determining Saturn's axes of rotation.)

A century later Jacobi predicted[8] for sufficiently rapid rotation that an axially symmetric oblate ellipsoid would become unstable resulting in a triaxial ellipsoidal shape. Cosmic triaxial shapes are observed in the smaller Martian, Jovian, and Saturian satellites[5,9-11] and asteroids[12,13] (Table 2 and Figure 3) and have been proposed for galaxies[6]. At even larger rotational frequencies a rotor is unstable with respect to binary fission[14]. Indeed, such an explanation has been proposed[5,15,16] for Hector, one of the Trojan asteroids.

The question of how cosmic systems obtain large angular momentum remains. One such hypothesis is illustrated in Figure 4. The escape velocity is not sufficiently large for all the fragments of a collision between two asteroids to escape. The remaining fragments reassemble. If the impact parameter of the collision is large enough, the angular momen-

Fig.1. Voyager 2 image of Saturn (from ref.11). The flattened poles characteristic of a rotating classical object can be seen.

Table 1. Ellipticity of the Planets[a]

Planet	Ellipticity[b]
Mercury	0.0
Venus	0.0
Earth	0.0034
Moon	0.002
Mars	0.0059
Jupiter	0.0637
Saturn	0.102
Uranus	0.024
Neptune	0.0266
Pluto	?

a) From ref.[5]
b) Defined as $(R_e - R_p)/R_p$.
R_e and R_p are equatorial and polar radii.

Table 2. Dimensions of some Triaxial Objects in the Solar System[a]

Martian Satellites	Dimensions[b]
Phobos	27.0 x 21.4 x 19.2
Deimos	15 x 12 x 11

Jovian Satellites	
Amalthea	135 x 85 x 77

Saturian Satellites	
1980S27	70 x 50 x 40
1980S26	55 x 45 x 35
1980S3	70 x 60 x 50
1980S1	110 x 100 x 80
1980S13	17 x 14 x 13
1980S6	18 x 16 x 15
Hyperion	205 x 130 x 110

Asteroids	
Metis (9)	122 x 95 x 55
Kalliope (22)	108 x 80 x 65
Laetitia (39)	128 x 75 x 42

a) From refs.[5,9-13]
b) Tabulated dimensions are radii in km

tum of the aggregated system may be adequate to form a triaxial object. Such violent collisions also may perturb an asteroid's orbit sufficient for it to be captured as a satellite by one of the planets. Thus the large number of triaxial objects among the smaller satellites of Jupiter and Saturn is accounted for. Indeed, evidence of violent collisions is preserved in large craters. It is easy to believe that if the collision responsible for the large crater on Mimas[10], shown in Figure 5 had been off center that this satellite would have fragmented. Likewise, the recent Voyager 2 photographs of Miranda[17], Figure 6, support such a picture of fragmentation and reaggregation.

Fig.2. Typical type E6 elliptical galaxy, NGC 205. This galaxy is the
smaller companion of the more familiar spiral Andromeda Galaxy
(M31). The figure is from ref.[7].

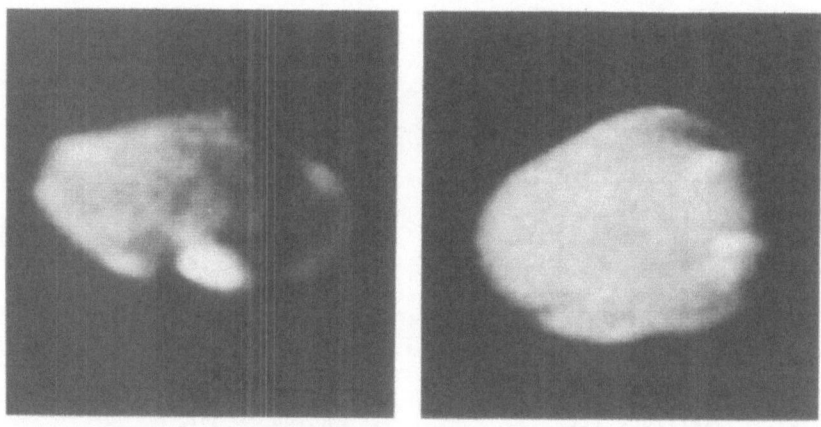

Fig.3. Two Voyager 1 images of the Jovian satellite Amalthea (from ref.[9])
at approximate right-angle aspect. The triaxial nature of this
satellite is indicated.

1.2 The Nuclear Quantum System

1.2.1. Different criteria for the shapes of classical and quantal
systems. In contrast to macroscopic bodies, the atomic nucleus is a
quantal system composed of a finite number of strongly-interacting fermions. The equilibrium shape depends on the detailed microscopic properties of the nuclear quantum system. Indeed, the fact that most stably-
deformed nuclei are prolate, not oblate like the earth, is such a quantal
property. A Nilsson model[18],[19] plot of the spectrum of nuclear states

Asteroids

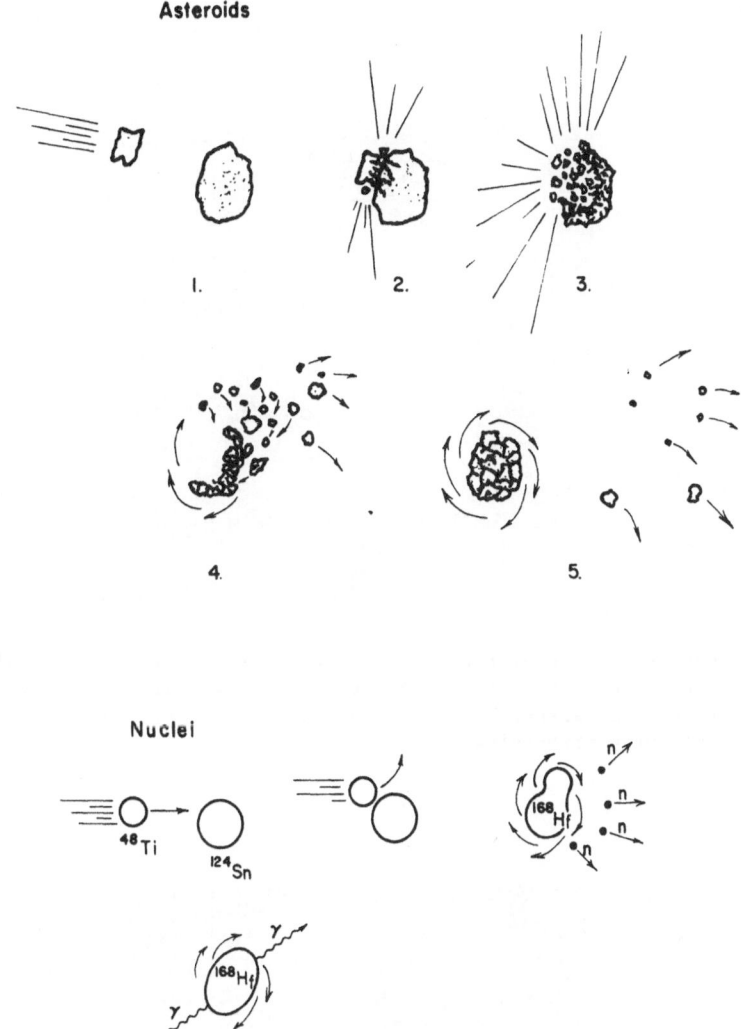

Nuclei

Fig.4. Pedagogical figure comparing the formation of rapidly rotating
classical objects (e.g. asteroids) and nuclei. It has been hypo-
thesized[9,12,13] that this mechanism can impart sufficient angular
momentum to a reaccelerated classical system for the object to
assume a triaxial shape or even to "fission[15,16]". The formation
of the rotating nucleus depicted has been studied at Daresbury
Laboratory[27]. Rapidly-rotating ^{168}Hf$_{96}$ nuclei were formed by
fusing 217 MeV ^{48}Ti with ^{124}Sn after the emission of four neutrons. A level scheme constructed from the gamma-ray decay of
^{168}Hf formed in this manner is shown in Figure 8.

as a function of quadrupole deformation often indicates a lower energy
for prolate shapes. For example, just above the N = 82 closed shell (see
Figure 7) there is a larger density of nuclear states for prolate deformations than for oblate. Thus the single-neutron energy is lowered for
prolate deformations relative to that of spherical or oblate shapes. Not
only the single-neutron and single-proton energies, but also the macro-

Fig.5. Voyager 2 image of the Saturian satellite Mimas (from ref.[10]). It is easy to imagine that had the impact responsible for the large crater been either a bit more violent or a little off center Mimas would have fragmented.

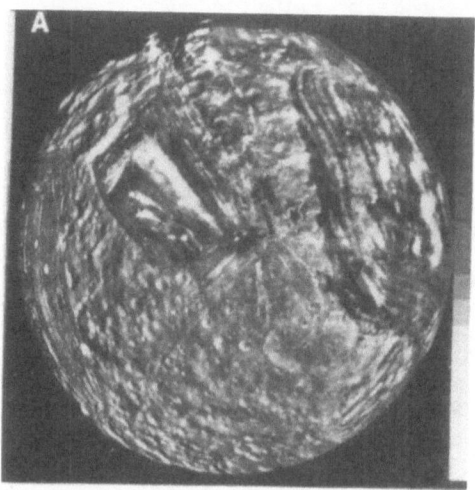

Fig.6. Recent (January 1986) Voyager 2 image of the Uranian satellite Miranda (from ref.[17]). Disruption followed by reaccreation has been suggested[17] as the source of the strikingly different color and cratering observed at various surface positions on Miranda.

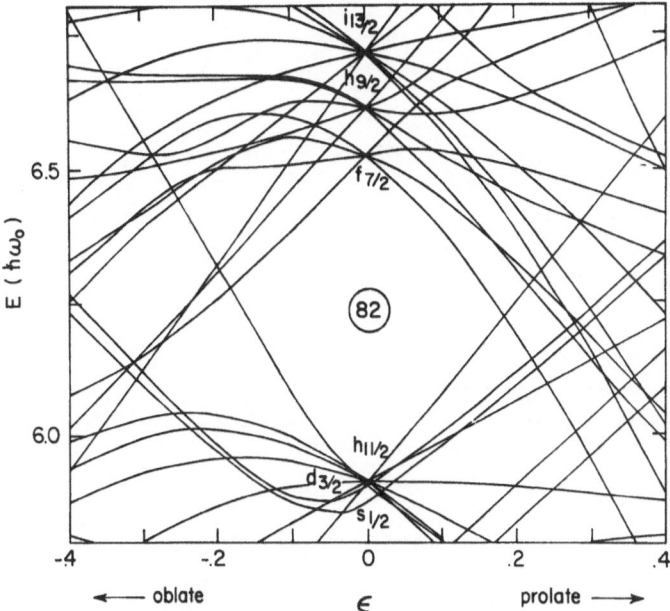

Fig.7. Calculated energy levels for independent neutron motion (Nilsson
model) in a spheroidal potential of quadrupole deformation ε. The
A = 165 parameters of ref.[19] were used. For reference the N = 82
closed shell and the spherical shell-model states are indicated at
ε = 0.

scopic binding energy[20] (liquid-drop energy), are important in determining the equilibrium nuclear deformation. The single-neutron and
single-proton spectrum of states favor prolate shapes in the lower and
middle portions of their respective shells and oblate shapes in the upper
portion of a shell (see Figure 7). The macroscopic binding energy, which
is maximized by a minimum number of nucleons on the nuclear surface,
favors spherical shapes. The balance between these effects is the basis
of the Strutinsky procedure[21,22] of calculating nuclear shapes.

1.2.2. Nuclear shapes. The loss of spherical symmetry in deformed
nuclei leads to the possibility of collective rotation*. The distinctive
features of collective rotation are a rotational band associated with
each intrinsic nuclear configuration and a large nuclear quadrupole moment. The energy associated with collective rotation varies regularly
producing a very different level scheme from a shell-model nucleus. The
excitation energy, E_x, of a quantum mechanical rotor is related to its
angular momentum, I, by:

$$E_x(I) = \frac{\hbar^2}{2\mathcal{J}} R(R+1) + E_j. \tag{1}$$

─────────────
* Collective rotations of a quantum mechanical sphere cannot be observed.
  Likewise, collective rotations about the symmetry axis of an axially-
  deformed quantal rotor cannot be observed.

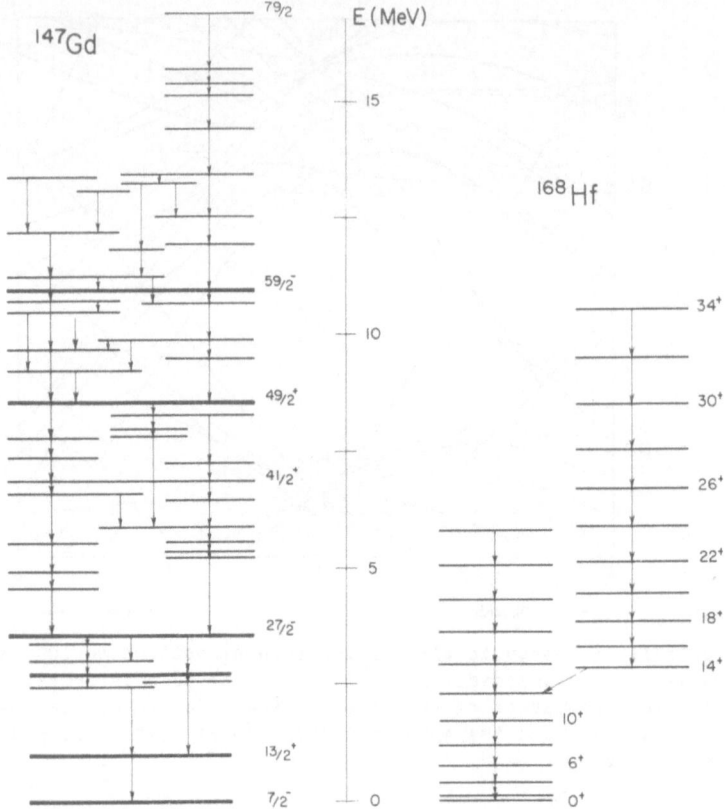

Fig. 8. Comparison of near yrast level schemes for a "collective rotational" nucleus, $^{168}Hf_{96}$ ref.[27], and a "single-particle" nucleus,
$^{147}Gd_{83}$ refs.[25,26]. Levels in ^{147}Gd with lifetimes $\geqslant 1$ nsec are
indicated by heavy lines.

Here ϑ is the moment of inertia, E_j is the excitation energy due to the
intrinsic nuclear excitation, and $\vec{R}$, the rotational angular momentum is
the difference between the total angular momentum, I, and the angular
momentum of the unpaired nucleons, j:

$$\vec{R} = \vec{I} - \vec{j}. \tag{2}$$

For the ground-state band of an even-even nucleus, where there is no
intrinsic excitation, $E_x(I)$ is given by the familiar parabolic expression*:

---

* This equation is obtained from the classical expression relating the
rotational energy, E_{rot}, and angular momentum, L:
$$E_{rot} = L^2/2\vartheta \tag{3}$$
by replacing L^2 with the corresponding quantum mechanical quantity
$I(I+1)\hbar^2$.

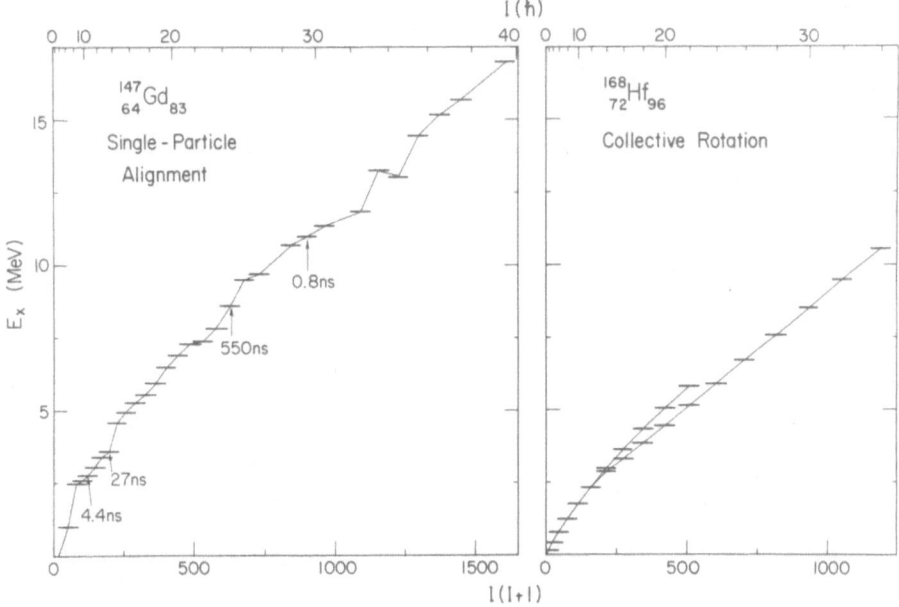

Fig.9. Comparison of excitation energy versus I(I+1) plots for collective
rotational, $^{168}Hf_{96}$, and "single-particle like," $^{147}Gd_{83}$, nuclei.
The corresponding decay schemes are shown in fig.8.

$$E_x(I) = \frac{\hbar^2}{2\vartheta} I(I+1). \qquad (4)$$

In contrast, $E_x(I)$ for a shell model nucleus is irregular and dependent upon the detailed spacing of the single-particle levels. These
different mechanisms for generating angular momentum in nuclei, summed
contributions of individual nucleons with aligned angular momentum vectors; and collective rotation of a deformed system, were mentioned in the
classic 1937 paper of Niels Bohr and Kalckar[23]. Nuclear rotational
bands, however, were not established[24] until 1953.

   Typical examples of near yrast level schemes for $^{147}Gd_{83}$
(refs.[25,26]), a single-particle nucleus near the N = 82 closed shell, and
$^{168}Hf_{96}$ (ref.[27]), a collective rotor, are compared in Figure 8. The
angular momentum dependence of the excitation energy of the yrast*
sequences of these two nuclei is compared in Figure 9. The decay sequences of $^{168}Hf_{96}$ illustrates the regular pattern of rotational bands
characteristic of the collective rotation of a deformed nucleus. $E_x(I)$
for the ground-state rotational band is a parabola with vertex at the
origin. The rotational band corresponding to an excited intrinsic confi-

───────────

*Yrast is a neologism[28] constructed as the superlative form of the Nordic
word "yr" meaning "dizzy." Thus the yrast sequence is the sequence of
the most "dizzy" nuclear states, or the states of maximum angular momentum for a given energy. This and many other specialized terms for the
spectroscopy of rotating nuclei are defined in the glossary of ref.[29].

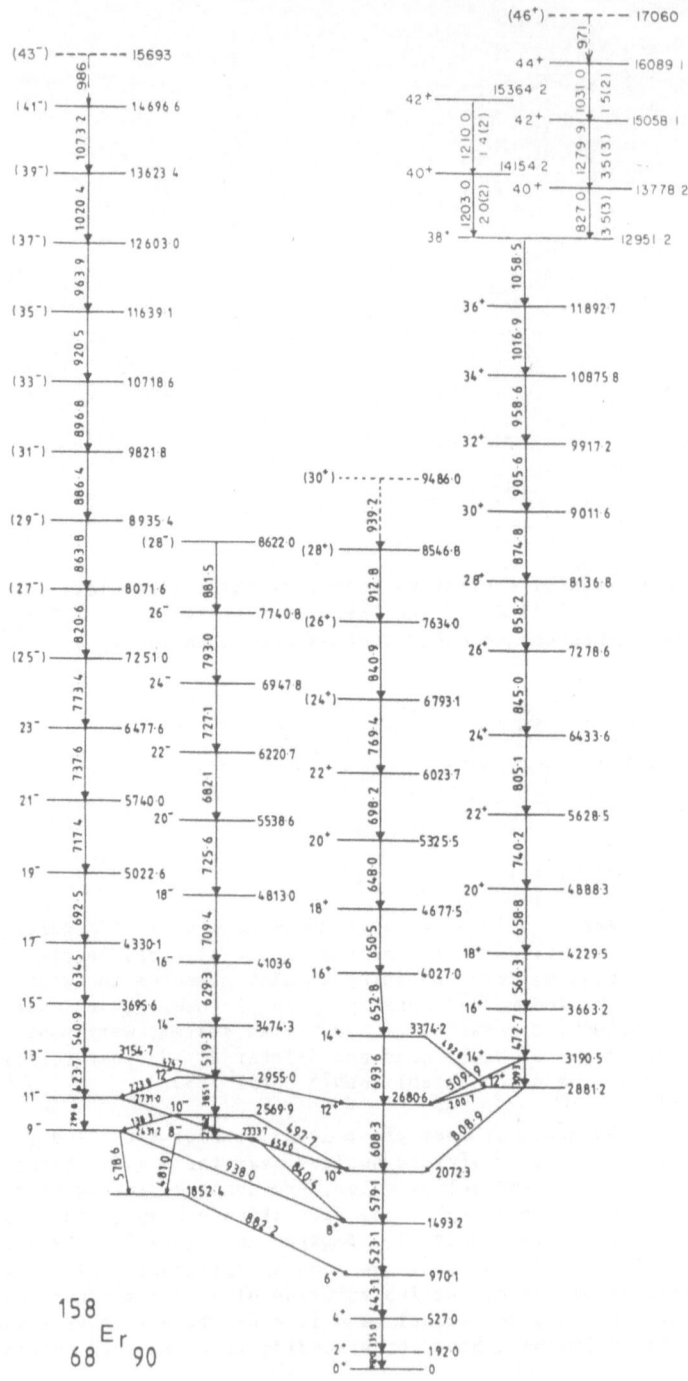

Fig.10. Level scheme for $^{158}Er_{90}$ established from the experimental
studies reported in refs.30,31.

guration (the S band - see sect.2) also has the expected parabolic dependence of $E_x(I)$ with the vertex displaced by (E_-j), see eq.(1). In contrast, the decay scheme of $^{147}Gd_{83}$, only one neutron removed from the
N = 82 closed neutron shell, is irregular and complicated by isomeric
states. Likewise the $E_x(I)$ of the ^{147}Gd yrast sequence is characteristic
of a system that continually changes configuration.

The contrasting decays, shown in Figures 8 and 9 and discussed in
the preceding paragraphs, can be described as either "single-particle" or
"collective rotational." Both structures, however, can coexist in the
same nucleus. The remainder of this section concentrates on two examples
of such shape coexistence: for $^{158}Er_{90}$ the shape of the yrast configuration changes as function of angular momentum; and for $^{152}Dy_{86}$, three
different shapes coexist at the same angular momentum.

1.2.3. <u>The angular momentum dependence of nuclear shapes.</u> A decay
scheme[30,31] of $^{158}Er_{90}$ is shown in Figure 10 and an $E_x(I)$ plot is shown
in the upper portion of Figure 11. Due to the compression of the excitation energy scale, a result of the large rotational energy associated
with these states, these data are replotted in the lower portion of
Figure 11 relative to a rotor with $\mathcal{J}$ = 70.4 MeV$^{-1}\hbar^2$. The yrast sequence
is typical of a deformed rotor up to I^π = 38$^+$. At larger angular momentum the yrast sequence divides into two sequences neither appearing rotational. Recent information, indicating that the lifetimes[31] of the nuclear states in the left-hand sequence are shorter (faster transitions)
than those of the right-hand sequence, leads to the interpretation[31,32]
of the left-hand sequence as a continuation of the rotation band, strongly perturbed by another sequence terminating with shell model states at
I^π = 40$^+$ and 46$^+$.

Low-lying I^π = 40$^+$ and 46$^+$ shell-model states can be constructed for
$^{158}Er_{90}$ by the single-fold occupation of the valence-shell large-Ω Nilsson configurations. (The detailed configuration of the 46$^+$ state of
^{158}Er, for example, is given in Table 3. The neutron configuration of
this level is illustrated in Figure 12.).

Table 3. Shell-model Configuration of the low-lying 46$^+$
Single-Particle State in $^{158}Er_{90}$

<u>Subshell</u>	$m_i = \Omega$	$\sum(m_i) = K^\pi$
$\pi(h_{11/2})^4$	11/2,9/2,7/2,5/2	32/2$^+$
$\nu(i_{13/2})^2$	13/2,11/2	24/2$^+$
$\nu(h_{9/2})^3$	9/2,7/2,5/2	21/2$^-$
$\nu(f_{7/2})^3$	7/2,5/2,3/2	15/2$^-$

total spin I^π=46$^+$

The angular momentum of such a state is the result of aligning the individual components of all the unpaired (i.e. those occupied by a single
nucleon) configurations - see Figure 13. Thus the total angular momentum
is constructed from the intrinsic motion of a "few" nucleons. Aligning
the angular momentum vectors of large-Ω orbitals restricts the intrinsic
motion of the active nucleons to the "waist" of the nucleus. The strong,
short-range attraction of a sufficient number of such unpaired large-Ω
nucleons will produce a "bulging nuclear waist" giving these "single-

particle" nuclei an oblate shape. The preference for oblate high-spin "single-particle" states in the lower portion of a major shell also can be seen from a Nilsson diagram (Figure 12). The large-Ω components of high-j configurations are low in energy for oblate shapes. Therefore, in contrast to the favoring of prolate shapes by the total density of states when angular momentum is ignored, oblate shapes are preferred for high-spin single-particle states.

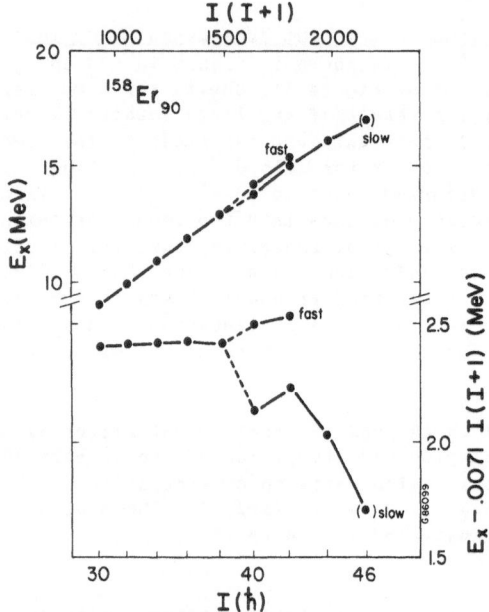

Fig.11. Excitation energy versus I(I+1) plot (top) for the large angular momentum, positive-parity, even-spin decay sequences in $^{158}Er_{90}$. The same data also is shown in the lower portion relative to a rigid rotor with moment of inertia of 70.4 MeV$^{-1}\hbar^2$. A level scheme for ^{158}er is shown in fig.10.

Yrast "high-spin" single-particle states, which have been known in "deformed" light nuclei for some time[35], have been predicted[36] for near transitional heavy nuclei such as $^{158}Er_{90}$ - see Figure 14. The surprise, however, is the strikingly-large interaction (as much as 20-30 keV) be-tween the "collective rotational" and "single-particle like" states. Not only are such states perturbed but both 40$^+$ states of ^{158}Er decay to the "rotational" 38$^+$ state - see Figure 10. Neither is this surprise limi-ted[32,37] to ^{158}Er. Some consequences of this apparent paradox are dis-cussed in refs.[32,37,38].

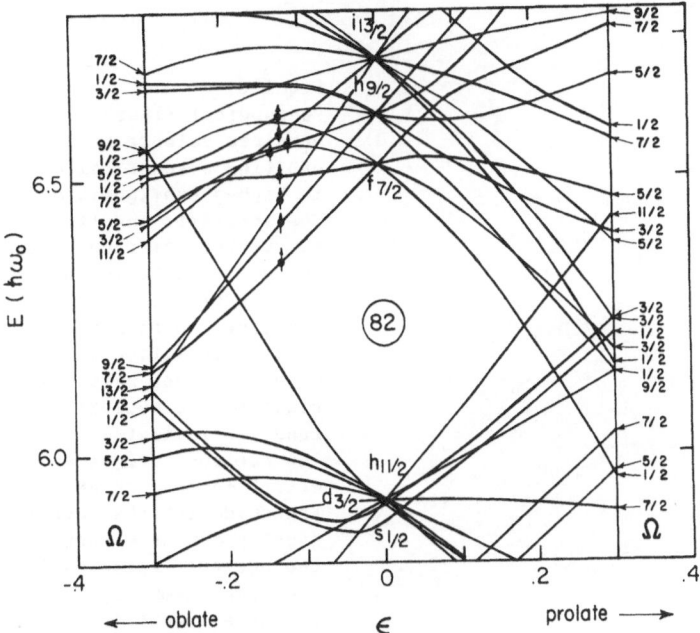

Fig.12. Calculated energy levels for independent neutron motion in a
spheroidal potential of quadrupole deformation (Nilsson model) as
shown in fig.7. The angular momentum projection on the nuclear
symmetry axis, Ω, is given for each level. The valence shell
single-neutron configuration, see also Table 3, corresponding to
the low-lying 46^+ "single-particle state" in $^{158}Er_{90}$ is indicated
by the arrows.

1.2.4. <u>Superdeformed nuclei.</u> For some time low-lying coexisting
"single-particle" and "collective rotational" shapes have been
known[33,39,40] in $^{152}Dy_{86}$ - see Figure 15. Recently gamma-rays interpreted as transitions between a sequence of coexisting "super deformed"
states also have been observed[34,41] in ^{152}Dy up to an angular momentum of
60 ± 2 units - see Figure 16. The technical achievement of establishing a
discrete state with an angular momentum of 60 units itself is spectacu-
lar. The largest angular momentum previously established[31] for a nuclear
state was 46 units.

The nearly-constant 47 keV separation of neighboring gamma-ray transitions in the "superdeformed" sequence (see Figure 16) corresponds to a
dynamic moment of inertia of 85 $MeV^{-1}\hbar^2$. (Expressions relating the kinematic and dynamic moments of inertia to the gamma-ray transition energy
and the angular momenta are derived in the Appendix.) This value for a
rigid nucleus of mass 152 corresponds to a quadrupole deformation of $\varepsilon \approx$
0.6 or a major to minor axis ratio (a:b) of 2:1. Indeed, strong single-
particle shell corrections are predicted[42]* both for N = 86 and Z = 66,
at this deformation (ε = 0.6). That is, $^{152}Dy_{86}$ is predicted to be a
"superdeformed double closed shell nucleus."

---

*Note the Nilsson deformation parameter, η, used by Strutinsky[42] is rela-
ted to the more standard parameterization of the quadrupole deformation
by $\eta \approx 16\varepsilon$.

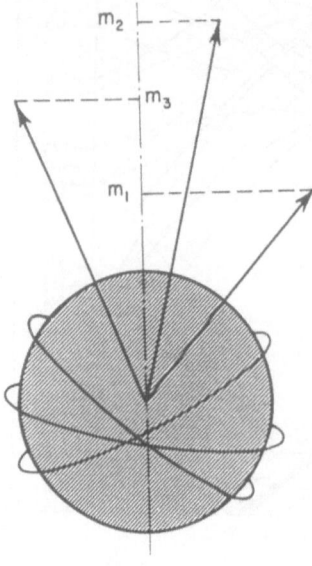

$$I = m_1 + m_2 + m_3 + \ldots$$

Fig.13.
Pedagogical figure of three
unpaired valence particles
orbiting a spherical nucleus
in high-j configurations.
The intrinsic angular momentum vectors of each is indicated. To take maximal advantage of the strong, attractive, short-range nu-
clear interaction these
high-j valence particles
cluster in a nuclear "waist
band" as indicated. Thus
the nucleus "bulges at the
waist" forming an oblate
spheroid with the intrinsic
angular momentum vectors of
the nucleus "aligning" along
a nuclear symmetry axis
giving a sizeable angular
momentum.

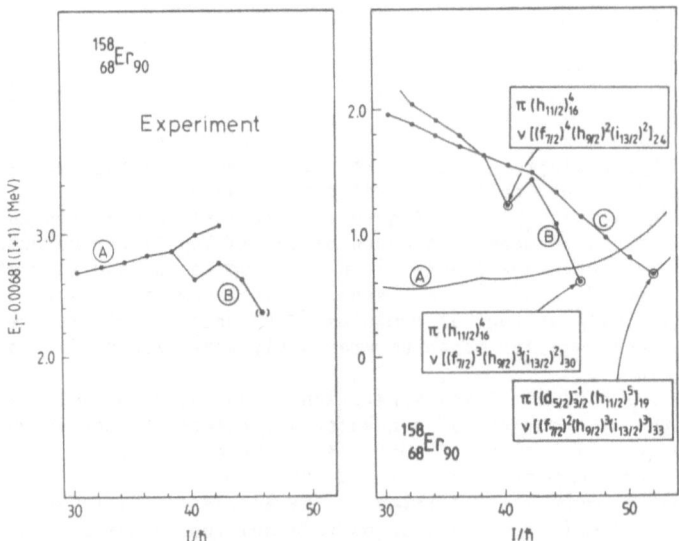

Fig.14. Comparison of high-spin experimental (left) positive-parity levels (see Figure 10) in $^{158}_{68}Er_{90}$ with the corresponding predicted[36] (right) values. These data are plotted relative to a rigid
reference with a moment of inertia of 73.5 $MeV^{-1}\hbar^2$. The valence-
shell single-particle configurations are given in the left-hand
portion for the band terminations. A more detailed description
of the configuration of the low-lying 46^+ state is given in Table
3.

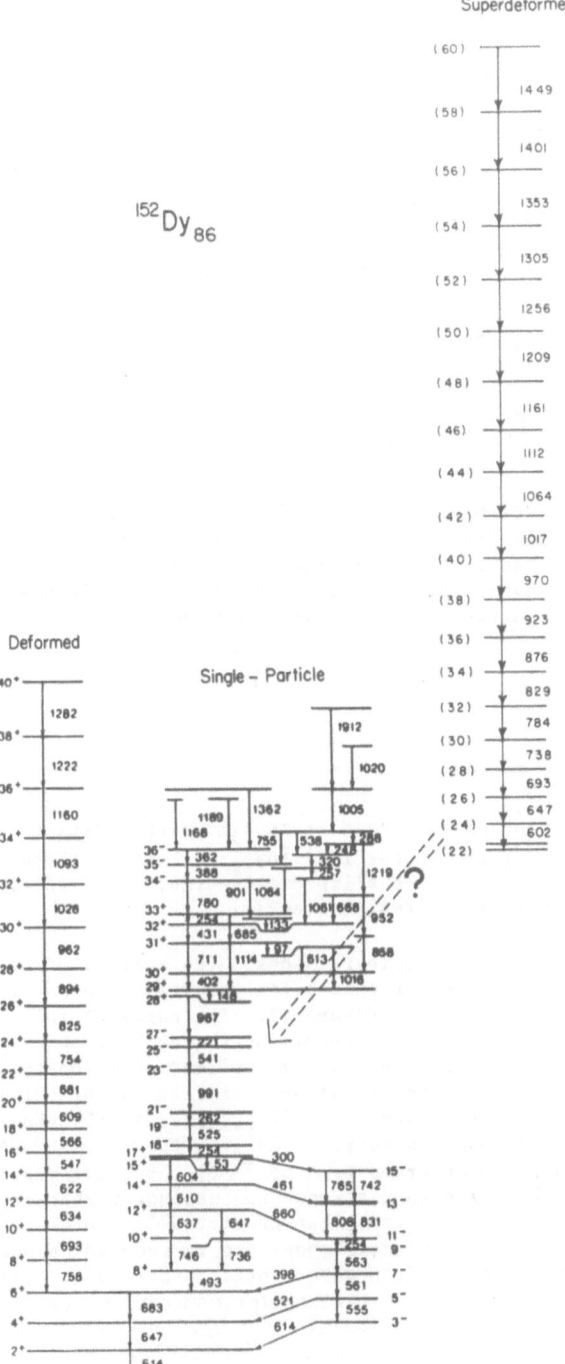

Fig.15. Partial [152]Dy[86] level scheme[33,34,39-41] depicting the coexisting single-particle, deformed collective, and superdeformed decay sequences. Neither the angular momentum nor the excitation energy of the superdeformed sequence is firmly established[34,41].

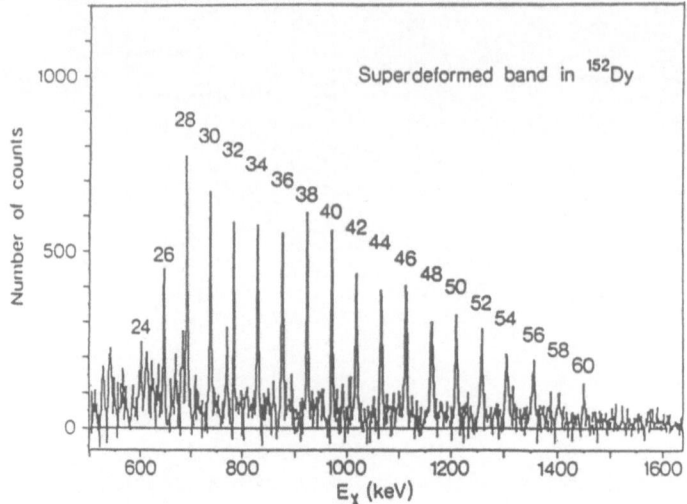

Fig.16. Gamma-ray energy spectrum[34,41] showing discrete transitions
between superdeformed states in $^{152}Dy_{86}$. The various gamma-ray
lines are labelled by the angular momentum associated with the
decaying level. The angular momentum is uncertain within two
units. (Courtesy of Daresbury Laboratory.)

The observation superdeformed state allows the detailed study of
three distinct coexisting shapes in $^{152}Dy_{86}$: slightly oblate ellipsoid
($\varepsilon \approx -0.15$; a:b$\approx$1.15:1); prolate ellipsoid ($\varepsilon \approx 0.20$; a:b$\approx$1.25:1); and superdeformed prolate ellipsoid ($\varepsilon \approx 0.6$; a:b$\approx$2:1).

1.2.5. <u>The variety of nuclear shapes.</u> The variety of predicted[43]
and observed[32,34,44-8] nuclear shapes for the yrast states of the dispro-
sium isotopes is summarized in Figure 17. The parallel roles of particle
number and angular momentum in determining the nuclear shape in "transitional nuclei" are emphasized by such a "nuclear shape phase diagram."
Angular-momentum dependent shape changes, similar to that of $^{158}Er_{90}$ (see
sect. 1.2.3), also are observed for $^{154}Dy_{88}$ (refs.44,46) and $^{156}Dy_{90}$
(refs.32,48). These shape changes, as well as the estimated[34,41] angular
momentum where the superdeformed band of $^{152}Dy_{86}$ becomes yrast, are indicated in this figure. The angular-momentum dependent shape changes are
observed about fifteen units less than predicted[31]. More recent calculations[37], in which the Z = 64 shell model gap is reduced, produce improved
agreement with experiment. Even more exotic shapes, e.g. triaxial el-
lipsoids, are predicted[43] at larger angular momentum where data is not
yet available.

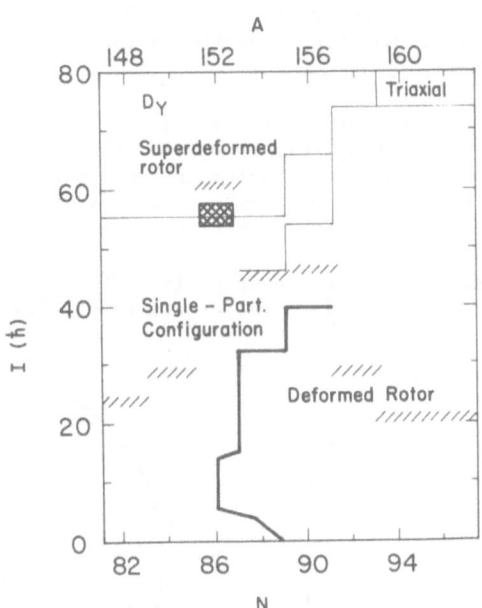

Fig.17. Diagram indicating the variety of predicted and observed shapes
of the yrast configuration of the even-even dysprosium isotopes
(Z = 66) as a function of neutron number, N, and angular momentum, I. Predicted[43] and observed[32,34,44-7] shape boundaries are
distinguished by light and heavy lines respectively. The best
estimate[34,41] of the angular momentum where the superdeformed
configuration of $^{152}Dy_{86}$ becomes yrast is indicated by the cross-
hatched area. The maximum angular momentum observed in each
isotope is denoted by the slashed lines.

2. PEDESTRIAN PAIR CORRELATIONS

2.1. The Pair Coupling Scheme

The atomic nucleus is unique amongst quantal systems. It consists of a finite number ($\leqslant 250$) of strongly interacting fermions. Two neutrons (or two protons) moving in the nucleus can best take advantage of the strong, short-range attractive nuclear interaction to minimize their energy by moving in time-reversed orbits - see Figure 18. The most efficient possibility, moving together in the same orbit, shown in the left-hand portion of this figure, violates the Pauli exclusion principle. Both nucleons have identical quantum numbers. Time-reversed motion, guaranteeing two interactions per orbital period, is the most efficient construction that does not violate the Pauli principle. The magnetic quantum numbers, m, of the two nucleons have opposite signs. Therefore, the net intrinsic angular momentum imparted to the nucleus is zero; i.e. the nucleons are paired.

The ground (minimum energy) states of nuclei are obtained by such pairwise occupation of the configurations with the lowest energies. Of course, for an odd number of neutrons (or protons) a single nucleon must be unpaired. This "pair coupling scheme" gives maximum consideration to correlations between each nucleon and one other nucleon.

The effect of paired protons added to the $^{208}Pb_{126}$ doubly-closed shell is illustrated in Figure 19. The experimental levels[50] of ^{210}Po

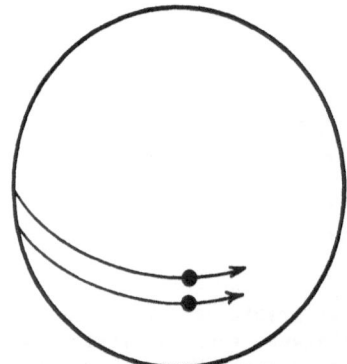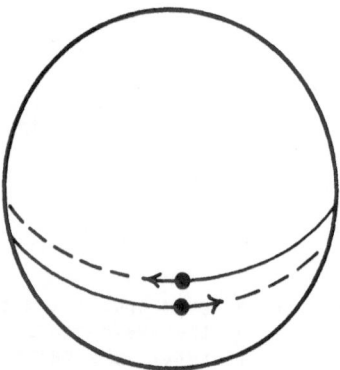

Lowest energy, not allowed by Pauli principle	Guarantees two interactions per orbit period

Fig.18. Pedagogical figure depicting two couplings in which a pair of nucleons moving in a nucleus take advantage of the strong, short-range, attractive nuclear interaction to minimize the energy of the system. The configuration shown to the left is the most efficient, since the two nucleons continuously interact. This configuration, however, is excluded by the Pauli principle. The "paired" configuration (shown to the right), guaranteeing two interacitons per orbital period, is the most efficient allowed coupling. Of course, nucleons must be considered as wavepackets, not point particles as indicated. However, the localization of the probability distributions of high-j configurations, shown in Figure 21, indicates the validity of the features of this figure. These simple arguments are taken from reference[49].

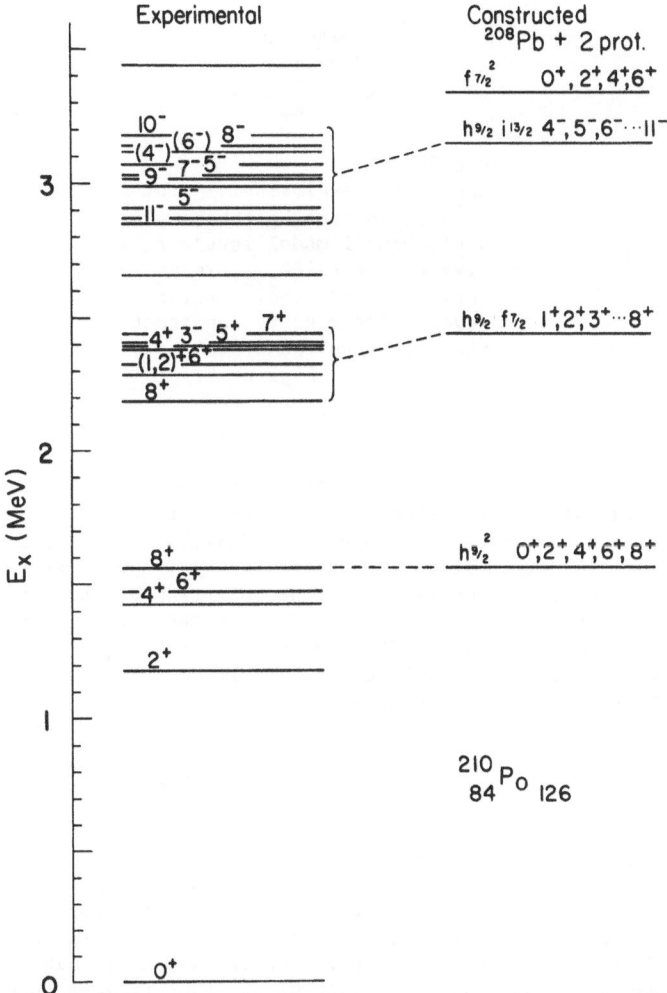

Fig.19. Comparison of the experimental spectrum of states[50] for $^{210}Po_{126}$ with that constructed from $^{208}Pb_{126}$ plus two protons in the various combinations of shell-model states taken from the single-proton spectrum[50] of $^{209}Bi_{126}$, see Figure 20. These two spectra of states are arbitrarily normalized at the 8^+ member of the $(h_{9/2})^2$ multiplet. The different behavior of the two-proton configuration in which both valence protons are in the same shell model configuration is attributed to pairing.

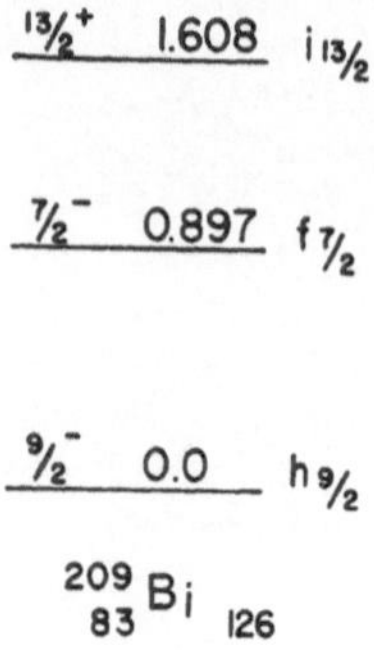

$\underline{\quad}^{13/2^+}\quad 1.608\quad i\,{}^{13}/_2$

$\underline{\quad}^{7/2^-}\quad 0.897\quad f\,{}^{7}/_2$

$\underline{\quad}^{9/2^-}\quad 0.0\quad h\,{}^{9}/_2$

$^{209}_{83}Bi_{126}$

Fig.20.
Low-lying level scheme[50] for $^{209}Bi_{126}$ depicting the single-proton shell model levels relative to the $^{208}Pb_{126}$ closed shell. These single-proton levels were used to construct the low-lying two-proton spectrum shown in Figure 19.

are compared to two-proton states constructed from tne $^{209}Bi_{126}$ single-proton spectrum of states. (The low-lying ^{209}Bi spectrum is shown in Figure 20.) The completely-paired 0^+ ground state of $^{210}Po_{126}$ is lowered in energy by 1.18 MeV relative to that of the next lowest state constructed from the $(h_{9/2})$ multiplet*. The splitting of the $(h_{9/2})$ multiplet is five times greater than those in which the two protons are in different shell-model states. These effects, largest for the completely paired 0^+ ground state, are attributed to pairing. Thus, sizeable pairing effects are observed in shell-model nuclei.

2.2. The Aligned Coupling Scheme and Deformed Nuclei

Pairing only considers two-nucleon correlations. However, exploiting the spatial localization of single-particle orbits and the strong, short-range attractive nuclear force, it is possible to construct a coupling scheme that includes many-particle correlations. For a specific shell-model state the localization is maximum for substates with m = |j| - see Figure 21. This localization increases with increasing orbital angular momentum. Thus the first pair of nucleons prefers to occupy the m = j and -j substates of a high-j orbital. The m = j-1 and -j+1 substates is the preferred occupation for the next pair of nucleons in the same large-j shell-model state. Not only is the spatial localization large, but these substates have the largest overlap with the occupied m = j and -j substates - see Figure 21.

---

*The observed preference for the 2^+, 4^+, ... states relative to larger angular momentum members of the $(h_{9/2})^2$ multiplet can be attributed[52] to higher multipoles in the pairng interaction. This same effect produces configuration dependent pair correlations discussed in section 2.4.

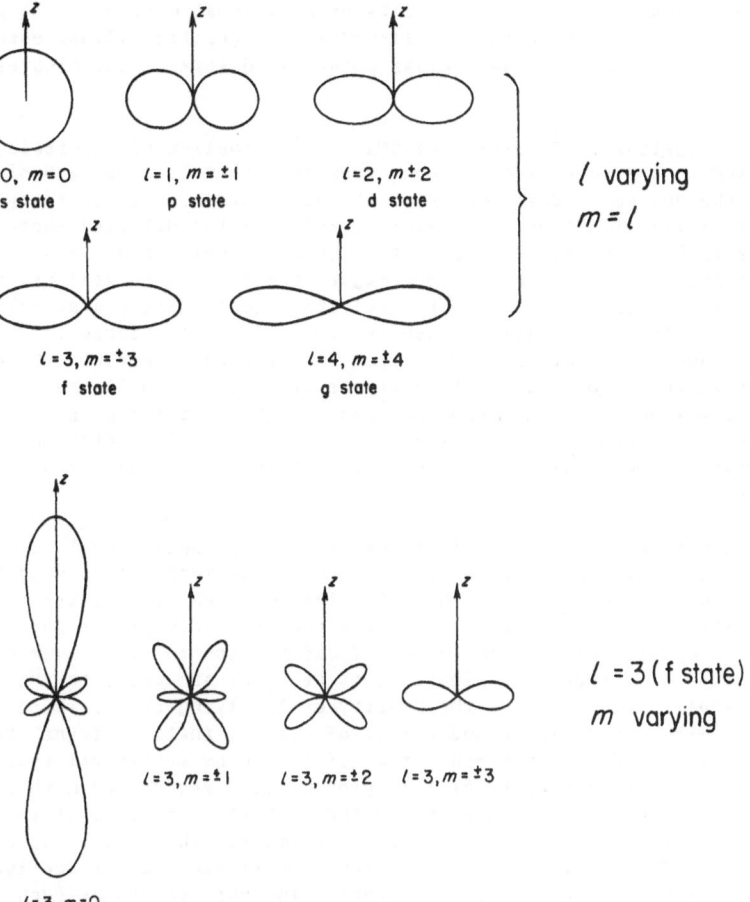

Fig.21. Probability distributions relative to the polar axis: (top) for
the m = ℓ component as a function of the orbital angular momentum, ℓ; and (bottom) for the ℓ = 3 (f state) as a function of the
magnetic quantum number, m. The aligned coupling scheme, see
text, results from placing the valence nucleons in these such
states in a manner that most efficiently takes advantage of the
strong, short-range attractive nuclear interaction. These distributions were taken from ref.[51].

The density distribution for such large-j, large-m orbitals is confined to a narrow band perpendicular to the direction of j. This component is lowered in energy if the nuclear potential is deformed, see e.g.
the Nilsson diagram shown in Figure 12. An equivalent statement is the
spatial localization of the large-j, large-m components distorts, or
polarizes, the nuclear matter distribution away from spherical symmetric

shapes[*]. (This, of course, is the Rainwater hypothesis[53] that led Aage
Bohr to reconsider[54] the possibility of rotational motion.) The occurrence of more than one high-j shell-model configuration allows even larger spatial overlaps as the nuclear potential deforms - see Figures 12
and 21.

The spatial localization of the large-j shell-model orbitals,
resulting in the so-called "aligned coupling scheme", is a mechanism to
deform the nucleus. Detailed calculations, however, are necessary to
determine: (i) if the nucleus deforms; and (ii) the detailed shape of the
nucleus as the large-j orbitals are occupied. Does the nucleus retain
axial symmetry? Is a prolate (one major- and two minor-axes) or an
oblate (two major- and one minor-axis) spheroidal shape preferred? The
latter question is addressed - see section 1.2.1. and Figure 12 - by
Nilsson model calculations (axial symmetry assumed). Prolate shapes are
favored in the lower and middle portions of a high-j shell, and oblate
shapes are favored in the upper portion.[**] The question of axial symmetry requires detailed calculations of, not only single-proton and -neu-
tron energies, but also the macroscopic binding energy of the nucleus -
see sect. 1.2.1.

Nuclear deformation provides the "element of anisotropy" allowing
the definition of the nuclear orientation. Therefore, it is possible to
observe rotational effects in nuclei. Indeed, such effects are observed
for midshell nuclei, where a sufficient number of occupied high-j orbitals deform the nucleus. The onset of deformation for even-even dysprosium isotopes is traced in Figure 22 as neutrons are added to the N = 82
closed shell. Not only does the density of low-lying states increase
dramatically with an increased number of valence shell neutrons, but the
distribution of $E_x(I)$ approaches that of a quantum mechanical rotor - eq.
(4). The increased density of low-lying states, associated with symmetry
breaking and the related increase in the degrees of freedom of the system, is a general feature of quantum systems[56]. The rotational band
shown for $^{158}Dy_{92}$ corresponds to a single intrinsic nuclear configuration
rotating with different angular momenta. The observation of such spectra
is associated with the rotational degree of freedom, a consequence of the
broken spherical symmetry of the system. Rotational spectra are not
observed for spherically symmetric quantum systems.

2.3. The Effect of Deformation on Pair Correlations

The loss of spherical symmetry, a result of the occupation of anisotropic valence-shell nuclear configurations, modifies both the spectrum
of independent-particle level spacing decreases as the (2j+1)-fold degenerate shell-model states divide into j+1/2 doubly-degenerate "Nilsson
states" in the deformed system - see Figure 12. (The spectrum of states

---

[*]In contrast, the occupation of large angular momentum atomic orbits
   does not distort the central atomic potential. The Coulomb force is
   relatively weak and long ranged. Thus deformed atoms and the associated rotational atomic spectra are not observed. Rotational molecular
   spectra, however, have been known for some time.[55] The atomic lattice
   of the molecules provides the element of anisotropy.

[**]In contrast, the requirement of large single-particle angular momentum
   (e.g. high-k states or "band terminations" such as the 46[+] state of
   $^{158}Er_{90}$) leads to the opposite conclusion: slightly oblate (prolate)
   shapes in the lower (upper) portion of major shells. The nuclear shape
   of such states is discussed in section 1.2.3.

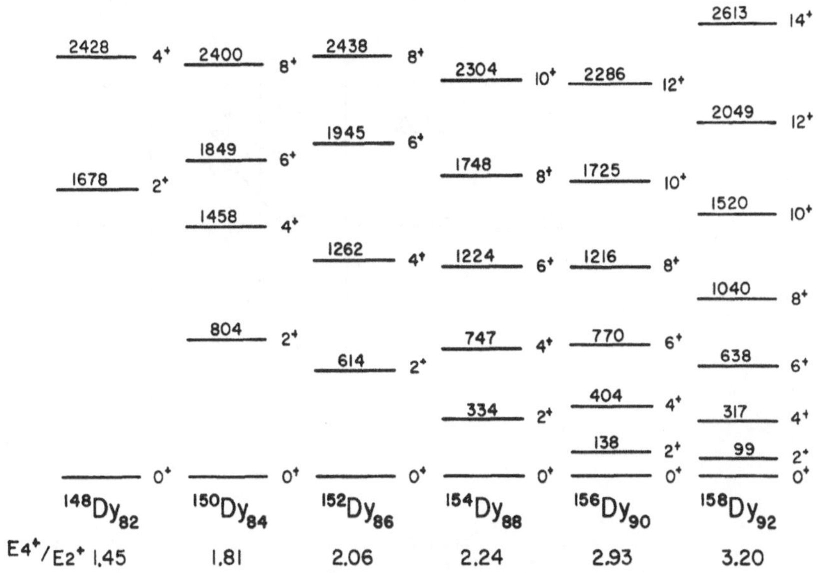

Fig. 22 energy-level diagram (energies in keV, with spin-parity J^π):

- $^{148}Dy_{82}$: 0^+; 1678 2^+; 2428 4^+
- $^{150}Dy_{84}$: 0^+; 804 2^+; 1458 4^+; 1849 6^+; 2400 8^+
- $^{152}Dy_{86}$: 0^+; 614 2^+; 1262 4^+; 1945 6^+; 2438 8^+
- $^{154}Dy_{88}$: 0^+; 334 2^+; 747 4^+; 1224 6^+; 1748 8^+; 2304 10^+
- $^{156}Dy_{90}$: 0^+; 138 2^+; 404 4^+; 770 6^+; 1216 8^+; 1725 10^+; 2286 12^+
- $^{158}Dy_{92}$: 0^+; 99 2^+; 317 4^+; 638 6^+; 1040 8^+; 1520 10^+; 2049 12^+; 2613 14^+

$E4^+/E2^+$: 1.45   1.81   2.06   2.24   2.93   3.20

Fig.22. Comparison of the low-lying spectrum of positive-parity states in even-even dysprosium isotopes[32,34,44,46-7] from $^{148}Dy_{82}$ to $^{158}Dy_{92}$. The ratio of the excitation energy of the lowest 4^+ state to that of the lowest 2^+ state also is given. The increased low-lying density of states as well as the increase in the E(4⁺)/(2⁺) ratio with increasing neutron number is taken as evidence for deformated nuclei. [E(4⁺)/E(2⁺) = 3.333... for a perfect quantum mechanical rotor, see eq. (4)].

and degeneracies associated with independent-particle motion in a variety of approximations to the nuclear potential is summarized in Figure 23.) If (i) the level spacing becomes sufficiently small, and (ii) the two-fold degeneracy of the levels (a consequence of time-reversal invariance) is maintained; then there is a sizeable probability in a quantum system to scatter pairs of time-reversed particles from one level to another. Not only is this scattering dependent on the energy difference between the two levels (monopole pairing), but in general there can be a dependence on the relative orientation of the nuclear orbitals as well. This orientational dependence, which is unique to nuclear physics, accounts for the overlap of the two orbitals and the angular dependence of the intranuclear scattering. In a few instances, see e.g. refs.[52,57-9], such effects have been considered by including configuration dependent terms in the pairing interaction.

For a sizeable pairing interaction and closely-spaced single-particle levels the scattering of pairs of particles between time-reversed states can be large enough to modify the occupation of the single-particle states. This scattering, and the resulting "smearing" of the Fermi surface, is depicted in Figure 24. The configuration of the resulting pair correlated state (e.g. the ground-state of a deformed even-even nucleus) does not dramatically change in the nucleus with two more or two less neutrons (or protons). Thus an enhancement is expected, and indeed is observed, for the addition or removal of two like nucleons[60]. Such "collective" contributions of a number of paired configurations can be considered in a manner analogous to the static deformation of the nuclear shape produced by the collective distortions of several anisotropic valence-shell orbitals. For example, both shape and pair correlations

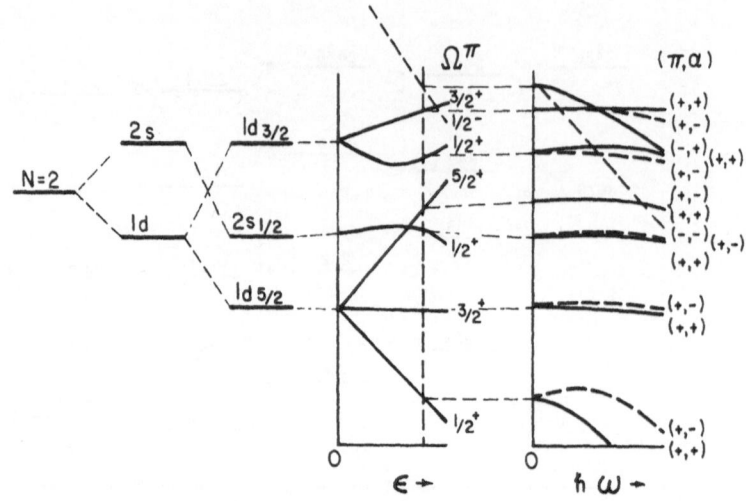

	Harm. Oscil.	Woods Saxon	Shell Model $\ell \cdot s$	axially deformed	Coriolis + Centrifugal $-\omega j_x$
Deg.	$(N+1)(N+2)$	$2(2\ell+1)$	$2j+1$	2	1
q.n labeling	N, π	N, ℓ, π	N, ℓ, j, π	$[Nn_3 \Lambda] \Omega^\pi$	$\alpha \quad \pi$

Fig.23. S-d shell spectra of energy levels corresponding to independent-particle motion in a variety of potentials which historically have been used to describe the nucleus. The degeneracies, as well as the quantum numbers used to label the energy levels are given for each potential.

can be separated into collective and intrinsic components. Thus we speak of a "deformation of the pair field". The "pair-gap parameter", Δ, is the appropriate deformation parameter for the static pair field.

The static deformation of the pair field also produces a major modification of the nucleonic motion. The resulting quasiparticle coupling scheme can be considered a generalized seniority coupling scheme. (For illustration corresponding quantities in the shell-model and quasiparticle coupling schemes are indicated in Table 4.) The elementary excitations, or quasiparticles, are hybrids or particles and holes, and the seniority, v, denotes the number of quasiparticle excitations. The quasiparticle energy, E_v, corresponding to the single-particle configuration v is given by

$$E_v = \sqrt{\Delta^2 + (\varepsilon_v - \lambda)^2} \tag{5}$$

where ε_v and λ are the single-particle energy of the state v and the Fermi level corresponding to the appropriate particle number.

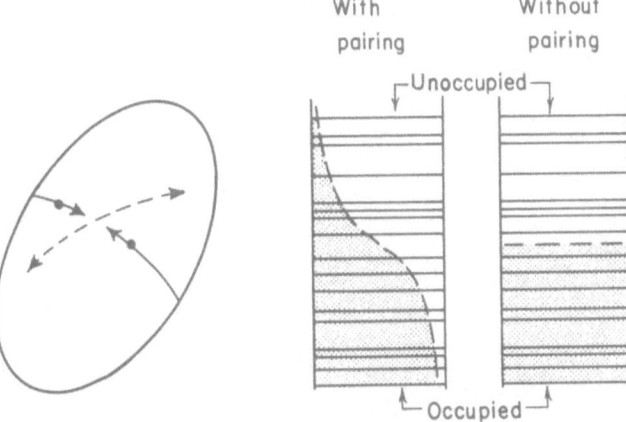

Fig.24. To the left the scattering of a pair of nucleons from one time-
        reversed orbital to another is depicted for a deformed system in
        which the level spacing is small. Of course, nucleons must be
        cosnsidered as wavepackets, not point particles as indicated. To
        the right the occupation for the diffuse Fermi surface of a pair-
        correlated state is compared with that of a sharp Fermi surface
        for an uncorrelated state.

Table 4. Seniority Coupling Schemes

Shell Model	Quasiparticle
Vacuum Configuration:	
Ground state of closed-shell nucleus	Even-even ground-states configuration
Basic Excitation:	
Single-particle and single-hole states in neighboring odd-A isotopes and isotones	Single-quasiparticle states in neighboring odd-A isotopes and isotones
Field:	
Spherical nuclear field	Deformed nuclear and deformed pair fields
Residual Two-body Interaction:	
Particle-particle, particle-hole, and hole-hole interactions	Residual quasiparticle interaction

The resulting modification of the spectrum of single-particle (-qua-
siparticle) states is illustrated in the left-hand and center portions of
Figure 25. The spectrum of single-particle (i.e. v=1) states is shifted
upward in energy by $E_{v=1,g.s.} \approx \Delta$ relative to the v=0 quasiparticle vacuum
(i.e. the ground state of the neighboring even-even nuclei). The spacing
of the lowest single-quasiparticle states also is compressed relative to
the corresponding spectrum of single-particle states in the absence of
pair correlations. For a mathematically complete treatment of pair correlations in nuclei and the quasiparticle coupling scheme the reader is
referred, for example, to ref.[61].

The effects of pairing is enhanced by the static deformation of the
pair field induced by the scattering of pairs of particles in deformed
nuclei. Compare, for example, the lowering of the completely paired
ground states in the presence (Figure 26) and absence (Figure 19) of
static pair correlations. The term pair correlations in the present article refers to a static deformation of the pair field producing effects
such as that depicted in Figure 26.

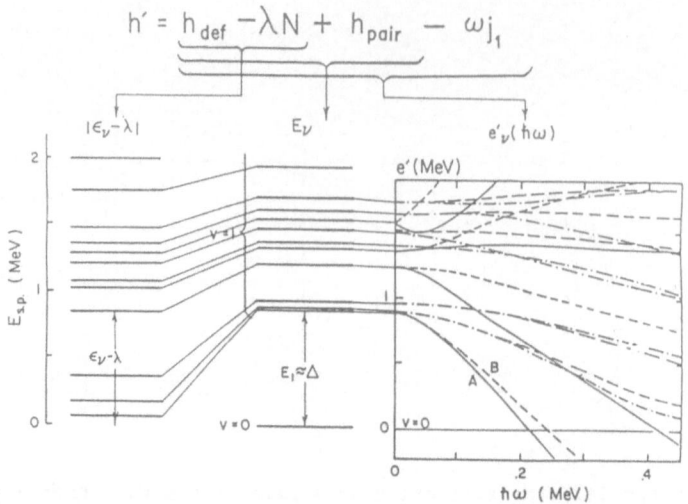

Fig.25. Spectra of Nilsson states (left), quasiparticle energies, E_ν, for
the Nilsson model plus pair correlations (center), and routhians,
$e'(\hbar\omega)$ (right) depicting the effects of pair correlations and
rotation on independent-particle motion in a deformed potential.
$(\pi,\alpha) = (+,1/2)$, $(+,-1/2)$, $(-,1/2)$, and $(-,-1/2)$ states are denoted by solid, short-dashed, dot-dashed, and long-dashed curves
respectively in the right-hand portion. The hamiltonian for
independent-particle motion in a rotating deformed potential is
given at the top of the figure, and the various terms associated
with each predicted spectrum are indicated. h_{pair} of eq.(7) is
written explicitly. The spectra were calculated assuming
$\varepsilon_2 = 0.242$, $\varepsilon_4 = \gamma = 0.0$, and $\Delta = 0.87$ MeV. They are appropriate for the
single-neutron and single-quasineutron spectra of $^{165}Yb_{95}$.

2.4. The Effects of Rotation on Intrinsic Nucleonic Motion

We earthlings can claim kin with the nucleons in a rotating nucleus.
We both "live" in a rotating reference frame influenced by Coriolis and
centrifugal forces. The resulting ocean currents and weather patterns
which, for example, make Northern Europe nearly habitable in winter and
the vortices formed during the draining of our bathtubs are accepted
consequences of these terrestrial rotationally induced forces.

Likewise, the Coriolis and centrifugal forces affect the intrinsic
structure of a rotating nucleus. Just as the terrestrial Coriolis force
has opposite signs for rockets fired to the east and west, the corresponding nuclear force will have opposite signs based on whether the nucleons
are moving in the direction of, or counter to, the rotation of the
nucleus, see Figure 27. Therefore, time reversal invariance is not conserved in a rotating nucleus. The nuclear rotation has defined a preferred direction of rotation.

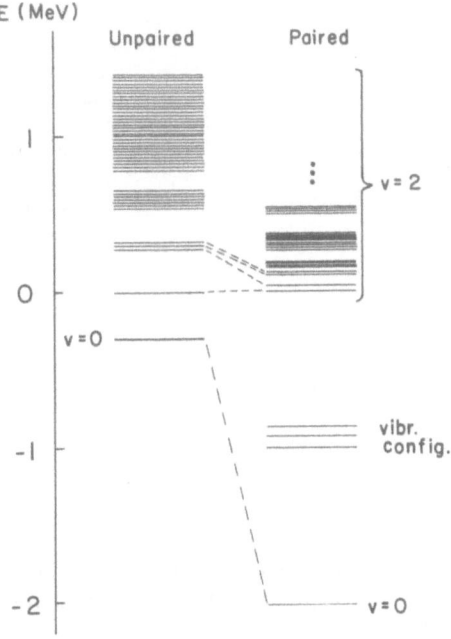

Fig.26. Comparison of the spectra of two-quasiparticle and two-particle
levels in a system with (right) and without (left) pair correlations for a spectrum of nearly equally-spaced single-particle
states. A level spacing of 300 keV and a pair gap of 1 MeV were
assumed. The spectra of levels are arbitrarily normalized at the
lowest two-quasiparticle and two-particle level. In a deformed
nucleus each intrinsic configuration depicted would be the intrinsic state for a rotational band. For completeness the low-
lying vibrational configurations also are included in the correlated picture, at an energy of about half the pair gap.

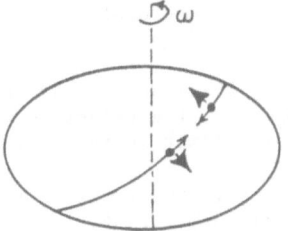

Fig.27.
Diagram illustrating the
Coriolis plus centrifugal
force vectors (large arrows)
for nucleons moving in time-
reversed orbits in a prolate
deformed nucleus rotating
about an axis perpendicular
to the nuclear-symmetry
axis.

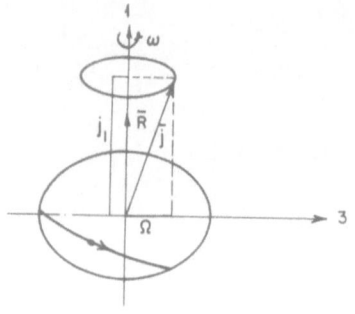

ROTATION ALIGNED

Fig.28.
Vector diagram depicting the angular momentum coupling of rotationally-aligned nucleons in an axially-symmetric nucleus rotating about an axis (labelled 1) perpendicular to the nuclear-symmetry axis (labelled 3). See also Figure 32.

Theoretically these rotationally-induced effects are included[29,62] by adding a Coriolis plus centrifugal term, $-\omega j_1$, to the independent-particle hamiltonian:

$$h' = h_{def} - \omega j_1. \qquad (6)$$

Here h_{def} is the hamiltonian for independent-particle motion in a deformed potential, e.g. the Nilsson potential[18,19]; ω is the angular frequency of rotation; and j_1 is the projection of the intrinsic nucleonic angular momentum on the axis of rotation - see Figure 28. Time-reversed states have opposite values of j_1; therefore, the two-fold degeneracy of the Nilsson levels is removed - see Figure 23. The resulting energy levels are labelled by the conserved quantum numbers: parity, $\pi=+$ or $-$, and signature*, $\alpha=+1/2$ or $-1/2$. The effects of the Coriolis plus centrifugal force is strongest for the configuration with the largest value of j_1, i.e. the configuration for which the intrinsic angular momentum is most nearly aligned with the rotational axis - see Figure 28. In the s-d shell the $\Omega = 1/2$ orbital derived from the $d_{5/2}$ subshell (i.e. the $1/2^+[220]$ Nilsson orbital), is the most alignable. Therefore, it is the most strongly affected by rotation - see Figure 23. Likewise, the $\Omega=1/2$ component of the $i_{13/2}$ neutrons, the most alignable rare earth configuration, is the most strongly influenced by rotation in this mass region - see Figure 29. Indeed, this highly-aligned orbital is responsible for the low-frequency band crossings or "backbends" in this mass region.

For heavier nuclei, where the pair-gap parameter, Δ, is larger than the average single-particle level spacing**, d, the effects of pair correlations, h_{pair}, also must be included explicitly in the hamiltonian:

---

*Signature, α, is the quantum number associated with the $R_1(\pi)$ symmetry, i.e. a rotation of 180° about the axis of nuclear rotation[62]. The relation between angular momentum and signature can be expressed as $I=\alpha$ Mod 2. In even-A systems $\alpha=0$ sequences contain even spins, and $\alpha=1$ sequences contain odd spins. For odd-A systems $\alpha=1/2$ sequences include $I = 1/2, 5/2, 9/2, 13/2,...,$ and $\alpha = -1/2$ sequences include $I = 3/2, 7/2, 11/2, 15/2,...$

**For example, in even-even deformed rare-earth nuclei $\Delta_n \approx 1$ MeV and $d_n \approx 0.3$ MeV. The subscript n denotes the neutron degree of freedom. The results of such values are shown in Figure 26.

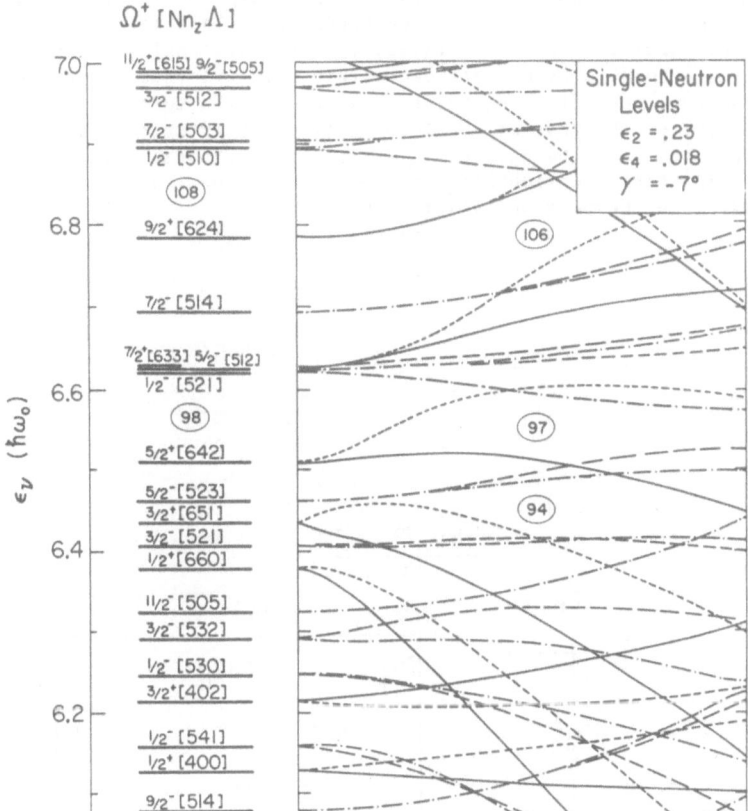

Fig.29. Calculated spectrum of N=4-6 single-neutron states in a rotating
nucleus in the absence of pair correlations. $(\pi,\alpha) = (+,1/2)$,
$(+,-1/2)$, $(-,1/2)$, and $(-,-1/2)$ states are denoted by solid,
short-dashed, dot-dashed, and long-dashed curves, respectively.
To the left the asymptotic quantum numbers, $\Omega^{\pi}[N,n_z,\lambda]$, are given
for each orbit. These values, of course, are only valid for $\hbar\omega=0$.
The deformations used (given in the upper right-hand portion of
the figure) correspond - to average values of minimum energy
deformations for the low-lying configurations of
$^{165,167,169}Yb_{95,97,99}$ at $\hbar\omega = 0.45$ MeV. Nilsson-model parameters
of ref.[19] were used. Predicted gaps at N = 94, 97, and 106 in
the rotating system are indicated.

$$h' = h_{def} + h_{pair} - \omega j_1 \qquad (7)$$

The effect of rotation on the spectrum of single-quasiparticle states is
illustrated in the right-hand portion of Figure 25. Just as in the absence of pair correlation, the Coriolis and centrifugal forces decrease
the energy of the aligned single-particle configurations (i.e. those with
large j_1), the energy of the aligned quasiparticle configurations also
is decreased in a rotating system. The low-lying highly-alignable single

quasiparticle configurations "fill" the pair gap and finally "plunge"
into the quasiparticle vacuum - see Figure 25. This is the nuclear equivalent of "gapless superconductivity". The rotational frequency where a
pair of quasiparticles becomes degenerate with the vacuum corresponds to
the frequency where the energy associated with the Coriolis and centrifugal force on the most alignable pair of quasiparticles equals the energy associated with the pair gap. At larger rotational frequencies the
two-quasiparticle configuration based on such a pair of highly-aligned
quasiparticles becomes the yrast configuration.

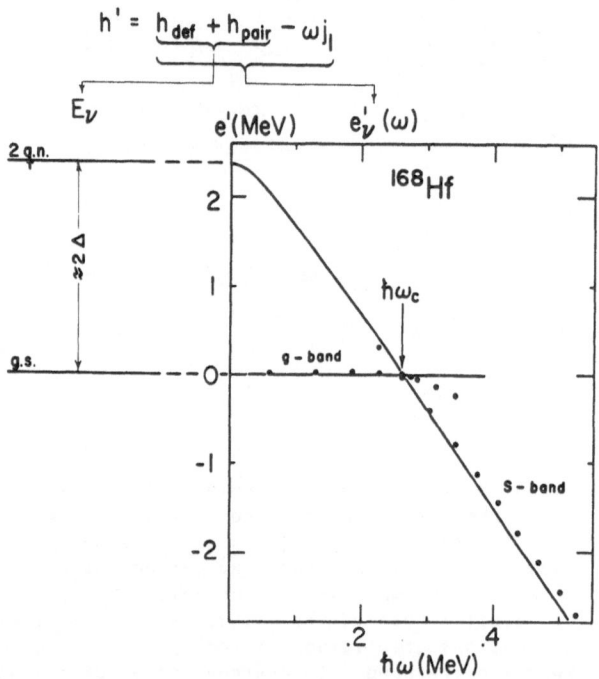

Fig.30. To the left the effects of pair correlations on a low-lying two-
quasiparticle configuration are indicated. To the right the
effects of the Coriolis plus centrifugal force are shown for a
highly-aligned high-j, low-Ω configuration. In this portion
"experimental" energies (points) for the yrast decay sequence[27]
of $^{168}Hf_{96}$ (see Figs.8 and 9), expressed in the intrinsic frame
are compared with cranking-model[62] calculations (lines). The
various terms in the hamiltonian associated with the left- and
right-hand parts are indicated. At the band-crossing frequency,
$\hbar\omega_c$, the Coriolis plus centrifugal forces on the pair of highly-
aligned $i_{13/2}$ quasineutrons compensate the effects of pair
correlations for this pair of aligned quasineutrons. Here the
energies of the ground-state band and that of the aligned
configuration, S-band, are degenerate.

This interpretation is shown more clearly in Figure 30, where only the most high-alignable low-lying two-quasineutron and the vacuum configurations are plotted for $^{168}Hf_{96}$. Pair correlations produce a completely-paired vacuum configuration, which is reduced in energy by about 2Δ relative to the lowest two quasipaticle states - see the left-hand side of Figure 30 and Figure 26. Coriolis plus centrifugal forces cause the highly-aligned two-quasineutron configuration to be depressed in energy when the system is rotated. If the most highly-alignable configurations are near the Fermi level, as is the case for the lower portion of a major shell in prolate deformed nuclei, then the pattern shown in the right-hand portion of Figure 30 is obtained. In this limit the crossing frequency, $\hbar\omega_c$, measures how rapid the nucleus must rotate for the Coriolis and centrifugal forces to compensate for the pair correlations. Thus the frequency of the band crossing corresponding to the alignment of the same pair of quasiparticles in various nuclear configurations is a relative measurement of the pair correlations.

The band crossing, or "backbend"[63-4] is seen in the decay scheme, see Figures 8 and 9, as a change in the intrinsic configuration of the yrast sequence. Indeed when the decay scheme data, $E_x(I)$, is changed to the intrinsic, or rotating, frame*, $e'(\hbar\omega)$, the band crossing frequency, $\hbar\omega_c$, is well defined - see Figure 30. Therefore, it is possible to determine on a relative scale the pair correlations associated with the aligning quasiparticles simply by measuring how rapid the nucleus must rotate for the Coriolis and centrifugal forces to compensate for the pair correlations.

The systematics of such band-crossing frequencies, corresponding to the alignment of the most alignable pair of $i_{13/2}$ quasineutrons, is summarized for light rare-earth isotopes in Figure 31. In most cases the band crossings are observed about 40 keV lower in rotational frequency for isotopes having an odd number of neutrons[68]. That is, the odd-N isotopes usually do not need to rotate as rapidly to compensate for pair correlations. The configuration of the odd quasineutron is "blocked," i.e. a pair cannot be scattered into it; therefore, it does not contribute to the pair correlations. Thus, pair correlations are reduced, and it is not necessary to rotate as fast to overcome them.

In a few instances, however, the crossing frequency in the odd-N isotopes is not reduced (see Figure 31). The unique feature of these seemingly anomalous cases is an oblate intrinsic configuration of the unpaired odd quasineutron[71,72]. (The orbital shapes** of intrinsic configurations are illustrated in Figure 32.) The explicit correlation between the shape of a valence quasineutron orbital and the shift in

---

*The rotating intrinsic frame is a natural coordinate system for the presentation of data. The single-particle hamiltonian, eq.(7), is valid only in this system. For example, other spectroscopic quantities of deformed nuclei, e.g. the magnetic moment, $\vec{\mu}$; the quadrupole moment, Q; and the particle addition and removal operators, a^+ and a, for stripping and pickup reactions, also are valid only in the intrinsic system. The Clebsch-Gordon coefficients multiplying these quantities for a deformed system are the result of a coordinate change from the intrinsic to the laboratory system[65]. The prescription for expressing the decay scheme data, $E_x(I)$, in the intrinsic frame, $e'(\hbar\omega)$, and referring it to the energy of the ground-state configuration, is given several places in the literature, see e.g. refs.[29,62,66]. The intrinsic frame excitation energy, e', often is referred to as the "routhian" because of its similarity with the classical quantity of the same name[67].

**For footnote, see next page.

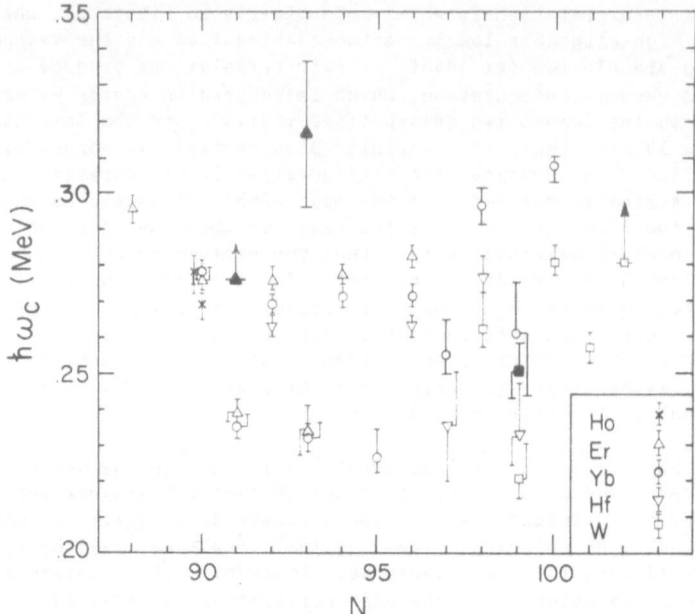

Fig.31. Band crossing frequency, $\hbar w_c$, systematics corresponding to the
alignment of a pair of $i_{13/2}$ quasineutrons in the yrast sequence
of even-N nuclei and for various negative-parity sequences in
odd-N nuclei. The solid points correspond to oblate valence
configurations. The definition of $\hbar w_c$ is illustrated in Figure
30. The original data sources are summarized in refs.[68-72].

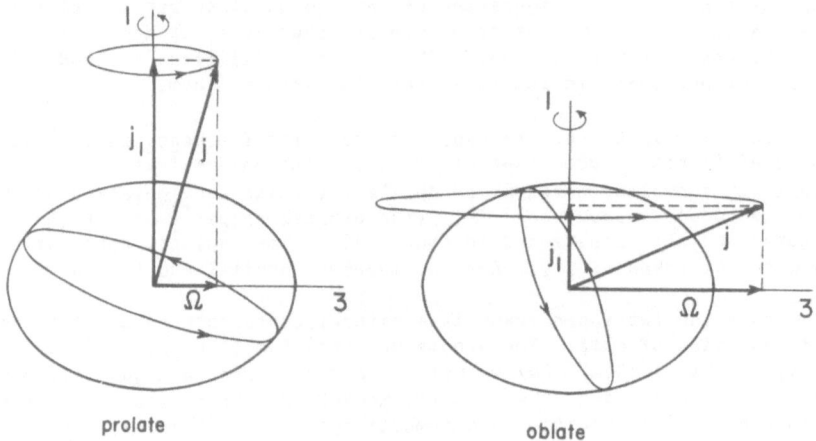

Fig.32. Schematic drawing of prolate and oblate nucleonic orbitals in
prolate deformed nuclei. The intrinsic angular momenta, j, for
particles moving in these orbits are indicated together with the
projections on the nuclear symmetry axis, Ω, and the rotational
axis, j_1.

**It is emphasized that these oblate or prolate shapes refer to the orbi-
tal shape of a specific intrinsic configuration, see Figure 32, not the
nuclear shape. Of course, it is the combined effect of such valence
shell orbitals that distort the nucleus.

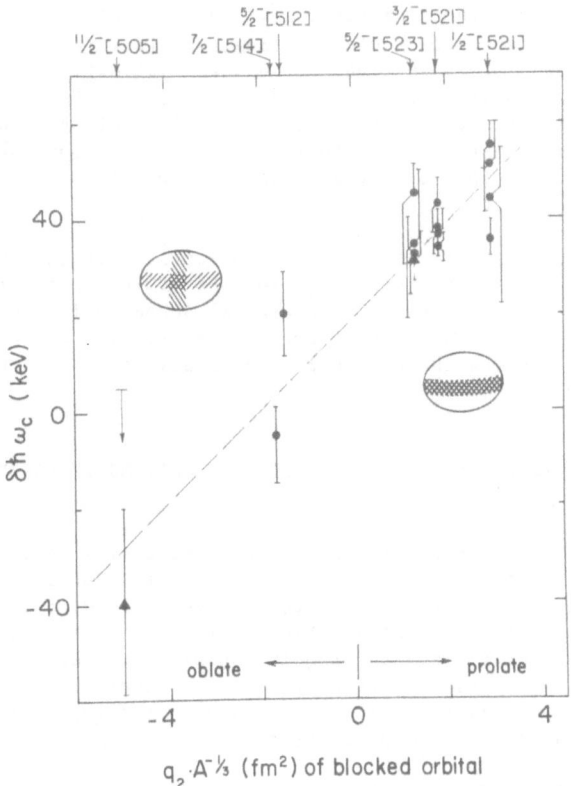

Fig.33. Correlation between the shift in band-crossing frequencies for neighboring even-N and odd-N nuclei, $\delta\hbar\omega_c = \hbar\omega_c(\text{even-N}) - \hbar\omega_c(\text{odd-N})$, and the shape (quadrupole moment) of the blocked quasineutron orbital in the odd-N nuclei for a series of Er, Yb, Hf, W, and Os isotopes. The original data are summarized in the caption to Figure 31. The quadrupole moments, defined[72] as positive for prolate and negative for oblate orbits - see Figure 32, are proportional to the negative slope of the energy on the Nilsson diagram - see Figure 12. The asymptotic quantum numbers of the various valence quasineutrons are indicated at the top of the figure. The minimal overlap between an oblate blocked orbital and the prolate aligning orbital, and the maximum overlap between prolate blocked and aligning orbitals are indicated for the limiting cases in the small diagrams.

band crossing frequencies between neighboring odd- and even-N isotopes is shown in Figure 33. The occupation (or blocking) of an oblate orbital does not reduce the pair correlations for the aligning quasineutrons, which are prolate. Thus the nucleus must rotate as rapidly to compensate for pair correlations as in the neighboring even-N isotopes.

Such data indicate that pair correlations are configuration dependent. This dependence requires quadrupole terms in the pairing interaction[52,57-9] accounting for the relative orientation of the nuclear orbitals - see the introduction to section 2.3.

The crossing-frequency data are a unique example of "shape-dependent pairing" in nuclei. Other examples, e.g. the excitation of low-lying states in near closed-shell nuclei[52], single-particle[73-4] and two-particle[57,75] transfer cross sections, band-head systematics[74], and moments of inertia[71] are sensitive to pairing contributions between a statistical ensemble of orbitals near the Fermi surface. In contrast, the band-crossing frequency data are specific to the pairing contributions between two configurations: i.e. that of the aligning quasiparticles and that of the valence quasiparticles.

The detailed analysis of such data indicates that configuration dependences of both the nuclear shape and the nuclear pair field must be included[72]. The configurations with orbital shapes differing with the average nuclear shapes, which contribute least to pair correlations, are just those that produce the strongest modifications of the nuclear shape.

3. TO BE OR NOT TO BE CORRELATED - PAIR CORRELATIONS IN RAPIDLY-ROTATING NUCLEI

Just as we earthlings, living on the surface of a rotating planet, are subject to Coriolis and centrifugal forces, the nucleons "living" in a rotating nucleus also are subject to such forces. More than a quarter of a century ago Mottelson and Valatin predicted[76] that in a rapidly-rotating nucleus these forces, not only inhibit pair correlations - see Figure 27, but should be sufficiently strong to completely quench them. Likewise, thermal excitation associated with increasing nuclear temperature also quenches[77-9] pair correlations.

This transition from the "superconducting" (paired) to the "normal" (unpaired) phase associated with the quenching of nuclear pair correlations is but one phase transition that can be studied in nuclear matter. Recently, other nuclear phase transitions, e.g. the transition from bound to unbound nuclear matter[80-1] and from hadronic matter to a quark-gluon plasma[82-3] have captured the imagination of our generation of nuclear, particle, and astrophysicists. Since these two phase transitions are of such current interest both at this school[81,83] and elsewhere, I would like to indicate the location of the pair phase transition on a somewhat modified version[84] of the "standard" ultra high energy nuclear matter phase diagram shown in Figure 34. Actually in this plot the pair phase transition becomes lost among the various exotica conjured from the depths of theorists' minds. Therefore, an enlarged diagram of the more traditional nuclear structure physicist "play ground" (indicated by the dashed region in Figure 34) is shown in Figure 35. The predicted[77-8] temparature of the pair phase transition is indicated, as is an expected small temperature increase with increasing nuclear density. An increased density decreases the nuclear level spacing, thereby increasing the scattering of pairs and the pair correlation. Thus it is necessary to go to larger temperatures to quench pair correlations.

Such a plot fails to reflect the quenching of pair correlations by the Coriolis and centrifugal forces in rotating systems - Figure 27 and sect. 2.4. Therefore, rotational frequency and temperature are more appropriate coordinates for considering the pair phase transition. Such a schematic pair phase diagram is compared in Figure 36 with a similar diagram illustrating an analogous process, the quenching of superconductivity by temperature and a magnetic field.

The two remaining questions are indicated schematically in Figure 37: (i) How do we know that pair correlations are quenched and that we

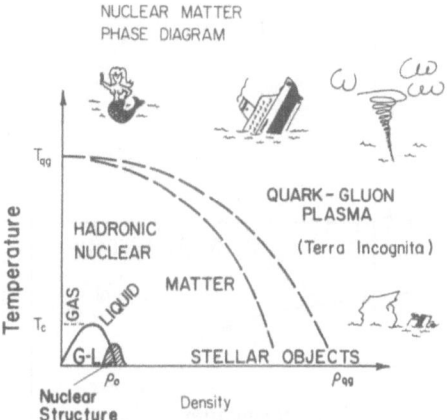

Fig.34. Nuclear matter phase diagram depicting the transition from bound to unbound nuclear matter (gas-liquid transition) and from hadronic matter to a quark-gluon plasma. The region of traditional nuclear structure, (shaded) is enlarged in Figure 35.

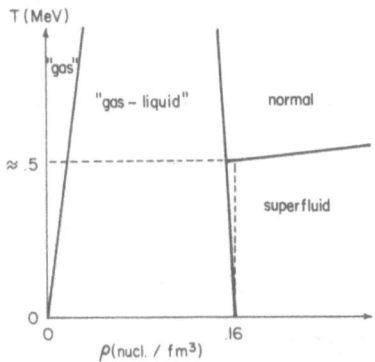

Fig.35. Enlargement of the nuclear structure portion of the phase diagram shown in Figure 34. The superfluid (paired) and normal (unpaired) phases are indicated.

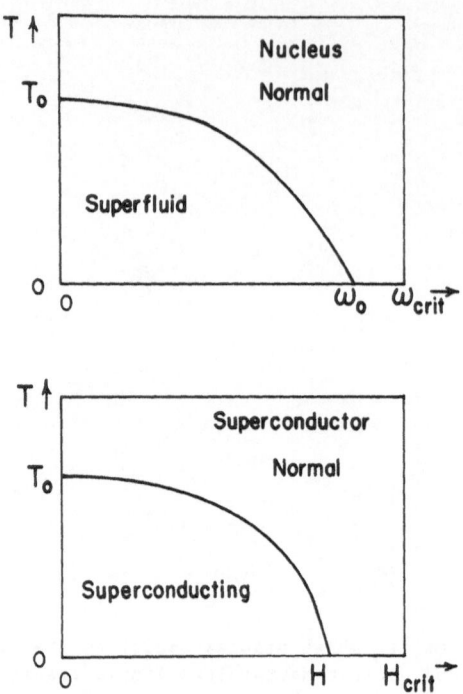

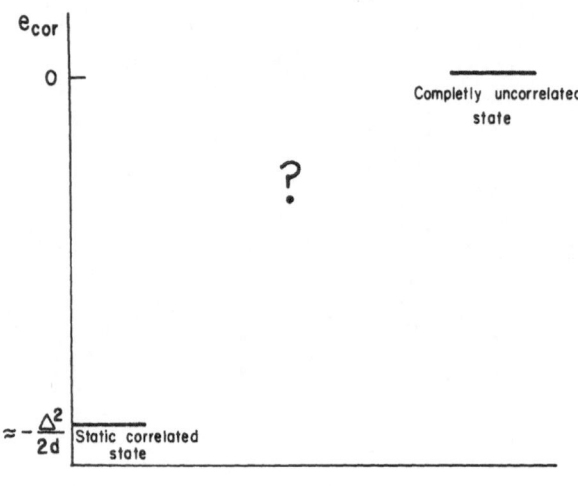

Fig.36. Comparison of the rotational- and temperature-dependent quenching
of nuclear pair correlations (nuclear superfluidity) and the mag-
netic- and temperature-dependent quenching of superconductivity.

Fig.37. Schematic diagram of the energy difference (correlation energy,
e_{corr}) between the static correlated state at rest and the completely uncorrelated state at large rotational frequencies. The
total correlation energy due to pair correlations in the limit of
equally-spaced energy levels[61] is indicated. The pertinent questions are: (i) how to determine that the nucleus is uncorrelated; and (ii) what is the nature of $e_{corr}(\omega)$ in the region of the
quenching of pair correlations.

are in the "completely uncorrelated state"? (ii) What is the nature of the phase transition? Is it sharp (first order) or gradual (higher order)? That is, what is the rotational frequency (also configuration and mass) dependence of the energy associated with pair correlations, e_{corr}?

3.1. Are Static Pair Correlations Quenched in Rapidly-Rotating Systems?

Besides the naive observation of a rigid-body moment of inertia[85] at least four observations have been taken as evidence for the quenching of static pair correlations at large angular momenta:

(i)   Configuration dependent and irregular variation of the Fermi level, λ_n or λ_p, with particle number N or Z, refs.[86-7];

(ii) Disappearance of band crossings[88-9];

(iii) No systematic preference for a specific configuration to be lowest in energy[87,90];

(iv) Disappearance of pair induced alignment[88].

In this subsection topics (i-iii) will be discussed.

3.1.1. The particle-number, or "gauge-space" alignment plot.
As so often in physics we resort to a technique used to establish another correlation – the quadrupole deformation of the nucleus. The parallel roles of the shape and pair fields[60,91-2] in the cranking hamiltonian are utilized to generalize an approach valid for the shape degree of freedom to the pair degree of freedom. Canonical variable plots[92] for the cranking hamiltonian, i.e. $I_x(\omega)$, the aligned angular momentum plot, and $N(\lambda_n)$, the particle-number plot, are shown in the left- and right-hand portions of Figure 38 respectively. These plots are given both in the limit of correlated (upper) and uncorrelated (lower) behavior of these systems with respect to the shape and neutron-pair degrees of freedom.

Just as the irregular variation of E_γ ($\approx 2\hbar\omega$) with aligned angular momentum, I_x, is evidence for noncorrelated, or single-particle, behavior of the near shell model nucleus, $^{150}Dy_{84}$, a similar pattern of the two-neutron binding energy B_{2n} ($=2\lambda_n$) is evidence for an uncorrelated behavior of rapidly-rotating odd-N ytterbium isotopes[87,93-101] with respect to the neutron-pair degree of freedom. The energy associated with the addition of the least bound pair of neutrons coupled to spin zero and positive parity in odd-mass ytterbium isotopes, i.e. λ_n, or with two units of angular momentum in ^{150}Dy, i.e. $\hbar\omega$, depends on the details of the single-particle spectrum of states as a function of N or I_x. Fluctuations occurring in the spacings of the single-particle states, a result of the finite number of constituent nucleons, give the irregular pattern. Indeed, the gaps in the single-particle levels near the spherical $^{150}Dy_{84}$ are larger than those for rotating deformed systems; therefore, the $I_x(\omega)$ plot for ^{150}Dy is more irregular than the $N(\lambda_n)$ plot for the rapidly-rotating deformed ytterbium isotopes.

In contrast, when the nuclear field is correlated with respect to a specific degree of freedom, the energy associated with that degree of freedom, e.g. λ_n or $\hbar\omega$, varies regularly as a function of the corresponding variable, N or I_x – see the upper portion of Figure 38. These regular patterns reflect the correlated nature of the intrinsic configurations, which remain constant for the addition of a pair of neutrons or two units of angular momentum. They are necessary conditions for the existence of a static deformation of the nuclear field with respect to the appropriate degree of freedom, i.e. the neutron pair field for $N(\lambda_n)$

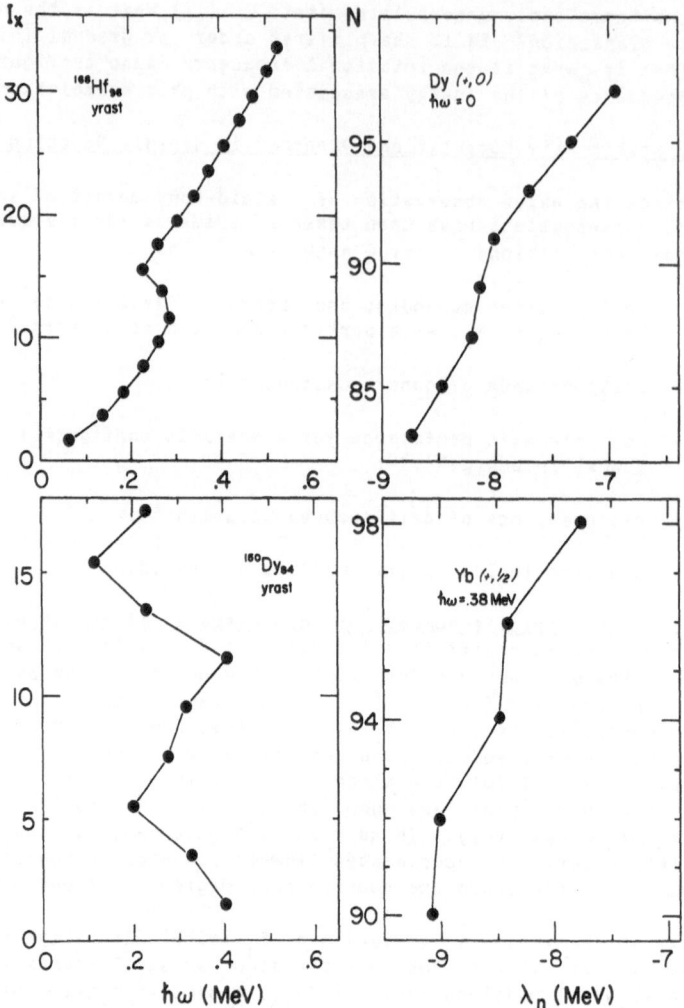

Fig.38. Comparison of configuration-space alignment plots, $I_x(\hbar\omega)$, (left-
hand side) and particle-number, or "gauge-space" alignment,
plots, $N(\lambda_n)$, (right-hand side). The plots shown in the top
(bottom) portion indicate collective (single-particle) behavior.

or the nuclear shape for $I_x(\hbar\omega)$. The discontinuities at specific values
of λ_n and $\hbar\omega$ in these plots indicate that $\Delta N=2$ neighboring isotopes or
states in a decay sequence differing by two units of angular momentum
have different correlations. For example, the singularity at N=89 in the
dysprosium particle-number plot is associated[45,102] with the transition
from nearly spherical to deformed ground-state shapes. Similarly, the
discontinuity in the aligned angular momentum plot for $^{168}Hf_{96}$ ref.[27] is
the "backbend" associated with the shift of the yrast sequence of this
nucleus to a different intrinsic configuration - see Figs.8, 9, and 30.

Particle-number plots for the lowest positive-parity configurations
both in odd- and even-N ytterbium isotopes and for the lowest negative-
parity configurations in odd-N ytterbium isotopes[70,87,93-101,103-5] are
compared at small and large rotational frequencies in Figure 39.

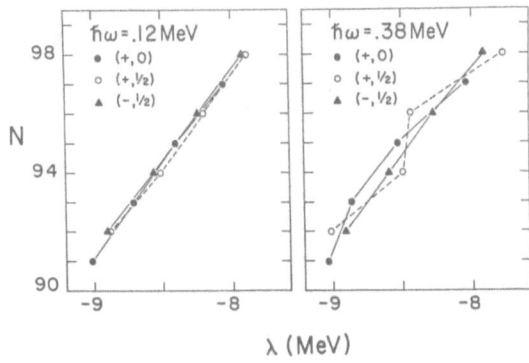

Fig.39. Particle number, or "gauge-space" alignment, plots for the lowest
$(\pi,\alpha) = (+,0)$, $(+,1/2)$, and $(-,1/2)$ decay sequences of ytterbium
nuclei at $\hbar\omega = 0.12$ MeV (left) and 0.38 MeV (right). The neutron
Fermi levels, λ, defined as in ref.[45], correspond to half the
separation energy of the least bound pair of neutrons coupled to
spin zero and positive parity for the specific configuration and
rotational frequency. The data are from ref.[70,87,93-101,103-5].

The change from the linear behavior of $N(\lambda_n)$ with minimal configuration
dependence at small rotational frequencies to an irregular behavior dependent on the configuration at large $\hbar\omega$ indicates a dramatic reduction
of the neutron pair correlations at large rotational frequencies. Such
irregularities in $N(\lambda_n)$ indicate that the static neutron pair field is
too weak to smooth the fluctuations in the single-neutron level spacings
(shown in Figure 29) by producing a neutron pair-correlated state. This
condition indicates that at large rotational frequencies the neutron
pair-gap parameter, λ_n, the deformation (or order) parameter in neutron-
number space, is less than, or of the order of, the average spacing of
time-reversed single-neutron level, d_n. For rare earth nuclei $d_n \approx 300$
keV, ref.[69,106].

3.1.2. The disappearance of band crossings. Two types of band crossings are known: (i) those based on the alignment of a pair of quasiparticles ("virtual" crossings); and (ii) those based on level crossings in
the single-particle spectrum of states ("real" crossings). The latter
type of band crossings[107] between configurations of the same number of
quasiparticles (or particle) corresponds to level crossing such as those
shown in Figure 29. These "real" crossings are independent of the existence of pair correlations. In contrast, the former (most common) type,
corresponding to crossings between bands differing in quasiparticle number by two, depends on the existence of pair correlations[62,68,88] - see
also sect.2.4. The disappearance of such "virtual" band crossings in the
absence of pair correlations is illustrated in Figure 40. With increasing rotational frequency the single-particle states no longer interact
pairwise with the vacuum, thereby increasing the particle number of the
vacuum by two units. In particular this occurs at $\hbar\omega=0.23$ MeV for the
positive-parity quasineutron levels labelled A and B and at the largest
values of $\hbar\omega$ shown for the negative-parity levels labelled E and F.

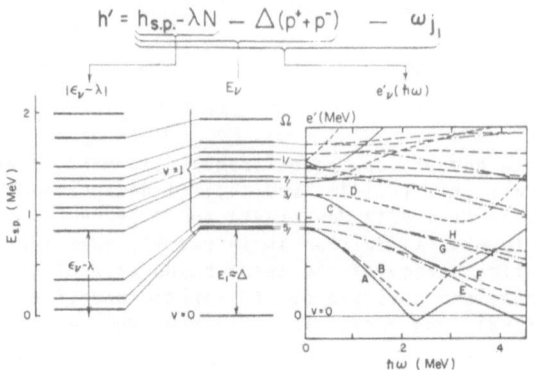

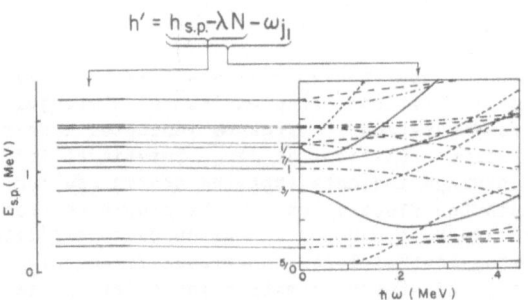

Fig.40. Comparison of single-particle levels in rotating systems with (top) and without (bottom) static pair correlations. In the upper portion the calcualted spectrum of Nilsson states (left), Nilsson-plus pairing states (center) and independent-particle states in a rotating system with pair correlations (right) are shown. In the lower portion the Nilsson spectrum of states is repeated to the left, and the independent-particle spectrum of states in a rotating system without pair corelations is shown to the right. The hamiltonians for each calculated spectrum of states are indicated. The associations of (π,α) with the various curves in the right-hand portion is the same as that of Figures 25 and 29. The parameters entering these calculations, appropriate for single-quasineutron and single-neutron states in $^{165}Yb_{95}$, are given in the caption to Figure 25.

Frauendorf[88] was the first to suggest that the disappearance of these "virtual" band crossings would provide evidence for the absence of pair correlations. Unfortunately, at the time of his suggestion the experimental data did not extend to sufficiently large rotational frequencies for this test to be made for the higher frequency band crossings. Sufficient progress, however, has been made in recent years to allow such tests. Empirical single-neutron diagrams can be constructed[108] by referring the routhians of odd-N isotopes to the (+,0) yrast configurations of the neighboring even-even systems - see Figure 41. (An average of the (+,0) routhians for the N-1 and N+1 isotopes helps to minimize systematic variations due to deformation changes.) Upbends in such a plot correspond to crossings in the (+,0) system that do not occur in the neighboring odd-A isotope, and downbends correspond to crossings in the odd-A system not occurring in the (+,0) configuration of the even-even neighbors. The first crossing (AB) in such plots can be used to define the zero level (vacuum) so that higher frequency crossings can be

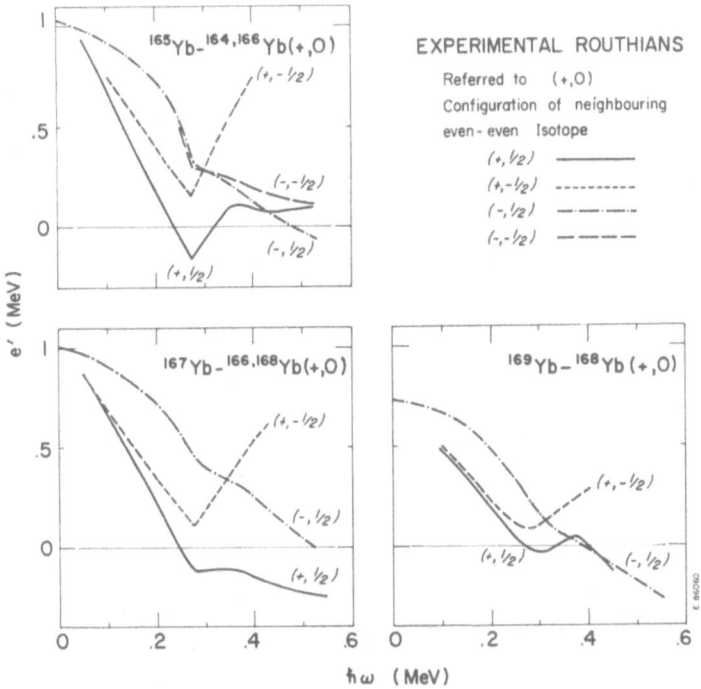

Fig.41. Empirical single-neutron diagrams for $^{165}Yb_{95}$, $^{167}Yb_{97}$, and $^{169}Yb_{99}$ constructed by referring the routhians of odd-N isotopes to the (+,0) configurations of the neighboring even-even systems as indicated. Upbends in such plots correspond to band crossings in the (+,0) system that do not occur in the neighboring odd-N isotope, and downbends correspond to crossings in the odd-A system not occurring in the (+,0) configurations of the even-even neighbors. The vacuum (zero) of these plots is defined from the lowest frequency crossing, i.e. the frequency where the (+,1/2) and (+,-1/2) orbitals upbend. The data are taken from refs.30,87,93,98-100,105.

searched for relative to this "vacuum. (This definition of the vacuum is the main contribution of Frauendorf as such polots of experimental data have been made for some time, see e.g. ref.[99].)

Such empirical plots should be comparable to the single quasiparticle plots in the presence of pair correlations if pair correlations exist. If, however, static pair correlations do not exist these higher frequency crossings will not exist in the even-even systems; therefore, the single-particle levels will not be reflected upward. Instead, they can "plunge into the vacuum." Indeed, such an analysis for $^{165,167,169}Yb_{95,97,99}$ indicates that the larger frequency crossing seems to disappear - see Figure 41. Therefore, the neutron pair correlations in the (+,0) configuration are not sufficiently strong to create the pair of quasineutrons necessary to produce a "virtual" crossing. Or in more technical terms[61] the pair condensate associated with the static deformation of the neutron pair field does not exist (is not strong enough to create a pair of quasineutrons.)

3.1.3. <u>Loss of preference for positive-parity configurations.</u> The experimental spectrum of single-particle states, the basic independent-particle excitations, contains information on the existence, or nonexistence, of correlations of all types, as well as on the details of the

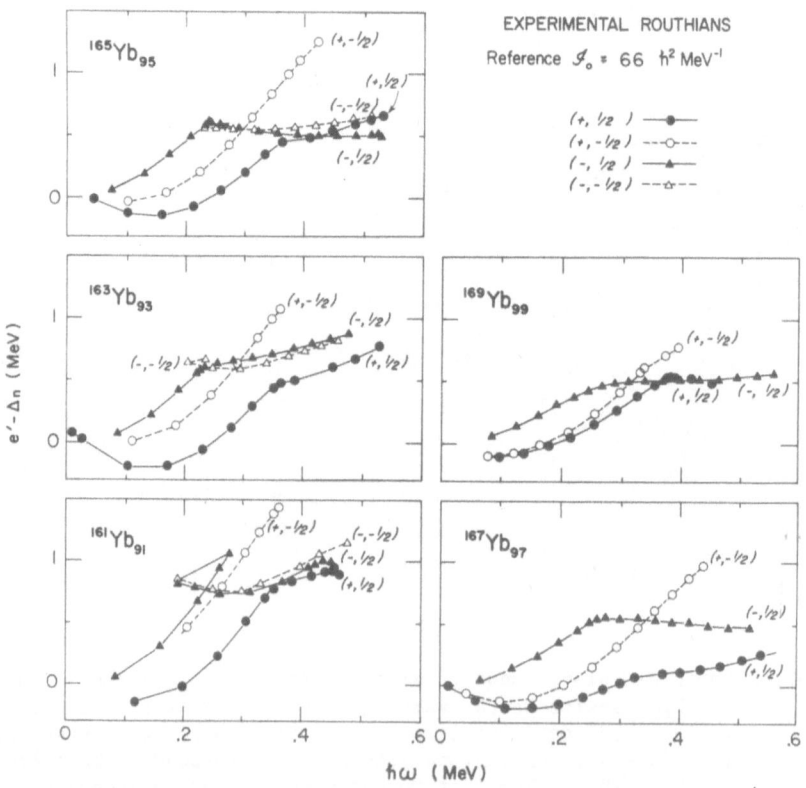

Fig.42. Rotating frame excitation energies, or routhians, e , as a function of ħω for high-spin decay sequences in $^{161,163,165,167,169}Yb_{91,93,95,97,99}$. The experimental data are from refs.[87,93-100]. See caption to Figure 43 for description of reference configuration.

single-particle potential. Such spectra of states for a series of odd-mass ytterbium isotopes[87,93-100] with from 91 to 99 neutrons are shown in Figure 42. A striking difference is observed between the low- and high-frequency portions of these spectra. At low frequencies, $\hbar\omega < 0.35$ MeV, the observed spectra of states for neighboring odd-N isotopes are similar. In these five isotopes an identical ordering of levels is observed: the $(\pi, \alpha) = (+, 1/2)$ configuration is lowest followed by the $(+, 1/2)$ and the $(-, 1/2)$ configurations. The level spacings gradually decrease with increasing mass number, reflecting the movement of the Fermi level, λ_n, away from the highly-alignable, low-Ω components of the $i_{13/2}$ neutrons. Such smooth systematic changes as a function of neutron number reflect the highly-correlated nature of these states and are characteristic of a system with strong neutron pair corelations. The favoring of the positive-parity single neutron states is understandable. The low-Ω components of the $i_{13/2}$ shell model configuration is the most alignable configuration in this shell. Pair correlations allow components of these configurations to mix into the positive-parity yrast states even when they are well below the Fermi level - see Figure 40.

In contrast, at large rotational frequencies, $\hbar\omega > 0.35$ MeV, a distinctly different spectrum of single-neutron states is observed. For each isotope the spectrum is unique. In $^{161}Yb_{91}$, $^{163}Yb_{93}$, and $^{167}Yb_{97}$ the yrast configuration has positive parity; whereas in $^{165}Yb_{95}$ and $^{169}Yb_{99}$ it has negative parity. Such a spectrum of states is characteristic of a system without neutron pair correlations. Each level scheme is uniquely characteristic of the sequence of single-neutron states occupied to give the correct spin, parity, and neutron number.

3.2. What is the Nature of the Pair Phase Transition?

The experimental data discussed in the preceding subsections answers the first of the two questions posed at the beginning of this section. The static neutron pair field is not sufficiently strong at large rotational frequencies to produce "virtual" band crossings or to smooth the fluctuations in the single-neutron levels. These facts argue that the static neutron pair gap, produced by the enhanced scattering of neutron pairs between closely-spaced, time-reversed orbits, is essentially quenched.

To address the nature of a phase transition, it is customary to plot the energy of the system as a function of the degree of freedom associated with this transition*. For the quenching of neutron pair corelations in a rapidly-rotating nucleus the appropriate plot[77] is $e'(\hbar\omega)$, e.g. Figure 30. Since we wish to isolate the energy associated with the various correlations, it is convenient to refer the rotating-frame experimental energies, e', to the least correlated configuration**. This is shown for $^{166}Yb_{96}$ in Figure 43 and for the odd-N ytterbium isotopes in Figure 42. Though at first these plots seem quite different from those expressed relative to a "correlated" reference, the relative energies of different configurations at a given energy are preserved. Compare, e.g. the routhians for the yrast configurations of the isotones $^{168}Hf_{96}$, shown referred to the most correlated (ground-state) configuration in Fig.30,

---

*Such a plot of the free energy of the system as a function of the intensive variable is standard for the study of phase transitions. The equivalent plot for the transition from hadronic matter to quark-gluon plasma is the energy density versus temperature. Such a plot is a product of lattice gauge calculations - see e.g. Figures 3-5 of ref.[109] and ref.[83].

**For footnote, see next page.

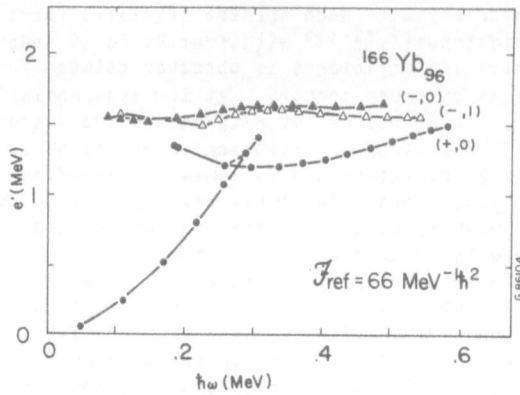

Fig.43. Rotating frame excitation energy (routhian), e', as a function of
the rotational frequency, $\hbar\omega$, for decay sequences in $^{166}Yb_{96}$.
These data[100,105] are "referred" to the least-correlated experimental states in this mass region[92], i.e. configuration with
constant moment of inertia $\mathscr{J}$ = 66 $\hbar^2MeV^{-1}$. Though these routhians appear somewhat different from those referred to the com-
pletely paired ground-state configuration, see Figures 30 and 41,
the relative separation of the various configurations is the
same.

and $^{166}Yb_{96}$, shown referred to the least correlated configuration in
Figure 43.

Not only are the relative energies more discernable when the data
are referred to the "uncorrelated" reference, but the correlation energy
also can be estimated as a function of rotational frequency on an "absolute scale" - see ref.[92]. At the lowest rotational frequencies the positive-parity configuration (i.e. the "completely-paired" ground-state
configuration) is strongly correlated[***] (occurs lower in energy) relative to the negative-parity configurations. With increasing rotational

---

**A constant moment of inertia equal to 66 $\hbar^2MeV^{-1}$ has been chosen[92] for
the reference configuration. This value, about 85% of the rigid-body
value for a deformed nucleus with a quadrupole deformation of ε_2=0.25,
corresponds to the average moment of inertia of the negative-parity
sequences at the largest rotational frequencies. The configurations
corresponding to these sequences are predicted[110-1] to be the least
correlated. The remaining 15% of the moment of inertia is associated
with residual proton pair correlations. Indeed, the average moment of
inertia of bands above the proton band crossings is the rigid-body
value.

***This is the same criteria used to establish correlations in shell-
model nuclei (Fig.19) and for band heads of rotational nuclei (Fig.26).

44

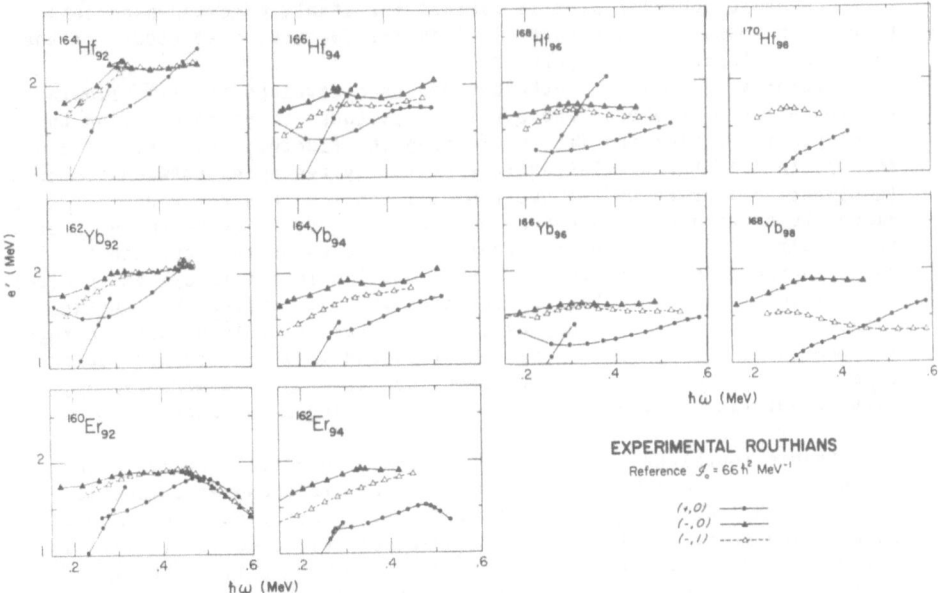

Fig.44. Rotating frame excitation energies, or routhians, e', as a function of rotational frequency, ℏω, for high-spin decay sequences in the N = 92, 94, 96, and 98 isotopes of erbium, ytterbium, and hafnium. The values of e' are defined as in ref.[62] except they are referred to a reference configuration with a constant moment of inertia of 66 ℏ²MeV⁻¹. The data are taken from refs.[27,70,87,94,98,100-1,103,116-19].

frequency the energy of this state increases rapidly with respect to the "unpaired reference," and the negative-parity configurations. Because of the strong pair correlations, the moment of inertia of the (+,0) configurations is greatly reduced. At ℏω=0.27 MeV the ground-state rotational band is crossed* by the S band. Above this band crossing, the (+,0) configuration remains correlated relative to the negative-parity configurations, which are nearly frequency independent above ℏω=0.30 MeV. The positive-parity routhian, however, continues to increase in energy up to ℏω=0.60 MeV where it approaches the energy of the negative-parity configurations. Likewise, the difference in slopes for the positive- and negative-parity configurations is minimized at the largest rotational frequencies. Except for the sharp band crossing at ℏω=0.27 MeV in the (+,0) decay sequence, the remaining features of the $^{166}Yb_{96}$ routhian plot (Figure 43) are smooth. Thus any other "phase transition" must be of higher order.

---

*Crossings between rotational bands based on different intrinsic configurations (e.g. crossings involving the excitation of a pair of quasiparticles - see sect. 2.4.) can be considered as phase transitions[77,108]. If the crossing is sharp, i.e. de'/dω is discontinuous, the transition is "first order."

The ^{166}Yb routhian plot is typical for stably-deformed even-even rare-earth nuclei - see Figure 44. Similar features also occur for the odd-mass isotopes - see Figure 42.

Recently an analysis technique has been developed[92]: (i) allowing the routhians of the odd- and even-mass isotopes to be compared on the same relative scale; and (ii) minimizing the fluctuations associated with the spacings of the single-particle states by averaging between isotopes (isotones) differing by two neutrons (protons). This procedure[*] is equivalent to a coordinate change in particle-number, or gauge space, just as the construction of the normal routhian corresponds to a coordinate change between the laboratory and the rotating intrinsic system in normal space - see sect. 2.4. Double routhians are well suited for studying quantities that vary smoothly as a function of particle number and angular momentum, e.g. shape and pair correlations. They also provide a smooth background which is convenient for "viewing" the more rapidly varying quantities, e.g. the single-particle level spacings.

Double routhians, $e''(\hbar\omega,\lambda_n)$, for the ground-state decay sequences of the even- and odd-mass ytterbium isotopes are shown as a function of $\hbar\omega$ for λ_n=-8.1 MeV (i.e. N=98) in Figure 45. This plot shows several interesting features: (i) for this choice of reference, $e''(-,1/2)$ is independent of rotational frequency for $\hbar\omega$>0.30 MeV; (ii) the (+,0) configuration is correlated (occurs lower in energy) relative to the (-,1/2) configuration; and (iii) the correlation of (+,0) relative to (-,1/2)

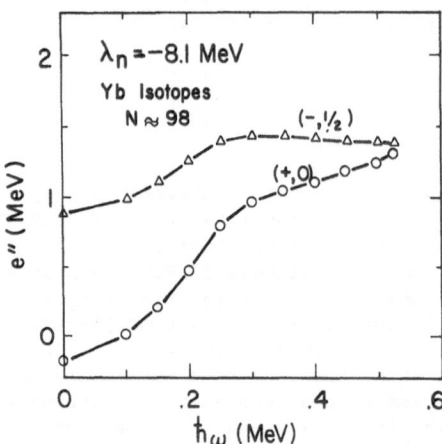

Fig.45. "Double" routhians[92], e'', for the ground-state decay sequences of even-N, (+,0), and odd-N (-,1/2), ytterbium isotopes. These values are plotted as a function of $\hbar\omega$ for λ_n = -8.1 MeV, corresponding to N≈98.

---

[*]Normal routhians, $e'(\hbar\omega)$ are converted[92] to double-routhians, $e''(\hbar\omega,\lambda)$, by subtracting the "gauge" space rotational energy, λN, and referring to the rotating liquid-drop energy, the appropriate gauge-space reference.

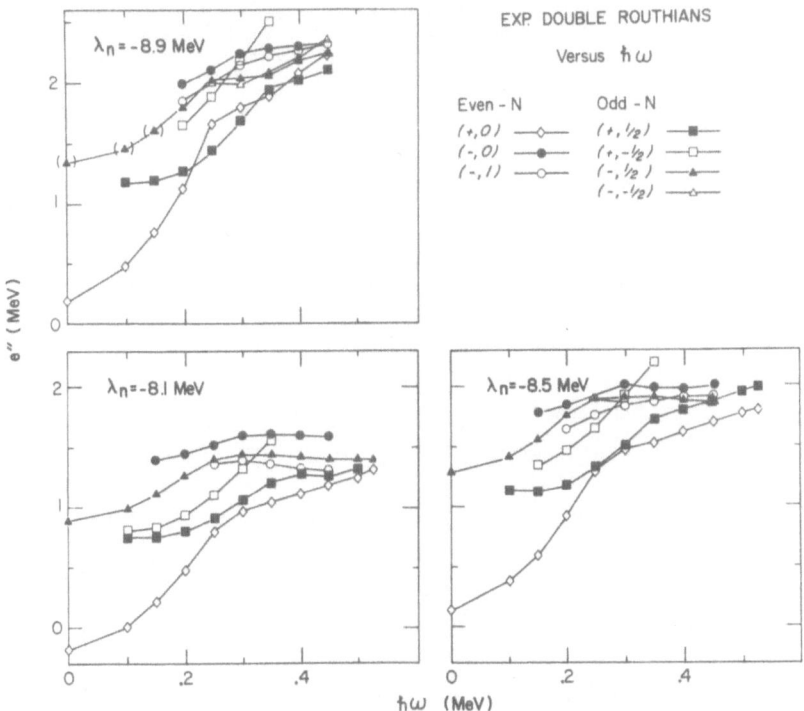

Fig.46. "Double" routhians[92], e'', for a series of ytterbium isotopes[70,87,93-101,103-5] plotted as a function of $\hbar\omega$ for constant
values of λ_n. The configurations are labelled by the conserved
quantum numbers (π,α).

decreases with increasing rotational frequency finally disappearing at
$\hbar\omega=0.525$ MeV where the data stops.

    Similar data also can be obtained for other configurations and va-
lues of the neutron Fermi level, λ_n. Systematic double routhians for the
ytterbium isotopes are shown as a function of $\hbar\omega$ for constant λ_n in
Figure 46 and as a function of λ_n for constant $\hbar\omega$ in Figure 47. The
correlation of the $(+,0)$ configuration relative to that of the negative-
parity configurations, discussed in the preceding paragraphs, is systematic. At the lowest rotational frequencies, where neutron pair correlations are expected to be large, this preference is maximum, >1 MeV. With
increasing rotational frequency these preferences persist but are greatly
reduced in magnitude. For $\hbar\omega=0.50$ MeV the average preference for the
$(+,0)$ configuration is reduced to as little as 100 keV relative to the
lowest negative-parity configuration. A similar, but smaller, correlation also is observed for the $(+,1/2)$ configuration.

    However, there is a paradox! Specific configurations appear to be
correlated (i.e. systematically occur lower in energy) up to $\hbar\omega\approx0.50$ MeV,
whereas static neutron pair correlations appear to be quenched at smaller
rotational frequencies - see subsect. 3.1. Neither is the variation of
energy with rotational frequency characteristic of a first-order phase
transition. There are no sharp discontinuities in the e'($\hbar\omega$) plots,
except at established band crossings. This apparent paradox and the
"smearing" of phase transition are addressed in subsect. 3.3. in terms of
dynamic pair correlations (pair vibrations).

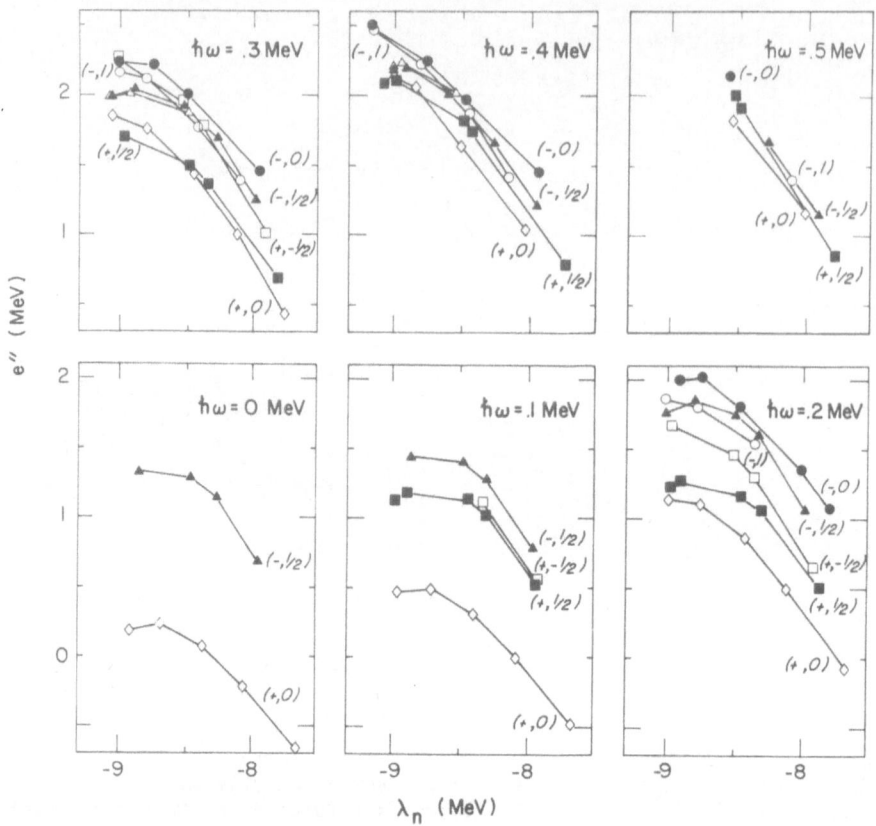

Fig.47. "Double" routhian[92], e'', for a series of ytterbium isotopes as a
function of λ_n for constant values of $\hbar\omega$. See also caption to
Figure 45.

3.3. Dynamic Pair Correlations.

Pair vibrations (resulting from fluctuations in Δ, the "gauge-space"
deformation or order parameter) are known[60-1,112] for shell-model nuclei
in the limit of $\Delta \lesssim d$. Such effects also are expected to be important in
rapidly-rotating nuclei at rotational frequencies where the Coriolis and
centrifugal forces are sufficiently large to reduce Δ to the magnitude of
the average level spacing in a rotating system.

The appropriate model for vibrations is the random phase approximation[113] (RPA). The appropriate basis of such a calculation in rotating
nuclei is the single-particle levels of a rotating deformed nucleus. The
calculation can be made both in the presence and in the absence of static
pair correlations, though until recently they have been confined to the
unpaired regime - see e.g. refs.[110,114-5].

Such calculations[111] for $_{166}Yb_{96}$ and $_{168}Yb_{98}$ are compared in Figure
48 with experimental data [87,100,105]. The effects of fluctuations both
in the presence and in the absence of static pair correlations are included. The corresponding predicted neutron static pair gap, Δ_n is shown in
Figure 49. An estimate of the contributions of pair fluctuations

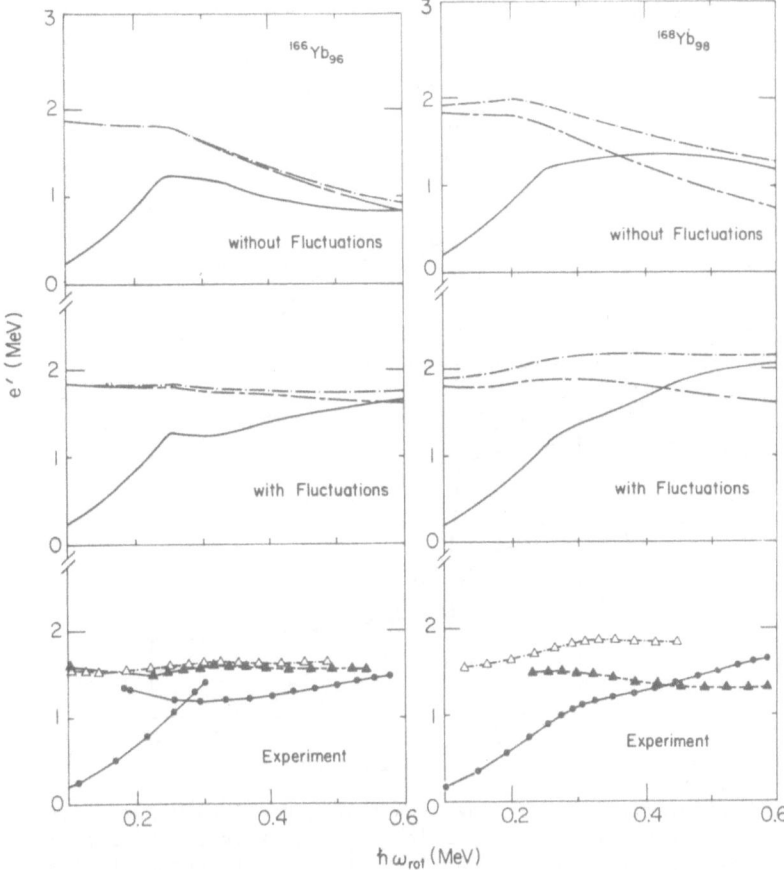

Fig.48. Comparison of calculated[111] routhians without (top) and with
(middle) the effects of dynamic pair fluctuations (RPA) and experiment routhians[87,100,105] (bottom) for $^{166}Yb_{96}$ and $^{168}Yb_{98}$. The
corresponding static pair gaps are shown in Figure 49. The calculations and experimental data are referred to reference configurations with constant moments of inertia equal to 62 and 66
$\hbar^2 MeV^{-1}$ respectively. $(\pi,\alpha)=(+,0)$ is denoted by solid lines and
points; $(-,0)$ by dot-dashed lines and open triangles; and $(-,1)$
by dashed lines and solid triangles.

$$\Delta_{dyn} = \sqrt{G(-\tilde{E}_{corr})} \qquad (8)$$

also is shown in this figure. Here, G is the monopole pair coupling
constant, and $\tilde{E}_{corr}$ is the correlation energy corresponding to pair fluctuations calculated from the RPA solutions[111]. When $\Delta_{dyn} \gtrsim \Delta$, the effects
of pair fluctuations are significant.

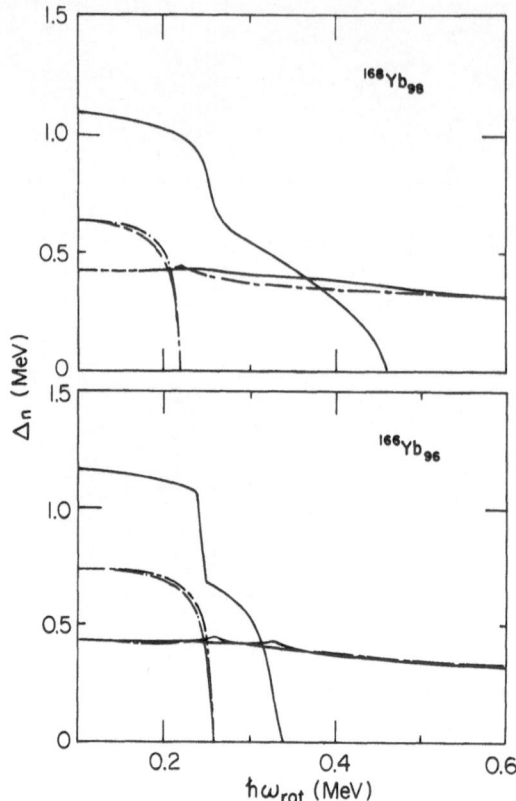

Fig.49. Comparison of static neutron pair gaps (lines approaching zero at
finite values of $\hbar\omega$) for $^{166}Yb_{96}$ and $^{168}Yb_{98}$. Values of the
dynamic neutron pair gap, Δ_{dyn} (lines that extend to large $\hbar\omega$),
as defined by eq.(8), also are shown for comparison. See also
caption to Figure 48.

In general, pair correlations decrease e', thereby increasing the
slope of $e'(\hbar\omega)$, and reducing both the aligned angular momentum[*] and the
moment of inertia. Pair fluctuations "smear" such features in the region
of a small static pair gap, extending this effect to larger rotational
frequencies. Thus pair fluctuations decrease the sharpness of an already
rather smooth predicted pairing "phase transition" - see Figure 48.

The calculated routhians are in excellent agreement with the $^{166}Yb_{96}$
and $^{168}Yb_{98}$ data - see Figure 48. The predicted isotopic dependence for

---

[*]The relation for aligned angular momentum, $i = -de'/d\omega$, can be derived
from eq.(7). $e' = \langle h' \rangle$, and $i = \sum j_1$.

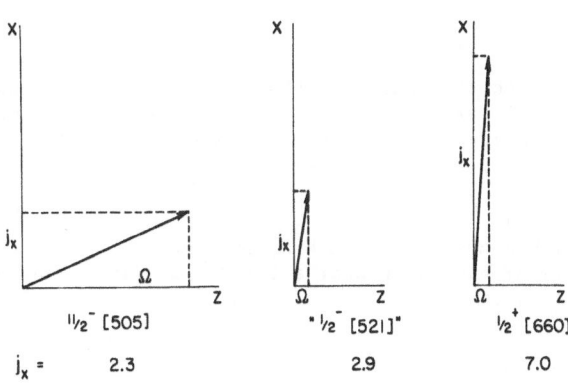

Fig.50. Alignments of the intrinsic angular momentum, j, with respect to
        the nuclear symmetry axis, z, and the rotational axis, x. Values
        are shown for the most-aligned neutron configuration, $1/2^+[660]$,
        and the two least-aligned configurations near the Fermi surface
        for the light ytterbium isotopes. $j=5/2$ and $\Omega=1/2$ was assumed
        for the "$1/2^-[521]$" configuration. Values of $j_x(j_1)$ also are
        given.

both static and dynamic pair correlations is understood. The variation
in the critical frequency associated with the quenching of the static
pair gap between $^{166}Yb_{96}$ and $^{168}Yb_{98}$ (see Figure 48) is the result of a
large pair matrix element for the "$1/2^-[521]$" neutron orbital. Its main
shell model components are $f_{5/2}$ and $p_{3/2}$; thus j_1 is small - see Figure
50. Therefore, it is necessary to rotate more rapidly for the Coriolis
and centrifugal forces ($=-\omega j_1$) to affect the time-reversal properties of
this configuration. This "hot orbital" is nearest the Fermi level for
N=98 - see Figure 29. It, however, is blocked in the negative-parity
configurations of $^{168}Yb_{98}$. Thus, it contributes strongly to both static
and dynamic neutron pair correlations of the (+,0) configuration of
$^{168}Yb_{98}$, less strongly to positive- and negative-parity configurations of
$^{166}Yb_{96}$, and not at all to the negative-parity configurations of
$^{168}Yb_{98}$.

     Other "hot" mid rare earth orbitals, requiring large rotational
frequencies to "unpair", include the $11/2^-[505]$ neutron orbital and the
$1/2^-[411]$ proton orbital. These are orbitals with small j_1 values resulting from either small j's or large Ω's - see Figure 50. However, quadrupole pairing effects[57-9], not included in the calculations of Figures
48 and 49 are expected to be important, especially for the $11/2^-[505]$
orbital[72-4].

     In summary, pair fluctuations (pair vibrations) produce significant
modifications in the correlations of rapidly-rotating, weakly-paired
systems. Features associated with the quenching of static pair correlations, e.g. increased alignment and moments of inertia, are "smoothed" to
larger rotational frequencies. Such effects are sensitive to the aligned
angular momentum j_1, of configurations near the Fermi level. The observed rather featureless variation $e'(\hbar\omega)$ near the pairing "phase transition" is reproduced by calculations that include pair fluctuations.

EPILOGUE

I hope that the students have had only half as much fun attending these lectures as I have in preparing them. Probably the most important consequence of the recent renaisssance of nuclear structure studies, is the group of enthusiastic young physicists that are entering the field. You are the future!

I close with a quote[120] from King Solomon:
"It is the glory of God to conceal a thing: but the honor of kings is to search out a matter."

ACKNOWLEDGEMENTS

The experimental nuclear structure program at the Niels Bohr Institute is a collaborative effort including staff members, guests, and collaborators from Daresbury Laboratory and the Universities of Liverpool, Lund, Jyväskylä, Manchester, and Oslo. In particular the participation of J. C. Bacelar, R. Bengtsson, T. Bengtsson, R. A. Broglia, R. Chapman, S. Frauendorf, M. Gallardo, I. Hamamoto, J. C. Lisle, J. N. Mo, W. Nazarewicz, M. A. Riley, H. A. Ryde, J. F. Sharpey-Schafer, Y. R. Shimizu, J. Simpson, G. Sletten, P. O. Tjøm, C.-H. Yu, J.-Y. Zhang, and S. Åberg is acknowledged. Most of the ideas presented have been formulated in collaboration with G. B. Hagemann and B. Herskind. This work has been supported by grants from the Commemorative Association for the Japan World Exposition and the Danish Natural Science Research Council.

APPENDIX

An experimental estimate of the moment of inertia of a rotating deformed nucleus can be obtained from either the first or second derivative of the excitation energy with respect to angular momentum. These two moments of inertia, labeled $\partial^{(1)}$ and $\partial^{(2)}$, are referred to as "kinematic" and "dynamic" moments of inertia respectively. These quantities are easily derived from the rotational formula (4):

$$
\begin{aligned}
E_I - E_{I-2} = E_\gamma &\\
&= \frac{\hbar^2}{2\partial^{(1)}}[I(I+1) - (I-2)(I-1)] \\
&= \frac{\hbar^2}{2\partial^{(1)}}(2I+1)
\end{aligned}
\tag{A1}
$$

and

$$
\begin{aligned}
E_{\gamma 2} - E_{\gamma 2} &= \frac{\hbar^2}{2\partial^{(2)}} \{(2I+1) - [2(I-2)+1]\} \\
&= 4\hbar^2/\partial^{(2)}
\end{aligned}
\tag{A2}
$$

The gamma-ray transition energies are as defined in Figure A1.

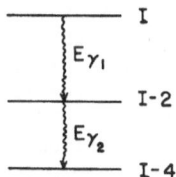

Fig.A1. Pedagogical gamma-ray decay sequence of two transitions defining the experimental quantities used to calculate the kinematic, $\partial^{(1)}$, and dynamic, $\partial^{(2)}$, moments of inertia.

Therefore, the kinematic and dynamic moments of inertia are given by:

$$
\frac{\partial^{(1)}}{\hbar^2} = (2I+1)/E_\gamma
\tag{A3}
$$

and

$$
\frac{\partial^{(2)}}{\hbar^2} = 4/(E_{\gamma_1} - E_{\gamma_2})
\tag{A4}
$$

respectively. These equations can be generalized to configurations other than that of the ground-states of even-even isotopes - see eqs. (1) and (2) - if the total angular momentum is replaced by I_1 (or I_x), the component of angular momentum on the rotational axis

$$
I_1 = \sqrt{(I+1/2)^2 - K^2}
\tag{A5}
$$

K is the component of the total intrinsic angular momentum on the nuclear symmetry axis. That is,

$$
K = \Sigma\Omega ,
$$

where the sum is over all nonpaired configurations. Note that K is not a conserved quantity in a rotating system.

References

1. T. Lucretius Carus, De Rerum Natura (On the Nature of Things), about 50 B.C.; Eng. trans. C. Bailey (Oxford University Press, 1936, Oxford).
2. See e.g. Galileo Galilei, A Dialogue on the Two Principal Systems of the World, 1632; Eng. trans. S. Drake (University of California Press, 1953, Berkeley).
3. I. S. Newton, Philosophiae Naturalis Principia Mathematica, (The Royal Society, 1687, London) Book III, Propositions XVIII-XX; Eng. trans. A. Motte (University of California Press, 1947, Berkeley).
4. An excellent account of the Lapland arc measurement is contained in I. Todhunter, History of the Mathematical Theories of Attraction and the Figure of the Earth (Constable, 1873, London), reprint edition (Dover, 1962, New York).
5. J. K. Beatty, B. O'Leary, and A. Chaikin, The New Solar System (Cambridge University Press, 1981, Cambridge).
6. J. Binney, Comments on Astrophysics 8:27 (1978).
7. P. Moore, The Atlas of the Universe (Beazley Ltd., 1970, London).
8. C. G. J. Jacobi, Poggendorf Annalen der Physik und Chemie 33 (1834) 229; reprinted in Gesammelte Werke 2 (Reimer, 1882, Berlin) p. 17.
9. B. A. Smith, L. A. Soderblom, T. V. Johnson, A. P. Ingersoll, S. A. Collins, E. M. Shoemaker, G. E. Hunt, H. Masursky, M. H. Carr, M. E. Davies, A. F. Cook II, J. Boyce, G. E. Danielson, T. Owen, . C. Sagan, R. F. Feebe, J. Veverka, R. G. Strom, J. F. McCauley, D. Morrison, G A. Briggs, and V. E. Suomi, Science 204:951 (1979).
10. B. A. Smith, L. Soderblom, R. Beebe, J. Boyce, G. Briggs, A. Bunker, S. A. Collins, C. J. Hansen, T. V. Johnson, J. L. Mitchell, R. J. Terrile, M. Carr, A. F. Cook II, J. Cuzzi, J. B. Pollack, G. E. Danielson, A. Ingersoll, M. E. Davies, G. E. Hunt, H. Masursky, E. Shoemaker, D. Morrison, T. Owen, C. Sagan, J. Veverka,. R. Strom, an V. E. Suomi, Science 212:163 (1981).
11. B. A. Smith, L. Soderblom, R. Batson, P. Bridges, J. Inge, H. Masursky, E. Shoemaker, R. Beebe, J. Boyce, G. Briggs, A. Bunker, S. A. Collins, C. J. Hansen, T. V. Johnson, J. L. Mitchell, R. J. Terrile, A. F. Cook II, J. Cuzzi, J. B. Pollack, G. E. Danielson, A. P. Ingersoll, M. E. Davis, G. E. Hunt, D. Morrison, T. Owen, C. Sagan, J. Veverka, R. Strom, and V. E. Suomi, Science 215:499 (1982).
12. P. Farinella, P. Paolicchi, E. F. Tedesco, and V. Zappala, Icarus 46:114 (1981).
13. S. J. Weidenschilling, Icarus 46:124 (1981).
14. J. Jeans, Problems of Cosmogony and Stellar Dynamics (Cambridge University Press, 1919, Cambridge).
15. A. F. Cook, in Physical Studies of Minor Planets, ed. T. Gehrels, NASA SP-267, p. 155.
16. S. J. Weidenschilling, Icarus 44:807 (1980).
17. B. A. Smith, L. A. Soderblom, R. Beebe, D. Bliss, J. M. Boyce, A. Brahic, G. A. Briggs, R. H. Brown, S. A. Collins, A. F. Cook II, S. K. Croft, J. N. Cuzzi, G. E. Danielson, M. E. Davies, T. E. Dowling, D. Godfrey, C. J. Hansen, C. Harris, G. E. Hunt, A. P. Ingersoll, T. V. Johnson, R. J. Krauss, H. Masursky, D. Morrison, T. Owen, J. B. Plescia, J. B. Pollack, C. C. Porco, K. Rages, C. Sagan, E. M. Shoemaker, L. A. Sromovsky, C. Stoker, R. G. Strom, V. E. Suomi, S. P. Synnott, R. J. Terrile, P. Thomas, W. R. Thompson, and J. Veverka, Science 233:43 (1986).
18. S. G. Nilsson, Mat.-Fys. Medd. Dan. Vid. Selsk. 29,no.16 (1961).
19. S. G. Nilsson, C. F. Tsang, A. Sobiczewski, Z. Szymanski, S. Wycech, C. Gustavsson, I. L. Lamm, P. Möller, and B. Nilsson, Nucl Phys. A131:1 (1969).

20. S. Cohen, F. Plasil, and W. J. Swiatecki, Ann. Phys. 82:557 (1974).

21. V. M. Strutinsky, Ark. Fys. 36:629 (1967).

22. See e.g. G. Andersson, S. E. Larsson, G. Leander, P. Möller, S. G. Nilsson, I. Ragnarsson, S. Åberg, R. Bengtsson, J. Dudek, B. Nerlo-Pomorska, K. Pomorski, and Z. Szymanski, Nucl. Phys. A268:205 (1976).

23. N. Bohr and F. Kalckar, Mat.-Fys. Medd. Dan. Vidensk. Selsk. 14,no.10 (1937); reprinted in Niels Bohr's Collected Works, ed. E. Rudinger (North Holland, 1986, Amsterdam) Vol.9, p.225.

24. Aa. Bohr and B. R. Mottelson, Phys. Rev. 90:717 (1953).

25. O. Bakander, C. Baktash, J. Borggreen, J. B. Jensen, J. Kownacki, J. Pedersen, G. Sletten, D. Ward, H. R. Andrews, O. Hausser, P. Skensved, and P. Tarus, Nucl. Phys. A389:93 (1982).

26. G. Sletten, S. Bjørnholm, J. Borggreen, J. Pedersen, P. Chowdhury, H. Emling, D. Frekers, R. V. F. Janssens, T. L. Khoo, Y. H. Chung, and H. Kortelahti, Phys. Lett. 135B:33 (1984).

27. R. Chapman, J. C. Lisle, J. N. Mo, E. Paul, A. Simcock, J. C. Willmott, J. R. Leslie, H. G. Price, P. M. Walker, J. C. Bacelar, J. D. Garrett, G. B. Hagemann, B. Herskind, A. Holm, and P. J. Nolan, Phys. Rev. Lett. 51:2265 (1983).

28. J. R. Grover, Phys. Rev. 82:557 (1951).

29. R. Bengtsson and J. D. Garrett, in Collective Phenomena in Atomic Nuclei, International Review of Nuclear Physics, Vol.2 (World Scientific, 1984, Singapore) p. 194.

30. J. Simpson, M. A. Riley, J. R. Cresswell, P. D. Forsyth, D. Howe, B. M. Nyako, J. F. Sharpey-Schafer, J. Bacelar, J. D. Garrett, G. B. Hagemann, B. Herksind, and A. Holm, Phys. Rev. Lett. 53:648 (1984).

31. P. O. Tjøm, R. M. Diamond, J. C. Bacelar, E. M. Beck, M. A. Deleplanque, J. E. Draper, and F. S. Stephens, Phys. Rev. Lett. 55:2405 (1985).

32. M. A. Riley, J. D. Garrett, J. F. Sharpey-Schafer, and J. Simpson, Phys. Lett. B177:15 (1986).

33. B. M. Nyako, J. Simpson, P. J. Twin, D. Howe, P. D. Forsyth, and J. F. Sharpey-Schafer, Phys. Rev. Lett. 56:2680 (1986).

34. P. J. Twin, B. M. Nyako, A. H. Nelson, J. Simpson, M. A. Bentley, H. W. Cranmer-Gordon, P. D. Forsyth, D. Howe, A. R. Mokhtar, J. D. Morrison, J. F. Sharpey-Schafer, and G. Sletten, Phys. Rev. Lett. 57:811 (1986).

35. See e.g. I. Ragnarsson, S. Åberg, and R. K. Sheline, Physica Scripta 24:215 (1981).

36. T. Bengtsson and I. Ragnarsson, Physica Scripta T5:165 (1983).

37. I. Ragnarsson, Zeng Xing, T. Bengtsson, and M. A. Riley, Physica Scripta 34:651 (1986).

38. J. D. Garrett, G. B. Hagemann, and B. Herskind, Ann. Rev. Nucl. Part. Sci. 36:419 (1986).

39. T. L. Khoo, R. K. Smithers, B. Haas, O. Hausser, H. R. Andrews, D. Horn, and D. Ward, Phys. Rev. Lett. 41:1027 (1978).

40. J. Styczen, Y. Nagai, M. Piiparinen, A. Ercan, and P. Kleinheinz, Phys. Rev. Lett. 50:1752 (1983).

41. P. J. Twin, in these Proceedings.

42. V. M. Strutinski, Nucl. Phys. A122:1 (1968).

43. C. G. Andersson, R. Bengtsson, T. Bengtsson, J. Krumlinde, G. Leander, K. Neergård, P. Olanders, J. A. Pinston, I. Ragnarsson, Z. Szymanski, and S. Åberg, Physica Scripta 24:266 (1981).

44. A. Pakkanen, Y. H. Chung, P. J. Daly, S. R. Faber, H. Helppi, J. Wilson, P. Chowdhury, T. L. Khoo, L. Ahmad, J. Borggreen, Z. W. Grabowski, and D. C. Radford, Phys. Rev. Lett. 48:1350 (1982).

45. R. Bengtsson, J.-Y. Zhang and S. Åberg, Phys. Lett. 105B:5 (1981).

46. H. W. Cranmer-Gordon, P. D. Forsyth, D. V. Elenkov, D. Howe, J. F. Sharpey-Schafer, M. A. Riley, G. Sletten, J. Simpson , I. Ragnarsson, Z. Xing, and T. Bengtsson, Nucl. Phys. A (in print).

47. S. Lunardi, M. Ogawa, H. Backe, M. Piiparinen, Y. Nagi, and P. Kleinheinz, in Proc. Sym. on High-Spin Phenomena in Nuclei, Argonne, 1979, ANL/PHY-79-4, p. 403.

48. M. A. Riley, N. J. Ward, P. D. Forsyth, H. W. Griffiths, D. Howe, J. F. Sharpey-Schafer, J. Simpson, J. C. Lisle, E. Paul, and P. Walker, in Proceedings of the Fifth Nordic Meeting on Nuclear Physics, Jyväskylä, Finland, March 1984, p. 353.

49. N. Rowley, Phys. Bull. 34:110 (1983).

50. C. M. Lederer and V. S. Shirley, Table of Isotopes, 7th edition (Wiley, 1978, New York).

51. L. Pauling and E. B. Wilson, Introduction to Quantum Mechanics (McGraw-Hill, 1935, New York).

52. R. A. Broglia, D. R. Bes, and B. S. Nilsson, Phys. Lett. 40B:213 (1974).

53. J. Rainwater, Phys. Rev. 79:432 (1950).

54. Aa. Bohr, Phys. Rev. 81:134 (1951).

55. N. Bjerrum, in Nernst Festschrift (Knapp, 1912, Halle) p. 90.

56. S. Bjørnholm, Aa. Bohr, and B. R. Mottelson, in Physics and Chemistry of Fission (IAEA, 1974, Vienna) Vol.I, p. 367.

57. D. R. Bes, R. A. Broglia, and B. Nilsson, Phys. Lett., 40B:338 (1972).

58. I. Ragnarsson and R. A. Broglia, Nucl. Phys. A263:315 (1976).

59. M. Diebel, Nucl. Phys. A419:221 (1984).

60. R. Broglia, O. Hansen, and C. Riedel, Adv. Nucl. Phys. 6:287 (1973).

61. Aa. Bohr and B. R. Mottelson, Nuclear Structure (Benjamin, 1975, Reading, Mass.) Vol.II, Chap. 6.

62. R. Bengtsson and S. Frauendorf, Nucl. Phys. A327:139 (1979).

63. A. Johnson, H. Ryde, and J. Sztarkier, Phys. Lett. 34B:605 (1971).

64. F. S. Stephens and R. S. Simon, Nucl. Phys. A183:257 (1972).

65. Aa. Bohr and B. R. Mottelson, see Ref. 61, Chap. 4.

66. S. Frauendorf, in Proc. of the Nucl. Phys. Workshop, Trieste, Italy, October 1981, ed. C. H. Dasso et al., (North-Holland, 1982, Amsterdam) p. 111.

67. H. Goldstein, Classical Mechanics (Addison-Wesley, 1950, Reading, Mass.) Chap. 7.

68. J. D. Garrett, O. Andersen, J. J. Gaardhøje, G. B. Hagemann, B. Herskind, J. Kownacki, J. C. Lisle, L. L. Riedinger, W. Walus, N. Roy, S. Jönsson, H. Ryde, M. Guttormsen, and P. O. Tjøm, Phys. Rev. Lett. 47:75 (1981).

69. J. D. Garrett and S. Frauendorf, Phys. Lett. 108B:77 (1982).

70. S. Jönsson, N. Roy, H. Ryde, W. Walus, J. Kownacki, J. D. Garrett, G. B. Hagemann, B. Herskind, R. Bengtsson, and S. Åberg, Nucl. Phys. A449:537 (1986).

71. J. D. Garrett, G. B. Hagemann, B. Herskind, J. Bacelar, R. Chapman, J.C. Lisle, J. N. Mo, A. Simcock, J. C. Willmott, and H. G. Price, Phys. Lett. 118B:297 (1982).

72. J. C. Bacelar, M. Diebel, O. Andersen, J. D. Garrett, G. B. Hagemann, B. Herskind, J. Kownacki, C.-X. Yang, L. Carlén, J. Lyttkens, H. Ryde, W. Walus, and P. O. Tjøm, Phys. Lett. 152B:157 (1985).

73. R. J. Peterson and J. D. Garrett, Nucl. Phys. A414:59 (1984).

74. J. D. Garrett, in High Angular Momentum Properties of Nuclei, ed. N. R. Johnson, Nuclear Science Research Conference Series, Vol.4 (Harwood, 1983, Chur) p. 17.

75. R. F. Casten, E. R. Flynn, J. D. Garrett, O. Hansen, R. J. Mulligan, D. R. Bes, R. A. Broglia, and B. Nilsson, Phys. Lett. 40B:333 (1972).

76. B. R. Mottelson and J. D. Valatin, Phys. Rev. Lett. 5:511 (1960).
77. A. L. Goodman, Nucl. Phys. A369:365 (1981).
78. K. Tanabe, K. Sugawara-Tanabe, and H. J. Mang, Nucl. Phys. A357:20 (1981); and K. Sugawara-Tanabe, K. Tanabe, and H. J. Mang, ibid. p. 45.
79. S. Levit, in these Proceedings.
80. See e.g. D. H. Boal, Nucl. Phys. A447:479c (1985); and refs. therein.
81. J. P. Bondorf, in these Proceedings.
82. See e.g. Proceedings of the Fifth International Conference on Ultra-Relativistic Nucleus-Nucleus Collisions, Nucl. Phys. A461 (1987).
83. G. Baym, in these Proceedings.
84. J. P. Bondorf, J. D. Garrett, C. Gregoire, H. H. Gutbrod, G. C. Morrison, S. Nagamiya, and J. Rafelski, Nucl. Phys. A447:655c (1985).
85. See e.g. I. Hamamoto, Physica Scripta T5:10 (1983).
86. J.-Y. Zhang, Nucl. Phys. A421:353c (1984).
87. J. C. Bacelar, M. Diebel, C. Ellegaard, J. D. Garrett, G. B. Hagemann, B. Herskind, A. Holm, C.-X. Yang, J.-Y Zhang, P. O. Tjøm, and J. C. Lisle, Nucl. Phys. A442:509 (1985).
88. S. Frauendorf, Nucl. Phys. A409:243c (1983).
89. J. D. Garrett, in Selected Topics in Nuclear Structure: Proc. of the XXI Winter School on Nucl. Phys., Zakopane, Poland, eds. R. Broda and Z. Stachura (Inst. of Nucl. Phys., 1986, Krakow) p. 15.
90. J. D. Garrett, Nucl. Phys. A421:313c (1984).
91. S. Åberg, R. Bengtsson, I. Ragnarsson, and J.-Y. Zhang, in Proc. of the Nucl. Phys. Workshop, Trieste, Oct. 1981, eds. C. Dasso et al. (North-Holland, 1982, Amsterdam) p. 273.
92. J.-Y. Zhang, J. D. Garrett, J. C. Bacelar, and S. Frauendorf, Nucl. Phys. A453:104 (1986).
93. C. Schück, N. Bendjaballah, R. M. Diamond, Y. Ellis-Akovali, K. H. Lindenberger, J. O. Newton, F. S. Stephens, J. D. Garrett, and B. Herskind, Phys. Lett. 142B:253 (1984).
94. J. Simpson, in Proc. XXIII Int. Winter Meeting on Nucl. Phys., Bormio, 1985, ed. I. Iori (Ricerca Scientifica ed Educazione Permanente, 1985, Milano) p. 187.
95. J. J. Gaardhøje, Thesis Univ. of Copenhagen (1980).
96. D. R. Haenni, H. Deijbakhsh, R. P. Schmitt, G. Mouchatz, L. L. Riedinger, and M. P. Fewell, Cyclotron Inst. Texas A&M Progress Report, 1983, p. 5.
97. J. Kownacki, J. D. Garrett, J. J. Gaardhøje, G. B. Hagemann, B. Herskind, S. Jönsson, N. Roy, H. Ryde, and W. Walus, Nucl. Phys. A394:269 (1983).
98. C. Schück, F. Hannachi, R. Chapman, J. C. Lisle, J. N. Mo, E. Paul, D. J. G. Love, P. J. Nolan, A. H. Nelson, P. M. Walker, Y. Ellis-Akovali, N. R. Johnson, N. Bendjaballah, R. M. Diamond, M. A. Deleplanque, F. S. Stephens, G. Dines, and J. Draper, in Proc. XXIII Int. Winter Meeting on Nucl. Phys., Bormio, Italy, 1985 (Ricerca Scientifica ed Educazuione Permanente, Milano, 1985) p. 294.
99. N. Roy, S. Jönsson, H. Ryde, W. Walus, J. J. Gaardhøje, J. D Garrett, G. B. Hagemann, and B. Herskind, Nucl. Phys. A382:125 (1982).
100. E. M. Beck, J. C. Bacelar, M. A. Deleplanque, R. M. Diamond, F. S. Stephens, J. E. Draper, B. Herskind, A. Holm, and P. O. Tjøm, Nucl. Phys. A464:472 (1987).

101. P. M. Walker, W. H. Bentley, S. R. Faber, R. M. Ronningen, R. B. Firestone, F. M. Bernthal, J. Borggreen, J. Pedersen, and G. Sletten, Nucl. Phys. A365:61 (1981).

102. G. G. Dussel, J. J. Liotta, and R. P. J. Perazzo, Nucl. Phys. A388:606 (1982).

103. J. N. Mo, S. Sergiwa, R. Chapman, J. C. Lisle, E. Paul, J. D. Garrett, M. A. Riley, G. Sletten, J. Hattula, and M. Jääskeläinen, Daresbury Ann. Report, 1985, p. 44; and Manchester-Jyväskylä-Daresbury-NBI Preprint, 1987.

104. L. L. Riedinger, O. Andersen, S. Frauendorf, J. D. Garrett, J. J. Gaardhøje, G. B. Hagemann, B. Herskind, Y. V. Makovetzky, J. C. Waddington, M. Guttormsen, and P. O. Tjøm, Phys. Rev. Lett. 44:568 (1980).

105. W. Waluś, N. Roy, S. Jönsson, L. Carlén, H. Ryde, J. D. Garrett, G. B. Hagemann, B. Herskind, Y. S. Chen, J. Almberger, and G. Leander, Physica Scripta 24:324 (1981).

106. J. D. Garrett, G. B. Hagemann, B. Herskind, Nucl. Phys. A409:259c (1983).

107. C.-X. Yang, J. Kownacki, J. D. Garrett, G. B. Hagemann, B. Herskind, J. C. Bacelar, J. R. Leslie, R. Chapman, J. C. Lisle, J. N. Mo, A. Simcock, J. C. Willmott, W. Waluś, L. Carlén, S. Jönsson, J. Lyttkens, H. Ryde, and P. O. Tjøm, Phys. Lett. 133B:39 (1983).

108. S. Frauendorf, in Proc. of the Fifth Nordic Meeting on Nucl. Phys., Jyväskylä, Finland, 1984, p. 19.

109. J. Engels, Nucl. Phys. A461:317c (1987).

110. R. A. Broglia, M. Diebel. S. Frauendorf, and M. Gallardo, Phys. Lett. 166B:252 (1986).

111. Y. R. Shimizu, J. D. Garrett, R. A. Broglia, M. Gallardo, and E. Vigezzi, NBI-Milano-Sevilla Preprint, 1987.

112. D. R. Bes and R. A. Broglia, Nucl. Phys. A80:289 (1966).

113. See e.g. D. Pines and P. Nozieres, The Theory of Quantum Liquids (Benjamin, 1966, New York).

114. R. A. Broglia and M. Gallardo, Nucl. Phys. A447:489c (1985).

115. Z. Szymanski, in Nuclear Structure 1985, eds. R. Broglia, G. B. Hagemann, and B. Herskind (North-Holland, 1985, Amsterdam) p. 343.

116. R. Chapman, J. O. Newton, J. C. Lisle, J. N. Mo, E. Paul, J. D. Garrett, G. B. Hagemann, B. Herskind, and A. Nelson, in Proc. of the International Nucl. Phys. Conf., Harrogate, U.K., 1986, Vol. 1, p. 161; and to be published.

117. Y. K. Agarwall, J. Recht, H. Hübel, M. Guttormsen, D. J. Decman, H. Kluge, K. H. Maier, N. Roy, J. Dudek, and W. Nazarewicz, Phys. Lett. 122B:207 (1983).

118. H. Hübel, K. P. Blume, K. H. Maier, A. Maj, H. Kluge, A. Kuhnert, J. Recht, and M. Guttormsen, Proc. XXIII International Winter Meeting on Nucl. Phys., Bormio, 1985 (Ricerca Scientifica ed Educazione Permanente, 1985, Milano) p. 255; and Nucl. Phys. A (in press).

119. J. C. Lisle, J. D. Garrett, G. B. Hagemann, B. Herskind, and S. Ogaza, Nucl. Phys. A366:281 (1981).

120. King Solomon, The Bible, Proverbs 25:2, about 950 B.C.; Eng. trans., King James Version, 1611.

SUPERDEFORMATION IN ^{152}Dy

P.J. Twin

SERC Daresbury Laboratory
Daresbury
Warrington WA4 4AD, U.K.

ABSTRACT

A discrete line band has been observed in ^{152}Dy extending up to 60$\hbar$. It has a quadrupole moment of 19 ± 3 eb and a moment of inertia of 83$\hbar^2$ MeV^{-1} in excellent agreement with predictions for a superdeformed shape.

INTRODUCTION

There have been many calculations[1]) which predict that ^{152}Dy will become prolate with a large deformation of $\beta \approx 0.6$ before it becomes unstable to fission. These structures are expected to become yrast at the highest spins. Once populated most of the decay intensity will remain in the superdeformed bands until it becomes sufficiently non-yrast for the decays out of band to compete with the strongly enhanced in-band E2 decays.

The first evidence[2]) for superdeformation was obtained at Daresbury using TESSA2 in which the E_γ-E_γ correlation technique was applied to the unresolved γ-ray continuum originating from states in ^{152}Dy above 40$\hbar$. The essence of the technique is that in the matrix of the energies of two coincident γ-rays the signature for rotational bands is the presence of ridges parallel to the $E_{\gamma 1} = E_{\gamma 2}$ diagonal. A narrow ridge was observed extending from 0.8 MeV to 1.35 MeV. The position of the ridge yielded a moment of inertia $\mathscr{J}^{(2)}_{band} = (85 \pm 2)\hbar^2$ MeV^{-1} which corresponded to the value expected for superdeformed collective bands.

A higher statistics experiment with 150 million events was recently carried out using the TESSA3 spectrometer[3]). The states in ^{152}Dy were populated using the ^{108}Pd(^{48}Ca,4n) reaction at 205 MeV with a target consisting of two 500 µg cm^{-2} self-supporting foils isotopically enriched to 95% in ^{108}Pd. A 15 mg^{-2} gold catcher foil was positioned 5 cm downstream of the targets such that it was outside the focus of the Ge array but within the full detection efficiency of the BGO ball. Four matrices of Ge-Ge coincidences were constructed. The "isomer" matrices were produced by demanding that three or more additional BGO ball elements fired between 30 ns and 200 ns after the main γ-γ event. The remainder of the events were placed in "prompt" matrices. The "low fold" and "high fold" matrices were obtained by setting sum energy and fold conditions which divided the ^{152}Dy events into two equal parts. All matrices included a narrow window 10 ns wide on the Ge-ball time interval to reject the slower neutron induced Ge events. The four matrices were composed of various fractions of ^{151}Dy(5n), ^{152}Dy(4n) and ^{153}Dy(3n) events as given in Table 1 and the total isomer matrix had a 56% efficiency in detecting ^{152}Dy events.

Table 1.

Final nucleus	Prompt		Isomer	
	Low fold	High fold	Low fold	High fold
^{151}Dy	43%	13%	30%	9%
^{152}Dy	40%	47%	70%	91%
^{153}Dy	17%	40%	–	–
Fraction of superdeformed events	0.3%	0.6%	0.5%	1.2%

A new band of 19 transitions extending from 602 keV to 1449 keV was observed in spectra generated from all four matrices. The spectra shown in Figs. 1 and 2 were obtained by summing gates on many of the γ-rays observed in the band. The signal-to-noise level in the four spectra in Fig. 1 show the variation with respect to both the fraction of ^{152}Dy events and the fraction of superdeformed events. Clearly the quality of the spectra improve by increasing both fractions.

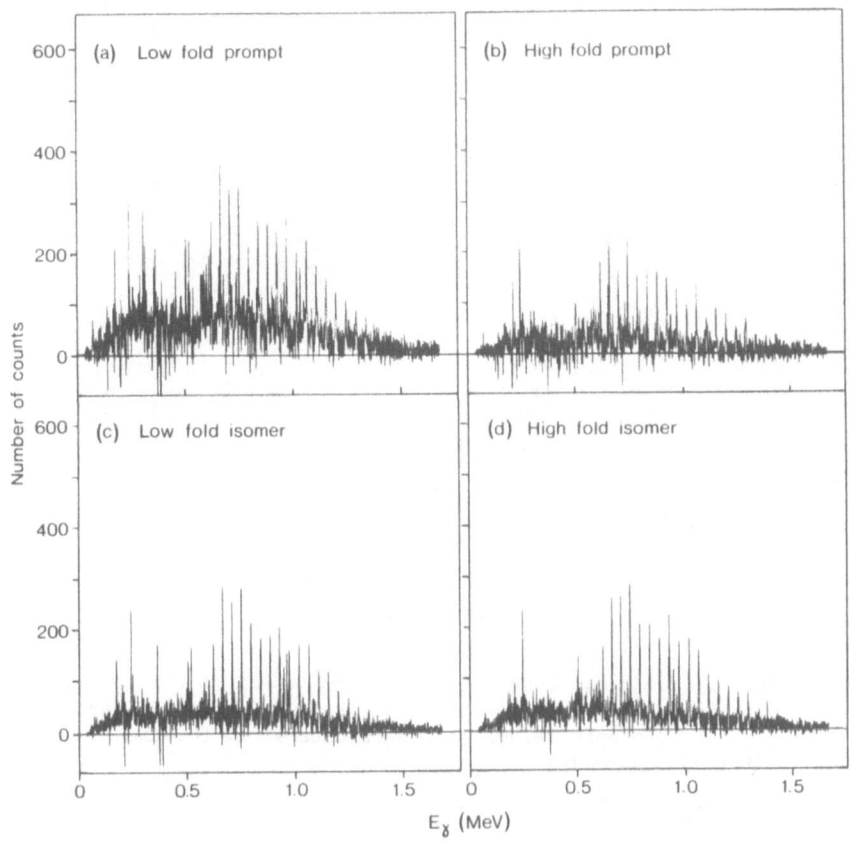

Fig. 1. The superdeformed band obtained by summing gates on the
647 keV, 692 keV, 737 keV, 829 keV, 923 keV, 1065 keV,
1160 keV, 1209 keV, 1257 keV and 1305 keV γ-rays. The spectra
are derived from (a) the "prompt" low fold matrix, (b) the
"prompt" high fold matrix, (c) the "isomer" low fold matrix
and (d) the "isomer" high fold matrix.

SPIN ASSIGNMENTS OF BAND MEMBERS

The spectra in Fig. 2 are the total "prompt" and "isomer" matrices.
The γ-rays indicated in the "prompt" matrix are the $2^+ \to 0^+$, $4^+ \to 2^+$ and
$6^+ \to 4^+$ decays in ^{152}Dy arising from that part of the decay out of the band
which bypasses the isomer and this has been determined to be 10 ± 3%.
The γ-rays indicated in the "isomer" matrix are transitions between
oblate states of spins 19^- and 25^- which are steps in the decay path out
of the superdeformed band. These give an average spin on entry to the
yrast oblate states of 21.7$\hbar$.

The superdeformed band must be yrast or close to yrast at spins
greater than 55$\hbar$ and therefore due to its high moment of inertia it will
be around 4 MeV above yrast when it de-excites, as illustrated in Fig. 3.

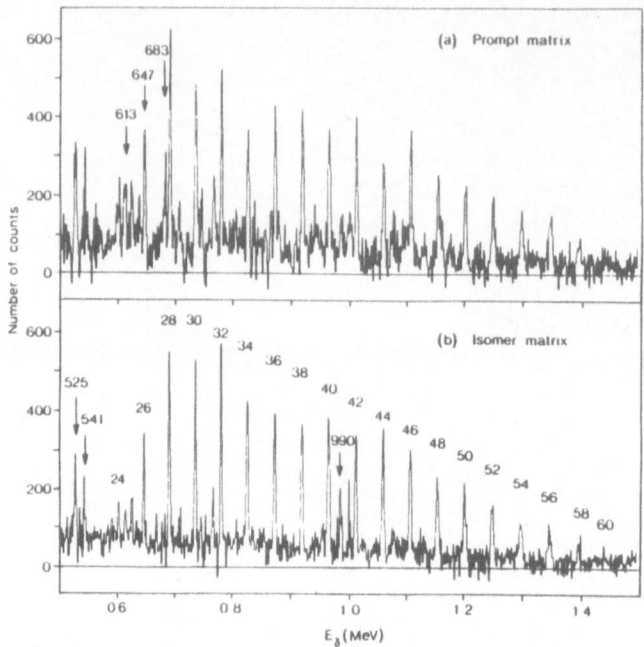

Fig. 2   The superdeformed band obtained by summation of the low and
high fold "isomer" (Fig. 2a) and "prompt" (Fig. 2b) spectra in
Fig. 1.   The spin values are those determined for the state
emitting the γ-rays.

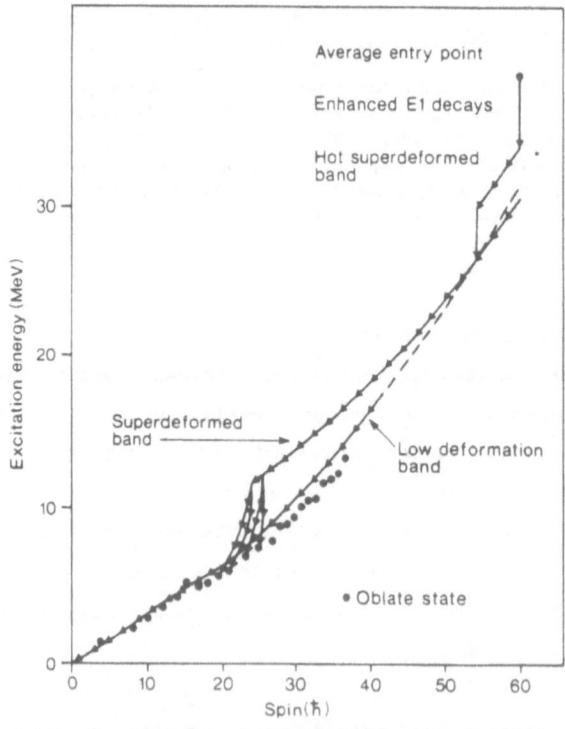

Fig. 3   The decay paths of the prolate collective bands in ^{152}Dy.
Decay paths feeding and de-exciting the superdeformed band are
illustrated.

It is proposed that this decay proceeds via statistical-like cascades of several dipole transitions. Assuming that the superdeformed band has even spins we assign the 647 keV transition to be (26^+-24^+) as this yields an average spin loss of $3.1\hbar$ for the statistical decays. Therefore the top transition in the band is the (60^+-58^+) transition at an excitation energy of about 30 MeV in ^{152}Dy. This extends the highest spin discrete level seen from about $46\hbar^4)$ to $60\hbar$ which is very close to the fission limit for the nucleus.

OTHER SUPERDEFORMED BANDS IN ^{152}Dy

The E_γ-E_γ correlation matrix has been examined using the techniques[2,5]) in which spectra of ΔE_γ were obtained for different regions of $\bar{E}_\gamma$. Peaks were observed at ΔE_γ = 47 keV corresponding to the superdeformed ridge in the "isomer" matrices. However, the intensities in these peaks were unaltered when the regions of $\bar{E}_\gamma$ were restricted to include only the known discrete line transitions in the superdeformed band and the remaining intensity showed no evidence of peaks corresponding to superdeformed ridges. Therefore it is concluded that the discrete line band accounts for the whole of the ridge structure observed in the γ-ray continuum measurements in ^{152}Dy. Thus, there is no evidence below $50\hbar$ for γ-rays from other superdeformed bands at higher excitation energies and any initial population of such bands must have fed into the yrast superdeformed bands above this spin.

THE QUADRUPOLE MOMENT OF THE BAND

The quadrupole moment of the superdeformed band has been deduced from an experiment using the Doppler Shift Attenuation Method. The target consisted of a single 1.3 mg cm^2 ^{108}Pd foil backed with 15 mg cm^{-2} of gold, so that the recoiling nuclei stopped in the target and backing. All the γ-rays associated with the oblate states in ^{152}Dy have lifetimes greater than a few picoseconds[6]) and, therefore, are not Doppler shifted. However, the decays from the higher spin superdeformed states are known to have lifetimes less than 100 fs[7]) and thus be Doppler shifted. These lifetimes were determined from measurements of the fractional Doppler shifts plus a knowledge of the slowing down time of the recoils in the target and its backing.

The gains of the germanium detectors were matched so that all the stopped transitions were aligned. The fast superdeformed γ-rays should then be Doppler shifted to higher and lower energies in the four forward detectors (θ = 35° cosθ = 0.82) and the four backward detectors

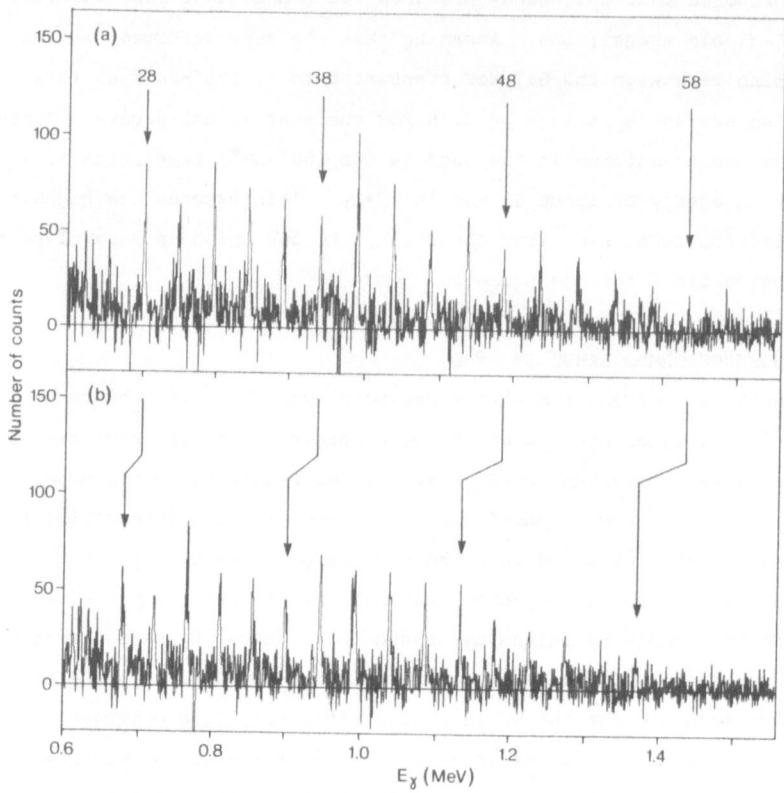

Fig. 4 Spectra of the superdeformed band with a backed target showing
the γ-rays Doppler shifted to (a) higher energies at θ = 35°
and (b) lower energies at θ = 145°.

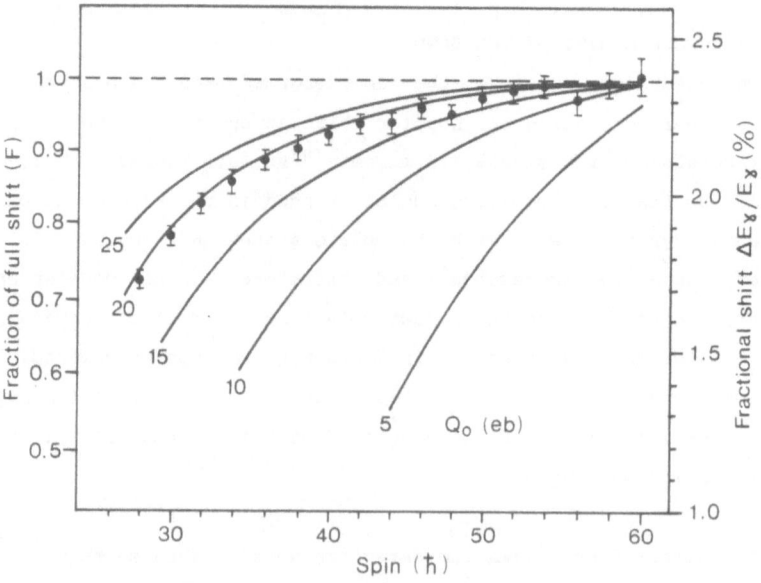

Fig. 5 The fractional Doppler shift as a function of the spin of the
emitting state.

($\theta = 145°$, $\cos\theta = -0.82$), respectively. The spectra shown in Fig. 4 were generated by summation of gates on the known superdeformed γ-rays. They clearly illustrate that the energies of the transitions are shifted and accurate measurements could be made of the average fractional shift $\Delta E_\gamma / E_\gamma = \langle \frac{v}{c} \cos\theta \rangle$, where v is the recoil velocity of nucleus. The fractional shift is plotted in Fig. 5 as a function of the spin of the emitting state. The data show that from an essentially full shift at the highest spin there is a gradual decrease in the fractional shift.

A knowledge of the feeding intensities of the superdeformed band is required to obtain lifetime values from the measured Doppler shifts. These intensities show that the band is 50% fed by spin 55ħ and 80% fed by spin 50h. From this data and the essentially full shifts observed at the highest spins it is concluded that there is no longer lived feeding of these states and that the γ-rays below 50ħ are completely fed via the superdeformed band. Therefore, in the analysis, the assumption was made that the band is fully fed at 60ħ and the effective lifetimes of each subsequent transition are the result of the cumulative feeding down the band. This produces an effective lifetime of each state which is approximately five times larger than the intrinsic lifetime. A second assumption was that the deformation remains constant throughout the band, a feature which is borne out by the constancy of the moment of inertia. Thus a single value of the intrinsic quadrupole moment was utilised.

The average time taken for the recoiling ions to traverse the target is 60 fs and therefore the γ-rays from the highest spin states are emitted before the recoils leave the target. Consequently in this case the slowing down process in the target plays a very important role. Also, over the time range of interest, the dominant process in both target and backing is electronic stopping. The Northcliffe and Schilling[8]) stopping powers were used after normalisation to the alpha stopping powers of Ziegler and Chu[9]). The nuclear stopping was also included following the method of Blaugrund[10]). The measured effective lifetimes are listed in Table 2.

Calculated fractional Doppler shifts are shown in Fig. 5 for quadrupole moments (Q_o) of 5, 10, 15, 20 and 25 eb. The data is fitted statistically by a Q_o of 19 ± 1 which is equivalent to a B(E2) strength of 2660 W.u. Taking into account an additional 15% error due to uncertainties in the slowing down process the value of Q_o becomes 19 ± 3 eb. This is very much larger than the measured values for Q_o of approximately

Table 2.

Spin $(\hbar)$	E_γ (keV)	Lifetimes τ_m(fs) Effective	Intrinsic^a	Spin $(\hbar)$	E_γ (keV)	Lifetimes τ_m(fs) Effective	Intrinsic^a
24	602.3±0.3			42	1017.0±0.2	33±7	5.5
26	647.2±0.2			44	1064.8±0.2	33±7	4.4
28	692.2±0.2	190±10	38	46	1112.7±0.3	22±6	3.5
30	737.5±0.2	135±10	28	48	1160.8±0.3	26±6	2.8
32	783.5±0.3	105±10	20	50	1208.7±0.3	14±7	2.3
34	829.2±0.2	82±10	15	52	1256.6±0.3	9±6	1.9
36	876.1±0.2	62±8	11	54	1304.7±0.3	5^{+7}_{-5}	1.5
38	923.1±0.2	54±11	9	56	1353.0±0.3	16^{+10}_{-8}	1.3
40	970.0±0.2	43±8	6.9	58	1401.7±0.4	3^{+10}_{-3}	1.1

^{a}The intrinsic lifetime is the calculated value assuming Q_o = 19 eb.

5 eb[11]) in rotational bands of normal deformation ε_2 = 0.2 in rare earth nuclei. It is consistent with the calculated value[12]) for the super-deformed quadrupole moment of ^{152}Dy of 18 eb.

THE MOMENT OF INERTIA OF THE BAND

The lifetime experiment also yielded important new data on some of the transitions in the superdeformed band. In the previous thin target measurement some of the γ-rays had similar energies to high intensity background γ-rays associated with oblate states in ^{152}Dy and, therefore, made it difficult to determine their energies accurately. However, in the backed target data these γ-rays were Doppler shifted into regions of lower intensity background and thus their energies could be accurately determined. The mean transition energies derived by averaging data from both experiments are listed in Table 2 and the resultant static ($\mathcal{J}^{(1)}_{band}$) and dynamic ($\mathcal{J}^{(2)}_{band}$) moments of inertia are illustrated in Fig. 6. The two moments are identical over the spin regime above 44$\hbar$, indicating that the nucleus is behaving like a rigid rotor. At lower spins the static moment slowly decreases with a corresponding increase in the dynamic moment. This feature may be related to an increase of fluidity or the reappearance of pairing as the nuclear rotation drops.

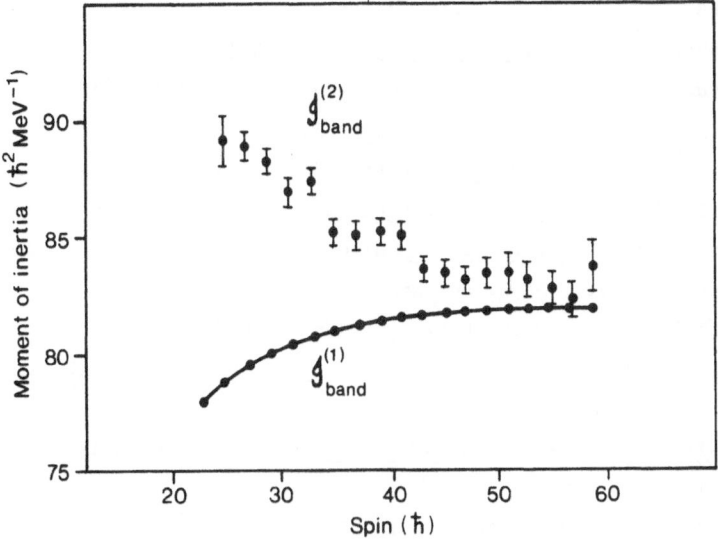

Fig. 6 The moments of inertia, $\mathscr{J}^{(1)}_{band}$ and $\mathscr{J}^{(2)}_{band}$, for the super-deformed band in ^{152}Dy.

CONCLUSION

In summary a discrete line band had been identified in ^{152}Dy extending from $22\hbar$ to $60\hbar$. It has the correct quadrupole moment and moment of inertia for a superdeformed prolate shape with $\varepsilon_2 = 0.6$ (an axis ratio of 2:1). The band exhausts the ridge structure and this indicates that it is the yrast superdeformed band separated by a considerable energy gap of probably 1 - 1.5 MeV from other bands of the same shape. The de-excitation from the band must occur several MeV above the yrast oblate states and it takes place within a couple of band transitions. The band is fed mainly between $50\hbar$ and $60\hbar$ and some phenomena such as enhanced giant dipole strength must be involved in order to cool the nucleus to yrast at this very high spin. These observations have opened up the whole field of superdeformed nuclei and many experimental groups are looking for other examples and theoretical groups are attempting to explain the present data. The months ahead will be exciting times as new data and theories are presented.

I thank my many colleagues who helped with the analysis and the experiments, particularly Mike Bentley, Barna Nyakö and Costas Kalfas.

REFERENCES

1. T. Bengtsson et al., Phys. Scripta 24: 200 (1981);
   J. Dudek and W. Nazarewicz, Phys. Rev. C 31: 298 (1985).

2. B.M. Nyakö et al., Phys. Rev. Lett. 52: 507 (1984).

3. P.J. Twin et al., Phys. Rev. Lett. 57: 811 (1986).

4. P.O. Tjom et al., Phys. Rev. Lett. 55: 2405 (1985).

5. B. Herskind, J. Phys. (Paris) Colloq. 41: C10 (1980).

6. T.L. Khoo et al., Phys. Scripta 24: 283 (1981).

7. P.J. Twin et al., Phys. Rev. Lett. 55: 1380 (1985).

8. L.C. Northcliffe and R.F. Schilling, Nucl. Data Tables A7: 233 (1970).

9. J.F. Ziegler and W.K. Chu, At. Data Nucl. Data Tables 13: 463 (1974).

10. A.E. Blaugrund, Nucl. Phys. 88: 501 (1966).

11. M. Oshima et al., Phys. Rev. C 31: 1988 (1986).

12. I. Ragnarsson and S. Absers, Phys. Lett. 180B: 191 (1986).

GIANT DIPOLE RESONANCES IN HOT NUCLEI

Kurt A. Snover

Department of Physics
Nuclear Physics Laboratory
University of Washington
Seattle, WA 98195

ABSTRACT

This paper represents a condensation of the lectures presented on the topic of giant dipole resonances in excited nuclei. It is meant to be self-contained; however, extensive reference is made to a recent review,[1] to which the reader is referred for further background and additional detail.

INTRODUCTION

We are currently experiencing a period of rapid growth in both our experimental and theoretical understanding of high energy gamma ray emission in heavy ion collisions. For heavy ion projectiles of mass $A \gtrsim 12$ and bombarding energies $E/A \lesssim 5\text{-}7$ MeV per nucleon, high energy γ-rays are produced by statistical emission from a thermally equilibrated, highly excited compound nucleus formed by fusion. The Giant Dipole Resonance is among the various excitation mode present in the equilibrated excited nucleus and it is the decay of the GDR which produces high energy gamma rays in the energy range $E_\gamma \sim 10\text{-}30$ MeV. Typically the GDR is excited along with other degrees of freedom, so that its decay leaves the nucleus in an excited state. An analysis of the γ-ray spectrum shape provides the average properties of the GDR built on excited states, which are sensitive to the size and shape of the excited nucleus. Indeed, much of the current interest in these studies stems from the information which the GDR provides on the shape (deformation) of excited nuclei.

A complementary area of study, which I mention only briefly here, is (p,γ) reactions populating excited nuclear states. Such reaction studies to low-lying discrete states were used many years ago to demonstrate the existence of GDRs built on excited states. More recently the systematics of the GDR have been studied built on a wide range of different individual states or groups of states up to $E_x \sim$ 20 MeV in nuclei in the mass range $12 \leq A \leq 51$. The (p,γ) reaction proceeds primarily by a doorway (nonstatistical) reaction mechanism, in which the GDR is induced by the isovector part of the entrance-channel interaction between the proton projectile and the target. The dipole multipole of this interaction drives target protons and target neutrons with opposite phase, generating the Giant Dipole vibration. Selective states in the residual nucleus are populated following γ-emission; namely, the same states that are formed in direct proton stripping, and the integrated (p,γ) strength to a given final state is proportional to the proton stripping spectroscopic factor. This proportionality may be understood in a simple schematic model in terms of the overlap between the proton-plus-target entrance channel and the microscopic particle-hole doorway character of the GDR built on a particular final state. Within a given nucleus the GDR's show rather regular features, occurring at roughly constant γ-ray resonance energies[2] and with widths that increase rapidly with increasing energy of the state upon which the GDR is built. Further details can be found in ref. 1.

Currently, most of the interest and activity in the field of GDRs built on excited states lies in statistical GDR decay studies in heavy ion reactions, which are possible over a wide range of conditions of nuclear mass, spin and excitation energy. It is precisely because the reaction mechanism of statistical decay is quantitatively well-understood that one can infer from such studies direct, quantitative information on the structure of highly excited nuclei. The remainder of this paper focusses on statistical decay.

GENERAL FEATURES OF STATISTICAL GDR DECAY

The GDR built on nuclear ground states is a general feature of all nuclei. Ground-state photoabsorption, for example, is dominated by a single compact resonance, the GDR, occurring at an energy of $E_D \sim 80 \ A^{-1/3}$ MeV in medium and heavy nuclei, and somewhat less in light nuclei. The contribution of multipoles other than E1 are relatively insignificant for reactions involving real photons. The GDR may be viewed macroscopically as the collective oscillation of all the protons against all the neutrons, with a dipole spatial pattern, and microscopically as a coherent particle-hole excitation. The restoring force is provided by the dipole multipole of the isovector $\vec{\tau}_1 \cdot \vec{\tau}_2$ part of the nucleon-nucleon interaction. The photoabsorption cross section is usually well-approximated by a one or two-component Lorentzian form

$$\sigma_{abs}\left(E_\gamma\right) = (60 \text{ MeV-mb})(2/\pi)(NZ/A) \cdot$$

$$\sum_{j=1}^{2} S_j \Gamma_j E_\gamma^2 \left[\left(E_\gamma^2 - E_j^2\right)^2 + E_\gamma^2 \Gamma_j^2\right]^{-1} \qquad [1]$$

where S_j, E_j and Γ_j are the resonance strengths (in units of the classical sum rule), energies and widths (E_D is the weighted mean of the energies E_j). Generally a one-component form is adequate for describing the GDR built on the ground states of spherical or near-spherical nuclei, whereas two components are necessary in the region of (axially symmetric) deformed nuclei with eigenenergies E_j approximately inversely proportional to the length of the principal axes of the deformed shape. The hydrodynamic model, which has been used extensively in analyzing the GDR built on ground states of deformed nuclei, relates the energy splitting and the strength ratio to the nuclear quadrupole deformation: $E_2/E_1 = 0.911 \, d + 0.089$, where d is the major-to-minor axis ratio, and $S_2 : S_1 = 2:1$ for a prolate shape and $1:2$ for an oblate shape. The more conventional nuclear deformation parameter[3] δ, the length difference of the major and minor axes divided by the radius of a sphere of equal volume, is given by $\delta = (d-1)d^{-1/3}$. Nuclear deformation determinations in cold nuclei (i.e., nuclear ground and low-lying excited states) from GDR shape analyses agree well with deformations inferred from other techniques such as nuclear quadrupole moment and B(E2) transition moment measurements. Deformation splitting of the GDR is one of its more dramatic features, and this plays an essential role in determining GDR shapes in hot nuclei.

Many of the important features of statistical decay of the GDR following heavy ion fusion are illustrated in Fig. 1 for the example of $^{12}C + ^{154}Sm \rightarrow ^{166}Er^*$ at $E(^{12}C) = 61.2$ MeV ($E_x = 49.2$ MeV). Calculations of fusion followed by statistical evaporation were made with the computer code CASCADE, now commonly used for γ-ray spectrum analyses. The top panel shows, in the upper insert, the approximate distribution of initial orbital angular momenta calculated in a smooth-cutoff strong absorption model as $\sigma_\ell = \pi \lambda^2 (2\ell+1) T_\ell$ where the transmission coefficient $T_\ell = \{1 + \exp[(\ell - \ell_0)/d]\}^{-1}$ and the compound fusion cross section

$\sigma_c = \sum_{\ell} \sigma_{\ell}$. Here d=2fm, typically, and ℓ_0 is the grazing angular momentum. Also shown in Fig. 1 are the yrast line and several different decay chains involving neutron and gamma emission (charged particle emission, which generally competes favorably with neutron emission, is in this heavy a nucleus relatively small). Statistical decay rates for both particle and gamma emission are calculated by use of the

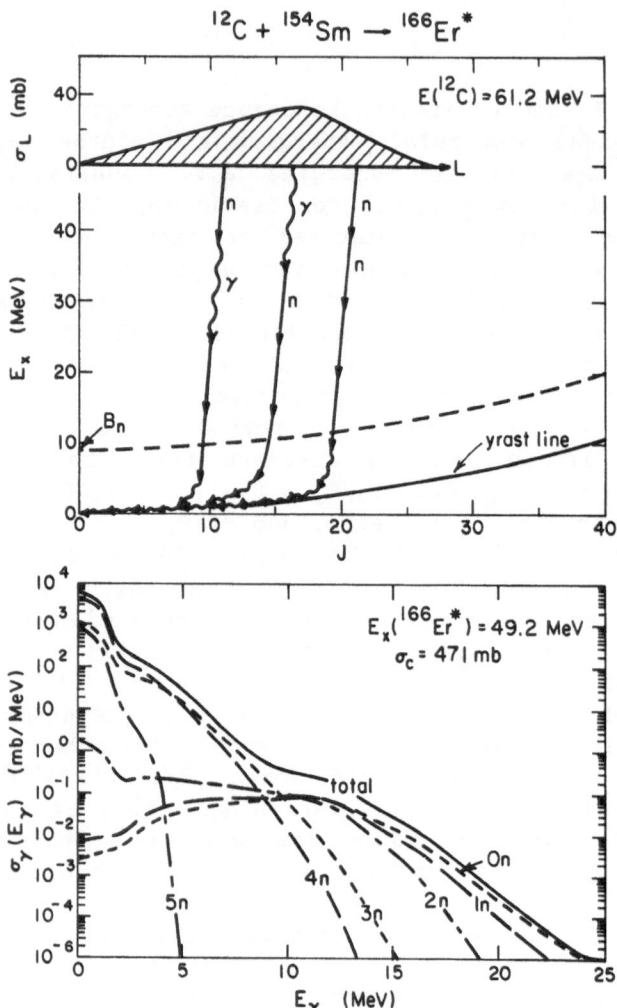

Fig. 1 Statistical decay following heavy-ion fusion in the reaction of 61 MeV $^{12}C+^{154}Sm\rightarrow^{166}Er^*$ (49 MeV). (Top) The inital spin distribution, the yrast line, and the competition between n and γ decay. (Bottom) The inclusive spectrum and the individual components labeled by the number of neutrons emitted preceeding the γ emission (with GDR parameters from Ref. 4).

reciprocity theorem relating decay to the inverse absorption process; the resulting particle decay widths are given by the usual expressions involving products of transmission coefficients and final-state level density, with the former calculated from averaged optical model parameters. The γ-decay width (per unit energy) calculated in the same manner is given by

$$\Gamma_\gamma^J \, \rho_J\left(E_i\right) = (\pi\hbar c)^{-2} \, \sigma_{abs}\left(E_\gamma\right) \left(E_\gamma^2/3\right)\rho_{J_f}\left(E_f\right) \qquad [2]$$

for decays between individual levels from an initial state of spin J (J_f = J or J±1) where $\rho_J(E_i)(\rho_{J_f}(E_f))$ is the density of initial (final) states of spin J(J_f). Here E_f = E_i − E_γ so that $\sigma_{abs}(E_\gamma)$ represents the average γ-ray absorption cross section for states at energy E_f and spin ≅ J. In practice, any measurement involves averaging over E_f and J, as well as neighboring nuclides (see below) and the usual assumption is that the γ-decay width for this process may be calculated from Eqs. (1) and (2) where S_{-1}, E_j and Γ_j represent GDR parameters which describe the absorption cross section averaged over the (final) states involved.

The bottom panel of Fig. 1 shows the calculated inclusive singles γ-ray spectrum (labelled "total") and its individual components labeled by the number of neutrons emitted preceeding the γ-ray. In the calculation, GDR parameters were used which are similar to ground-state photoabsorption (the actual values were determined from fitting ^{12}C + ^{154}Sm data,[4] as discussed below). Low energy $E_\gamma \lesssim$ 1-2 MeV γ decays, predominantly E2, occur late in the decay chain, when the nucleus has cooled to an energy $\lesssim$ E_{yrast} + B_n and is effectively bound. At higher E_γ the decays are predominantly E1. Here the decay probabilities depend on the relative size of E_γ compared to K_x where $K_x \sim$ 10-12 MeV is the total (binding plus kinetic) energy carried away by competing particle emission x. For $E_\gamma > K_x$ as is the case for $E_\gamma \sim E_D$, gamma emission is favored to occur at the earliest stage. Hence the 0n channel, corresponding to γ-decay from the initial compound nucleus before any particles are emitted, is the strongest single component in Fig. 1 (bottom). On the other hand, γ-emission for which $E_\gamma < K_x$ is most likely to occur near the end of the decay chain. These simple features arise because the γ-decay probability Γ_γ/Γ_x involves a ratio of level densities of final states populated by γ-decay to final states populated by particle decay and this ratio decreases with decreasing E_i when $E_\gamma >$ K_x and increases with decreasing E_i when $E_\gamma < K_x$. The region of the spectrum which is sensitive to the GDR parameters is the region $E_\gamma > K_x$. The convex "bump" which the calculated spectrum (and all measured spectra!) exhibit in the energy region $E_\gamma \sim E_D$ is an inherent consequence of the role of the GDR in the γ-emission process. On the other hand, the extraction of GDR parameters from measured spectra can only be done by comparison (ideally, by fitting) with detailed statistical model calculations, and hence are model-dependent.

The absolute γ-decay probability for the various steps may also be found from Fig. 1. For example, the integrated γ-ray production cross section $\int \sigma_\gamma(E_\gamma)dE_\gamma$ for the 0n channel is $\cong 0.6$ mb which, when compared to the calculated fusion cross section $\sigma_c = 471$ mb yields a decay probability of $\cong 1.3 \times 10^{-3}$. Also, one sees in this example that γ-decays of ^{166}Er, ^{165}Er and ^{164}Er contribute at $E_\gamma = E_D \cong 14$ MeV in the ratio of $\cong 47:35:18\%$, reflecting the decreasing probability of high energy γ emission as the nucleus de-excites, as discussed above.

With this information, the mean energy $\bar{E}_f$ of the ensemble of final states populated by γ-emission with $E_\gamma \sim E_D$ may be calculated. The approximate temperature to which this correspond is

$$T_f = \left[\frac{\bar{E}_f - \bar{E}_{rot}}{a}\right]^{1/2}$$

where $\bar{E}_f = \bar{E}_i - E_D$, $\bar{E}_{rot} \cong \bar{J}_f(\bar{J}_f+1)\hbar^2/2I_{rigid}$, $I_{rigid}/\hbar^2$ (MeV^{-1}) $= A^{5/3}(1+0.32\ \beta)/72$, and a $\cong A/8$ is the mass parameter in the level density. This expression follows from the statistical mechanics definition of temperature $T \equiv [d\ ln\rho/dE]^{-1}$ neglecting energy dependences in the level density other than the exponential factor $exp(2\sqrt{a(E-E_{rot})})$. Here $\bar{J}_f$ is the mean spin of the populated ensemble, which may also be estimated from the statistical model, and turns out for most cases to be well approximated by $\bar{J}_f = 2\ell_0/3 \cong \bar{J}_i$. Of course, this applies to the inclusive singles spectrum - more restrictive conditions on spin may be provided by gating with a multiplicity filter or sum energy spectrometer. On the other hand, both spin and energy (temperature) may be varied somewhat independently in singles by studying reactions with different projectile-target mass asymmetry which form the same compound nucleus.

The hypothesis that high energy, γ emission arises only from statistical decay following thermal equilibration for heavy ion collisons in the energy range discussed here has been tested in a variety of different experiments (see ref. 1). Below I present some examples of experimental results for statistical GDR decay obtained at Seattle using a large NaI spectrometer.

EXAMPLES OF STATISTICAL DECAY

Comparison of Deformed and Spherical Cases

Fig. 2 (right side) shows results[5] for the ^{12}C + ^{64}Ni reaction forming $^{76}Se^*$ at E_i = 45.2 MeV and initial spin $\sim$ 14 $\hbar$ and the ^{18}O + ^{72}Ge reaction near the Coulomb barrier, populating $^{90}Zr^*$ at E_i = 49.9 MeV and a low spin ($\sim$ 5 $\hbar$). The top panel shows the measured spectrum shape and the Cascade statistical model calculation(folded with the detector response) which was least-squares fitted to the data in the high energy region by varying the GDR parameters. The fits represent an excellent description of the measured spectrum shapes over more than six orders of

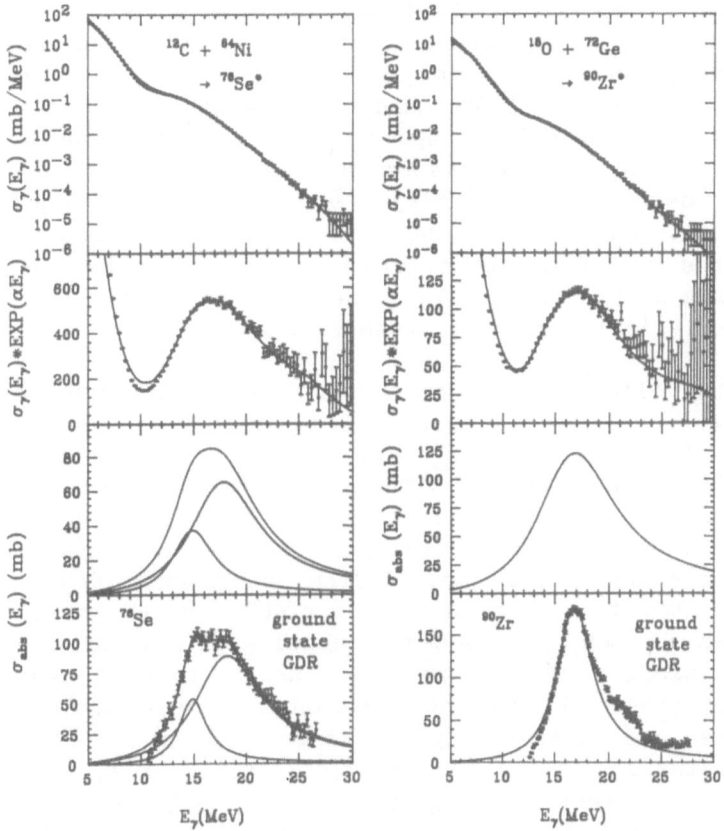

Fig. 2 Top panel: measured spectra and fitted Cascade calculations for the indicated reactions. Second panel: spectra and fitted calculations multiplied by exp (αE_γ), α^{-1} = 1.60 MeV for $^{76}Se^*$ and 1.71 MeV for $^{90}Zr^*$. Third panel: fitted $\sigma_{abs}(E_\gamma)$. Bottom panel: g.s.-GDR data and fitted shapes for $\sigma_{abs}(E_\gamma)$.

magnitude! The fitted $\sigma_{abs}(E_\gamma)$ is shown in the third panel
- in the case of ^{90}Zr* a single component Lorentzian works
just fine, whereas the ^{76}Se* results require two
components. Now the quality of the fit and the sensitivity
of the data to the shape of $\sigma_{abs}(E_\gamma)$ is impossible to judge
on the compressed semi-log plots shown in the top panels.
Fig. 2 (second panel) shows a linear plot of both data and
spectra multiplied by $\exp(\alpha E_\gamma)$ for display purposes - the
exponential factor α has been adjusted so that the shape of
the multiplied spectra is similar to the shape of $\sigma_{abs}(E_\gamma)$
determined from the fit to the data. The actual fitted
$\sigma_{abs}(E_\gamma)$ is shown in the third panel, while the bottom panel
shows $\sigma_{abs}(E_\gamma)$ results for the g.s.-GDR.

Now ^{90}Zr and neighboring nuclides are spherical at low
energy. The g.s.-GDR in ^{90}Zr shows a narrow ($\Gamma \sim 4$ MeV)
single-Lorentzian shape (the high energy shoulder is a $T_>$ =
$T_3 + 1$ peak resulting from isospin splitting - see below).
The decays of ^{90}Zr* (49.9 MeV) show a strongly broadened
(Γ=8.75 MeV) single-Lorentzian GDR with about the same
resonance energy and strength. On the other hand, ^{76}Se is a
nucleus for which strong quadrupole deformation is important
in the low-energy level spectra, and for which the g.s.-GDR
is split into 2 components due to deformation. The GDR
shape deduced from decays of ^{76}Se* also requires 2
components and, remarkably, agrees within experimental error
with the g.s.-GDR, not only in overall shape but in detail
in terms of the properties - energy, width and strength - of
the individual components (see Fig. 2).

The effect of strong deformation on the GDR shape in
excited nuclei is clearly evident in the decays of ^{166}Er*
(49.2 MeV) and ^{160}Er* (43.2 MeV) nuclei formed in
^{12}C+^{154,148}Sm reactions.[5] These rare earth nuclei behave at
low energy like good rotators, with prolate deformations
δ=0.30 and 0.25, respectively. Again, within experimental
error the GDR shapes deduced from decays at high excitation
energy are the same as the g.s.-GDR shapes measured in
photoabsorption on rare-earth nuclei with similar low-energy
deformation. These results have several important
implications: 1) the deformation of these nuclei remain
prolate and essentially unchanged in magnitude up to
temperatures $T \cong 1.2$ MeV and spins $J \sim 0$-25 $\hbar$, 2) all GDR
properties, such as centroid energy, strength and width
remain unchanged as well. In particular, the spreading
width due to mixing of the GDR with more complicated
configurations has not changed appreciably compared to the
g.s.-GDR, even though the excited nucleus decays involve
much higher excitation energy. Recent GDR spreading width
calculations[6] in ^{90}Zr at comparable temperature are
consistent with this conclusion.

Systematics of the GDR at Moderate Temperature and Spin

A summary[7] of some of the basic features of the GDR in excited nuclei from A=46 to 166 is shown in Fig. 3. Most of the compound nuclei were formed with spins $\sim$ 0-25 $\hbar$ and initial excitation energies $E_i \sim$ 40-52 MeV, corresponding to mean final-state temperatures (following γ-decay) ranging from $\sim$ 1.2 MeV in the heavy cases to $\sim$ 2 MeV in the lightest case. Spectra which were fitted with a single Lorentzian are indicted by an S; other cases required a double Lorentzian as indicated by a D. The plotted quantities are the total GDR strength, the mean resonance energy and the FWHM (full width at half maximum) of the fitted $\sigma_{abs}(E_\gamma)$ (for a single Lorentzian fit these are S, E_D and Γ).

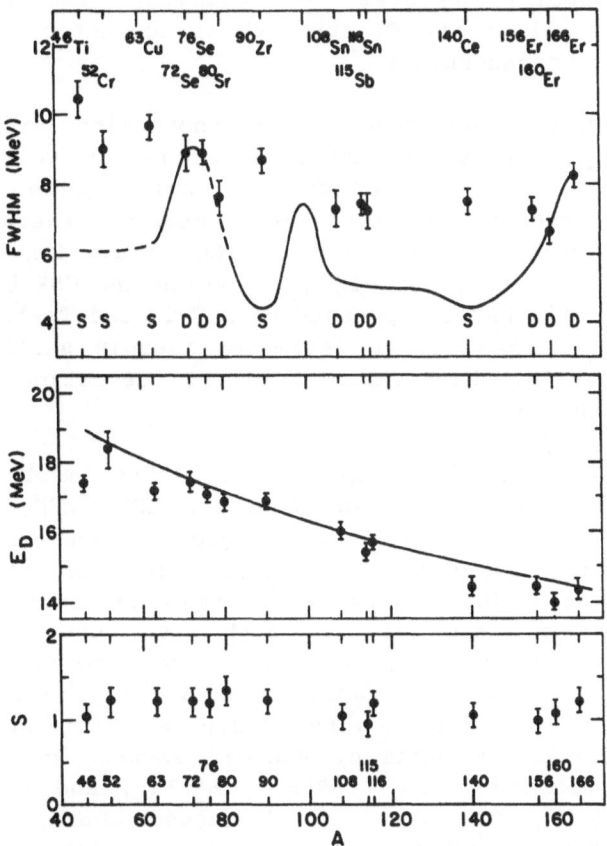

Fig. 3 Systematics of the GDR in excited nuclei at T=1-2 MeV and J=0-25 $\hbar$ as a function of mass. Bottom: resonance strengths in units of the classical E1 sum rule. Middle: resonance energies together with a curve representing g.s.-GDR energies. Top: resonance widths (FWHM) together with a curve approximating the g.s.-GDR width for nuclei of similar mass and isotope. S and D indicate whether a single or double Lorentzian shape was required to fit the excited-nuclei data. The measured compound nuclei are shown.

In the high-energy region of the γ-ray spectra, the yield scales as the product $\sigma_c \cdot S$ of the fusion cross section σ_c and the sum rule strength fraction S. The uncertainty in the extracted strengths is estimated to be ±15-20%, due to ±10-15% uncertainty in σ_c and ±10% for detector efficiency. The strengths for A=46 to 166 have been multiplied by factors of 1.6 to 1.07 to account for the neglect of isospin in the statistical model fits, as estimated from calculations with an isospin-dependent version of CASCADE [7]. The measured strengths are essentially all consistent with 1 classical dipole sum rule, in good agreement with ground state systematics, indicating that within experimental error, these resonances fully equilibrate before decay [this is an assumption in the application of the reciprocity theorem in deriving Eq. (2)], and essentially all of the E1 strength is accounted for by the assumed Lorentzian shapes.

The GDR resonance energies are shown along with a curve which fits ground-state GDR energies fairly well in this mass region. The measured energies agree quite well with the curve except for the light masses, where isospin splitting becomes important. The heavy-ion entrance channels always have isospin $T=T_3$, while the GDR built on a $T=T_3$ level will have components with $T=T_3$ and T_3+1. Ground-state systematics lead to estimated isospin shifts of ∼ 1 MeV for A=46-63, which explains most of deviations from the curve for the light masses.

In the width plot (top of Fig. 3) the smooth curve is an approximation to the ground-state GDR FWHM, which is well-known to be narrow in the region of spherical nuclei and broadened primarily by quadrupole deformation in other mass regions. The observed GDR FWHM in excited nuclei varies more or less smoothly with mass, even though the single- and the double-Lorentzian shapes are significantly different in the various nuclei. The widths shown in Fig. 3 are generally broader than the g.s.-GDR widths observed in the same or in neighboring nuclei, except for the three strongly deformed cases - $^{166}Er^*$, $^{160}Er^*$, and $^{76}Se^*$. The $^{166}Er^*$ and $^{160}Er^*$ cases, as discussed above, are clear examples of the persistence of ground-state-like deformation at high energies. Thus it is tempting to infer that nuclear quadrupole deformation plays a fundamental role in determining the GDR shapes in other excited nuclei as well.

Strong support for this conjecture is contained in the calculations of Gallardo et al.[8] The basic idea is that the ensemble of excited nuclear states characteristic of a nucleus in thermal equilibrium at finite temperature contains a distribution of deformations, and thermal averaging over such a deformation distribution leads to substantial broadening of the GDR. An essential element in this argument is that the nuclear potential energy as a function of the quadrupole shape degrees of freedom becomes rather flat at moderate temperature and spin for nuclei which are spherical or near-spherical at low energies. This appears to be the case for $^{108}Sn^*$, where the calculated[8] and measured[9] broadening appear comparable. More recently, calculations[10] have been made of the effects of thermal broadening in $^{90}Zr^*$ and $^{164}Er^*$ at temperatures and spins similar to those of the experiments described above. The regular show little or no broadening in the case of $^{164}Er^*$, while the $^{90}Zr^*$ GDR is broadened significantly (factor of 1.4) compared to the g.s.-GDR, in qualitative agreement with experiment. However, the magnitude of the calculated width increase in $^{90}Zr^*$ is substantially less than that observed experimentally (factor of 2). This raises the interesting question of whether the nuclear potential energy as a function of the quadrupole shape degrees of freedom at finite temperature in a spherical nucleus like $^{90}Zr^*$ is substantially flatter than given by present theory, or whether thermal fluctuations in other shape degrees of freedom (e.g., octupole) are important.

For all of the double-Lorentzian examples in Fig. 2, the fitted strength ratios favor prolate-like shapes (S_2/S_1 > 1). On the other hand, higher angular momenta should tend to drive the nuclear shape toward oblate deformation. For the shapes represented in Fig. 3, where angular momentum probably plays a minor role, the gradual increase in the FWHM as A decreases appears to be related to the higher temperatures for the light cases. The issue of determining excited nuclear shapes (prolate vs. oblate) is discussed in more detail below.

The results shown in Fig. 3 for the GDR strength and energy are in good agreement with recent theoretical calculations.[11] The GDR strength is given essentially by the classical dipole sum rule, which is determined by the number of neutrons and protons in the nucleus, and hence is not expected to change in an excited nucleus. The exchange current enhancement factor in the sum rule is calculated to depend very weakly on temperature. From finite temperature RPA calculations, the GDR energy is expected to drop with increasing T, due primarily to the softening of the nuclear surface. However, the calculated drop is only 2-3% at T=2 MeV.

All results for GDR parameters in excited nuclei depend on assumed level densities. Our results have been obtained using a prescription similar to that of Puhlhofer[12] and are sensitive primarily to the assumed level density in the "high energy" region $E_x > 80$ $A^{-1}/^3$ MeV, for which the Δ's are taken from the Myers droplet model mass formula without the Wigner term and with shell and pairing corrections removed, and the a parameter is given by a=A/8. In spite of years of study, the nuclear level density in the energy range of interest is not firmly established. However, the above choice of parameters is a common one, and in the present work it leads to consistent results for GDR properties in excited nuclei over a wide mass range. Given the growing theoretical consensus that the GDR energy and strength should be stable over a wide range of temperatures and spin, one can turn the problem around and use the agreement between excited-state and g.s.-GDR strengths and/or energies to place limits on acceptable level density parameters. For example, in the heavier nuclei a change of a to A/7 or A/9 leads to $\sim \pm0.5$ MeV change in E_D and $\pm 30\%$ change in S, and both of these changes are about at the limit of acceptability in comparison with g.s.-values. One should note that any level density with the same slope $d\rho/dE$ as that used in the present analyses would yield similar results.

The GDR as a Function of Spin and Temperature in $^{63}Cu^*$

A different perspective on systematics is afforded by examining the properties of the GDR as a function of temperature and spin within a given nucleus. As an example of this type of study, I present some recent results obtained at Seattle[13] on decays of the compound nucleus $^{63}Cu^*$.

The purpose of this experiment was to search for a possible dependence of the GDR energy, width and strength on nuclear temperature and/or spin. Use of different entrance channels permitted the formation of $^{63}Cu^*$ with a variety of different spins and initial excitation energies, ranging from ^{4}He+^{59}Co forming $^{63}Cu^*$ at E_i=22.5 MeV and $\ell_0 \cong 4.5$, to ^{18}O+^{45}Sc at E_i=77 MeV and ℓ_0=34, where ℓ_0 is the grazing ℓ-value. ^{6}Li+^{57}Fe and ^{12}C+^{51}V reactions were also employed, with several reaction pairs that formed the compound nucleus at the same energy but different spin.

Results of least-squares fitting of Cascade statistical model calculations to the measured spectra are shown in Fig. 4. All spectra were consistent with statistical decay except for ^{4}He+^{59}Co and ^{6}Li+^{57}Fe reactions at the highest bombarding energies, which were omitted from the figure. All spectra were fitted adequately by a single Lorentzian, for which the deduced strengths, energies and widths are shown in Fig. 4 as a function of mean final excitation energy $\widetilde{E}_{xf} \equiv \overline{E}_f - \overline{E}_{rot}$. Also shown are the parameters for the g.s.-GDR. To a good approximation the GDR strengths and energies are the same for all reactions and energies, and are in agreement with ground-state values. Thus these quantities are also independent of spin. The width, however, increases smoothly by about a factor of two in going to the highest excitation energy, which corresponds to a temperature $T_f = [\widetilde{E}_{xf}/a]^{1/2} \cong 2$ MeV for a = A/8 = 7.9.

In order to get a better idea of the relative importance of temperature and spin in broadening the GDR, the width is shown in Fig. 5 as a function of (final-state) temperature T_f^2 and initial grazing angular momentum ℓ_0. This plot shows that at large $T_f \sim 1.6-1.9$ MeV, the contours of constant Γ are only weakly dependent on ℓ_0 suggesting that the dominant contribution to the broadening is due to temperature.

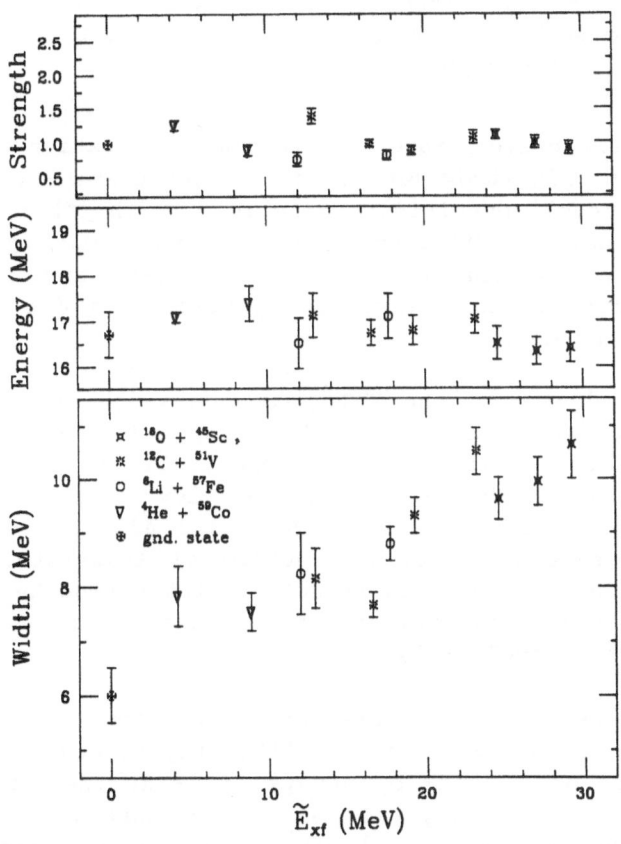

Fig. 4 Fitted GDR energies, strengths and widths for decays of the $^{63}Cu^*$ compound nucleus, as a function of final excitation energy above the yrast line.

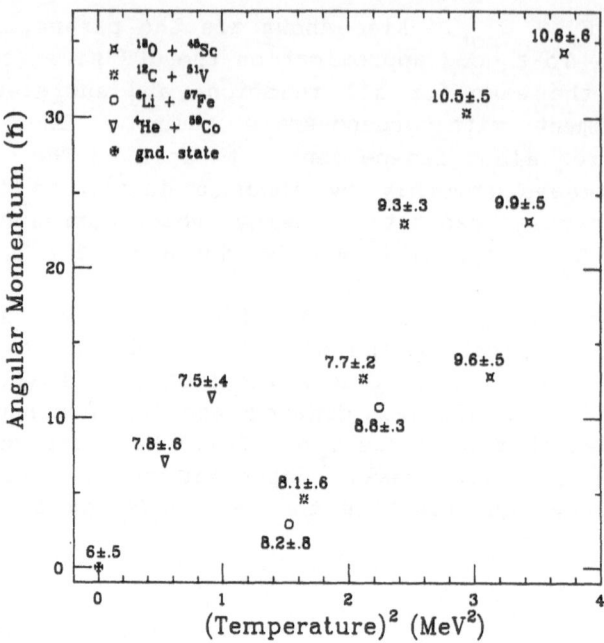

Fig. 5 Fitted GDR width as a function of (final state) temperature squared T_f^2 and initial grazing angular momentum ℓ_0 ($\bar{J} \approx 2\ell_0/3$) for decays of $^{63}Cu^*$.

The extensive range of initial excitation energies examined in this study permitted a closer examination of the possible level densities compatible with the measured spectra. Several different prescriptions[13] were judged to be adequate over the energy range studied, and the error bars in Figs. 4 and 5 include the variations due to these different choices.

A broadening of the GDR resonance width as a function of temperature, similar to that shown in Fig. 4, was found in ref. 9 for decays of Sn^* isotopes. If one excludes the broad, strongly deformed cases at low temperature, the remainder of the cases shown in Fig. 3 also correspond to about the same width as a function of temperature as is shown in Fig. 4. Thus all these data taken together suggest a rather universal temperature - dependent broadening, independent of nuclear mass.

The absence of any appreciable variations in S or E_D as a function of spin in decays of $^{63}Cu^*$ may be contrasted with the results of a study of decays of $^{162}Er^*$ performed with the Heidelberg NaI crystal ball.[14] In that experiment, a strong downward shift of E_D of $\approx$ 2 MeV was observed as a function of increasing angular momentum from 10 to 50 $\hbar$. The

Heidelberg experiment is the only one of its kind, and the result is incompatible with our current theoretical understanding of the physics of rapidly rotating nuclei. Although the $^{63}Cu^*$ decay experiment does not extend to as high a spin as the $^{162}Er^*$ study (<40 $\hbar$ compared with 50 $\hbar$), the rotational frequencies in the $^{63}Cu^*$ study are over twice as large as in the $^{162}Er^*$ study, so that the effects of angular momentum on the properties of $^{63}Cu^*$ might be expected to be at least as great as in $^{162}Er^*$.

Shape Changes at Finite Spin and Temperature

An exciting application of statistical GDR analyses is the search for nuclear shape changes at finite temperature and spin. At high enough angular momentum, nuclei may be expected to be oblate deformed, due to the combined effects of Coriolis and centrifugal forces, and irrespective of their deformation (e.g., spherical or prolate) at low spin. Finite temperature is expected to hasten this process, due to the softening of the potential energy surface. Evidence has been presented[15] for a prolate-oblate shape change in $^{166}Er^*$ and $^{158}Er^*$ at temperatures T ~ 1.2-1.4 MeV and spins $\bar{J}$ ~ 13-26 $\hbar$. More recently, apparent prolate shapes have been deduced from GDR studies in excited Pb^* isotopes, and the authors propose that these shapes have their origin in the superdeformed configurations as have been predicted[8] for moderate spin ~ 30 $\hbar$ and low temperature T < 0.8 MeV for ^{196}Pb.

The fundamental limitation in these studies arises from the fact that the individual components of a split GDR are overlapping and generally not resolved. The result is that the decomposition of a measured GDR shape in terms of two components with a definite strength ratio is problematical in the best of circumstances. It depends on the assumed (Lorentzian) lineshape, and the use of only two components in the fitting process. This latter assumption is more easily justified in the analysis of g.s.-GDR shapes, where axial symmetry is a good approximation, than in the analysis of decay of excited nuclei, where thermal fluctuations are generally important and thus triaxial shapes must certainly contribute. Even so, the data certainly would not support an analysis in terms of more than six parameters. However, even under the assumption of a 2-component Lorentzian shape, prolate-like and oblate-like fits to the data in which six parameters are independently varied (energy, width and strength of each component) are usually difficult to distinguish. What is needed is some additional constraint in the fitting process to reduce the ambiguity.

A reasonable constraint appears to be the width ratio Γ_2/Γ_1 which is strongly correlated with the apparent strength ratios S_2/S_1 in GDR analyses.[4] For the g.s.-GDR, one always finds $\Gamma_2 > \Gamma_1$ for (prolate) deformed nuclei; in fact, the experimental ground-state systematics are well described by a power law $\Gamma_2/\Gamma_1 \, \alpha \, (E_2/E_1)^{1.9}$. This variation is believed to be due primarily to the energy dependence of the spreading width, which is expected in infinite nuclear matter to vary quadratically with the resonance energy, and somewhat more slowly in finite nuclei.[17] This is consistent with calculations of the splitting of the (degenerate) upper component of the g.s.-GDR due to nuclear structure effects (see e.g., ref. 18), which show only a small splitting. Hence a strongly oblate-deformed nucleus would also be expected to have $\Gamma_2 > \Gamma_1$ (unfortunately there are no such ground-state examples in nature). Based on such a constraint, the authors of ref. 16 conclude that their Pb* GDR shapes are most likely prolate. Such a constraint on the analysis of the data of ref. 15 would also imply more prolate-like shapes. Thus it is clear that more work is necessary, both theoretically and experimentally, to understand the shapes - prolate vs. oblate- of excited nuclei.

The GDR at Very High Energies

At bombarding energies $E/A \gtrsim 7$ MeV per nucleon, heavy ion collisions may produce high energy gamma rays by several different mechanisms. One of these mechanisms is nuclear bremsstrahlung, now well-established as a process which produces an exponential-like high energy tail extending out to $E_\gamma \gtrsim 100$ MeV for collisions with $E/A \gtrsim 20$ MeV per nucleon. Although important aspects of the emission process still need to be clarified, such as absolute cross sections and angular anisotropies, the basic mechanism appears to be incoherent emission from neutron-proton collisions at an early stage in the reaction process.[19] Most of these experiments do not measure the spectrum below $E_\gamma \sim 20$ MeV, in the region of the GDR. In work done at LBL[20] on 7 to 17 MeV per nucleon ^{12}C and 19 MeV per nucleon ^{19}F colliding with various targets, the statistical GDR bump is seen superposed on the bremsstrahlung tail. A bump is also evident[21] near the GDR energy in γ-ray spectra from the collisions of 44 MeV per nucleon $^{40}Ar + {}^{158}Gd$. With tags on peripheral and on central collisions, these authors estimate that in the former case the GDR bump arises mainly from decay of target-like fragments excited to initial energies $\sim$ 90 MeV, while for central collisions, γ-rays in the GDR region come from decay of a broad mass distribution of mostly lighter fragments. It seems clear that to isolate and measure the statistical GDR decay from regions of very high initial excitation energy, such as from fusion-like processes, in high energy heavy ions collisions is a difficult experimental challenge.

CONCLUSIONS

I hope I have conveyed to you a sense of the current excitement and interest in the topic of Giant Dipole Resonances in hot nuclei, along with a feeling for some of the interesting physics which is emerging from these studies. The interested reader is invited to explore the theoretical aspects of these problems in the contributions to this conference by P.F. Bortignon and by S. Levit.[22]

I am indebted to my Seattle colleagues including J.A. Behr, G. Feldman, C.A. Gossett, J.H. Gundlach, and M. Kicinska-Habior for their invaluable contributions to the work presented here.

REFERENCES

1. K. A. Snover,   Ann. Rev. Nucl. Part. Sci.   36:545 (1986).
2. Actually, in N=Z nuclei recent detailed analyses of the resonances shapes (G. Feldman et al., to be published) show evidence for a drop in the GDR energy as one goes from low to high final-state energy, due apparently to isospin splitting discussed later in this text.
3. $\delta \cong 0.95 \beta$ where β and γ are the generalized quadrupole shape parameters of Bohr and Mottelson.
4. C. A. Gossett, K. A. Snover, J. A. Behr, G. Feldman, and J. L. Osborne, Phys. Rev. Lett. 54:1486 (1985).
5. J. H. Gundlach, K.A. Snover, J.A. Behr, C.A. Gossett, G. Feldman, and E.F. Garman, to be published.
6. P. F. Bortignon, R. A. Broglia, Proc. topical Meet. on Phase Space Approach to Nucl. Dynamics, Trieste, ed., M. DiToro. Singapore: World Scientific (1985); P. F. Bortignon, R. A. Broglia, G. F. Bertsch, and J. Pacheco, to be published (1986). the whole reference?
7. C. A. Gossett,   K. A. Snover,   J. H. Gundlach, G. Feldman, J. Behr, and M. Kicinska-Habior, to be published.
8. M. Gallardo, M. Diebel, T. Døssing and R. A. Broglia, Nucl. Phys. A 443:415 (1985).
9. J. J. Gaardhoje, C. Ellegaard, B. Herskind, D. Diamond, M.A. Delaplanque, G. Dines, A.O. Macchiavelli, and F.S. Stephens, Phys. Rev. Lett 56:1783 (1986).
10. M. Gallardo, F. J. Luis, and R. A. Broglia, to be published.

11. See e.g., H. Sagawa and G. F. Bertsch, Phys. Lett. B 146:138 (1984); see also ref. 1 for additional references.

12. F. Puhlhofer, Nucl. Phys. A 280:267 (1977).

13. M. Kicinska-Habior, K. A. Snover, C. A. Gossett, J. Behr, G. Feldman, H. Glatzel, J. H. Gundlach, and E. F. Garman, to be published.

14. D. Schwalm, Proc. of the 1984 Erice Summer School on Heavy Ion Physics, Erice, Italy.

15. J. J. Gaardhøje, C. Ellegaard, B. Hersking, and S. G. Steadman, Phys. Rev. Lett 53:148 (1984); J. J. Gaardhoje, "Nuclear Structure 1985," ed. R. A. Broglia, G. Hagemann, B. Herskind, Amsterdam: North Holland (1985) p. 519.

16. D. R. Chakrabarty, M. Thoennessen, N. Alamanos, P. Paul, and S. Sen, Phys. Rev. Lett. 58:1092 (1987).

17. G. F. Bertsch, P. F. Bortignon and R. A. Broglia, Rev. Mod. Phys. 55:287 (1983); G. F. Bertsch, private communication.

18. G. Maino, A. Ventura, L. Zuffi, and F. Iachello, Phys. Rev. C 30:2191 (1984).

19. G. F. Bertsch, contribution to this conference.

20. C. A. Gossett, J. A. Behr, J. H. Gundlach, K. A. Snover, K. T. Lesko, and E. B. Norman, to be published.

21. R. Hingmann, W. Kühn, V. Metag, R. Muhlhans, R. Novotny, A. Ruckelshausen, W. Cassing, H. Emling, R. Kulessa, H.J. Wollersheim, B. Haas, and J.P. Vivien, Phys. Rev. Lett. 58:759 (1987)

22. See also Y. Alhassid, S. Levit, and J. Zingman, Phys. Rev. Lett. 57:539 (1986); S. Levit, Proc. of the Internat. Nuc. Phys. Conf., Harrogate, UK, 1986.

HOT NUCLEI - THEORY AND PHENOMENA

Shimon Levit

Department of Nuclear Physics
Weizmann Institute of Science
Rehovot 76100
ISRAEL

Lectures delivered at the International School of Heavy Ion Physics, 2nd Course: The Response of Nuclei under Extreme Conditions, Erice, 12-22 October 1986.

ABSTRACT

I review the basic theoretical ideas and techniques used in the description of hot nuclei. I then discuss, in some detail, the shape transitions and the upper temperature limits of nuclear stability.

TABLE OF CONTENTS

1. INTRODUCTION

A large nucleus can store hundreds of MeV of the excitation energy in its dense "reservoir" of excited states. If properly delivered this energy equilibrates and leads to formation of a hot drop of nuclear matter containing two hundreds and more nucleons. With sufficient ingenuity one is able to study the physical properties of such a drop, e.g. the equation of state of the inside material, the temperature dependence of its surface tension, the disappearance with the temperature of the quantum mechanical effects (shell, pairing, etc.), the responses to rotations, collective and coherent vibrations, etc. Recent experiments with heavy ions begin to open these exciting new directions of the nuclear physics research.

Heavy ion reactions provide an almost unique way to heat nuclei in laboratory. Although such reactions can in principle follow different reaction mechanisms it appears reasonable to expect that a substantial part of the reaction cross section goes via a formation of an equilibrated nuclear complex. Experimental efforts are presently becoming more and more elaborate and precise in the attempts to isolate and study such an equilibrated component of the nuclear cross section (Refs. 1-4).

Numerous phenomena can be identified and studied in hot nuclei. The so called continuum γ-rays from the deexcitation of these nuclei provide information on the temperature dependence of the properties of the nuclear giant resonances (Refs. 5-12) which can be compared with the available theory (Refs. 13-19). The most studied giant dipole component depends quite sensitively on the shape of the hot nucleus. The first results on measuring in this way the shape as a function of the temperature and spin were reported in Refs. (7,8). Such measurements (Ref. 20) are relevant for the studies of various shape related, phase-transition like phenomena in hot nuclei. In the regime $T \sim 0.5 - 1$ MeV these are the "depairing" transitions (Refs. 21,22) and for higher temperatures ($T \sim 1$-3 MeV) - the shape changing transitions (Refs. 23-28).

Above $T \sim 5$ MeV one is expected to approach a limit of maximum excitation energy (heat) which a fully equilibrated nucleus can absorb and still remain a self-bound system. Experimental evidence showing a possible detection of such a limit were presented recently (Refs. 2,3,29,30). Theoretical considerations (cf. below, Section 3) indicate that measuring the temperature T_{lim} of this limit in different nuclei is expected to give valuable information of the hot nuclear matter equation of state and the temperature dependence of the nuclear surface tension.

Only partially equilibrated nuclei exist above T_{lim}. Studying their formation, expansion and fragmentation is expected to "probe" different parts of the nuclear matter equation of state at high temperatures. This is especially interesting in relation to the critical point of the coexistence of two phases of nuclear matter, liquid and gas, found in all available theoretical calculations (Refs. 29-47).

2. THEORETICAL IDEAS AND TECHNIQUES IN THE DESCRIPTION OF HOT NUCLEI

2.1 Equilibration Assumption, Possible Experimental Realizations and Tests of the Equilibration

Perhaps the most important conceptual ingredient in the common description of the highly excited nuclei formed in heavy ion reactions is the assumption of equilibration. This assumption is not always clearly formulated, may be of a more or less restrictive form but appears to be essential if one wishes to reduce the problem of dealing with high density of levels, complicated wave functions, strong nuclear interactions, etc., to a global description in terms of few macroscopic parameters. Even if the equilibration is not really achieved in a given high energy nuclear phenomenon, the consequences of assuming it form a convenient baseline, starting point, etc. with respect to which a more elaborate description can be constructed and the deviations from which form a meaningful subject of theoretical and experimental studies.

It is the easiest to understand and formulate the global equilibration assumption. If one knows the excitation energy E^*, the linear and angular momenta P and J, etc. at which the highly excited nucleus was formed, one assumes the equal probability distribution of all the states with these quantum numbers. For a given energy E^* this means that the probability to find the nucleus in a state i is $w_i = (R_i/\Delta N) = (\Delta E/\Delta N)(R_i/\Delta E) = \rho^{-1}(E^*)\delta(E_i - E^*)$, where E_i is the energy of the state i, R_i is zero for E_i outside and one for E_i inside the interval $(E^*, E^* + \Delta E)$, ΔN is the number of states in this interval, ρ - the density of states and ΔE is let go to zero. For known P, J, etc. one multiplies w_i by $\delta(P_i - P), P_J$ (projection on J), etc. The resulting probability distribution is called microcanonical in the standard terminology of statistical mechanics. To see how it can be used consider a simple example of transitions between highly excited nuclear states caused by an operator Q. If all the initial states at an energy E' were equally populated (here enters the equilibration assumption) and the only information on the final states was their energy E'' then the transition probability is

$$dp(\epsilon = E' - E'')/d\epsilon \sim \rho^{-1}(E') \sum_{i,f} |< f \mid Q \mid i >|^2 \, \delta(E' - E_i)\delta(E'' - E_f) \quad (2.1a)$$

which can be written as the Fourier transform (H is the Hamiltonian)

$$\int_{-\infty}^{\infty} \frac{dt}{2\pi} e^{i\epsilon t} Tr\delta(E' - H)Q^+(t)Q(\circ)/Tr\delta(E' - H) \quad (2.1b)$$

of the "microcanonical response" of the system at energy E' to the perturbation Q. Here $Q(t) = exp(iHt)Qexp(-iHt)$ and one relates (2.1a) to (2.1b) by evaluating the traces in the basis of the eigenstates of H and also inserting these states as intermediate between $Q^+(t)$ and $Q(\circ)$. It is usually difficult to work with the δ-function distribution (it lacks multiplicatively) and in theoretical discussions it is common to replace it (approximately) by more convenient exponential forms which I will discuss shortly.

Intuitively one expects that the global equilibration may have occurred during the evolution of a highly excited nuclear system if its deexcitation followed routes which were very different from the one by which the energy was initially deposited and if sufficient time delays have occurred between the initial and final stages. Statistically one may also expect that the total cross section of a complicated heavy ion reaction always contains a part which goes via a formation of a fully equilibrated nuclear complex. It is an experimental task to isolate this part and to determine how significant it is. It is natural to assume that the equilibrated component forms a significant fraction of the *fusion part* of the total cross section. Recent experimental studies (Refs. 1-4) seem to indicate that this component persists up to very high excitation energies of several MeV per nucleon ($\sim$ 5 MeV temperature).

Global equilibration means that the parts of the nuclear system had enough time together to reach the complete relaxation i.e. to "explore" all the regions of the phase space available under the given values of the excitation energy, angular momentum and other conserved additive quantum numbers. One may introduce more restricted forms of the equilibrium in which only "fast" degrees of freedom had time to relax whereas the slow ones (shape, average particle density, etc.) continue to evolve dynamically leading to the picture of the system going through a continuous sequence of *constrained* equilibria. An example of a theoretical treatment of such a situation is given in Ref. (45). It forms the conceptual basis for the works (cf. Refs. (34-35)) on the time evolution of nuclear systems along the isentropes of the appropriate equation of state.

When the number of the constraints under which one assumes the instantaneous equilibrium increases, say from the average particle density to the entire density (and current) distribution one goes over into the hydrodynamical type of the description with local equilibration taking place in each "point" (small volume cell) of the system.

While experimental tests of the global equilibrium are relatively clear, it is difficult to conceive of direct experimental verification of the constrained equilibrium not to speak of the local hydrodynamical equilibration.

2.2 Introducing the Temperature

The average level density of large enough nuclei is exponentially large and of the form $\rho(E^*) = \rho_o e^{S(E^*)}$ for energies several MeV above the ground state. According to the discussion in Sec. 7 of Ref. (52) $S(E^*)$ has the interpretation of the entropy and ρ_o is the inverse of a certain characteristic energy interval. One can introduce a convenient variable, the temperature (cf. Sec. 9 of Ref. (52)),

$$T^{-1}(E^*) = \frac{d}{dE^*} ln \left(\frac{\rho(E^*)}{\rho_o} \right) = \frac{dS}{dE^*} \tag{2.2}$$

which is a measure of how fast the level density at given E^* grows with E^*. Since both $S(E^*)$ and E^* are extensive, $T(E^*)$ is an intensive variable. It is on the average monotonic in E^*. The most frequent use of $T(E^*)$ variable is in replacing the δ-function microcanonical distribution by more convenient exponentials in the expressions of the type (2.1). Consider the function $\rho(E^*)e^{-\beta E^*}$ depending on a parameter β. For large nuclei this is a product of a fast rising and an exponentially decreasing functions and, therefore, it is sharply peaked at some E^* which is determined by the value of β via the relation $\beta = dS/dE^*$, i.e. $\beta = T^{-1}(E^*)$. We can, therefore, use the so called canonical distribution

$$\rho(E^*) exp[\beta (F(\beta) - E^*)] \tag{2.3}$$

with $exp(\beta F(\beta)) = \sum_i e^{-\beta E_i} = \int_o^\infty \rho(E^*) exp(-\beta E^*) dE^*$, as a properly normalized replacement of the δ-function microcanonical distribution. In (2.1b) this replacement is achieved by substituting $exp[\beta F(\beta) - \beta H]$ instead of $\delta(E' - H)/Tr\delta(E' - H)$ and tuning the value of β to give the maximum of (2.3) at $E^* = E'$. In this way the correlation function in (2.1b) becomes

$$Tr\, e^{-\beta H} Q^+(t) Q(o) / Tr\, e^{-\beta H} \tag{2.4}$$

i.e. the temperature dependent response of the system. The formal relation between the canonical and the microcanonical distributions is established by the use of the saddle point approximation in evaluating the traces over the states, say

$$Tr\ Oe^{-\beta H} = \sum_i <i\,|\,O\,|\,i> e^{-\beta E_i} = \int_0^\infty dE^* O(E^*)\rho(E^*)e^{-\beta E^*} \qquad (2.5)$$

with the saddle point value of E^* given as above by $\beta = d\ ln(\rho/\rho_o)/dE^*$. This is familiar from the standard discussion of the averaged nuclear level density (cf. Ref. (54)). In (2.5) the fixed value of the inverse temperature β means fluctuations of E^* represented by the integral in the r.h.s. It is useful sometimes to think about the inverse relation

$$Tr\ \delta(E^* - H) = (2\pi i)^{-1} \int_{-i\infty}^{i\infty} d\beta Tr exp(\beta E^* - \beta H) \qquad (2.6)$$

as representing the *fluctuations of the inverse temperature* β for a fixed value E^* of the energy. In this representation the saddle point evaluation of the integral in (2.6) gives the average β for a given E^*.

It is hoped that the above discussion answers the frequently asked questions of whether the introduction of the temperature implies that a nucleus is in a contact with a heat bath, etc. The real physical assumption is the equilibration and the two probability distributions $\rho^{-1}(E^*)\,\delta(E_i - E^*)$ and $exp\,(\beta F(\beta) - \beta E_i)$ are two approximately equivalent descriptions of the equilibration related to each other by the saddle point evaluation of appropriate integrals.

2.3 Rotations

Formation of highly excited nuclei in heavy ion reactions is usually accompanied by transfer of measurable amount of the angular momentum J from the relative to the intrinsic motion. The equilibrium distribution is, therefore, $\rho^{-1}(E^*,J)\delta(E_i - E^*)P_J$ where P_J is the projection on the given J and $\rho(E^*,J)$ is the density of levels at this J. Transformation to the temperature representation exactly analogous to the discussed above replaces this distribution by $P_J exp(-\beta H)/Z(T,J)$ where $Z(T,J) \equiv exp[-\beta F(\beta,J)] = Tr\ P_J exp(-\beta H)$ is the nuclear partition function with fixed J.

It is convenient to transform from the angular momentum J to its intensive partner ω - the angular velocity of rotations. This can be done as follows. The use of the probability distribution $P_J exp(-\beta H)/Z(T,J)$ is always for calculating averages of the observables, etc., i.e. traces over the states of the system. Consider as an example the partition function itself. It can be written as

$$Z(T,J) = Tr\ P_J e^{-H/T} = \sum_\nu (2J+1)\, e^{-E^*(\nu,J)/T} = \int_0^\infty dE^* \rho(E^*,J)e^{-E^*/T} \qquad (2.7)$$

where ν denotes all the quantum numbers of nuclear levels except J and its projection M. Introducing $\rho(E,M)$, the density of levels with a given M and using

$$\rho(E,J) = \rho(E,M = J) - \rho(E,M = J+1) \approx -(\partial/\partial M)\rho(E,M)\,|_{M=J+1/2} \qquad (2.8)$$

one can rewrite (2.7) as

$$Z(T,J) \cong -(\partial/\partial M) \int_0^\infty dE^* \rho(E^*, M) e^{-E^*/T} \,|_{M=J+1/2}$$

$$= -(\partial/\partial M) \int_0^{4\pi} (d\phi/4\pi) e^{-i\phi M} Tr e^{-(H-i\vec{\omega}\cdot\vec{J})/T)} \,|_{M=J+1/2} \qquad (2.9)$$

where the direction of $\vec{\omega}$ is arbitrary and $|\vec{\omega}| = \phi T$. The last equality in (2.9) can be easily verified by evaluating the trace in common eigenstates of H, J^2, and the projection of $\vec{J}$ on $\vec{\omega}$. The integral over ϕ then projects on M. The upper limit 4π of the ϕ integration takes care of both the integer and half-integer possible values of M.

Carrying out the differentiation in (2.9) one obtains

$$Z(T,J) = \int_0^{4\pi T} (i\omega d\omega/4\pi T^2) \, exp\left\{-\left[i(J+1/2)\omega + F(T,i\vec{\omega})\right]/T\right\} \qquad (2.10)$$

where

$$F(T,\vec{\omega}) = -T \, ln \, Tr \, e^{-(H-\vec{\omega}\cdot\vec{J})/T} \qquad (2.11)$$

is the free energy of the nucleus in a frame which rotates with the angular velocity $\vec{\omega}$. The appearance of the imaginary $i\vec{\omega}$ in the free energy in (2.10) does not cause problems. For a time reversal H the free energy $F(T,\vec{\omega})$ in (2.11) is an even function of ω and the continuation to $i\vec{\omega}$ is done trivially by replacing ω^2 to $(-\omega^2)$.

The arbitrariness of the choice of the direction of $\vec{\omega}$ in (2.9-10) is, of course, related to the rotational invariance of the *exact* H which appears in these expressions. Later on when discussing the mean-field approximations for H it will be clear how the orientation of the nuclear shape in deformed situations is to be related to the direction of $\vec{\omega}$.

The integral in (2.10) should be interpreted as the sum over the fluctuations of ω for a given value of J. The average ω is found by the stationary point condition of the integrand, i.e.

$$J + 1/2 = -\partial F(T,\vec{\omega})/\partial\omega \qquad (2.12)$$

It is worth commenting that a more precise variant of the formula (2.10) i.e. without the approximation (2.8), is obtained by using the exact form of the projection operator P_J

$$P_J = \frac{2J+1}{2\pi} \int_0^{4\pi} sin(\phi/2) sin\left[(J+1/2)\phi\right] exp\left[i\vec{\phi}\cdot\vec{J}\right] d\phi \qquad (2.13)$$

in (2.7). Replacing the product of sines in (2.13) by the difference $[cos J\phi - cos(J+1)\phi]/2$ and going over to complex exponentials one recovers (2.9) as an approximation to the difference.

2.4 Mean-Field Approximations - Variational and Functional Integral

Theoretical discussions of the properties of hot nuclei are based on various versions of the mean field approximation. One seeks to replace the equilibrium probability distributions of the previous sections which depended on the *exact* Hamiltonian H by approximate distributions with an optimal (self-consistent) one-body Hamiltonian H_o. Here again the canonical distribution (2.3) is much more convenient than the microcanonical δ-function. In looking for the criterion for the optimization of H_o one has two main approaches, variational and functional integral. Both have their advantages and drawbacks and I will briefly discuss them.

The exact canonical distribution $D = exp(-\beta H)/Z(T)$ can be obtained by minimizing the free energy

$$F(T, D) = Tr\ DH + T\ Tr\ D ln\ D \qquad (2.14)$$

with respect to all possible density matrices D under the constraint $Tr\ D = 1$. Reducing the space of variations and restricting D to be an exponential of a suitably parametrized one-body Hamiltonian i.e. $D \rightarrow D_o \sim exp(-\beta H_o)$, calculating $F(T, D_o)$ in (2.14) and minimizing w.r.t. the parameters of H_o one finds the optimal H_o and D_o within the given parametrization. For instance choosing H_o of the form $H_o = \sum_{\alpha,\beta} h_{\alpha\beta} a_\alpha^+ a_\beta$ and minimizing (2.14) with respect to $h_{\alpha\beta}$ one obtains (cf. Appendix A of Ref. 25 for details) the temperature dependent generalization of the Hartree-Fock equations. The pairing effects can be included by adding $\sum_{\alpha\beta} g_{\alpha\beta} a_\alpha^+ a_\beta^+ + c.c.$ to this H_o and minimizing w.r.t. $g_{\alpha\beta}$ as well.

When interested in the shape evolution with temperature a simpler though more restrictive choice of H_o is that of the Nilsson Hamiltonian supplemented by the Strutinsky procedure in order to account for the correct dependence of the average free energy on the deformation. At higher temperatures when the shell effects are not important the temperature dependent version of the Thomas-Fermi approximation can be useful. If only the global aspects of the hot nuclei behaviour are of the interest (stability, etc.) the hot liquid drop model proves to be a useful approximation (cf. below).

Mean-field approximations for hot nuclei can also be derived using the functional integral representation for the nuclear partition function. For an exact Hamiltonian of the form $H = tr\ [t\hat{\rho} + (1/2)\hat{\rho}V\hat{\rho}]$ with t-kinetic energy, V-two body interaction and $\hat{\rho}_{\alpha\beta} = a_\alpha^+ a_\beta$ one can write this representation as (Ref. 48).

$$Z(T) = e^{-F(T)/T} = \int D[\sigma] exp\left[1/2 \int_o^\beta tr\sigma V\sigma\right] exp\left[-F(T, \sigma)/T\right] \qquad (2.15)$$

where the integration variable σ is a "density function" $\sigma_{\alpha\beta}(\tau)$ which depends on the imaginary time-like parameter $\tau, 0 \leq \tau \leq \beta = T^{-1}$. For static, i.e. τ independent σ the functional $F[T, \sigma]$ is $-T\ ln\ Tr\ exp(-H_\sigma/T)$ with $H_\sigma = tr(t\hat{\rho} + \hat{\rho}V\sigma)$, a one-body Hamiltonian describing the motion in the σ-dependent one-body potential $tr\hat{\rho}V\sigma$.

Expression (2.15) shows that the exact Z is a weighted sum (i.e. the functional integral) of the partition functions of the same system placed in the one-body H_σ's with all possible values of the "density" σ. The mean-field approximation is obtained by minimizing the integral in (2.15) i.e. by minimizing $F(T,\sigma) - (1/2)tr\sigma V\sigma$. This leads to the temperature dependent Hartree equations for σ. One of the advantages of the form (2.15) is that it clearly shows that this Hartree value of σ is "the most probable" one and that going beyond the mean-field approximation means to include the fluctuations in σ, i.e. to allow for all possible values of σ each with the weight $exp\{-[F(T,\sigma) - (1/2)tr\sigma V\sigma]/T\}$.

2.5 Mean-Field and Rotations, the Moment of Inertia

Rotation of hot nuclei is described by replacing H by $H - \vec{\omega} \cdot \vec{J}$ and integrating over $|\vec{\omega}|$ as discussed in Section 2.3. The choice of the direction of $\vec{\omega}$ was arbitrary there due to the rotational invariance of the exact H. As is well known the replacement of H by a trial one-body Hamiltonian H_o and subsequent minimization may lead to the mean-field H_o which is not invariant under rotations, i.e. to a deformed nuclear shape. In the absence of the rotations ($\omega=0$) this shape can have an arbitrary orientation and one has continuous degeneracy of variously oriented shapes.

When $\vec{\omega}$ is not zero the choice of its direction eliminates this degeneracy and determines the orientation of the deformed nuclear shape relative to $\vec{\omega}$. Although the choice of the direction of $\vec{\omega}$ *was* arbitrary for the exact H, once it is made and H replaced by H_o this choice fixes the orientation of the shape by the same minimization procedure which selects the optimal H_o. This minimization with respect to the orientation can be carried out in a very general way. Consider the free energy $F(T,\vec{\omega})$ for some choice of the deformed H_o and the direction of $\vec{\omega}$. Expanding one finds

$$F(T,\vec{\omega}) = F(T,\omega = 0) - 1/2 \sum_{i,j=1}^{3} I_{ij}(T)\omega_i\omega_j + ... \tag{2.16}$$

where I_{ij} are interpreted (cf. below) as the components of the moment of inertia tensor. In the intrinsic frame (x',y',z') of the principle axes of I_{ij} the sum in (2.16) is $\omega_{x'}^2 I_{x'x'} + \omega_{y'}^2 I_{y'y'} + \omega_{z'}^2 I_{z'z'}$ and the components of ω in this system are $\omega_{x'} = -\omega cos\phi sin\theta$, $\omega_{y'} = \omega sin\phi sin\theta$, $\omega_{z'} = \omega cos\theta$. Inserting these expressions in (2.16) and minimizing with respect to the orientation angles ϕ and θ one finds (cf. Appendix B of Ref. 27) that the minimum occurs when one of the principle axes of the nuclear shape is oriented along $\vec{\omega}$ and moreover the moment of inertia around this axis is the largest. For instance when the nucleus is prolate it must have its symmetry axis perpendicular to $\vec{\omega}$, when it is oblate - parallel, etc.

We emphasize that this fixing of the orientation of the shape relative to $\vec{\omega}$ is a part of the mean field approximation in the sense that it is found by minimization of the free energy w.r.t. to the parameters of H_o, i.e. its orientation relative to $\vec{\omega}$. Allowing for the fluctuations around the mean-field values one must include the fluctuations of the orientation relative to $\vec{\omega}$ as well (cf. below, Section 2.7).

The actual value of the nuclear moment of inertia determines the relation between the angular momentum and the average angular velocity of the rotations (cf. Eqs. (2.12) and (2.16)). Using (2.11) with H replaced by the mean-field H_o one finds (the details of the following can be found cf. in Appendix A of Ref. 27).

$$I_{ij} \equiv -\partial^2 F/\partial \omega_i \partial \omega_j \mid_{\omega=0} = \int_o^\beta < J_i(\tau) J_j(o) > d\tau \qquad (2.17)$$

where $\beta = 1/T$ and $< J_i(\tau) J_j(o) > = Z^{-1} Tr \left[e^{-\beta H} (e^{H\tau} J_i e^{-H\tau} J_j) \right]$. The classical limit of (2.17) is $< \sum_\alpha m(r_\alpha^2 \delta_{ij} - r_{\alpha i} r_{\alpha j}) >$, the classical expression for the moment of inertia. Quantum mechanical evaluation of (2.17) is easiest when the restriction of the fixed particle number is dropped and instead a chemical potential term $-\mu N$ is added to the Hamiltonian. One obtains for I_{zz} for instance

$$I_{zz} = \sum_i |< i \mid j_z \mid i >|^2 \, (-\partial f_i/\partial \epsilon_i)$$
$$+ \sum_{i \neq k} |< i \mid j_z \mid k >|^2 \, \frac{f_k(1 - f_i) - f_i(1 - f_k)}{\epsilon_i - \epsilon_k} \qquad (2.18)$$

where $\mid i >$ and ϵ_i are the s.p. states and energies and f_i - their thermal occupations. The two parts of (2.18) have clear physical meaning. When the nuclear shape is symmetric around the axis of rotation (chosen as z in this case), e.g. for an oblate shape, the second term is zero whereas the first describes the contribution to the moment of inertia I_{zz} by the only possible mechanism of the quantum rotation around a symmetry axis - the single particle alignment near the Fermi energy. At zero temperature $-\partial f_i/\partial \epsilon_i \rightarrow \delta(\epsilon_i - \mu)$ giving the expression discussed in Ref. (54). When the symmetry axis is perpendicular to z only the second term in (2.18) is not zero and it describes the contribution due to the collective rotation of the entire shape. At $T \rightarrow O$ it goes to the familiar cranking model result.

2.6 Elimination of the Continuum States

Consider the description of a nucleus as nucleons moving in a given one-body (shell model) potential and let us calculate the free energy of such a system at a non-zero temperature T. It is convenient to calculate with a chemical potential term $-\mu N$ and one obtains

$$\Omega(T, \mu) \equiv -T \, ln \, Tr \, exp\left[-\beta \left(H_o - \mu N \right) \right]$$
$$= -T \left(\sum_{boundstates} + \int_o^\infty deg(\epsilon) \right) ln \, \{1 + exp\left[(\mu - \epsilon)/T \right]\} \qquad (2.19)$$

The sum is over the bound single particle states and $g(\epsilon)$ denotes the level density of the continuum single particle states which exist since the potential vanishes at large distances. The presence of this continuum results in the dependence of $\Omega(T, \mu)$ on the volume in which the whole system is placed. It is easiest to see this by calculating $g(\epsilon)$ from the asymptotic form of the wave functions $sin(kr + \delta)$

with δ - the phase shifts (for details, cf. Ref. 49). The dependence of Ω on the volume is spurious and should be eliminated. It leads for instance to the inability to fix the number of particles *in the nucleus* by the standard relation $N = -\partial \Omega / \partial \mu$. Indeed the r.h.s. of this relation depends on the volume and for a *fixed* N one would find a volume dependent chemical potential which becomes positive and large as the volume increases to infinity. The dependence of (2.19) on the volume reflects a simple and basically correct physical fact that this expression describes a nucleus in the equilibrium with the gas of the surrounding particles, i.e. with its vapor. In order to isolate the part corresponding to the nucleus it is necessary to subtract the vapor contribution. This can be readily achieved if one observes that $g(\epsilon)$ in (2.19) can be rigorously written as $g_o + \Delta g$ where g_o is the volume dependent s.p. level density of the free system (with no central nuclear potential present) and Δg is entirely due to the effect of the potential. In the spherically symmetric case $\Delta g(\epsilon) = (4/\pi) \sum_{l=0}^{\infty} (2l + 1)[d\delta_e(\epsilon)/d\epsilon]$ where $\delta_e(\epsilon)$ is the phase shift of the l-th partial wave of the continuum s.p. states. Replacing $g(\epsilon)$ by $\Delta g(\epsilon)$ in (2.19) gives a physically meaningful definition of the nuclear free energy which is finite, volume independent and satisfies the requirements which are usually associated with such a quantity. In practice such a replacement is done by replacing $\Omega(T, \mu)$ by the difference $\Delta\Omega(T, \mu) = \Omega(T, \mu) - \Omega_o(T, \mu)$ where Ω_o is calculated without the nuclear potential.

In the self-consistent mean-field theory the nuclear potential is not fixed and depends on T and μ. The particle density $\rho(\vec{r}) = \sum_i f_i \psi^2(\vec{r})$ extends outside the nuclear volume and becomes constant at large distances due to the population of the continuum states. Elimination of the volume dependence in this case and the proper definition of the nuclear free energy is based upon an observation (Ref. 48) that the self-consistent mean field equations posses *two distinct solutions* for a given T and μ. One solution corresponds to a nucleus plus the surrounding vapor and another to the uniform vapor only. We show an example of these two solutions in Fig. 1. The natural definition of the nuclear part of the free energy is

$$\Delta\Omega_{nucleus}(T, \mu) = \Omega_{nucleus+vapor}(T, \mu) - \Omega_{vapor}(T, \mu) \qquad (2.20)$$

and the particle number in the nucleus should be fixed by $N = -\partial\Delta\Omega/\partial\mu$ which is obviously volume independent. The generalization of (2.20) to include the long range Coulomb potential and the results of the applications of the entire procedure are amply discussed in Refs. (49,50).

2.7 Fluctuations

There are many "sources" of fluctuations in theoretical schemes used in the description of hot nuclei. I will try to indicate the most important of them. The functional integral formula (2.15) shows that the mean-field potential is only the most probable one and that in principle all potential shapes must be admitted each with the probability given by an analog of a Boltzmann factor, the integrand in (2.15). The simplest way to account for these fluctuations is to approximate these Boltzmann factors by a Gaussian in $\sigma - \sigma_{mean-field}$ with the width given by the second variation of the exponential in (2.15). This was done in Ref. (48)

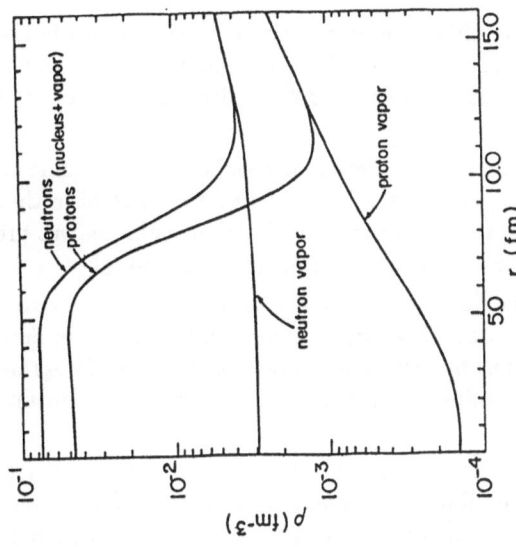

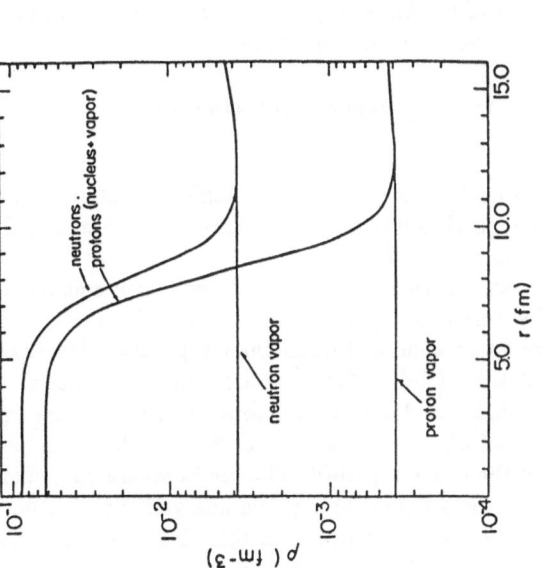

Fig. 1

Radial dependence of the neutron and proton densities of nucleus + vapor and vapor solutions of the thermal Hartree-Fock equations at T = 7 MeV for uncharged (left) and normal, charged (right) 208Pb. The particle numbers in both solutions for a constant chemical potential μ are volume dependent. μ was fixed to give the volume independent number of protons (82) and neutrons (126) in the nucleus which was defined using the "nucleus+vapor minus the vapor" subtraction procedure discussed in the text and Refs (48-50).

with the results for the partition function $Z = Z_{mean-field} \cdot Z_{fluctuation}$. with

$$Z_{fluctuations} = \frac{\prod_{j>k} 2 sinh\left(\Delta\epsilon_{jk}/2T\right)}{\prod_{\omega_\nu>0} 2 sinh\left(\omega_\nu/2T\right)} \qquad (2.21)$$

where $\Delta\epsilon_{jk} = \epsilon_j - \epsilon_k$ are the differences of the single particle states in the Hartree potential and ω_ν are the eigenfrequencies of the temperature dependent RPA equations. The simple physical meaning of (2.21) is discussed on pp. 1035-36 of Ref. (48).

The Gaussian approximation is definitely not sufficient when some of the "directions" around $\sigma_{mean-field}$ are "soft". In such cases the integral in these directions should be evaluated exactly. An example of such calculation was presented in Ref. (19) and discussed in the lectures of Bortignon and Broglia.

Another "source" of the fluctuations are the fluctuations of the temperature for the fixed value of the excitation energy. These were mentioned in Section 2.2 and can be taken into account by using the expression (2.6), i.e. integrating over all temperatures. I am not aware of any published attempt to implement this procedure.

When a rotating nucleus is described by the cranking type of formulas, i.e. by adding $-\vec{\omega} \cdot \vec{J}$ term to the Hamiltonian one must consider, for a given value of the angular momentum J, the fluctuations of the nuclear orientation relative to the direction of $\vec{\omega}$ as well as fluctuations of the magnitude $|\vec{\omega}|$. These issues were discussed in Sections 2.3 and 2.5 and the relevant expressions were given in Eqs. (2.10) or (2.13). To the best of my knowledge no account of the work on these fluctuations in hot nuclei appeared in the literature.

3. SHAPE TRANSITIONS IN HOT ROTATING NUCLEI

3.1 Mean-Field Results

Existence of deformed nuclei is a quantum mechanical effect related to the shell structure of the single particle levels in the mean-field potential and their filling up to a sharp Fermi energy in accordance with the Pauli principle. For non-zero temperature the occupations of the levels do not form a sharp distribution, the shell effects become less important and one expects that the deformed nucleus gradually changes into a spherical when the temperature increases. All available mean-field calculations (Refs. 23,24) confirm this expectation. Moreover, they show (Ref. 25) that the change of a deformed nucleus into a spherical is not an asymptotic process fully completed only at $T \to \infty$ but occurs at finite well defined values of T of the order 1-3 MeV. The temperature dependence of the deformation of the mean-field potential is a non-analytic function and in analogy with other statistical systems one can refer to this phenomenon as a shape transition. On Fig. 2 an example of the equilibrium deformation of the mean-field of ^{168}Yb is plotted as a function of the temperature. One clearly observes the non-analytic behaviour of β_{eq} very similar to, say, the average magnetization in ferromagnets.

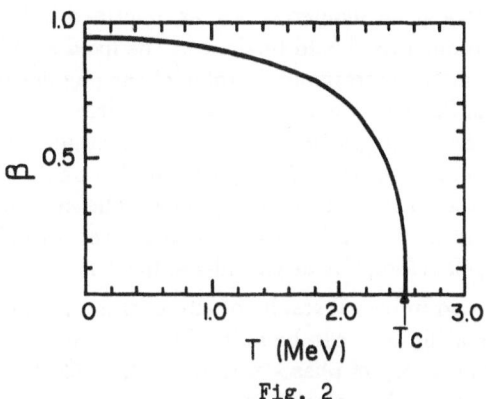

Fig. 2

Equilibrium deformation of [168]Yb vs. temperature. The
dimensionless β is related to the mass quadrupole moment
as Q/(20 barns) (Ref. 25).

It must be stressed that the analogy with the phase transitions is valid only at the level of the mean-field description. The fluctuations in finite nuclei will smooth the non-analytic mean-field behaviour. Yet the mean-field analysis forms a baseline of the discussions of the shape transition upon which more refined theories should be based.

Experimentally hot nuclei are produced in heavy ion reactions where together with the excitation energy large amounts of the angular momentum are transferred from the relative to the intrinsic motion. It is well known that spinning prolate nuclei change their shape to triaxial and eventually to oblate when the angular momentum increases. As will be shown below this change of the *mean-field shape* from triaxial to oblate is also a non-analytic shape transition. Since the temperature tends to drive nuclei to spherical and rotations to oblate one expects an interesting behaviour of the nuclear shapes when studied in the plane of the temperature or the excitation energy E^* vs the spin J.

3.2 Universal Features of Shape Transitions in Hot Rotating Nuclei - the Landau Theory

The shape of the mean-field of a hot nucleus is found by minimizing a trial mean-field free energy at a given temperature T as described in Section 2.4. In the presence of the rotations this should be done at the fixed angular momentum J or more conveniently at the corresponding value of the angular velocity $\vec{\omega}$ of rotations as described in Sections 2.4 and 2.5. The trial free energy depends on the variational parameters of the particular mean-field method used, e.g. single-particle wave functions in the Hartree-Fock approximation or the shape parameters of the potential in the Nilsson-Strutinsky scheme. The nuclear shape transitions occur at temperatures where pairing effects do not seem to play an important role and they will, therefore, not be considered below.

The most relevant and universal results of any microscopic mean-field theory of shape transitions at finite T can be anticipated on the basis of a unified framework of the Landau theory of phase-transitions (Ref. 52, Par. 139) in statistical systems. This framework is perfectly adequate for nuclear problems since it is mean-field in nature. Its advantages are in allowing to isolate those features which do not depend on the details of one or another microscopic mean-field scheme but have a more universal character. These features can be displayed on the relevant phase diagrams and compared with other statistical systems. The Landau approach also offers a useful and economical parametrization of the results of the microscopic calculations and singles out a small number of the most relevant combinations of the parameters upon which the behaviour of the equilibrium shape depends.

The details of the Landau theory of nuclear shape transitions were presented in Refs. (25-28) and I will only outline the main features of the approach. Let us denote by $F(T, \vec{\omega}, \rho)$ the trial free energy at a given temperature T and rotational frequency $\vec{\omega}$. Here ρ denotes the set of parameters with respect to which the minimization should be performed. For studying the equilibrium nuclear shapes the most crucial are $\alpha_{2\mu}(\mu = -2, ..., 2)$, the quadrupole deformation parameters. One considers, therefore, instead of $F(T, \vec{\omega}, \rho)$ an effective reduced free energy $F(T, \vec{\omega}, \alpha_{2\mu})$. Formally this reduction can be made following the discussion of Section 2.1 of Ref. 25. In the terminology of the Landau theory $\alpha_{2\mu}$ are called the order parameters the non-zero values of which characterize the degree of a symmetry breaking. In nuclear case the symmetry which is broken is of course the rotational invariance of the intrinsic nuclear mean-field Hamiltonian which has a non-spherical potential in deformed nuclei.

In the presence of rotations the rotational symmetry in the intrinsic frame is broken explicitly by the preferred direction of the rotational axis, i.e. the direction of $\vec{\omega}$. In analogy with an external magnetic field applied to a ferromagnet, the nuclear rotation acts in the intrinsic frame as an external field controlled by the magnitude of $\vec{\omega}$. The free energy of the entire system, including $\vec{\omega}$ must be rotationally invariant, therefore, $F(T, \vec{\omega}, \alpha_{2\mu})$ can depend on $\alpha_{2\mu}$ and $\vec{\omega}$ only via their invariant combinations. The $\vec{\omega}$ independent such invariants are $[\alpha \times \alpha]^{(0)}$ and $[[\alpha \times \alpha]^{(2)} \times \alpha]^{(0)}$. The lowest invariants involving $\vec{\omega}$ are quadratic $[\omega \times \omega]^{(0)}$, $[[\omega \times \omega]^{(2)} \times \alpha]^{(0)}$, $[[\omega \times \omega]^{(2)} \times [\alpha \times \alpha]^{(2)}]^{(0)}$, etc. Assuming analyticity of $F(T, \vec{\omega}, \alpha_{2\mu})$ one can expand in powers of the invariants retaining only the physically most relevant terms. The minimization of the resulting expression with respect to $\alpha_{2\mu}$ gives the equilibrium shape at the given T and ω. The minimization w.r.t. the "orientation" of $\alpha_{2\mu}$ (i.e. of the intrinsic frame in which $\alpha_{20} = \beta\cos\gamma$, $\alpha_{2,\pm2} = (\beta/\sqrt{2})\sin\gamma$ and $\alpha_{2,\pm1}=0$) relative to the direction of $\vec{\omega}$ can be done in a general form as described in Section 2.5. After this one arrives at the following expression for the fee energy (cf. Section 2.1 of Ref. 27 for details)

$$F(T, \omega, \alpha_{2\mu}) = F_o(T) + A(T)\beta^2 - B(T)\beta^3\cos3\gamma + C(T)\beta^4 - \frac{1}{2}I_{zz}(\beta, \gamma)\omega^2 \quad (3.1a)$$

where z is chosen along the direction of $\vec{\omega}$ and $I_{zz}(\beta, \gamma)$ is the corresponding component of the moment of inertia. Its general dependence on β and γ is

$$I_{zz}(\beta, \gamma) = I_o(T) - 2R(T)\beta\cos\gamma + 2I_1(T)\beta^2 + 2D(T)\beta^2\sin^2\gamma + O(\beta^3). \quad (3.1b)$$

Only those minima of (3.1a) are admitted for which I_{zz} is the largest component of the inertia tensor, i.e.

$$I_{zz} > I_{xx}, I_{yy} \quad (3.2)$$

The coefficient functions $A(T), B(T)$, etc entering the above expressions can not be determined at this level of the theory and must be considered as phenomenological parameters which should be found microscopically or, better still, constrained experimentally. However, for the general analysis it is sufficient to determine the expected qualitative dependence of these coefficients on T. Studies of Refs. (26-28) showed that in most situations this dependence is of the following general character which is typical for a statistical system undergoing phase transitions. The coefficient $A(T)$ is negative at low T and changes sign at some $T = T_c$. $B(T)$ and $C(T)$ are > 0 for nuclei with prolate ground states. $R(T), I_1(T)$ and $D(T)$ are positive, do not depend strongly on T in the vicinity of T_c and are fairly well approximated in this region by their rigid body values (cf. Eqs. (2.7b) in Ref. 27). $F_o(T), I_o(T)$ are not relevant for the minimization.

The minimization of (3.1) w.r.t. β and γ is most easily done using the variables $a_o = \beta \cos \gamma, \sqrt{2}a_2 = \beta \sin \gamma$. One first finds all the minima of (3.1) in the entire (β, γ) plane and then selects only those which obey (3.2). An important symmetry of (3.1) is $\gamma \to -\gamma$ which corresponds to interchanging x and y principal axis. The shapes related by this symmetry describe the same physical situation and therefore, it is sufficient to consider only $0° \geq \gamma \geq -180°$ region. The minimization of (3.1) can be carried out analytically and the evolution of the equilibrium with changing T and ω can then be studied.

The results depend very sensitively on the relative magnitudes of the coefficients R and D in (3.1b). In the rigid body moment of inertia R dominates over D whereas $I_o = R = 0$ and D dominates in the moment of inertia for irrotational flow. Since the relevant case in $T \sim 1 - 3$ MeV region is closer to the rigid body situation, it is useful to discuss first the extreme limit of dominant R and $D=0$. In this case, the equilibrium deformation for $B > 0$ depends on two combinations of the parameters τ and ω/ω_c where

$$\tau = AC/B^2, \qquad \omega_c = (9/10)(B/C)(B/R)^{\frac{1}{2}}. \tag{3.3}$$

The combination τ vanishes at $T = T_c$ and is monotonic and practically linear in the difference $T - T_c$ in all cases studied so far (cf. below). The evolution of the equilibrium shape in the shape transition region is conveniently represented on the *phase diagram* of τ vs. ω/ω_c shown in Fig. 3 on which the lines of constant deformations β (in units of B/C) and γ are plotted. Note that we work with the convention in which the rotation is along the z axis so that $\gamma = -120°$ represents a prolate shape rotating perpendicular to its symmetry axis while $\gamma = -180°$ is oblate shape with symmetry axis along that of the rotations. The shape is triaxial between these limits.

On Fig. 3 one sees that the equilibrium shape changes rapidly from almost prolate to oblate when the temperature increases in the vicinity of T_c (i.e. $\tau=0$) for fixed ω/ω_c. The transitions of the equilibrium mean-field shape from triaxial to oblate are always non-analytic. For $\omega < \omega_c$ they have the abrupt, first order character, i.e. the equilibrium shifts discontinuously from triaxial with $\gamma \neq -180°$ to oblate, $\gamma = -180°$, when τ increases above the values on the dashed-dotted line. For $\omega > \omega_c$ the transitions from triaxial to oblate are continuous, second order when τ increases above the solid line. The lines of the first and second order transitions join at the point $\tau = 63/128$ and $\omega/\omega_c=1$. This point is called tricritical in the standard terminology of statistical mechanics called so because of the coincidence of three minima of the free energy at this point. It is important to emphasize that it is the $\gamma \to -\gamma$ symmetry mentioned above which is responsible for this phenomenon, (cf. Ref. (27)). A phase diagram similar to Fig. 3 is found in e.g. liquid crystals in electric fields and antiferromagnets in magnetic fields (Ref. 53). In finite nuclear systems the order of the transitions is relevant only for the extreme mean-field description. A more relevant aspect of the phase diagram of Fig. 3 is the overall pattern of the behaviour of the shape and the large fluctuations in the vicinity of the tricritical point.

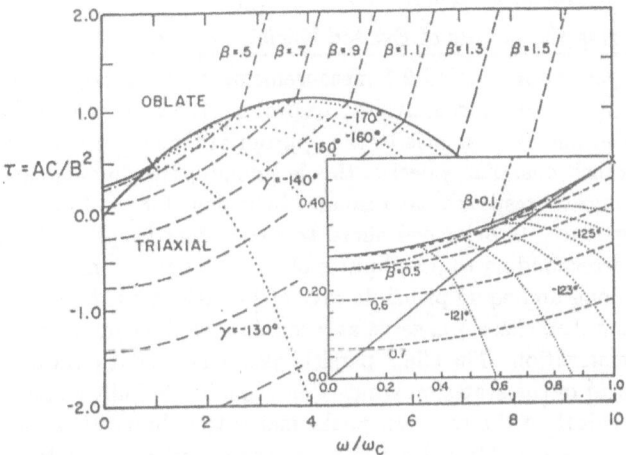

Fig. 3

The universal phase diagram of nuclear shape transitions in the reduced
temperature τ and ω/ω_c variables, Eg.(2.3), for D=0 limit. The tricritcal
point at ω/ω_c = 1, τ = 63/128 is denoted by X. To its left the dash-dotted
line denotes the first order transitions. The solid line to the right of
X denotes the second order transitions. Also shown are the lines of β = const
(dashed, β in units of B/C)and γ = const (dotted).
Note that Z is the axis of rotation (cf. text). In the upper part of the
figure γ is equal $-180°$ representing oblate shapes with the symmetry axis
parallel to that of rotation. The solid lines to the left of X denote the
boundary of the coexistance region of the triaxial and oblate "phases"
(cf. Ref. 26).

Departures from the $D = 0$ limit can also be analysed analytically. The phase diagram in this case has similar qualitative features as in Fig. 3 but has important quantitative differences especially in the large ω region (cf. Ref. 27). The transformation from the temperature and ω variables to the experimentally relevant E^* (excitation energy) and J (spin) is done easily within the present approach. One must, however, know the quantitative dependence of $\tau, B/C$, etc. on T. This dependence is presently being investigated microscopically for many nuclei and one of the examples is shown in Figs. 4 and 5 where we display the results of the calculations of the phase diagram for ^{166}Er. The region $J \simeq 10 - 20\hbar, E^* \simeq 60$ MeV where ^{166}Er was recently studied experimentally (Refs 7,8) is close to the tricritical point and the equilibrium is very sensitive to the values of E^* and J offering a possible explanation of the observed rapid transitions from almost prolate to oblate shape.

3.3 Measuring the Shapes of Excited Nuclei

As described in Section 2.1 measurements of the deexcitation of an equilibrated excited nucleus is equivalent to probing the Fourier transform of the microcanonical nuclear response to the appropriate operator which causes the deexcitation transitions. For γ decays this is (mainly) the dipole operator. The corresponding nuclear response has resonant behaviour for the frequency of the giant dipole resonance. In deformed nuclei this resonance is split into two or three components corresponding to possibilities of collective vibrations with different frequencies along non-equal principle axes of the deformed shape. The magnitude of the splitting can, therefore, serve as a measure of the character and the size of the nuclear deformation. The effect persists also when the excitation energy (nuclear temperature) of the states on which the resonance is built increases although the increasing width of the resonant peaks makes the observations more difficult than in the ground state. More discussions on the experimental determination of nuclear shapes at finite T and J are found in the lectures of Kurt Snover in this school.

4. UPPER TEMPERATURE LIMITS OF NUCLEAR STABILITY

4.1 Mean-Field Results

Experimentally one can form and study the same compound nuclei at higher and higher excitation energies E^*. How far up in E^* can one go and still have a selfbound nuclear drop and not just a puff of dispersing vapor? How do the nuclear properties change on the way to losing the binding effects? Is the disappearance of the binding a gradual process or a sudden phase transition like limit? The recent temperature dependent Hartree-Fock calculations provided this information (Ref. (49,50)). In order to extend these calculations to the relevant temperature region $T \sim 5 - 10$ MeV it is necessary to consistently deal with the single particle states in the continuum as described in Section 2.6. On Fig. 6 an example of the vapor subtraction procedure is shown. Despite the significant dependence of the single particle potential on the temperature for $T > 5$ MeV the deviations of the dependence of the entropy, energy, chemical potentials, etc. on T from the simplest Fermi gas relations is not significant. This is explained by a

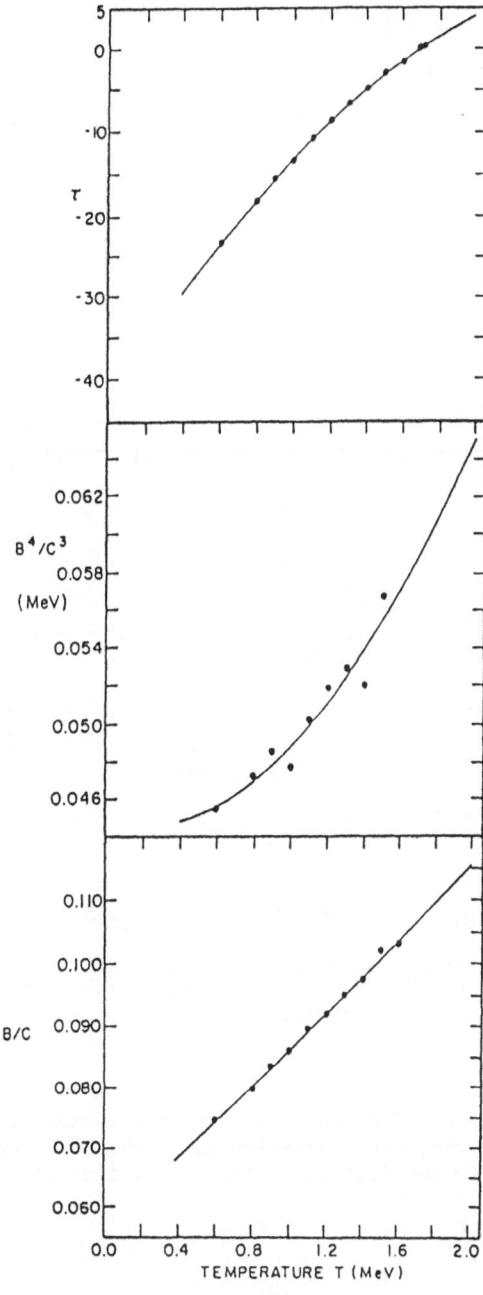

Fig. 4

166Er nucleus - calculated (dots) and fitted [Eq. (4.10) of Ref.(27)] temperature dependences of the parameters: $\tau = AC/B^2$ (the dimensionless reduced temperature), B^4/C^3 (the scale of the deformation dependent part of the free energy) and B/C (the scale of the deformation).

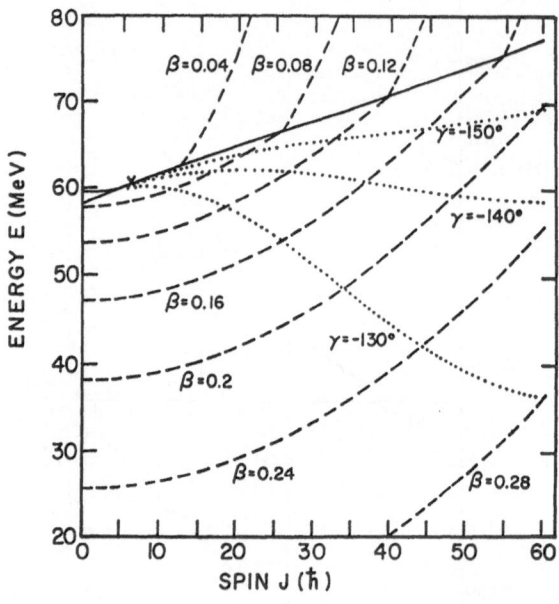

Fig. 5

The phase diagram of ^{166}Er in the plane of the excitation energy-angular momentum variables. Notation as in Fig. 4 except β is not scaled by B/C. For the details of the calculation cf. Ref. 27.

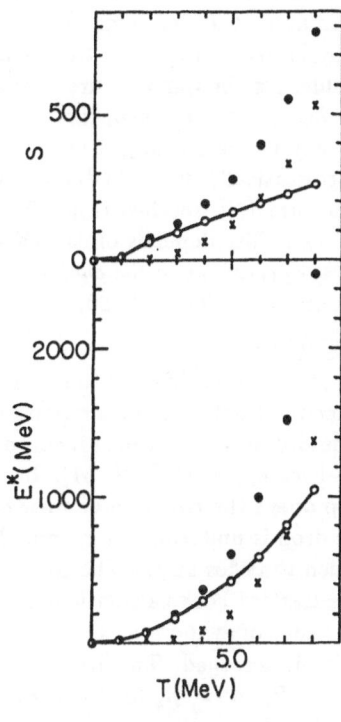

Fig. 6

Temperature dependence of the entropy S and the excitation energy E* of the nucleus+vapor (dots) and vapor (crosses) solutions and the subtracted quantities (circles) for ^{208}Pb (cf. Ref. 50). Despite significant changes of the single particle potential for T> 5 MeV the subtracted quantities continue to show the Fermi gas behaviour, S~T, E*~T^2. Above T = 8 MeV the nucleus becomes unstable (cf. Fig. 7 and Ref. 50) delicate cancellation of several counteracting effects (Ref. 49).

In order to study the limiting temperature of nuclear binding it is useful to compare the temperature dependences of the Hartree-Fock solutions for a normal nucleus and for its uncharged partner in which the Coulomb interaction is artificially switched off. This comparison is presented on Fig. 7. The uncharged nucleus exists up to $T=12.5$ MeV but becomes less and less dense with very diffuse surface spreading over larger radii. It disappears above $T=12.5$ MeV and only uniform vapor exists at $T=13$ MeV.

This merging of the nucleus and vapor solutions in the uncharged system corresponds to reaching the critical temperature T_c of liquid-gas coexistence found in the nuclear matter calculations (Refs. 31,32). Indeed the Hartree-Fock solution for the nucleus + vapor system at $T < T_c$ (cf. Fig. 1) explicitly exhibits this coexistence of the more dense liquid-like phase inside the nucleus and the less dense gas-like phase, i.e. vapor outside. As the temperature is raised towards T_c, the difference between the phases decreases and disappears at and above T_c. Such behaviour, however, is featured only in the uncharged nuclear system. The normal charged nucleus becomes electrostatically unstable due to the Coulomb repulsion long before the critical temperature T_c is reached (Fig. 7). The value T_{lim} of the temperature of this Coulomb instability depends on the charge-to-mass ratio of the nucleus and is, moreover, *very sensitive to the choice of the effective interaction* used in the Hartree-Fock equations (cf. Ref. 50).

4.2 A Simple Thermodynamic Model

The Coulomb instability and its sensitivity to the choice of the effective interaction can be better understood within the framework of a simple finite-temperature version of the liquid-drop model which incorporates the main macroscopic features of the Hartree-Fock approach (Ref. 51). In this model the matter inside the hot nuclear drop obeys the equation of state of the liquid phase of the bulk nuclear matter. The drop is uniformly charged and has a temperature-dependent surface tension which vanishes at T_c. The conditions at the surface are chosen in accordance with the Hartree-Fock calculations by demanding the equilibrium with the surrounding vapor of evaporated particles. For simplicity the vapor is assumed to be completely screened. The main equations of the model are the coexistence conditions $P_n = P_v, \mu_n = \mu_v$ for the pressures and the chemical potentials of the matter inside the nucleus and the outside vapor that has the same temperature. For the charged nuclear drop $P_n = P_{on} + P_{coul} + P_{surf}$ and $\mu_n = \mu_{on} + \mu_{coul}$, where the subscript zero denotes the *bulk* quantities given by the *liquid phase part* of the nuclear matter equation of state. The surface and the Coulomb contribution are found in a standard way. The vapor P_v and μ_v are given by the *gas phase part* of the bulk matter equation of state.

An example of the graphical solution of the coexistence equation is shown in Fig. 8,9. As the temperature increases the differences $P_{on} - P_v, \mu_{on} - \mu_v$ and P_{surf} decrease and vanish at T_c. The Coulomb contributions, P_{coul} and μ_{coul}, on the other hand, do not vanish as $T \to T_c$. At some limiting temperature T_{lim} the coexistence point reaches the upper boundary of the vapor phase and no coexistence is possible above T_{lim}. The value of T_{lim} (6.57 MeV in this example) is

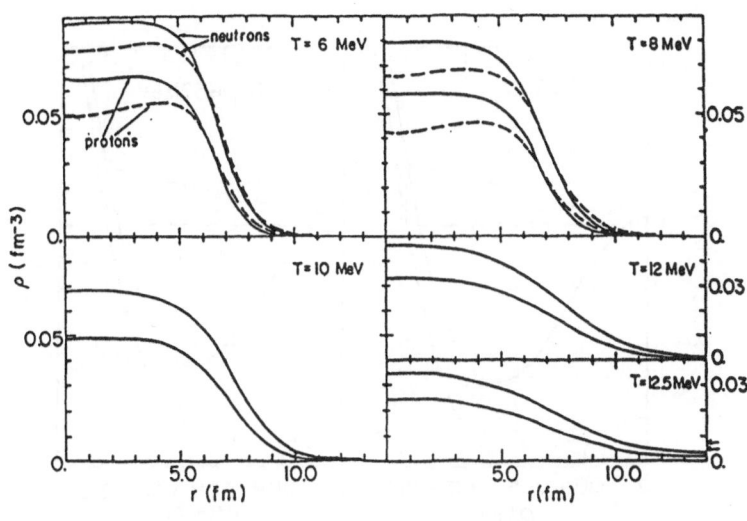

Fig. 7

Subtracted proton (lower curves) and neutron (upper curves) density
distributions at different temperatures for uncharged 208Pb (full lines)
and normal, charged 208Pb (dashed lines) calculated with SKM interaction
in R= 16 fm box (Ref. 50).
Normal 208Pb is unstable above T=8 MeV. Above T=12.5 MeV there is
only a uniform solution of the Hartree-Fock equations for uncharged
208Pb signifying the regime above the critical temperature of the
liquid-gas coexistance. The arrows show the proton and neutron density
of the uniform solution

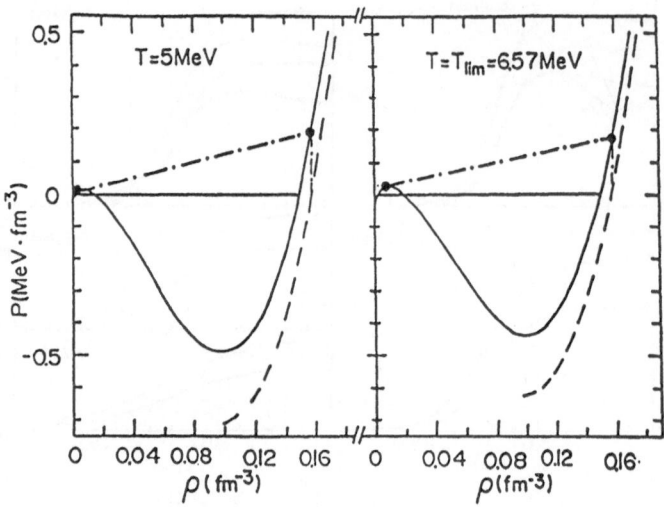

Fig. 8

Modified Maxwell construction for the nuclear charged drop-vapor
coexistence (heavy dots connected by dashed-dotted lines) for
^{109}Ag for temperatures below (T=5 MeV) and at the limiting value
(T=6.57 MeV). Full drawn curves are the bulk isotherms with
horizontal lines depicting the conditions of the coexistence of
the bulk materials (normal Maxwell construction). Dashed lines
show the full pressure $P_{On} + P_{surf}$ vs. the density inside the
charged drop (Ref. 51).

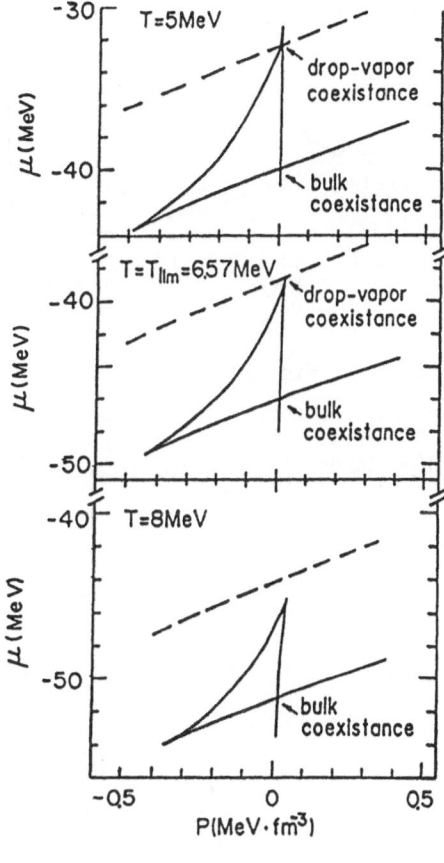

Fig. 9

Dependence of the chemical potential on the pressure for temperatures
below (T= 5 MeV), at (T = 6.57 MeV) and above (T = 8 MeV) the limiting
temperature. Full drawn curves denote the dependence for the bulk
quantities, dashed curve – the dependence of μ_{on} + μ_{coul} on P_{on} + P_{coul}
+ P_{surf} inside a charged drop representing [109] Ag. No coexistance is
possible above T_{lim} (Ref. 51)

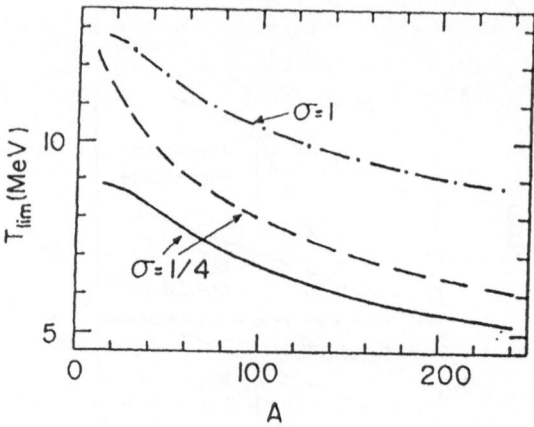

Fig. 10

Dependence of the limiting temperature on the nuclear mass A, for nuclei along the β-stability line for different equations of state and temperature dependence of the surface tension. [For detail, cf. Ref. (51).]

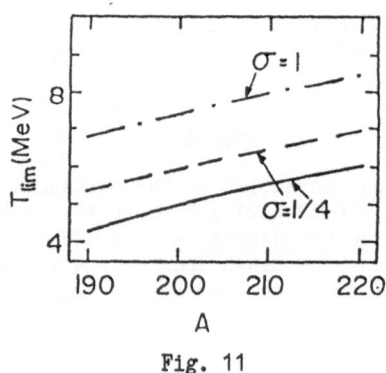

Fig. 11

Dependence of the limiting temperature of mass A for fixed charge $Z=82$ [cf. Ref. (51)].

much below T_c (17.22 MeV). It is obvious that contrary to the universal character of T_c, the limiting temperature is not universal and depends on the nucleus under consideration. The theoretical dependence of T_{lim} on atomic weight A, charge to mass ratio Z/A, the equation of state and on the temperature dependence of the surface tension was studied in detail in Ref. 51. The results are summarized on Fig. 10,11. The stiffer equation of state leads to higher values of T_{lim}. This dependence can be correlated with the dependence of T_{lim} on the effective interactions used in the microscopic Hartree-Fock calculations of Refs. 49,50.

The physical significance of T_{lim} is simple. Above the excitation energy which corresponds to T_{lim} a given nucleus cannot reach a complete equilibrium and will decay in a more violent way than by slow particle evaporation. In recent analyses (Refs. 2,3,27,28) evidence was found for the existence of the maximum energy that a nucleus can receive without breaking. In view of the theoretical picture discussed above studies of this energy and its dependence on the charge-to -mass ratio of formed nuclei may provide valuable information on the basic characteristics of nuclear matter - its equation of state and the temperature dependence of its surface tension.

Acknowledgement

My knowledge about the subjects of this review was mostly aquired during my collaborations with Yoram Alhassid, Paul Bonche, Arthur Kerman, John Manoyan, T. Troudet, Domenique Vautherin, and Jon Zingman, to all of whom I am sincerely greatful.

5. REFERENCES

1. S. Song et al., Phys. Lett. 130B, 14 (1983).
2. B. Borderie et al., Z. Phys. Z316, 243 (1984).
3. D. Jacquet et al., Phys. Rev. Lett. 53, 2226 (1984).
4. R.J. Charity et al., Phys. Rev. Lett. 56, 1354 (1986).
5. J.O. Newton et al., Phys. Rev. Lett. 46, 1383 (1981).
6. J.J. Gaardhoje et al., Phys. Lett. 139B, 273 (1984).
7. J.J. Gaardhoje et al., Phys. Rev. Lett. 53, 148 (1984).
8. C.A. Gossett et al., Phys. Rev. Lett. 54, 1486 (1985).
9. W. Hennerici et al., Nucl. Phys. A396, 329 c (1983).
10. B. Haas et al., Phys. Lett. 120B, 79 (1983).
11. A.M. Sandorfi et al., Phys. Lett. 130B, 19 (1983).
12. J.J. Gaardhoje et al., Phys. Rev. Lett. 56, 1783 (1986).
13. K. Neergard, Phys. Lett. 110B, 7 (1982).
14. D. Vautherin and N.V. Vinh Mau, Phys. Lett. 120B, 261 (1983).
15. P. Ring, L.M. Robledo, J.L. Egido, and M. Faber, Nucl. Phys. A419, 216 (1984) and the references therein for the earlier work of this group.
16. O. Civitarese, S. Furni, M. Ploszajczak, and A. Faessler, Nucl. Phys. A408, 61 (1983).
17. H. Sagawa and G.F. Bertsch, Phys. Lett. 146B, 138 (1984).
18. O. Civitarese, R.A. Broglia, and C.H. Dusso, Ann. Phys. 156, 142 (1984).
19. M. Gallardo, M. Diebel, T. Dossing, and R.A. Broglia, Nucl. Phys. A443, 415 (1985).
20. K. Snover, lectures in this school and the review on Giant Resonances in Excited Nuclei, Annual Review of Nucl. and Part. Aci. V36, 1986.

21. A.L. Goodman, Nucl. Phys. A352, 30, 45 (1981); A369, 365 (1981); A370, 90 (1981).

22. K. Tanabe, K. Sagawara-Tanabe, and M.J. Mang, Nucl. Phys. A357, 20, 45 (1981).

23. L.G. Moretto, Nucl. Phys. A182, 641 (1972); M. Brack and P. Quentin, Phys. Scripta, A10, 163 (1974), Phys. Lett. B52, 159 (1974); P. Quentin and H. Flocard, Ann. Rev. Nucl. Sci. 28, 523 (1978).

24. A.K. Ignatiuk, I.N. Mikhailov, L.H. Molina, R.G. Nazmutdinov, and K. Pomorsky, Nucl. Phys. A346, 191 (1980).

25. S. Levit and Y. Alhassid, Nucl. Phys. A413, 439 (1984).

26. Y. Alhassid, S. Levit, and J. Zingman, Phys. Rev. Lett. 57, 539 (1986).

27. Y. Alhassid, J. Zingman, and S. Levit, Nucl. Phys. A, in press.

28. Y. Alhassid, J. Manoyan, and S. Levit, in preparation.

29. E.C. Pollacco et al., Phys. Lett. 146B, 29 (1984).

30. X. Campi, J. Desbois, and E. Lipparini, Phys. Lett. 138B, 353 (1984); 142B, 8 (1984), Nucl. Phys. A428, 327c (1984).

31. U. Mosel, P.-G. Zint and K.M. Passler, Nucl. Phys. A236, 252 (1974); G. Sauer, M. Chanda, and U. Mosel, Nucl. Phys. A264, 221 (1976).

32. W.A. Kupper, G. Wegmann, and E.R. Hilf, Ann. Phys. 88, 454 (1974).

33. M. Jaqaman, A.Z. Mekjian, and L. Zamick, Phys. Rev. C27, 2782 (1983).

34. M.W. Curtin, M. Toki, and D.K. Scott, Phys. Lett. 123B, 289 (1983).

35. G. Bertsch and P.J. Siemens, Phys. Lett. 126B, 9 (1983); J.A. Lopez, and P.J. Siemens, Nucl. Phys. A431, 728 (1984).

36. J.E. Finn et al., Phys. Rev. Lett. 49, 1321 (1982).

37. R.W. Munich et al., Phys. Lett. 118, 458 (1982).

38. M.M. Gutbrod et al., Nucl. Phys. A387, 177c (1982).

39. C.B. Chitwood et al., Phys. Lett. 131B, 289 (1983).

40. A.D. Pangiotou et al., Phys. Rev. Lett. 52, 496 (1984).

41. P.J. Siemens, Nature 305, 410 (1983).

42. H. Schulz et al., Phys. Lett. 147B, 17 (1984).

43. H. Sagawa and G.F. Bertsch, Phys. Lett. 155B, 11 (1985).

44. A. Vicentini, G. Jacucci, and V.R. Pandharipande, Phys. Rev. C31, 1783 (1985).

45. W. Bauer, D.R. Dean, U. Mosel, and U. Post, Phys. Lett. 150B, 53 (1985).

46. J.P. Bondorf et al., Phys. Lett. 150B, 57 (1985).

47. D.M.E. Gross, Zhang Xiao-ze and Xu Shu-Yan, Phys. Rev. Lett. 56, 1544 (1986) and the references therein.

48. A.K. Kerman and S. Levit, Phys. Rev. C24, 1029 (1981);A.K. Kerman, S. Levit, and T. Troudet, Ann. Phys. 148, 436 (1983).

49. P. Bonche, S. Levit, and D. Vautherin, Nucl. Phys. A427, 278 (1984).

50. P. Bonche, S. Levit, and D. Vautherin, Nucl. Phys. A436, 265 (1985).

51. S. Levit and P. Bonche, Nucl. Phys. A437, 426 (1985).

52. L.D. Landau and E.M. Lifshitz, Statistical Physics (Pergamon, NY 1970).

53. A. Aharony, in "Critical Phenomena", Lecture Notes in Physics, 186,207 (Springer Verlag, New York 1983); R. Hornreich, Phys. Lett. 109A, 232 (1985).

54. A. Bohr and B. Mottelson, Nuclear Structure, Vol. 1 (Benjamin, New York, 1969) C42, App. 2.

RELAXATION OF NUCLEAR MOTION

P. F. Bortignon

Department of Physics, University of Padova,
and INFN LNL, Legnaro, Italy

R.A. Broglia

Department of Physics, University of Milano,
and INFN Milano, Milano, Italy
and
The Niels Bohr Institute, University of Copenhagen,
DK–2100 Copenhagen Ø, Denmark

1. INTRODUCTION

Heavy-ion collisions make it possible to measure the properties of nuclei at high excitation energy, where a statistical description of the nuclear response becomes necessary. In the lectures of S.Levit[1], it is shown how this concept can be introduced using the methods of the micro– and the grand–canonical ensembles.

Based on some of these results, we generalize, in the present lectures, a microscopic model of the damping of simple nuclear excitations, which has proved successful near the ground state[2,3], to the finite temperature situation[4].

The physical assumptions of the model are that Hartree–Fock (HF) theory of the single–particle motion and the Random Phase Approximation (RPA) theory of vibrational motion provide a satisfactory lowest order description of the nuclear properties, and that an important relaxation mechanism arises from the coupling of particles to low–lying surface vibrations. The coupling is treated in terms of Matsubara finite temperature Green functions, containing the zero–temperature case as a limit. Applications of the model are compared with empirical data. In particular, gamma– and particle–decay experiments are shown to test the doorway–state approach adopted for the spreading. Finally, the newly discovered phenomenon of rotational damping is discussed.

2. DAMPING OF SINGLE–PARTICLE AND VIBRATIONAL MOTION.

2.a) Finite Temperature Green Functions.

At finite temperature T, a many–body system is statistically distributed over all of its excited states*. Therefore, the "ground state average", which is used

---

* We do not discuss here under which conditions very excited nuclei can eventually reach statistical equilibrium.

to calculate the zero–temperature propagators of single–particle and vibrational motion, must be replaced by an appropriate average over (for example) a grand canonical ensemble[1,5].

In the grand canonical ensemble, the system is immersed in a bath held at a fixed temperature. The system can exchange particles or energy with the bath and viceversa. In the fermion case, the system Hamiltonian, H, is canonically replaced by the operator $H - \mu N$, where μ is the chemical potential and N the number operator. Introducing the distribution operator $\hat{\rho}$

$$\hat{\rho} = e^{\beta(H - \mu N)}, \tag{1}$$

and the partition function Z

$$Z = tr\hat{\rho}, \tag{2}$$

where $\beta = 1/T$ and the symbol "tr" denotes trace, the average value of any operator $\hat{O}$ is given by

$$\langle \hat{O} \rangle = \frac{tr\hat{O}\hat{\rho}}{Z}. \tag{3}$$

Introducing the single–particle creation and destruction operators in the Heisenberg picture and the Wick time–ordering operator, the zero temperature Green function for a fermion can be written as an expectation value of the operator $\hat{T}\{c_k(t_2)c_k^+(t_1)\}$ in the ground state. Hence the finite temperature Green function may be obtained simply by averaging this operator over a grand canonical ensemble by means of (3). A drawback of this definition is that the Hamiltonian appears in both the fermion operators and in the distribution operator $\hat{\rho}$. Thus, the actual calculation of the single–particle propagator requires two perturbation expansions. To avoid this inconvenience Matsubara[6] treats time as a complex temperature. The details of the method are discussed in many textbooks[5] and we report below the expressions for the unperturbed fermion and boson Green functions. For fermions one obtains[5]

$$G_\alpha^{(0)}(i\rho_n) = \frac{1}{i\rho_n - \epsilon_\alpha} \tag{4}$$

where the imaginary frequency has the values $i\rho_n = i(2n+1)\pi T$ with n an integer and the single–particle energy ϵ_α is measured from the chemical potential. The boson unperturbed Green function has the form

$$D_\lambda^{(0)}(i\omega_n) = \frac{1}{i\omega_n - \omega_\lambda} - \frac{1}{i\omega_n + \omega_\lambda} \tag{5}$$

where $i\omega_n = i2n\pi T$.

The graphical rules[5] for evaluating diagrams with these Green functions are essentially the same as for zero temperature except that the integrals over frequency are replaced by sums over the discrete imaginary frequencies $i\rho_n$ and $i\omega_n$. These sums generate the fermion and boson occupation factors $n_F(\alpha) = [exp(\epsilon_\alpha/T) + 1]^{-1}$ and $n_B(\omega_\lambda) = [exp(\omega_\lambda/T) - 1]^{-1}$ in the proper combination. This point is easily understood. In fact, let the function $f(z)$, product of Green functions, have poles at z_i with residues a_i and consider an integral of the form

$$I = \lim_{R \to \infty} \int \frac{dz}{2\pi i} f(z) n_B(z) \tag{6}$$

where the contour is a large circle of radius R which tends to ∞. In this limit, the integral vanishes and we obtain

$$\sum_{i\omega_n = i2n\pi T} f(i\omega_n) = -\beta \sum_i a_i n_B(\beta z_i) \tag{7}$$

The same procedure is used to evaluate fermion sums, that is,

$$\sum_{ip_n = i(2n+1)\pi T} f(ip_n) = \beta \sum_i a_i n_F(\beta z_i). \tag{8}$$

2.b) The surface coupling model

As anticipated in the introduction, we discuss the relaxation mechanism arising from the coupling of particles, described in the HF theory and by the Green function (4), to low–energy surface vibrations described in the RPA approximation and by the Green function (5).

For excitation energies below about $15 MeV$, surface vibrations seems to be the most important degrees of freedom the HF field can couple to, and the single–particle motion is damped by exciting them[2]. Consistently, the empirical absorptive potential is surface peaked below $15 MeV$ neutron scattering energy[7].

The first correction to the unperturbed single–particle Green function (4) is the self–energy associated with the graph of fig. 1, which is given by

Fig. 1. Self–energy diagram for the single–particle state. An arrowed line describes the particle propagation while a wavy line describes the vibration. Adapted from ref.[4].

$$\left(\sum^{(1)}(1, ip_n) \right) = -T \sum_{2,\lambda} V^2(1,2;\lambda) \sum_{i\omega_n} \frac{1}{i(p_n - \omega_n) - \epsilon_2}$$
$$\times \left(\frac{1}{i\omega_n - \omega_\lambda} - \frac{1}{i\omega_n + \omega_\lambda} \right), \tag{9}$$

where $V(i,j;\lambda)$ is the particle–vibration coupling strength. Performing the sum over the boson frequency $i\omega_n$ as in (7), the following expression for the self–energy is obtained[4],

$$\left(\sum_{ret}^{(1)}(\epsilon + iI) \right)_p = \sum_{2,\lambda} V^2(1,2;\lambda) \left\{ \frac{1 + n_B(\lambda) - n_F(2)}{\epsilon + iI - \epsilon_2 - \omega_\lambda} + \frac{n_B(\lambda) + n_F(2)}{\epsilon + iI - \epsilon_2 + \omega_\lambda} \right\}.$$
$$\tag{10}$$

Note that an analytical continuation in the frequency variable ip_n has been carried out and the averaging parameter I introduced.

The structure of eq. (10) may be easily understood recognizing the four time–ordered $T = 0$ processes corresponding to the diagram of fig. 1 at $T \neq 0$.

Two terms are present because the intermediate state can be formed either by creation or annihilation of a vibrational quantum. Moreover, the Green function (4) includes both particle and hole propagation.

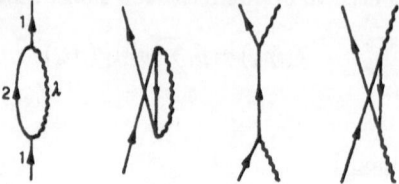

Fig. 2. The four time–ordered $T = 0$ processes corresponding to the diagram of fig. 1, at $T \neq 0$. Adapted from ref.[4].

The lowest order correction to the unperturbed vibration Green function (5) is given by the processes shown in figs. 3 and 4.

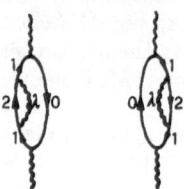

Fig. 3. Single–particle self–energy corrections to the RPA Green function. Adapted from ref.[4].

Fig. 4. Lowest order vertex correction to the RPA Green function. adapted from ref.[4].

The corresponding expressions are[4]

$$\left(P_{ret}^{(2)}(GR, \omega + iI)\right)_p = \sum_{1,0} V^2(1, 0; GR) \sum_{2,\lambda} V^2(1, 2; \lambda)$$

$$\times \left\{ \frac{n_B(\lambda) + 1 - n_F(2)}{\Delta\omega_{\lambda 20}} \left[\frac{n_F(0) - n_F(1)}{(\Delta\omega_{01})^2} + \frac{n_F(1)}{(\Delta\epsilon_{21})^2} \right] \right.$$

$$\left. - \frac{n_B(\lambda)n_F(2)}{\Delta\omega_{\lambda 20}(\Delta\epsilon_{21})^2} + \frac{(n_B(\lambda) + 1 - n_F(2))n_F'(1)}{\Delta\omega_{01}\Delta\epsilon_{21}} \right\}$$

$$- \left\{ \text{same as above with} \quad \omega_\lambda \to -\omega_\lambda, n_B(\lambda) \to n_B(-\lambda) \right.$$

$$= -(1 + n_B(\lambda)) \right\}$$

$$(11a)$$

$$\left(P_{ret}^{(2)}(GR, \omega - iI)\right)_h = \left(P_{ret}^{(2)}(GR, -\omega - iI)\right)_p, \qquad (11b)$$

and

118

$$
\left(P_{ret}^{(2)}(GR, \omega + iI)\right)_v = \sum_{12,30,\lambda} V(1,0;GR)V(12,30;\lambda)
$$

$$
\times \left\{ n_B(\lambda) \left[\frac{n_F(0)}{\Delta\omega_{01}\Delta\omega_{\lambda 20}\Delta\epsilon_{30}} - \frac{n_F(1)}{\Delta\omega_{01}\Delta\omega_{-\lambda 13}\Delta\epsilon_{21}} \right. \right.
$$

$$
+ \frac{n_F(2)}{\Delta\omega_{32}\Delta\omega_{\lambda 20}\Delta\epsilon_{21}} + \left. \frac{n_F(3)}{\Delta\omega_{32}\Delta\omega_{-\lambda 13}\Delta\epsilon_{30}} \right]
$$

$$
+ \frac{1}{\Delta\omega_{32}\Delta\omega_{01}} \left[-\frac{n_F(1)[1-n_F(2)]}{\Delta\epsilon_{21}} - \frac{n_F(0)[1-n_F(3)]}{\Delta\epsilon_{30}} \right.
$$

$$
\left. \left. - \frac{n_F(0)[1-n_F(2)]}{\Delta\epsilon_{\lambda 20}} + \frac{n_F(1)[1-n_F(3)]}{\Delta\omega_{-\lambda 13}} \right] \right\}
$$

$$
- \{\text{same as above with } \omega_\lambda \to -\omega_\lambda,
$$

$$
n_B(\lambda) \to n_B(-\lambda) = -(1 + n_B(\lambda))\}
\tag{11c}
$$

$$
\Delta\omega_{\pm\lambda jj'} = \omega + iI - (\pm\omega_\lambda + \epsilon_j - \epsilon_{j'}),
$$

$$
\Delta\omega_{jj'} = \omega + iI + \epsilon_j - \epsilon_{j'},
$$

$$
\Delta\epsilon_{jj'} = \omega_\lambda + \epsilon_j - \epsilon_{j'},
$$

More transparent expressions are obtained by keeping only the temperature dependence of the boson occupation factors, the specifics of the single–particle occupation factors being unimportant for a very collective giant mode. Furthermore, we keep only the poles associated with the vanishing of the real part of $\Delta\omega_{\pm\lambda ki}$, where k identifies a level above the Fermi surface and i one below it. The reduced expression for (11a) and (11c) are then

$$
\left(P_{ret}^{(2)}(GR, \omega + iI)\right)_p = \sum_{1,0} \frac{V^2(1,0;GR)}{(\Delta\omega_{01})^2} \sum_{12,\lambda} V^2(1,2;\lambda) \left(\frac{1 + n_B(\lambda)}{\Delta\omega_{\lambda 20}} + \frac{n_B(\lambda)}{\Delta\omega_{-\lambda 20}} \right),
\tag{12a}
$$

$$
\left(P_{ret}^{(2)}(GR, \omega + iI)\right)_v = - \sum_{12,34,\lambda} \frac{V(1,0;GR)V(12,30;\lambda)V(2,3;GR)}{\Delta\omega_{01}\Delta\omega_{32}}
$$

$$
\times \left\{ \left[\frac{n_B(\lambda)+1}{\Delta\omega_{\lambda 20}} + \frac{n_B(\lambda)+1}{\Delta\omega_{\lambda 13}} \right] + \left[\frac{n_B(\lambda)}{\Delta\omega_{-\lambda 20}} + \frac{n_B(\Lambda)}{\Delta\omega_{-\lambda 13}} \right] \right\}.
\tag{12b}
$$

This result is just what one would expect from figs 3 and 4 and eq. (10). The process in which a phonon is created in the intermediate state is enhanced by a factor $(1 + n_B(\lambda))$. In addition, a physical state may be created by annihilating a phonon already present. This appears in terms proportional to $n_B(\lambda)$ and shifts the strength at lower excitation energy.

The diagrams of figs. 3 and 4 contain five internal lines and there will be five occupation factors in the corresponding expressions of the "usual" perturbation theory, while not more than two appear in eqs. (11a), (11b) and (11c). In ref.[8], it was verified that all terms with three or more occupation factors cancel out when the sum is carried out over all time orderings, and that the final result agrees with Matsubara's expressions. Clearly, the calculations with the Green functions (4) and (5) are more economic.

3. CALCULATIONS AT ZERO TEMPERATURE

3.a) Résumé

The main results obtained in the surface coupling model for relaxation of single–particle and vibrational motion at zero temperature are reviewed in refs.[2,3,9]. The conclusion is that the model describes the damping to better than a factor of two accuracy at low excitation energy – say, below $15 - 20 MeV$ – where the surface absorption dominates in the optical potential. The coherence between particle and hole in a collective giant vibration can drastically modify and generally reduce[10] the spreading of the mode as compared to the damping of the single–particle states.

3.b) Exclusive Experiments

In the previous sections we discussed the relaxation of the single-particle and vibrational motion due to the coupling of the mean field to selected doorway states, consisting of collective vibrations and single-particle states. Exclusive experiments like γ-decay of single-particle (hole) states populated in stripping (pick-up) reactions, γ- and particle-decay of giant vibrations excited in inelastic scatterings can test this picture.

γ-decay

a). The γ-decay of deeply–bound neutron holes states in ^{111}Sn (that is $1g_{9/2}$, $2p_{1/2}$ and $2p_{3/2}$) has recently been investigated [11] via the reaction $^{112}Sn(^{3}He, \alpha\gamma)$. The main part of the decay (up to 80%) occurs through the feeding of quasiparticle–plus–vibration coupled states, in agreement with the doorway picture. The previously deduced strength functions are extended to higher energy as shown in fig. 5. In particular, a second bump predicted in ref.[2] at $\sim 7 MeV$ in the case of the $g_{9/2}^{-1}$ hole, is identified, although much broader (cf. fig. 3 of the first of ref.[2]).

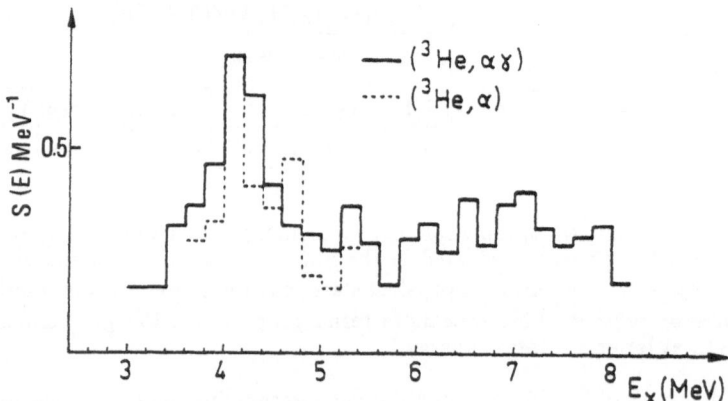

Fig. 5. Adapted from ref.[11]. Comparison of the experimental $(1g_{9/2})_n^{-1}$ strength functions obtained in two experiments.

b) In a study [12] of the γ-decay of the isoscalar giant quadrupole resonance in ^{208}Pb, the low–lying octupole vibration, although predicted to be dominant in the spreading of the giant mode, was hardly populated. This result is understood[13] in terms of the quenching of the *isovector* E1 transition between two states having predominantly *isoscalar* character. Moreover, the energy of the transition ($\approx 8.5 MeV$) is such, that the coupling to the giant dipole resonance reduces the square of the effective charge by a factor of 15, as shown by the relation

$$e_{eff} = e[1 + \chi(E_\gamma)], \qquad (13)$$

where the dipole polarizability is [14]

$$\chi(E_\gamma) = -\chi_0(\hbar\omega_D)^2 \times [(\hbar\omega_D + E_\gamma + i\Gamma_{D/2})(\hbar\omega_D - E_\gamma - i\Gamma_{D/2})]^{-1}. \qquad (14)$$

The values of the parameters entering in these equation are in the case of ^{208}Pb $\chi_0 = 0.76, \hbar\omega_D = 14 MeV$ and $\Gamma_D = 4 MeV$. The situation is reversed for the very high energy E1 transition between the isovector giant quadrupole resonance ($E_z \approx 22 MeV$ in ^{208}Pb) and the low-lying octupole state. In this case, being $E_\gamma \gg \hbar\omega_D$, the effective change is enhanced, $\mid 1 + \chi \mid^2 \approx 3$ and the transition is estimated [13] to be a factor $\approx 10^3$ stronger than in the previous case leading to a γ–width of $\approx 6 keV$. This offers the possibility of a good experimental determination of the properties of the giant mode as well as of a fruitful test of the theory. In fig. 6, the strength function of this mode is shown, as calculated in the model of ref.[2].

Neutron decay of giant resonances in heavy nuclei

Particle-decay experiments are a powerful tool in the study of giant vibartions, especially in light nuclei, where the giant modes are usually very fragmented and the decay of charged particle is allowed. Recently, very interesting information on the neutron decay of the giant quadrupole, dipole and monopole resonances energy region in heavy nuclei, have become available[15,16]. A common qualitative feature of these data is the enhancement of the neutron cross section, with respect to the statistical decay of the compound nucleus, not only to the low-energy single-hole levels of the A-1 system, but also to the hole-plus-vibration states at higher excitation energy. This result seems to confirm the central role played by the doorway states in the decay of the giant modes.

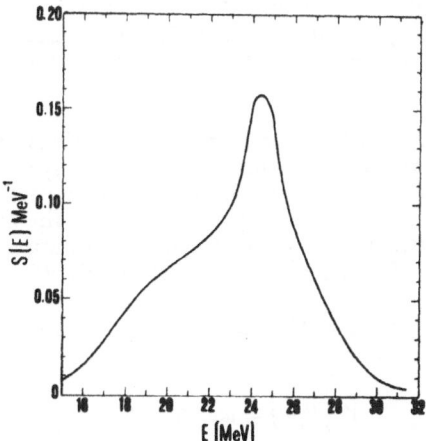

Fig. 6. Strength function of the isovector giant quadrupole resonance calculated in the model of the first of ref.[2].

An accurate quantitative analysis is necessary, both from the experimental and the theoretical side, to assess this point quantitatively.

4. CALCULATIONS AT FINITE TEMPERATURE

4a. Mean Field at Finite Temperature

For temperatures below $T \approx 4-5 MeV$, the Hartree–Fock solutions of medium and heavy spherical nuclei are rather stable with respect to changes in the temperature while the HF energies ϵ_α in eq. (4) are essentially temperature independent[1,17]. The main effect the temperature has on the Hartree–Fock solutions is the change of the single–particle occupation factors $n_F(\alpha)$ and the associated smoothing of the Fermi surface on an energy interval of the order of T.

At the level of the RPA, the isovector dipole strength function of spherical nuclei is rather T–independent in the same temperature range, while the isoscalar strength functions are more affected in the low–energy part[18].

These results are valid for the case of the spherical HF–solution of hot nuclei. However, the formation of highly excited nuclei in heavy ion reactions is usually accompanied by transfer of large amount of angular momentum. The consequences of this effect in terms of deformation of the mean field are discussed in the sect. 4.c)

4.b. Single–particle and Giant Resonance States at Finite Temperature in "Spherical" Nuclei

In ref.[4], the imaginary part $W(\epsilon, T)$ of the single–particle self–energy has been calculated for the valence neutron particle and hole states of ^{208}Pb, according to eq. (10), as a function of the energy ϵ and the temperature T. For $T \neq 0$, the main effect is that $W(\epsilon, T)$ is different from zero also at the Fermi energy, implying a shorter mean–free path λ of the nucleons in the hot nucleus. For infinite systems the mean–free path can be written in terms of W as[2,3]

$$\lambda = \frac{\hbar k}{m_k} \frac{\hbar}{2W} \tag{15}$$

where m_k is the momentum–dependent effective mass (cf. ref.[3] and sect. 4.d for more details). In fig. 7, the results obtained[4] for the case of ^{208}Pb averaged over the valence orbitals are displayed for a particle at the Fermi energy as a function of the temperature.

As already found at $T = 0$[2,3], the largest contributions to W arise from the coupling to low–lying octupole vibrations. This is especially true at $T \neq 0$ because of the presence of octupole states at very low energies with particularly large boson factors. Corrections due to Pauli principle violations in the graph of fig. 1, and implying boson occupation factors at the unperturbed particle–hole energies, cf. ref.[19], have been evaluated and amount to less than 10%[4,20].

A simple parametrization of the results at $T \neq 0$ and $\omega = \epsilon - \epsilon_F \neq 0$ is provided by[4]

$$W(T, \omega) = a\omega^\alpha + bT^\alpha \tag{16}$$

with $\alpha \approx 1$, $a \approx 0.2$ and $b \approx 0.5$, for values of T and ω of few MeV. This very approximated expression neglects the strong shell structure usually displayed by the results[2,3].

The linear dependence on ω and T is at variance with the results for infinite systems, found in ref.[21], where a quadratic dependence was obtained, but consistent with those of surface–dominated systems[22]. This is because the unperturbed particle–hole response function depends linearly on the excitation energy, while the collective response function, summed over many multipolarities in finite nuclei, is rather constant with energy, (cf. the appendix in ref.[4])

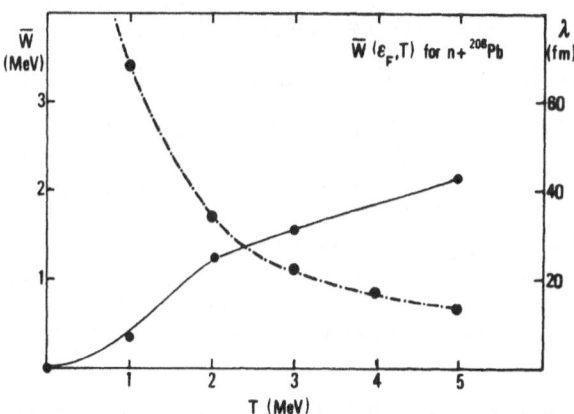

Fig. 7. The imaginary part $W = Im\left(\sum_{ret}^{(1)}(T, \epsilon + iI)\right)$ of the self–energy of neutron single–particle states in ^{208}Pb for $\epsilon = \epsilon_F$. The quantity W is averaged over the valence orbitals[4]. The dot–dashed line shows the mean–free path λ calculated according to (15).

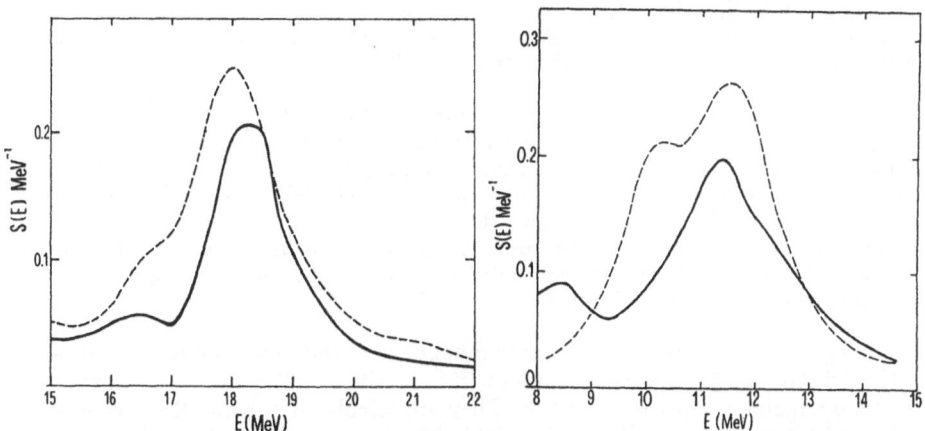

Figure 8. Strength functions for the Giant Dipole Resonance in ^{90}Zr and for the isoscalar Giant Quadrupole Resonance in ^{208}Pb at $T = 0$ (dashed line) and $T = 3MeV$ (full line). Adapted from ref.[4].

In fig. 8, the strength functions $S(E)$ are displayed for the isovector giant dipole resonance in ^{90}Zr (GDR) and for the isoscalar giant quadrupole (GQR) in ^{208}Pb, as calculated in ref.[4]. The strength function $S(E)$ is written in the doorway approximation in the form[2]

$$S_a(E) = \frac{S_0}{\pi} \frac{\Gamma(E+iI)/2 + I}{\left[E_a - E - \Delta E_a(E+iI)\right]^2 + \left[\Gamma(E+iI)/2 + I\right]^2} \tag{17}$$

where

$$\Delta E_a(E+iI) = Re\big(P_{ret}^{(2)}(E+iI)\big)$$

$$\Gamma(E+iI) = 2 \times Im\big(P_{ret}^{(2)}(E+iI)\big)$$

are the real and imaginary part of the self–energy of the giant resonance a.

Even at $T = 3MeV$, the position of the peaks in fig. 8 hardly moves and their widths have become, if anything, smaller. In both cases however, the area under the peak has been reduced by about 30%. This strength is now found at lower excitation energies, over an energy range of several MeV. This is because at $T \neq 0$, both $\big(P_{ret}^{(2)}(\omega + iI)\big)_{p,h}$ and $\big(P_{ret}^{(2)}(\omega + iI)\big)_v$ (cf. eqs. 11a,b,c) display poles at $\omega = (\epsilon_p - \epsilon_h) + \omega_\lambda$ and $\omega = (\epsilon_p - \epsilon_h) - \omega_\lambda$. While the first energy coincides with the peak of the giant resonance at $T = 0$, the second value of ω is shifted down from this value by $2\omega_\lambda$. This result implies that the damping of the full strength function has increased.

Before concluding this paragraph, it is noted that the properties of the single–particle self–energy at finite temperature in the Brueckner–Hartree-Fock and in the Thomas–Fermi approximation have been recently discussed in refs.[23].

4.c. The Giant Dipole Resonance in Highly Excited Nuclei

As discussed by K.A. Snover[24], a larger change in the width of the Giant Dipole Resonance, than that shown in fig. 8, has been observed in excited nuclei, with temperatures in the range $1 < T < 2MeV$. For systems that are spherical in the ground state the width practically doubles, while minor changes are observed in the case of nuclei which have deformed ground states, and which consequently display large widths of the GDR already at $T = 0$.

This is because in the heavy–ion reactions leading to the hot nuclei a large amount of angular momentum is transferred and the system become strongly deformed, in general displaying oblate shape. Consequently, the main breaking of the dipole strength can be obtained by calculating the GDR within the RPA in the deformed particle–hole basis. Thus, the vibrations along the different axis find sizable differences in the restoring force. There seems to be consensus[14] in this damping mechanism which is the adiabatic limit of the coupling to the quadrupole vibration described by figs. 3 and 4.

In fig. 9, examples of dipole strength function are displayed for different values of the deformation parameters ϵ and γ in ^{108}Sn excited at $T = 1.5MeV$ and with the total angular momentum $I = 40\hbar$. They are calculated[25] in the RPA using a deformed Nilsson potential[14] as mean field. The average photo–absorption cross section $< \sigma_{lab} >$ shown in the inset was calculated making use of the relation

$$< \sigma_{lab}(E,I,T) > = \frac{\int \epsilon d\epsilon d\gamma P(\epsilon, \gamma, I, T) \sigma_{lab}(E, \epsilon, \gamma, I, T)}{\int \epsilon d\epsilon d\gamma P(\epsilon, \gamma, I, T)} \tag{18}$$

where

$$P(\epsilon, \gamma, I, T) \propto exp\big[-F(\epsilon, \gamma, I, T)/T\big] \tag{19}$$

the quantity F being the free energy (cfr. ref.[1]).

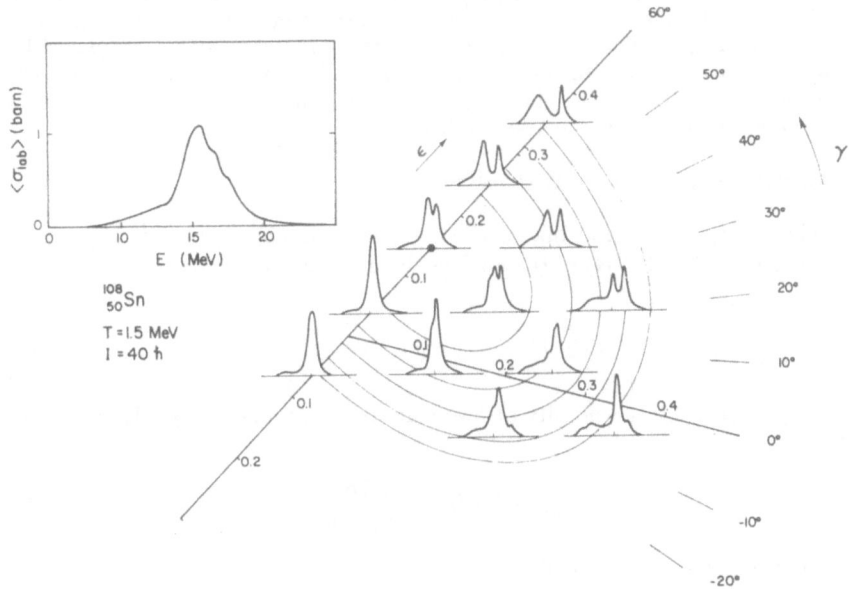

Fig. 9. Adapted form ref.[25]. See text.

Because of the finite temperature, the system explores the (ϵ, γ) potential energy surface with the probability (19). The comparison of the results obtained in this model for ^{108}Sn with experimental data is displayed in fig. 10.

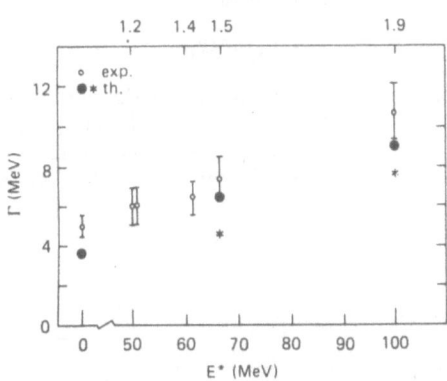

Fig. 10. Adapted from ref.[8]. Calculated spreading width of the GDR in Sn nuclei in the adiabatic approximation (stars) and taking into account in an average way also the effects of vibrations around the deformed shape (solid dots) in comparison with the data[26].

These calculations do not take into account the fact that as the temperature increases the compound nucleus spends shorter and shorter period of time at the different points of the (ϵ, γ) plane. The associated phenomenon of motional narrowing (cf. Section 5) is expected to play an important role. Estimates making use of the Fermi Gas Model will indicate that the effect observed in Fig. 10 is strongly quenched[42]. If this is the case, other effects that go beyond the mean field picture

125

have to be invoked. In particular, as the temperature increases, the mean–free path becomes shorter, and particle–particle collisions may play a role. However, it is still an open problem whether the very strong increase with T displayed by the many- quasiparticle width (cf. Section 5), and which essentially controls the degree of motional narrowing affecting the damping width of the giant resonances is well described by the Fermi Gas Model.

4d. Effective Masses and Level Density

The density of levels at high excitation energy $E^*(E^* \gg \epsilon_F A^{-1})$ is given, in the Fermi Gas Model (FG) of the nucleus, by a formula of the type [27]

$$\rho(E^*) \propto \exp\left\{2\left(\frac{\pi^2}{6}g_0 E^*\right)^{\frac{1}{2}}\right\} \tag{20}$$

where $g_0 \equiv g(\epsilon_F)$ is the density of levels at the Fermi energy ϵ_F. Its FG value therefore is

$$g_0 \approx \frac{3}{2}A/\epsilon_F = \frac{3Am^*}{(\hbar\kappa_F)^2} \tag{21}$$

Traditionally, the quantity

$$a \equiv \frac{\pi^2}{6}g_0 \tag{22}$$

is introduced. Using for the effective mass m^* the value 1 and for the Fermi momentum κ_F the saturation value $\kappa_F \simeq 1.37 fm^{-1}$, it is obtained

$$a \approx A/15(MeV)^{-1} \tag{23}$$

This value is much smaller than that obtained from fittings to neutron resonance and proton scattering data[27,28] which leads to

$$a_{emp} \approx A/8(MeV)^{-1} \tag{24}$$

The missing physics in the FG is essentially the surface of the nucleus. On the surface, the Fermi momentum κ_F is small. The coupling of the HF field to the low–lying surface vibrations increases[3] the effective mass both at the Fermi energy and at the nuclear surface by a factor $\sim 1.3 - 1.5$, (cf. fig. 11).

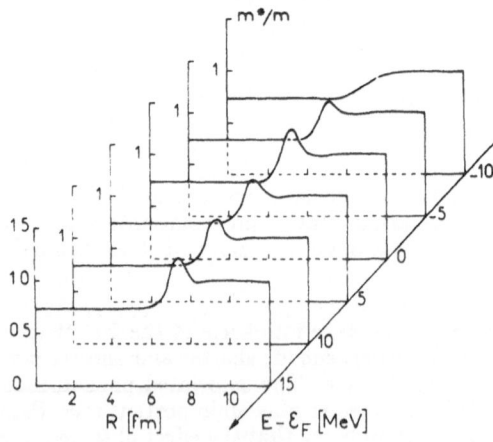

Fig. 11. Adapted from ref.[29]. Calculated dependence upon R and $E - \epsilon_F$ of the effective mass m^* in the case of protons in ^{208}Pb.

As discussed in ref.[3], the effective mass m^*/m can be written as the product of two quantities

$$\frac{m^*}{m} = \frac{m_\kappa}{m} \times \frac{m_\omega}{m} \qquad (25)$$

The κ-mass m_κ/m is connected to the non-locality (momentum dependence) of the single-particle potential. Its value is different from unity (already) in the HF approximation.

The ω-mass m_ω/m reflects the coupling of the static HF field to other degrees of freedom, in particular to the low-lying collective surface vibrations of the nucleus. The coupling produces a genuine energy-dependence of the real part $\Delta V(\omega, r)$ of the self-energy corrections to the HF field, as displayed in fig. 12a.

The function $\Delta V(\omega, r)$ is connected[3] to the imaginary part $W(\omega, r)$ of the single-particle field through a dispersion relation

$$\Delta V(\omega, r) = \frac{P}{\pi} \int \frac{W(\omega', r)}{\omega - \omega'} dz \qquad (26)$$

The ω-mass can be then defined as[3]

$$\frac{m_\omega(\omega, r)}{m} = 1 + \Delta m_\omega = 1 - \frac{d}{d\omega} \Delta V(\omega, r) \qquad (27)$$

and from the curve of fig. 12a the clear enhancement at the Fermi energy shown in fig. 12b is obtained.

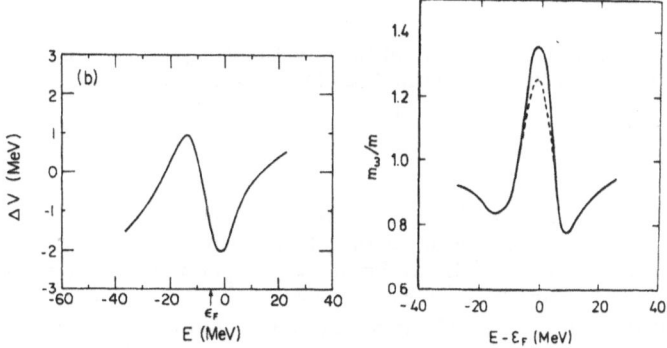

Fig. 12a and 12b. Adapted from ref.[3]. Example of the real part of the self–energy corrections to the HF field as calculated in the second of ref.[3] and of the corresponding effective mass m_ω/m, cf. eq. 27.

Results like those of fig. 11 and 12 can explain the discrepancy between the values (23) and (24).

We may ask now how this picture evolves as a function of the temperature. The results of a recent investigation[30] are summarized in fig. 13 for the case of neutron states in ^{208}Pb. The enhancement at the Fermi energy of the ω-mass disappears at a temperature of the order of 3-4 MeV*. The reason for this behaviour is that at high temperature the distinction between occupied and unoccupied levels tends to disappear (all occupation factors are equal at $T \rightarrow \infty$) and the characteristic shape in fig. 12a tends to become flat.

---

* Similar results have been obtained by other authors [31].

According to eqs. (21) and (22), the decreasing with temperature of the effective mass implies a decrease of the a parameter in the same temperature range.

Empirical evidence for this behaviour has been recently presented in ref. [32], where the level density parameter for $A \approx 160$ nuclei have been determined at excitation energies of 100 to 400 MeV from coincidence measurements between heavy residues, light particles and gamma–rays. While the low energy data are consistent with the value $a = A/8MeV^{-1}$, at high temperature, $T \approx 6MeV$, a smaller value $a = A/13MeV^{-1}$ is needed. The transition occurs in the range $2.5 < T < 4.5MeV$.

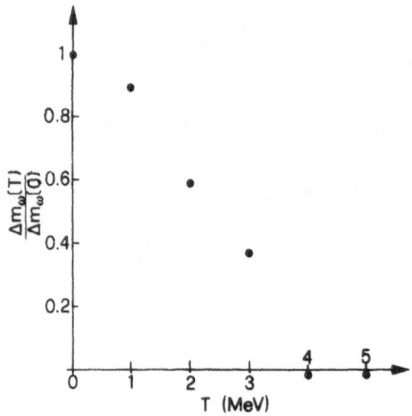

Fig. 13. Adapted from ref.[30]. Values of the quantity Δm_ω, defined in eq. (27), as function of the temperature and normalized to the zero–temperature value.

As discussed by K.Snover[24], precise determination of the empirical value of a is obtained in the study of the properties of the giant dipole resonance in very excited nuclei. The available data are limited to temperatures below 2 MeV and are consistent with the standard value $a = A/8MeV^{-1}$. This is in no contradiction with the predictions of fig. 13 and the data in ref.[32]. Experiments which explore the higher temperature regimes will be of extreme interest to check the model.

5. DAMPING OF ROTATIONAL MOTION AT FINITE TEMPERATURE

At an excitation energy of few MeV above the yrast line, rotational bands become very closely spaced. Any of them can be viewed as a collective state imbedded in a background of other states, to which it may couple by a residual interaction. The admixture between a band and the background leads to the damping of the rotational motion[33−36]. Three distinct mechanisms, namely differences in rotational frequencies and in quadrupole moments and spin dependence of the interaction, can contribute to this damping, each reflecting a property which can make rotational bands different. In fact, if a manifold of rotational bands all having the same quadrupole moment and the same frequency at a given angular momentum are mixed by a spin-independent residual interaction, the rotational pattern of the original manifold will be preserved (cf. fig. 14).

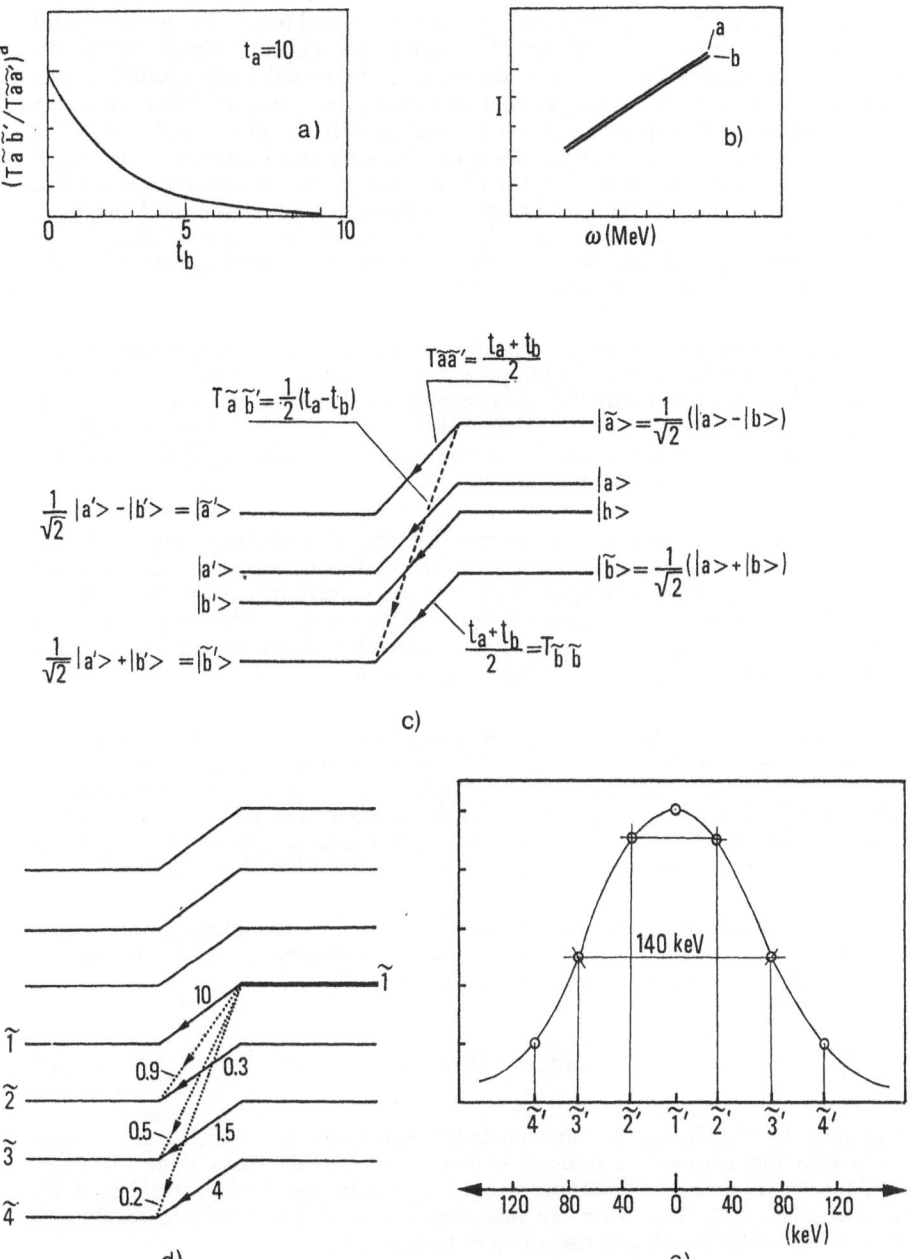

Fig. 14. Schematic representation of the persistence of the rotational pattern, when rotational bands with the same rotational frequencies are mixed (cf. (b) and (c)), unless an unrealistic dispersion in the intrinsic quadrupole moments and then in the quadrupole transition amplitudes (cf. (a), (d) and (e)) is assumed.

The response of the rotating compound nucleus to electromagnetic quadrupole decay has an analogue[37] in the line width observed in the phenomenon of nuclear magnetic resonance (NMR) in condensed matter[38-40]. Each nuclear spin at a given site in the lattice precessing with a Larmor frequency determined by the external magnetic field $\vec{B}_0$, is the analogue to a rotational band. The whole crystal at a given temperature and taking into account the magnetic dipole interaction between nearest neighbours, is the analogue to the mixed band manifold. The response of the system to an external time-dependent magnetic field sweeping through the crystal perpendicular to the constant field B_0, is the analogue to the stretched electromagnetic quadrupole decay between two consecutive rotational states. The line width observed in the absorbance process is due to the spread in Larmor frequencies and reflects the spatial inhomogeneity of the local magnetic field. Its analogue in the nuclear case is the rotational damping width Γ_{rot} of the electromagnetic quadrupole decay, arising correspondingly from the spread in the rotational frequencies of the manifold of bands.

While the damping width of collective motion of a many-body system is as a rule an increasing function of the temperature, the nuclear magnetic resonance line shape becomes narrower with T. This phenomenon, known as motional narrowing, arises because higher temperatures imply shorter diffusion times between different lattice sites. This is also predicted to be the case for the damping of the rotational motion, at least within some temperature interval.

From the above discussion it emerges that the rotational damping width Γ_{rot} is controlled by two quantities: the spread in rotational frequencies $\Delta\omega_0$ and the time τ (hopping time) a compound nucleus state stays in any single rotational band (diffusion time of the atoms between lattice sites). This last quantity is connected to the range of energies Γ_μ over which the rotational bands are mixed by the residual interaction, according to the relation $\tau = 1/\Gamma_\mu$.

In the case where the hopping time is very large as compared to the spread in frequencies, the rotational damping will be propositional to $\Delta\omega_0$. In the opposite situation the decay of the rotational motion takes place as a random walk process. The steps having a length $l = 2\Delta\omega_0/\Gamma_\mu$, the number of steps being $T_2\Gamma_\mu$, where T_2 is the time it takes for the dephasing angle between the phases of the spinning members of the rotational manifold to become of the order of one radian.

If one assumes that all bands are parallel ($\Delta\omega_0 = 0$), but allows the associated quadrupole moments to the different, the rotational transitions will acquire a broad halo which is distributed over an energy rangy Γ_μ. In fact

$$\overline{M_{E2}^2} = (\bar{M}_{E2})^2 + \sigma_{E2}^2, \tag{28}$$

the quantity M_{E2} being the quadrupole transition associated with a given band. This halo will however be difficult to detect, as unrealistically large variations of M_{E2}, of the order of 2–20, are needed to obtain any sizable breaking of the intensity (cf. Fig. 14). One can then conclude that the rotational damping is essentially controlled by the spreading in frequencies.

At low rotational frequencies, the nucleus rotates as a whole. Because of the marked shell structure, which can be viewed as magnetic impurities in the case of condensed matter, the rotation, which plays the role a magnetic field does in the case of infinite systems, couples differently with the motion of different nucleons. Those moving in high–l orbitals like $i_{13/2}$ (neutrons) and $h_{11/2}$ (proton) decouple from the rotating core, giving rise to the phenomenon of backbending[41] (gapless superconductivity in condensed matter).

One can thus write

$$\Gamma_{rot} = \begin{cases} 2(2\Delta\omega_0), & \Gamma_\mu \ll 2\Delta\omega_0, \\ 2\dfrac{(2\Delta\omega_0)^2}{\Gamma_\mu}, & \Gamma_\mu > 2\Delta\omega_0. \end{cases} \tag{29}$$

The range of energy Γ_μ over which the unperturbed rotational bands are mixed to form a compound nucleus state, depends on temperature through the seniority quantum number v. This quantity labels the average number of particles which at a given temperature are excited through the Fermi surface. In the Fermi Gas Model, and for a medium heavy nucleus of mass $A \approx 160$, one obtains

$$v = g_0 T 3(U)^{1/2} \quad (MeV) \tag{30}$$

In deriving this equation the relation $U = aT^2$ between excitation energy U and temperature T has been used. The energy U is measured relative to the yrast line. The quantity a has been discussed in paragraph 4d.

Making use of the empirical relation[2] $\Gamma_{v=1} = (1/15MeV)U^2$ for the damping width of a one-quasiparticle state, the inverse of the hopping time $\Gamma_\mu = 1/\tau$ among states of the compound nucleus can be written as

$$\Gamma_\mu(U) \approx v\Gamma_{v=1}\left(\frac{U}{V}\right) \approx \frac{1}{45}U^{3/2}(MeV). \tag{31}$$

The total angular momentum of rotational bands can thus be expressed through the classical relation $I = \mathcal{F}\omega + i$. Here the first term is due to the collective rotation of the nucleus as a whole with rotational frequency ω. The second term stems from the partial alignment of the single particle angular momenta along the axis of rotation. The variation of the rotational frequency at a given angular momentum can thus arise from variations in the alignment i, and in the moment of inertia $\mathcal{F}$ associated with the collective rotation.

To the extent that we can treat fluctuations in the alignment and in the moment of inertia independently we obtain

$$\Delta\omega_0 = \omega_0\sqrt{\left(\frac{\Delta i}{I}\right)^2 + \left(\frac{\Delta\mathcal{F}}{\mathcal{F}}\right)^2}, \tag{32}$$

where ω_0 is the average frequency at angular momentum I.

In the case of a compound nuclear state with v quasiparticles, the alignment is given by the random sum of v contributions, $(\Delta i)^2 = v < (i_{v=1})^2 >$. The average single particle alignment associated with the motion of a nucleon in a spherical nucleus is proportional to the radius R of the nucleus, that is $< (i_{v=1})^2 > \sim R^2 \sim A^{2/3}$.

This value is, however, strongly reduced in the case of deformed nuclei and will eventually vanish in the limit of infinite deformation. This is because the breaking of the rotational symmetry leads to a strong coupling of the single-particle motion to the deformed body which rotates as a whole (Goldstone mode). For a typical quadrupole deformation where the ratio between larger and smaller axis is 1.3 : 1, and at an angular momentum $I = 40$ one obtains $< (i_{v=1})^2 > \approx 1.5$. Hence,

$$(\Delta i)^2 \approx v < (i_{v=1})^2 > \approx 2.8 \cdot 10^{-3}(U)^{1/2}. \tag{33}$$

Fluctuations in the moment of inertia $\mathcal{F}$ are connected with fluctuations in the nuclear shape. For cold deformed nuclei, the most important surface fluctuations are associated with quadrupole vibrations. The coordinates used to describe these degrees of freedom are the relative elongation of the nuclear shape, β, and the deviation from axial symmetry, γ. In terms of these coordinates, the moment of inertia can be written as [14]

$$\mathcal{F} = \mathcal{F}_{sph}\left(1 + \left(\frac{5}{4\pi}\right)^{1/2}\beta\cos(\gamma - \pi/3)\right), \tag{34}$$

where $\mathcal{F}_{sph}$ is the rigid body moment of inertia of the spherical shape. Thermal excitation of the vibrations leads to fluctuations in the moment of inertia equal to

$$\left(\frac{\Delta\mathcal{F}}{\mathcal{F}}\right)^2 \approx T\frac{5}{4\pi}\left(\frac{1}{4}\frac{1}{C_\beta} + \frac{3}{4}\frac{\beta^2}{C_\gamma}\right) \approx 2 \times 10^{-4}U^{1/2}. \tag{35}$$

The quantities C_β and C_γ are the stiffness constants of the β – and γ – quadrupole vibrations, which can be determined from the corresponding energies and electromagnetic transition probabilities. Experimental values for a nucleus of mass $A \approx 160$ have been used in the above estimates as well as

$$E_\beta \approx E_\gamma \approx 1MeV, \qquad B(E2)_\beta \approx 1s.p.u., \qquad B(E2)_\gamma = 10s.p.u.$$

A rapid rotation of the nucleus may drastically alter the stiffness coefficients, and also lead to situations where the quadratic approximation to the deformation energy implied by equation (35) breaks down. A guideline for the magnitude of the stiffness coefficients is provided by the liquid drop model of the nucleus. For the nuclear mass region considered, the liquid drop stiffness coefficients are approximately 10 times smaller than those extracted from the experimental data. Consequently $(\Delta\mathcal{F}/\mathcal{F})^2 \approx 2 \cdot 10^{-3}U^{1/2}$ seems of inertia. This quantity is to be compared with $(\Delta i/I)^2 \approx 2.8 \cdot 10^{-3}U^{1/2}$.

Collecting the different terms together one obtains

$$\Delta\omega_0 \approx 35U^{1/4}keV, \tag{36}$$

where typical values $I = 40$ and $\mathcal{F}/\hbar \approx 70MeV^{-1}$ were used.

We can now calculate the two expressions appearing in eq. (29), as well as the energy U_c which divides their range of validity. From the relation $(\Gamma_\mu/U_c) = 2\Delta\omega_0(U_c)$ one obtains

$$U_c = 2.5MeV. \tag{37}$$

Consequently, for $A \approx 160$ and $I \approx 40$, one can write

$$\Gamma_{rot} = \begin{cases} 140 & U^{1/4}keV \quad (U \ll 2.5MeV), \\ 400 & U^{-1}keV \quad (U \gg 2.5MeV). \end{cases} \tag{38}$$

At sufficiently high excitation energies, the density of quasiparticle states increases so much that Γ_{rot} will eventually increase with energy. No analogue to this phenomenon seems to exist in the case of NMR because of the strong interaction between the unperturbed rotational band, as opposed to the weak collision regime governed by magnetic scattering.

The rotational width can, in the high temperature regime, be calculated making use of the Golden rule

$$\Gamma_{rot,\Delta V} \approx 2\pi\bar{V}^2 P_2. \tag{39}$$

Making again recourse to the Fermi gas model, the density of states P_2 around the Fermi surface connected by a two-body interaction at temperature T is

$$P_2 = \frac{2\pi^2}{3}\frac{T^2}{D} \approx 45U^{3/2}\,MeV^{-1},\tag{40}$$

where the value $D \approx 200keV$ typical of deformed nuclei was used.

With the help of a schematic pairing force quenched by the rotational motion, that is,

$$\bar{V}^2 \approx G/I,\tag{41}$$

one obtains, for typical values $G \approx 25/A$ and for $A = 160$ and $I = 40$,

$$\Gamma_{rot,\Delta V} \approx 4U^{3/2}keV \qquad (U > 6MeV).\tag{42}$$

The lower limit for which this expression is valid, was calculated equating $\Gamma_{rot,\Delta V}$ and the second espression of eq. (38).

There is quantitative evidence for the soundness of the simple estimates (38), as can be seen from fig. 15. Better resolution experiments are needed to check the predictions of the model concerning the predicted phenomenon of motional narrowing.

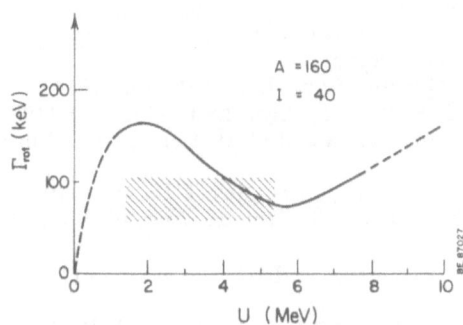

Fig. 15. The rotational damping width Γ_{rot} described by eqs. (38) and (42) as a function of the excitation energy U above the yrast line. The experimental information[35] is represented by the shaded area.

Arguments similar to those used in this section, have been applied to the contribution to the spreading width of giant vibrations arising from thermal fluctuations. A regime of motional narrowing is also expected in this case[42].

ACKNOWLEDGEMENT

A.R. Spalla carefully typed these lecture notes.

REFERENCES

1. S. Levit, these proceedings.
2. P. F. Bortignon and R. A. Broglia, Nucl. Phys. A371:405 (1981).
   G. F. Bertsch, P. F. Bortignon and R. A. Broglia, Rev. Mod. Phys. 55:287 (1983).
3. C. Mahaux, P. F. Bortignon, R. A. Broglia and C. H. Dasso, Phys. Reports 120:1 (1985).
   P. F. Bortignon, R. A. Broglia, C. H. Dasso and C. Mahaux, Phys. Lett. 140B: 163 (1984).
4. P. F. Bortignon, R. A. Broglia, G. F. Bertsch and J. Pacheco, Nucl. Phys. A460:149 (1986).
5. G. Mahan, "Many–particle physics" Plenum, New York (1981).
6. T. Matsubara, Pr. Theor. Phys. 14:351 (1955);
   N. P. Landsman and Ch. G. van Weert, Phys. Reports 145:141 (1987).
7. J. Rapaport, Phys. Reports 87:25 (1982).
8. R. A. Broglia, D. R. Bes, P. F. Bortignon and M. Gallardo, "Nuclear Structure, Reactions and Symmetries", eds. R. A. Meyer and V. Paar, World Scientific, Vol. II: 882 (1986).
9. P. F. Bortignon, "Nuclear Structure 1985", eds. R. A. Broglia, G. B. Hagemann and B. Herskind, North–Holland,: 535 (1985).
10. P. F. Bortignon, R. A. Broglia and C. H. Dasso, Nucl. Phys. A398: 221 (1983).
11. F. Azaiez et al., Nucl. Phys. A444: 373 (1985).
12. J. R. Beene et al., "Nuclear Structure 1985", eds. R. A. Broglia, G. B. Hagemann and B. Herskind, North–Holland,: 503 (1985).
13. P. F. Bortignon, R. A. Broglia, G. F. Bertsch, Phys. Lett. 148B: 20 (1984).
    J. Speth, D. Cha, V. Klemt and J. Wambach, Phys. Rev. C31: 2310 (1985).
    P. F. Bortignon and R. A. Broglia, to be published.
14. A. Bohr and B. R. Mottelson, "Nuclear Structure", Vol. II, Benjamin, (1975).
15. A. van der Woude, "Nuclear Structure 1985" eds. R. A. Broglia, G. B. Hagemann and B. Herskind, North–Holland,: 489 (1985).
16. A. Bracco, these proceedings.
17. M. Brack and P. Quentin, Phys. Lett. 52B: 159 (1974).
18. H. Sagawa and G. F. Bertsch, Phys. Lett. 146B: 138 (1984).
    O. Civitarese, R. A. Broglia and C. H. Dasso, Ann. Phys. (N.Y.) 156: 142 (198.).
19. A. K. Kerman and S. Levit, Phys. Rev. C24: 1029 (1981).
    A. K. Kerman, S. Levit and T. Troudet, Ann. of Phys. 154: 456 (1984).
    See also ref[1], eq. (2.21).
20. G. F. Bertsch, P. F. Bortignon, R. A. Broglia and C. H. Dasso, Phys. Lett. 80B: 161 (1979).
21. P. Morel and P. Noziéres, Phys. Rev. 126: 1909 (1962).
22. H. Esbensen and G. F. Bertsch, Phys. Rev. Lett. 52: 2257 (1984).
23. A. Lejeune, P. Grange, M. Matzolff and J. Cugnon, Nucl. Phys. A453: 189 (1986).
    A. H. Blin, R. W. Hasse, B. Hiller and P. Schuck, Phys. Lett. 16B: 221 (1985).
24. K. A. Snover, these proceedings and Ann. Rev. of Nucl. and Part. Sci.: 36: 545(1986) and references therein.
25. M. Gallardo, M. Diebel, T. Døssing and R. A. Broglia, Nucl. Phys. A443: 415 (1985).
26. J. J. Gaardøje et al., Phys. Rev. Lett. 53: 148 (1984) and 56: 1783 (1986).
27. A. Bohr and B. R. Mottelson, "Nuclear Structure", Vol. I, Benjamin, (1969).
28. J. R. Huizenga and L. G. Moretto, Ann. Rev. Nuc. Sci. 22: 427 (1972).
29. N. Van Giai and P. Van Thieu, Phys. Lett. 126B: 421 (1983).
30. P. F. Bortignon and C. H. Dasso, Phys. Lett. 189B: 381 (1987).
31. R. W. Hasse and P. Schuck, Phys. Lett. 179B: 313 (1986).
    N. Vinh Mau and D. Vautherin, Nucl. Phys. A445: 245 (1985).
32. G. Nebbia et al., Phys. Lett. 176B: 20 (1986).

33. B. Lauritzen, T. Døssing and R. A. Broglia, Nucl. Phys. A457: 61 (1986).

34. T. Døssing,"Nuclear Structure 1985", eds. R. A. Broglia, G. B. Hagemann and B. Herskind, North-Holland,: 379 (1985).

35. J. C. Bacelar, G. B. Hagemann, B. Herskind, B. Lauritzen, A. Holm, J. C. Lisle and P. O. Tjøm, Phys. Rev. Lett. 55: 1858 (1985).

36. F. Stephens, in Proceedings of the Second International Conference on Nucleus–Nucleus Collisions, Sweden 10–14 June 1985, Nucl. Phys. A447: 217c (1986).
    J. E. Draper, E. L. Dines, M. A. Deleplanque, R. M. Diamond and F. S. Stephens, Phys. Rev. Lett. 56: 309 (1986).

37. R. A. Broglia, T. Døssing, B. Lauritzen and B. R. Mottelson, Phys. Rev. Lett. 58: 326 (1987).

38. N. Blombergen, E. M. Purcell and R. V. Pound, Phys. Rev. 73: 679 (1948).

39. C. P. Poole and H. A. Farrach, "Relaxation in Magnetic Resonance", Academic Press, 1971.

40. J. Reisse, The Multinuclear Approach to NMR Spectroscopy, eds. J. B. Lamberg and F. G. Riddell, D. Reidel Publ. Co., 63 (1983).

41. F. S. Stephen and R. S. Simon, Nucl. Phys. A 183: 257 (1972).

42. B. Lauritzen, R. A. Broglia, R. Ormand and T. Døssing, Phys. Rev. Lett. (to be published).

NEUTRON DECAY OF GIANT RESONANCES IN ^{208}Pb

A. Bracco

Dipartimento di Fisica, Universita di Milano, 20133 Milano, Italy

and

J.R. Beene, F.E. Bertrand, M.L. Halbert, R.L. Auble, D.C. Hensley, D.J. Horen, R.L. Robinson, and R.O. Sayer

Oak Ridge National Laboratory, Oak Ridge, Tennessee 37830, USA

ABSTRACT

The neutron decay of the giant multipole resonance region between 9 to 15 MeV in ^{208}Pb has been studied. The giant resonances were excited by inelastic scattering of ^{17}O at 380 MeV. Neutrons from ^{208}Pb and γ rays from ^{207}Pb* were detected in the ORNL Spin Spectrometer and the ^{17}O in ΔE-E silicon detector telescopes. The neutron branching ratios for the decay to the ground state and to the low lying excited states of ^{207}Pb were measured as a function of the of excitation energy of ^{208}Pb and compared to Hauser-Feshbach calculations. Evidence for non statistical neutron decay to selected single-hole states, and to hole-surface vibration and hole-pairing vibration coupled states was found.

1. INTRODUCTION

A giant vibration can be viewed as a correlated state of a particle above the Fermi surface and a hole in the Fermi sea. These states are located in general at excitation energies above the particle binding energies and therefore they decay predominantly by charged particle and neutron emission in light nuclei and, due to the Coulomb barrier, only by neutron emission in heavier nuclei. Gamma decay is also possible, but the relative decay branch goes from 10^{-3} to 10^{-5}. The coupling of the 1p-1h state to the continuum gives rise , for the excitation of closed shell nuclei, to the direct decay into the hole states of the (A-1) nucleus with partial width given by the escape width $\Gamma^{\uparrow}$ (cfr. figure 1). The 1p-1h states can couple through the residual interaction to more complicated states present in the vicinity of the resonance. Two particle-two hole states, referred to as doorway states, represent the first step of the damping process of the resonance torward the compound nucleus. For closed shell nuclei, the 2p-2h states containing a pair of uncorrelated fermions and low lying surface vibrations or collective pairing modes are expected to be important doorway states (refs. 1,2). The 2p-2h states couple again to more complicated states until a completely equilibrated system is reached. Particle decay can thus occur either directly from the giant resonance or at any of the intermediate levels of the damping process or from fully equilibrated compound states. As a consequence, the comparison of the experimental neutron decay branching ratio with the statistical model predictions for states of the A-1 nucleus having different nuclear structure can be used to study both the microscopic structure and the damping process of giant resonances.

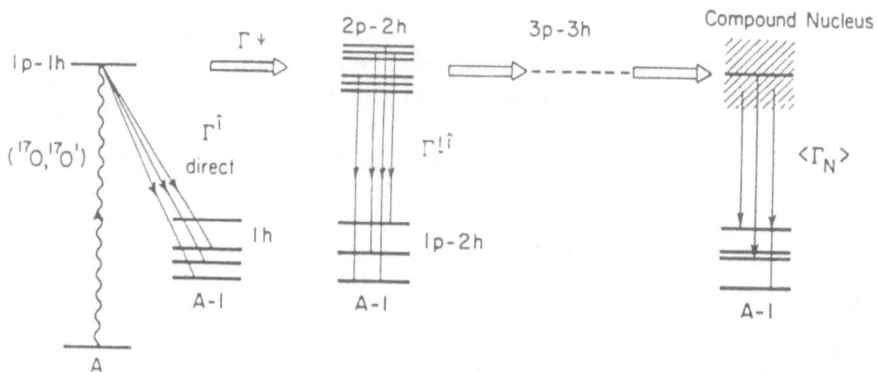

Fig.1.Illustration of neutron decay from a giant resonance state. $\Gamma^{\uparrow}$ is the direct escape width, $\Gamma^{\downarrow}$ is the spreading width, $\Gamma^{\downarrow\uparrow}$ is the escape width from intermediate states of the damping process of the giant resonance. $\langle\Gamma_N\rangle$ is the decay width from the conpound nucleus.

The experiment reported here is concerned with the measurements of the neutron decay of the giant multiple resonance region (9-15 MeV of excitation energy of) ^{208}Pb. The neutron branching ratios for the decay to the low-lying hole states and particle vibration coupling states (1p-2h) were measured as a function of ^{208}Pb excitation energy and were compared to the Hauser-Feshbach predictions.

In previous studies of neutron decay from the giant resonance region of ^{208}Pb (refs. 3,4,5,6) the comparison of the neutron energy spectra with statistical model calculations indicates the presence of a small non statistical neutron branch in the giant monopole resonance region. However, a more recent statistical model analysis (ref. 7) of the data of reference 6 suggests the absence of non-statistical decay. The lack of good statistics and energy resolution of these measurements did not allow the study of the neutron branching decay ratios to individual states in ^{207}Pb as a function of ^{208}Pb excitation energy.

The present experiment differs in two important aspects from the previous ones; first, by using inelastic scattering of heavy ions the GQR and GMR are strongly excited, and second the decay studies were made for a number of well resolved 1h and 1p-2h final states.

2. THE EXPERIMENT

In the present study, the resonances were excited by inelastic scattering of 380 MeV ^{17}O. This reaction resulted in a large excitation cross section for the the the quadrupole and monopole resonances in comparison with the underlying continuum.

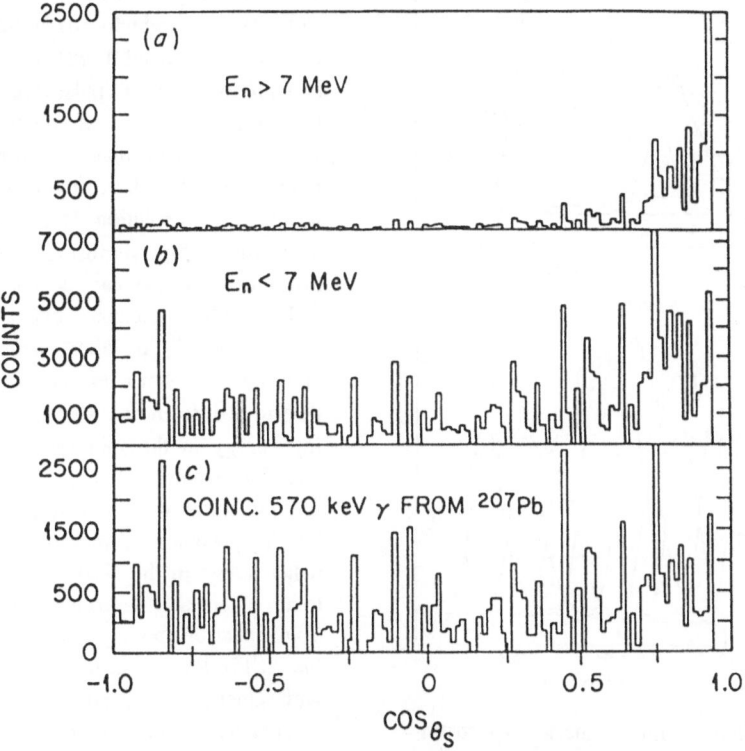

Fig.2.Neutron angular distributions for different neutron kinetic energies E_n (parts (a) and b)). θ_s is the neutron angle with respect to the scattered ^{17}O. The angular distribution shown in part c is that of neutrons decaying from ^{208}Pb at excitation energy 9-15 MeV and populating the first excited state (570 KeV) in ^{207}Pb.

The inelastically scattered ^{17}O was detected in six cooled Si surface-barrier telescopes arranged symmetrically around the beam at an angle of 13 ^{0} and subtending $\Delta\theta = 3$ ^{0} and $\Delta\varphi = 9$ ^{0} each and a total solid angle of 22.6 msr. The telescopes consisted of two elements of thickness 0.5 mm and 1. mm, respectively. The energy resolution was 800 KeV and the mass resolution sufficient to separate ^{17}O from adjacent oxygen isotopes. Neutrons and gamma rays were detected in 70 elements of the ORNL Spin Spectrometer (ref. 8). The spectrometer consists of a spherical shell of NaI 17.8 cm thick, divided into 72 independent modules surrounding the target chamber.

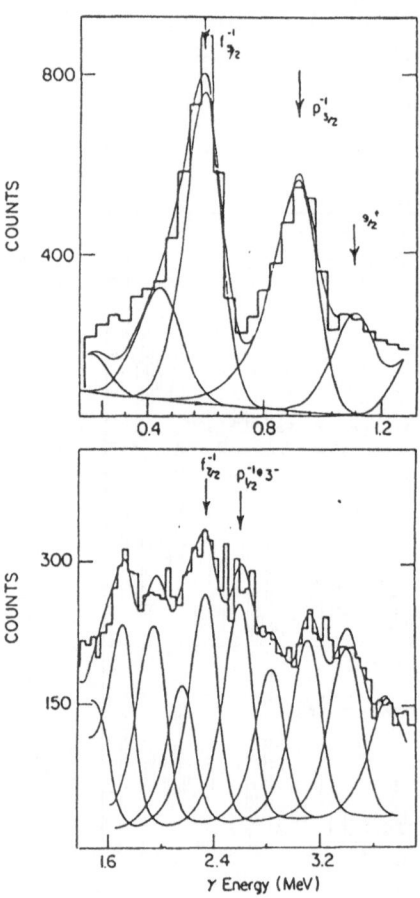

Fig.3.Sum Gamma energy spectrum for ^{208}Pb excitation energy of 12-13 MeV. The states indicated by the arrows are the ones having known decay branchings to the ground and isomeric states.

The raw data obtained from the spectrometer consisted of pulse heights from the individual NaI elements and times of these pulses relative to the inelastically scattered ^{17}O with which they were in coincidence. The total gamma ray pulse height was obtained by summing all those pulses which occured in a prompt time window. This window (which was a function of pulse height) was narrow enough to eliminate pulses resulting from detection of neutrons with energy less than ~ 7 MeV. Because of the short flight path, the energy resolution from neutron time of flight is insufficient to distinguish individual levels in ^{207}Pb. The residual excitation energy in ^{207}Pb after neutron emission is , however, accurately determined from the sum gamma ray energy in the spin spectrometer. The events of interest for the present study were thus identified by the presence of neutrons (i.e. delay pulses in the NaI), and by a total γ energy deposited in the Spin Spectrometer equal to one of the ^{207}Pb level energies and ^{17}O with kinetic energy corresponding to excitation energy of ^{208}Pb above one neutron emission threshold (7.4 MeV). Althogh redundant, the detection of neutrons is used for discriminating against contributions from processes other than inelastic excitation followed by neutron decay. The investigation of these processes was made by considering angular distributions. Figure 2a shows the angular distribution with respect to the direction of the scattered particle of the pulses from the spin spectrometer with an apparent gamma energy larger than 7 MeV (at these energies is not possible to distinguish between neutrons and gammas by time of flight). The fact that these events are tightly distributed around the direction of the scattered particles suggests that these are neutrons from the (^{17}O,^{18}O*) neutron pick up process in which the ^{18}O decays by neutron emission while in flight toward the detector (sequential decay). Such a process should produce neutrons moving approximately at the beam velocity (E_n ~ 21 MeV) which would not be identified by time of flight and which would often deposit large energies in the NaI detectors. The angular distribution of neutrons (i.e. delayed pulses) for ^{208}Pb excitation energy larger than 8 MeV and with energy less than ~ 7 MeV is shown in figure 2b. A forward-backward asymmetry is seen in the data . In the case of the angular distribution of neutrons corresponding to population of the first excited state in ^{207}Pb, and for ^{208}Pb excitation energy from 9 to 15 MeV, the forward-backward asymmetry is

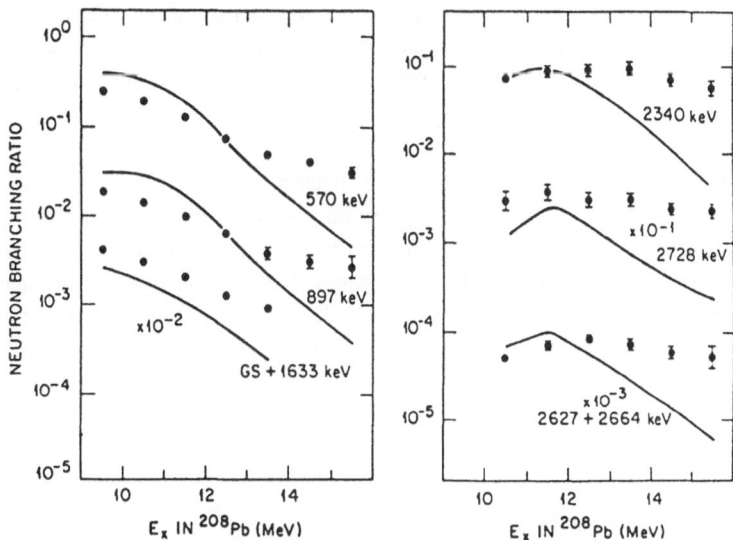

Fig.4. Neutron branching ratios obtained with the present measurements as a function of ^{208}Pb excitation energy for given excited states in ^{207}Pb. The curves are the Hauser Feshbach predictions described in the text.

much less enhanced (cfr. fig. 2c). This asymmetry is typical of non statistical decay. However, it is important, as a first step of the study, to restrict the analysis to a region free of any contaminations from scattered high energy neutrons from the neutron pick up and sequential decay mechanism, which might be present in the forward hemisphere. Another potential mechanism for production of neutrons unrelated to giant resonance decay is the knockout reaction. Such a process is expected to produce neutrons peaked near the target recoil direction (ref. 9). No evidence for a significant yield of such neutrons (at ~ 85 ^{0} deg in this case) was found.

The present analysis is therefore based on a determination of the initial ^{208}Pb excitation energy from the scattered ^{17}O energy, the detection of a neutron in the Spin Spectrometer (identified by time of flight) in the backward hemisphere, and the determination of the residual ^{207}Pb excitation energy by total gamma energy emitted as prompt radiation, detected by the Spin Spectrometer. A sample of total gamma energy spectrum is shown in figure 3, corresponding to an excitation energy of 12-13 MeV in ^{208}Pb. The spectrum was fitted with gaussians whose centroids are either at the energy of known excited states in ^{207}Pb or at the energy difference of known states from 1.633 MeV, the energy corresponding to the isomer state i13/2 of ^{207}Pb. Since the isomer could not be distinguished from the ground state, the deexcitation the ^{207}Pb levels which could go through the isomeric state were included in the analysis in this way. The peaks indicated by the arrows in figure 3 are the ones that are better separated from the others and correspond to states in ^{207}Pb for which the branching decay ratio to the ground state and to the isomeric state is well known. In addition, the nuclear structure of these states is known. The neutron branching ratios at a given ^{208}Pb excitation energy for the decay to a particular ^{207}Pb state was obtained from the area of the corresponding gaussian in the sum energy spectrum divided by the number of ^{17}O single events at the same ^{208}Pb excitation energy. Corrections for neutron and gamma efficiency were applied.

The efficiency of the Spin Spectrometer for neutron detection (45% on average up to 7 MeV neutrons) was measured by the ratio of the gamma yields from the first excited state of ^{207}Pb with and without the presence of neutrons (delay pulses) at different ^{208}Pb excitation energy bites. The gamma total and photopeak efficiencies were measured detecting gamma events from ^{207}Bi, ^{88}Y, and ^{60}Co sources in the same experimental configurations. The obtained values are in agreement with the ones reported in reference 8.

3. RESULTS AND DISCUSSION

The measured neutron branching ratios for the decay to the ground plus the isomeric states, to the 570 KeV (5/2$^-$), the 890 KeV (3/2$^-$), and the 2340 KeV (7/2$^-$) states , to the surface vibration coupling 2623 and 2665 KeV doublet, and to the pairing vibration coupling 2728 KeV state are shown in figure 4. The error bars contain the statistical and peak fitting uncertainties. For the 570 and 890 KeV states, the yield of events corresponding to the deexcitation of higher excited states through the emission of two or more gamma rays and with the detection of only the 570 or 890 KeV gamma rays were subtracted. These corrections amount to only a few percent.

The Hauser-Feshbach predictions are also presented in figure 4. The calculations were made using the optical model potential of reference 10, the experimentally known levels of ^{207}Pb up to 4 MeV and, above this energy, the level density of reference 11 which gives a good account of the experimental data in the relevant spin and excitation energy range. The calculations are linear combinations of neutron branching ratios for ^{208}Pb states of multipolarity 0,1,2,3,4. The strength distribution for multipolarities up to 4 was taken from (p,p') data (ref. 12). Making use of the fraction of the EWSR reported in reference 12, the cross sections for (^{17}O,^{17}O') at 13^0 were calculated using DWBA. Except for an almost constant underlying continuum, this distribution of L=0-4 strengths provides a good account of the experimental inelastic spectrum for ^{208}Pb excitation energies in the 9 to 15 MeV region of interest here. The effects of the decay of this continuum is not included in the calculation. The strength distribution used and the calculated DWBA cross sections are given in table I.

The general trend of the present measurements is shown in figure 5, where the ratios of the experimental and calculated neutron decay branching ratio are plotted. The differences between experiment and statistical predictions in the ^{208}Pb excitation energy region < 12 MeV are smaller than the ones in the higher energy region. Near to the neutron emission threshold, the statistical calculations are very sensitive to the

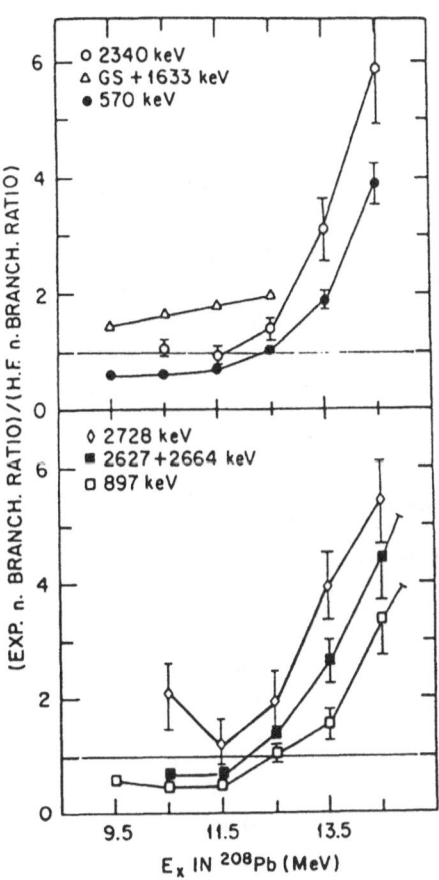

Fig.5. Ratio of the experimental and calculated neutron branching ratios as a function of ^{208}Pb excitation energy for given low lying excited states in ^{207}Pb. Straight lines are drawn between the points to guide the eyes.

values of the transmission coefficient which at these very low neutron energies have significant sizes only for small neutron orbital angular momenta. The calculations of the transmission coefficients obtained making use of the neutron strength functions of reference 13 differ from the ones obtained with the optical potential only by few percent and cannot improve the agreement between experiment and statisitical calculations in the [208]Pb excitation energy region < 12 MeV. However, if addition of L=6 strength is made , as previously suggested (refs. 2 and 14), the calculations describe better the data up to 12 MeV [208]Pb excitation energy with the exception of the first excited state (ref. 15). For [208]Pb excitation energies from 12 to 15 MeV there is a net excess of the experimental branching ratios over that calculated. At these excitation energies the statistical model calculations are very sensitive to the value used for the level density of [207]Pb, the neutron transmission coefficient being almost one. The experimental neutron branching ratios can be reproduced by the statistical model calculations only if a level density at least a factor of 5 smaller is used. Such a small value is inconsistent with the measured level density. In addition, the enhanced neutron branching ratio does not depend on the nature of the [207]Pb states.

The neutron decay to various low lying states in [208]Pb leads to different information concerning the giant resonances in [208]Pb. The neutron branching ratios to [207]Pb hole states (G.S.,570 KeV, 890 KeV, 1633 KeV, 2340 KeV) could be used to obtain information about specific particle-hole amplitudes of the wave function describing the giant resonances. On the other hand, the neutron branching ratios to the 2600 KeV and 2728 KeV [207]Pb particle vibration states can be used to extract information on the 2p-2h doorway states by which the giant resonances are expected to couple to the compound nucleus.

Table 1. Energies, widths, multipolarities, and fractions of the energy weighted sum rules of the giant resonances in [208]Pb. In the last column the cross sections for excitation of the giant resonances with [17]O are given.

[208]Pb E^* (MeV)	Γ (MeV)	L	%EWSR	σ (mb)
8.11	0.4	4	3	3.85
8.35	0.4	4	4	3.79
8.86	0.4	2	7	5.73
9.34	0.4	2	5	3.42
10.6	2.	2	70	30.3
12.0	2.4	4	10	7.41
13.6	4.0	1	100	4.36
13.9	2.9	0	100	10.0

The collective vibration for the two states are the low lying octupole vibration and the pair removal mode of [208]Pb, respectively (cf. refs. 16 and 17). It will therefore be interesting to compare the non-statistical decay of the giant resonances in ^{208}Pb for particular ^{207}Pb final states to theoretical predictions, when they become available. The latter comparison will provide a powerful tool to study the microscopic structure of giant resonances (ref. 18) at different levels of complexity in its decay torward the compound nucleus.

REFERENCES

1. G.F. Bertsch, P.F. Bortignon and R.A. Broglia, Rev. Mod. Phys. 55, 287(1983)

2. P.F. Bortignon and R.A. Broglia, Nucl. Phys. A371,405(1981)

3. A. van Der Woude, Nuclear Structure 1985, R. Broglia, G.B. Hagemann and B. Herskind editors, p.489

4. W. Eyrich, A. Hofmann, U. Scheib, S. Schneider, F. Vogler, and H. Rebel, Phys. Rev. Lett. 43,1369(1979).

5. H. Steuer, W. Eyrich, A. Hofmann, H. Ortner, U. Scheib, R. Stamminger, D. Steuer, and H. Rebel, Phys. Rev. Lett. 47,1702 (1981).

6. W. Eyrich, K. Fuchs, A. Hofmann, U. Scheib, H. Steuer, and H. Rebel, Phys. Rev. C 29,418(1984)

7. H.Dias and E. Wolynec, Phys. Rev. C30,1164(1984)

8. M. Jaaskelainen, D.G. Sarantites, R. Woodward, F.A. Dilmaman, J. T. Hood, R. Jaaskelainen, D.C. Hensley, M.L. Halbert, and J.H. Barker, Nucl. Instrum. Meth. 204,385(1983).

9. H. Ejiri, Journal de Physique 45,C4-135(1984).

10. C.M. Perey and F.G. Perey, Atomic Data and Nucl. data Tables 17,1(1985).

11. A. Gilbert and A. G. W. Cameron, Can. Jour. of Phys. 13,1116(1963).

12. F. E. Bertrand, E.E. Gross, D.J. Horen, R.O. Sayer, T.P. Sjorenn, D. K. McDaniels, J. Lisantti, J.R. Tinsley, L. W. Swenson, J.B. McClelland, T.A. Carey, K. Jones, and S.J. Seestrom-Morris, Phys. Rev. C34,45(1986).

13. D.J. Horen, J.A. Harvey, and N.W. Hill, Phys. Rev. C18,722(1978).

14. H.P. Morsch, P. Decowski, M. Rogge, P. Turek, L. Zemco, S.A. Martin, G.P.A. Berg, W. Hurlimann, J. Meissburger, and J.G. Romer, Phys. Rew. C28, 1947(1983).

15. J.R. Beene, A. Bracco, F.E. Bertrand, M.L. Halbert, D.C. Hensley, R.L. Auble, D.J. Horen, R.L. Robinson, T.P. Sjoreen, R.O. Sayer, Phys. Div. Prog. Rep. for Period Ending Sept. 30,1986, ORNL and to be published.

16 D.R. Bes and R.A. Broglia, Phys. Rev. C3,2389(1971).

17. I. Hamamoto, Phys. Rep.10,63(1974).

18 R.A. Broglia and P.F. Bortignon, Giant Multiple Resonances, Ed. F.E. Bertrand, Harwood Ac. Publisher, New York (1979)317.

A SEMICLASSICAL THEORY OF NUCLEAR EXCITATION

BASED ON THE VLASOV EQUATION [*]

A. Dellafiore and F. Matera

Università di Firenze, Dipartimento di Fisica
and INFN, Sezione di Firenze
Largo E. Fermi 2; 50125 Firenze, Italy

The Vlasov equation is an equation for the single-particle probability density $f(\vec{r},\vec{p},t)$. Formally it resembles the Boltzmann equation without collision term[1]. The physical hypothesis underlying Vlasov's approximation is that, on the average, the effect of collisions between particles can be simulated by a mean field in which each particle is allowed to move independently. This mean field is related in a self-consistent way to the particle density.

The Vlasov equation is a classical equation of motion. It was originally developed for studying plasma physics. The possibility of applying it to nuclei, where quantum effects play an important role, was first discussed by Bertsch[2]. He pointed out that the Liouville theorem (representative points in phase space moving as an incompressible fluid) plays an important role in assuring that the Pauli principle (one point per cell h^3 in phase space) is not badly violated during the motion. Thus the Vlasov equation can be used to describe nuclear excitations in a semiclassical approximation.

Recently a semiclassical theory of giant resonances based on the Vlasov equation has been developed[3]. This theory, contrary to previous attempts, does not require an additional scaling hypothesis on the phase-space density. In[3] the solution of the linearized Vlasov equation has been expressed in terms of a responde function $D(\vec{r},\vec{r}',\omega)$. This function satisfies the following RPA-type integral equation

$$D(\vec{r},\vec{r}',\omega) = D^\circ(\vec{r},\vec{r}',\omega) + \int d\vec{x} \int d\vec{y}\ D^\circ(\vec{r},\vec{x},\omega)\ u(\vec{x},\vec{y})\ D(\vec{y},\vec{r}',\omega)\ , \qquad (1$$

where $u(\vec{x},\vec{y})$ is the interaction between any two constituents of the many-body system and D° is the response function for particles moving in a fixed equilibrium potential $U_\circ(\vec{r})$. The strength function associated with an external driving field $Q(\vec{r})$ is given by

---

[*] Talk presented by A.Dellafiore

$$S(\omega) = -\frac{1}{\pi} \; \text{Im} \int d\vec{r} \int d\vec{r}' \; Q^*(\vec{r}) \; D(\vec{r},\vec{r}',\omega) \; Q(\vec{r}') \; . \tag{(2}$$

If the fluctuation of the mean field induced by the external force is neglected, then $D = D^\circ$ and $S = S^\circ$. When the nuclear response is evaluated in this approximation, the frequencies and the strengths of the system eigenmodes are determined in a scheme which corresponds to the usual shell model (fixed mean field). Once this is done, the effect of the mean field fluctuation can be taken into account and the new eigenfrequencies and strengths of the normal modes can be obtained by solving eq.(1).

We briefly illustrate these steps by assuming simple models for both the mean field and the interaction $u(\vec{x},\vec{y})$. We shall see that the method based on the Vlasov equation gives results which are remarkably similar to those given by analogous quantum calculations. Apart from being computationally much simpler than the usual quantum approach, the method has the merit of unveiling the classical skeleton underlying the structure of nuclear spectra in terms of simple properties of the classical orbits such as the period of radial motion and the precession frequency of the orbit. Moreover the connection with quantum mechanics can be made quite clear[4]: in the limit of large quantum numbers the method based on the Vlasov equation gives essentially the same results as the WKB approximation to the Schrödinger equation. Besides being conceptually very satisfactory, this connection between the classical and quantum methods can be exploited to introduce, if necessary, quantum corrections to the results given by the Vlasov equation. It also allows for a better quantitative understanding of the unavoidable limits of a semiclassical description of the quantum nuclear system.

For a spherical nucleus, the L partial-wave component of the strength function S° is given by[3,4]

$$S_L^\circ(\omega) = \frac{4}{4\pi} \sum_{\ell} (2\ell+1) \sum_{\Delta n=-\infty}^{+\infty} \sum_{\Delta\ell=-L}^{+L} (\Delta n + \frac{\Delta\ell}{\nu_\gamma(\ell)}) \frac{4\pi}{2L+1} \left| Y_{L,\Delta\ell}(\tfrac{1}{2}\pi,\tfrac{1}{2}\pi) \right|^2 \cdot$$

$$\tag{3.a}$$

$$|Q(\Delta n,\Delta\ell)|^2 \; \delta(\hbar\omega - (\Delta n + \frac{\Delta\ell}{\nu_\gamma(\ell)}) \hbar\omega_\circ(\ell)) \; .$$

In this equation the quantity $\omega_\circ(\ell)$ is the frequency of radial motion for particles on the Fermi surface with angular momentum $\lambda = (\ell + \tfrac{1}{2})\hbar$, while the ratio $(\omega_\circ/\nu_\gamma)$ gives the precession frequency in the plane of the orbit for the same particles. For a fixed central potential these two parameters alone determine all the eigenfrequencies of the system:

$$\omega(\Delta n,\Delta\ell) = (\Delta n + \frac{\Delta\ell}{\nu_\gamma(\ell)}) \; \omega_\circ(\ell) \; . \tag{3.b}$$

The strength of each eigenmode is proportional to the Fourier coefficients $Q(\Delta n,\Delta\ell)$ which are the classical limit of the quantum radial matrix elements $<n+\Delta n, \ell+\Delta\ell|Q|n,\ell>$ (see appendix A of[4]). Hence Δn corresponds to the change in the radial quantum number, while $\Delta\ell$ gives the variation of the angular momentum quantum number.

In order to get a feeling about the way in which the shape of the mean field $U_o(r)$ determines the features of the excitation spectrum, it is useful to compare the response of an oscillator potential $U_o(r)=\frac{1}{2}m\Omega r^2$ to that of a square well potential. The nuclear potential, which is better approximated by a Saxon-Woods shape, will give results in between these two extreme cases.

Because of its high degree of degeneracy the oscillator potential is a very special case. Both quantities ω_o and ν_γ are independent of ℓ ($\omega_o = 2\Omega$ and $\nu_\gamma = 2$), the orbits are closed and each eigenmode is actually a superposition of (L+1) degenerate eigenmodes corresponding to different ($\Delta n, \Delta\ell$) combinations[3]. For example for isoscalar quadrupole excitations there are three degenerate eigenfrequencies corresponding to the combinations ($\Delta n=0, \Delta\ell=2$), ($\Delta n=1, \Delta\ell=0$) and ($\Delta n=2, \Delta\ell=-2$) which all contribute to the $2\hbar\Omega$ transition. There are also three solutions at zero frequency corresponding to the combinations ($\Delta n=-1, \Delta\ell=2$), ($\Delta n=0, \Delta\ell=0$) and ($\Delta n=1, \Delta\ell=-2$) which however do not contribute to the transition strength, as can be seen directly from eq. (3.a).

For a different shape of the central potential the oscillator degeneracies will be removed for two reasons:
1) in general the period of radial motion (and hence ω_o) will depend on the angular momentum of the particles;
2) the classical orbits will not be necessarily closed (ν_γ is no longer an integer and can depend on ℓ).
For a square well potential of radius $R = r_o A^{1/3}$ the λ-dependence of both ω_o and ν_γ is easily calculated:

$$\hbar\,\omega_o(\lambda) = \hbar\,\frac{\pi\,\bar\lambda}{mR^2}\,\frac{1}{\sqrt{1-(\lambda/\bar\lambda)^2}} \qquad (4.a$$

$$\nu_\gamma(\lambda) = \frac{\pi}{\arccos(\lambda/\bar\lambda)} \quad , \qquad (4.b$$

where $\bar\lambda = (9\pi A/8)^{1/3}\hbar$ is the maximum value of λ. We can see from eq. (4.a) that for, say, A=40 and r_o=1.4 fm, $\hbar\,\omega_o$ increases from $\hbar\,\omega_o(\ell=0)\approx$ 29 MeV to $\hbar\,\omega_o(\ell=1)\approx 31$ MeV and to $\hbar\,\omega_o(\ell=2)\approx 34$ MeV for increasing values of the particle angular momentum. Correspondingly the quantity ν_γ given by eq. (4.b) changes from $\nu_\gamma(\ell=0)\approx 2.13$ to $\nu_\gamma(\ell=1)\approx 2.45$ and to $\nu_\gamma(\ell=2)\approx 2.93$. As a consequence the contribution of nucleons with, say, $\ell=2$ to the giant quadrupole excitation, instead of being concentrated, like for the harmonic oscillator, at an energy $2\hbar\Omega \approx 24$ MeV, is now split into a triplet at frequencies $\hbar\,\omega_o\approx 34$ MeV and $\hbar(\omega_o\pm\Delta\omega)$, with $\hbar\Delta\omega=(1-2/\nu_\gamma)\hbar\,\omega_o$ ≈ 10 MeV. Similarly particles with different values of ℓ give eigenfrequencies which can be easily calculated from eqs. (3.b) and (4).

The strength associated with the giant quadrupole triplet is not equally shared between all members of the multiplet. By explicitly evaluating the Fourier coefficients $Q(\Delta n, \Delta\ell)$ it can be seen that the eigenmode ($\Delta n=0, \Delta\ell=2$), which is lowest in energy, takes most of the transition strength associated with the multiplet. The rest of the strength is shared

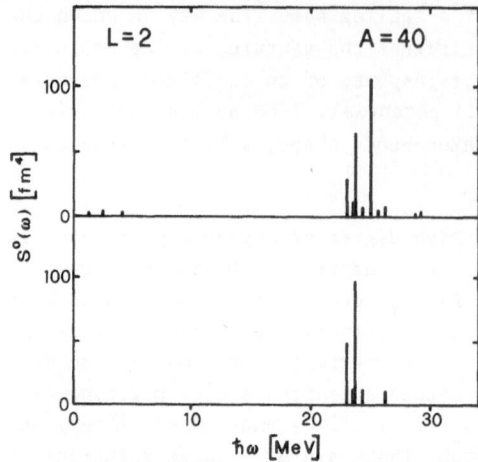

Fig. 1. Isoscalar quadrupole strength function for a
Saxon-Woods potential. The upper part does not
include quantum corrections. The lower part includes quantum corrections evaluated in[4].

between the other two members of the triplet, always according to the rule
that the excitation strength decreases with increasing energy.

The upper part of Fig.1 shows the spectrum of isoscalar quadrupole excitation for a somewhat more realistic Saxon-Woods potential. Like for
the square well, the fragmentation of levels is due to the angular momentum
dependence of both ω_o and ν_γ .

A further consequence of taking a potential other than harmonic oscillator is that the modes ($\Delta n=1, \Delta \ell =-2$), which do not contribute to the
oscillator strength because they correspond to zero frequency, can give now
a nonvanishing contribution at low energy. This is the origin of the very
small excitations displayed in the upper part of Fig.1 at low energy. Since
we can relate each eigenmode given by the semiclassical Vlasov theory to
the corresponding quantum transitions of the shell model (see[4]) we can see
that these modes correspond to transitions $\ell \rightarrow \ell -2$ within an oscillator
shell of the shell model. For A=40 this shell is fully occupied and these
transitions are actually forbidden by the Pauli principle. This and other
quantum effects can be easily incorporated into the semiclassical theory
based on the Vlasov equation . The lower part of Fig.1 shows the effect of
such quantum corrections. Moreover it has been shown in[4] that the effect
of a spin-orbit potential can be easily described in the framework of the
Vlasov theory.

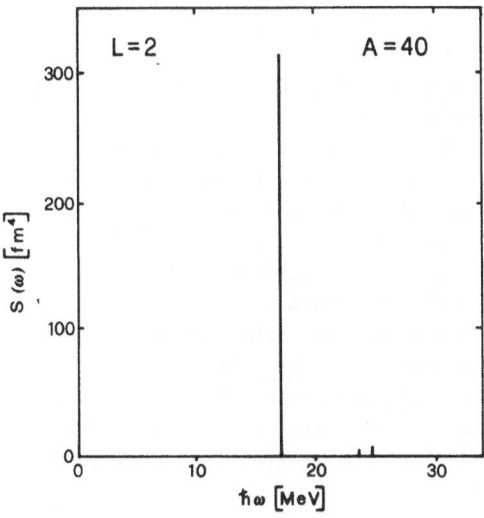

Fig. 2. Effect of the residual interaction on the
isoscalar quadrupole strength for A=40.

Up to now we have studied the nuclear response for a fixed mean field
$U_o(r)$. However the external field $Q(r)$ modifies the nuclear density and
consequently also the nuclear mean field. This has the effect of changing
both the eigenfrequencies and the relative excitation strengths of the normal modes given by S^o.

In order to take into account the fluctuation of the mean nuclear field
induced by the external field we have to solve the integral equation (1).
For illustrative purpose it is sufficient to assume a simplified separable
interaction of the multipole-mutlipole type. Figure 2 shows the redistri-
bution of the isoscalar quadrupole strength obtained in this way[5]. The
result is strikingly similar to that of much more complicated quantum RPA
calculations[6].

In conclusion we want to stress that the semiclassical theory of nuclear excitation based on the Vlasov equation offers a key for understanding
the features of seemingly complicated nuclear spectra in terms of a few
simple classical quantities. Thanks to the unambiguous connection which can
be established between the semiclassical and quantum excitation spectra,
quantum corrections can be easily introduced into the Vlasov theory. In this
way the theory becomes a tool, suitable also for quantitative predictions,
which has the advantage of greatly reducing the numerical effort required
by a fully quantum RPA approach.

REFERENCES

1. P. Ring and P. Schuck, "The Nuclear Many-Body Problem",
   Springer, Berlin (1980).
2. G. F. Bertsch, Dynamics of heavy ion collisions, in : "Nuclear
   Physics with Heavy Ions and Mesons", R. Balian, M. Rho and
   G. Ripka, ed., North Holland, Amsterdam (1978).
3. D. M. Brink, A. Dellafiore and M. Di Toro, Solution of the
   Vlasov equation for collective modes in nuclei,
   Nucl. Phys. A456:205 (1986).
4. A. Dellafiore and F. Matera, Semiclassical kinetic equation and
   the WKB approximation, Nucl. Phys. A460:245 (1986).
5. G. F. Burgio, Thesis, Università di Catania, unpublished.
6. T. S. Dumitrescu, C. H. Dasso, F. E. Serr and Toru Suzuki,
   Collective excitations in spherical nuclei, Jour. Phys.G
   12:349 (1986).

ON THE PRODUCTION OF SUPERHEAVY ELEMENTS AND THE LIMITATIONS TO GO BEYOND

P. Armbruster

GSI Darmstadt, P.O. Box 11 05 41, D-6100 Darmstadt, FRG

THE NEW ISOTOPES OF THE HEAVIEST ELEMENTS

All the work I present has been achieved by a group of physicists, the 'SHIPPERS' working together since about 10 years. To the work contributed during the years more than 20 physicists. To mention all names is impossible but I will give a few names, and I acknowledge that without each of them this paper would not have been presented: G. Münzenberg, F.P. Heßberger, S. Hofmann, W. Reisdorf, K.-H. Schmidt, and the Ph.D. students C.C. Sahm and J.G. Keller.

All the experiments of the group are performed using the strong beams of the 'UNILAC' accelerator to produce evaporation residues (EVR), the velocity selector 'SHIP' to separate the EVR's from the primary beam, and an implantation technique allowing to identify single atoms by their α-decay genetics. A comprehensive review of our work was given recently[1].

Figure 1 compiles our knowledge of the heaviest isotopes. Since 1981 we produced 14 isotopes of the elements 104 to 109 by cold fusion, $E^* =$ (15–25) MeV, of the target nuclei ^{207}Pb, ^{208}Pb, and ^{209}Bi and the projectiles ^{50}Ti, ^{54}Cr, and ^{58}Fe. For 13 isotopes α-decay has been established, that is α-energies and α-branchings have been measured. Only for one of these isotopes, ^{257}104, α-decay was known before. Spontaneous fission was found for 5 isotopes only, ^{260}106, ^{257}105, ^{255,256,258}104. For all the isotopes except the last one, α-decay as a competing decay mode was established. Surprisingly, for the 5 known isotopes of the heaviest elements ^{266}109, ^{264,265}108, ^{261,262}107 no spontaneous fission has been detected. In the chains N–Z = 47–49 spontaneous fission is strongest for the elements 104 and 105, the corresponding isotopes of

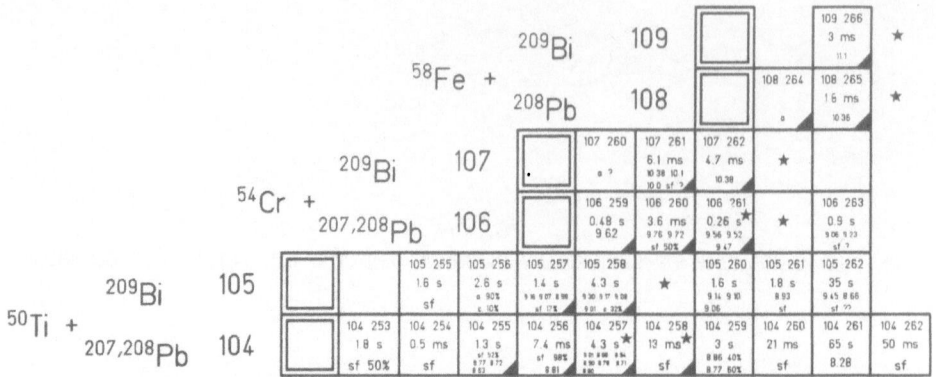

Fig. 1. The heaviest isotopes known 1986. ✦ indicates compound systems leading to EVR's. The 14 isotopes of elements 104-109 seen at SHIP are indicated by small triangles.

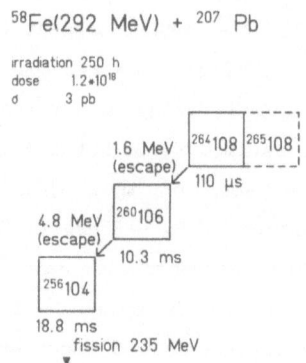

^{58}Fe(292 MeV) + ^{207}Pb

irradiation 250 h
dose 1.2•10^{18}
σ 3 pb

Fig. 2.

The only decay chain of ^{264}108 known,[2], observed in the reaction ^{207}Pb(^{58}Fe,n) at 5.04 MeV/u with a cross section of ∿ 3 pb.

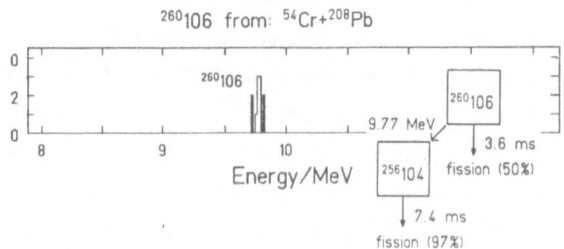

^{260}106 from: ^{54}Cr+^{208}Pb

Fig. 3. The decay of ^{260}106 [3]. Spontaneous fission of ^{256}104 follows α-decay (9.76 MeV) of ^{260}106.

154

elements 106-109 have smaller fission branches. For the isotope $^{264}108$ a single decay chain, Fig. 2, has been found in a recent experiment ^{207}Pb $(^{58}\text{Fe,n})^{264}108$, [2]. α-Decay was established followed by fission in the second generation at $^{256}104$. The even-even isotope $^{260}106(T_{1/2} = 4$ ms) has a 50 % fission branch[3], whereas its daughter fissions at least with a 97 % probability. The partial fission halflife is not increased going from $^{256}104$ to $^{260}106$. Figure 3 proves for $^{260}106$ the decay genetics, α-decay $\rightarrow$ spontaneous fission, which was never seen before and was proposed to be a characteristic of superheavy element decay. The α-energy, 9.76 MeV, observed in 8 decay chains followed by fission of $^{256}104$ becomes such a unique decay mode due to the background-free detection of fission, that the α-spectrum does not contain a single accidental count.

ON THE NUCLEAR STRUCTURE OF THE HEAVIEST ISOTOPES

The measured values of α-energies, α-decay halflives, and fission halflives can be used to establish trends of these quantities up to element 109 and to extrapolate the trends to higher proton numbers. Comparison with theoretical predictions of mass excesses, shell corrections, and fission barriers becomes possible.

Mass Excesses, Shell Correction Energies, and Fission Barriers

For the even elements, Z = 104, Z = 106, and Z = 108, α-energies of the even-even isotopes with N-Z = 48 have been determined. Figure 4 displays decay chains showing the α-decay of $^{256}104$,[4] $^{260}106$,[3] and $^{264}108$,[2]. There was only one chain found for the $^{256}104$ α-decay with $E_\alpha = 8.81 \pm 0.02$ MeV giving an α-branch of smaller than 3 %. Assuming a reduced α-width of 1, as has been found for $^{260}106$ and the other known doubly even isotones of N = 156, from the measured α-halflife of $^{264}108$ an α-decay energy for this isotope of $(11.0^{+0.8}_{-0.3})$ MeV has been derived. The three α-energies assuming groundstate to groundstate decay allow us to determine from the experimentally known mass excess of $^{252}102$ [5], the mass excesses of $^{256}104$, $^{260}106$, and $^{264}108$. The mass excesses of $^{256}104$, $^{260}106$, and $^{264}108$ are (94.2$\pm$0.1) MeV, (106.6$\pm$0.1) MeV, and (120.0$\pm$0.3) MeV, respectively. These values are compared in Table 1 to different mass excess predictions[6-11]. Best agreement is obtained with the prediction of Liran and Zeldes [10] and the very recent mass tables of Møller et al.[11]. All other approaches fail to reproduce the experimental values by about 1 MeV, the nuclei being more bound by this amount than predicted.

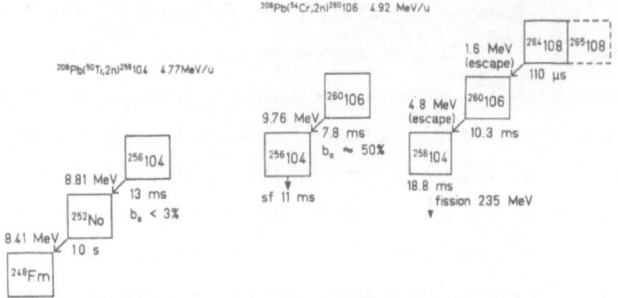

Fig. 4. Decay chains of the even-even nuclei $^{256}104$, $^{260}106$, and $^{264}108$ showing α decay $^{2-4}$. From the α energies, mass excesses of the isotopes are determined.

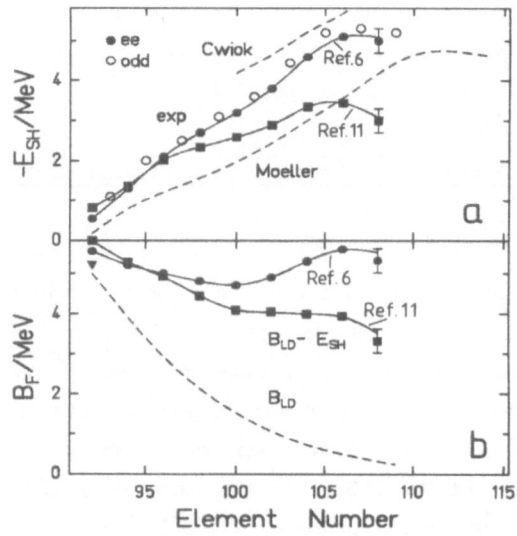

Fig. 5. For isotopes of the sequence N-Z = 48 from Z = 92-109 shell correction energies (a) and fission barriers (b) are given. Full lines refer to data derived from the experimental mass excess values, broken lines to theoretical predictions. The macroscopic mass values[6,11] are used to determine experimental shell corrections and fission barriers, respectively. These are compared to theoretical predictions for shell corrections[12,13] and to the expectation of the liquid drop model[14].

All isotopes of the N-Z = 48 sequence are known up to ^{266}109. Taking for the odd elements the highest α-energies observed, mass excesses were determined as for the even elements. For odd elements in case of transitions to excited states, the mass excesses obtained would be too large. Table 2 gives the new α-energies, mass excesses, and shell correction energies, $\Delta E^{shell} = M_{exp} - M_{macro}$, for the N-Z = 48 isotopes beyond Z = 102. The macroscopic mass excesses are taken from the tables published by Møller and Nix [6] or the latest mass tables of Møller et al. [11], which give macroscopic mass excess values corresponding to a stronger binding of the spherical reference nuclei. This leads to reduced values for shell corrections and fission barriers compared to the evaluation relying on the 1981-mass tables, which we used in the discussion of Ref.1.

Table 1. Mass excess values for ^{256}104, ^{260} 106 and ^{246}108

Nucleus	Ref. (6)	(7)	(8)	(9)	(10)	(11)	(1,2)
^{264}108	121.4	121.28	120.40	120.4	120.27	118.8	120.0±0.3
^{260}106	108.13	108.27	107.29	107.7	106.94	106.4	106.62±0.06
^{256}104	95.77	95.90	94.84	95.6	94.37	94.3	94.23±0.05

Table 2. The α energies, mass excesses, shell corrections, and fission barriers for N-Z = 48 isotopes between Z = 102 and Z = 109

Isotope	^{252}102	^{254}103	^{256}104	^{258}105	^{260}106	^{262}107	^{264}108	^{266}109	Ref.
E_α^{max} (MeV)	8.42	8.46	8.81	9.30	9.76	10.38	11.0	11.10	(1,2)
M^{exp} (MeV)	82.87	89.66	94.23	101.52	106.62	114.48	120.0	128.17	(1,2)
M^{macro} (MeV)	86.67	94.11	98.84	106.58	111.63	119.65	125.04	133.45	(6)
	85.75	93.03	97.60	105.16	110.03	117.86	123.03	131.13	(11)
E_{shell}^{exp} (MeV)	− 3.8	− 4.4	− 4.6	− 5.1	− 5.0	− 5.2	− 5.0	− 5.3	(6)
	− 2.9	− 3.4	− 3.4	− 3.6	− 3.4	− 3.4	− 3.0	− 3.0	(11)
E_{shell}^{cal} (MeV)	− 2.4	− 3.0	− 3.1	− 3.5	− 3.5	− 4.1	− 4.1	− 4.8	(12)
	− 2.6	− 3.4	− 3.3	− 3.8	− 3.6	− 4.3	− 4.2	− 4.8	(11)
B_f^{macro} (MeV)	1.1	0.9	0.7	0.6	0.5	0.4	0.3	0.2	(14)
B_f^{exp} (MeV)	4.9	5.3	5.3	5.7	5.5	5.6	5.3	5.5	(6)
	4.0	4.3	4.1	4.2	3.9	3.8	3.3	3.2	11)
B_f^{cal} (MeV)	3.5	3.9	3.8	4.1	4.0	4.5	4.4	5.0	(12)
	3.7	4.3	4.0	4.4	4.1	4.7	4.5	5.0	(11)

Figure 5a (full lines) presents the shell correction energies for all N-Z = 48 isotopes between Z = 91 and 109 using the mass values as given in Table 2 or taken from the latest Wapstra tables [5]. There is a small systematic odd-even difference of about 0.2 MeV between odd and even elements all over the range of elements concerned. The shell correction increases up to ^{258}105 and then stays about constant up to ^{266}109. Using the 1986-macroscopic masses of Møller et al. [11], the experimental shell correction are reduced by up to 2 MeV for the heaviest isotopes. Besides in Fig. 5a (broken lines) the calculated shell corrections [12,13] are shown. Most nuclei are more stabilized than predicted by Ref. 12, whereas Ref. 13 overestimates their stabilization. The uncertainties in macroscopic mass values determine the accuracy of the experimental shell corrections. The experimental errors in M_{exp} are small compared to the systematical errors entering via M_{macro}. However, differences between theoretical predictions[12,13] are of comparable size. The shell corrections both experimentally and theoretically are uncertain at least by $\pm$ 1 MeV. Within those limits, the predicted increase of shell corrections for heavier elements is established experimentally at least in its general trend.

The fission barrier B_f is the sum of the macroscopic liquid drop fission barrier and the shell corrections of the groundstate and saddle point masses. The latter is difficult to measure and calculate, but it is known to be a small contribution and is neglected. The macroscopic fission barrier could be taken from different calculations [6,7] or from an analysis of experimental fission barriers. The experimental macroscopic barriers .from Ref. 14 are used in the following and are given for the N-Z = 48 isotopes in Table 2. Figure 5b shows the experimental fission barriers, $B^{exp} = B^{macro} - E^{shell}$, and the experimental macroscopic barriers for the N-Z = 48 isotopes of element Z = 92 to Z = 109. Again we give fission barriers obtained via the macroscopic mass values of Refs. 6 and 11. Surprisingly the barriers stay high, in spite of the strongly decreasing macroscopic barrier, e.g. for ^{260}106 only 0.5 MeV stem from the macroscopic contribution, whereas (4.3$\pm$1) MeV are due to the shell correction energy of the ground state mass. For ^{232}U with a similar fission barrier the contributions are inverse, almost no shell corrections but a macroscopic barrier giving the main contribution.

The fission barrier protecting the isotopes of the new elements is high because of the strong groundstate stabilization of these nuclides. The stabilization shows an increasing trend in the entire range of the heaviest elements. The measured α-energies compared with the α-energies expected for nuclei having no shell correction show a difference in α-energies which is always smaller than 0.4 MeV, demonstrating that the surface as a whole is shell stabilized. The shell stabilization drastically changes the fission barrier, but only negligibly changes the α-energies. The α-energies reflect small local deviations of the mass surface, but they do not reveal the slowly changing trends toward higher shell correction energies. Only mass excess measurements disclose the increasing shell stabilization of the entire range of nuclides.

Halflives for α-Decay and Spontaneous Fission

The measured partial α-lifetimes of the even-even N-Z = 48 isotopes were compared to lifetimes corresponding to the macroscopic α-energies. The α-halflives were calculated according to a prescription given in Ref. 15. All reduced α-widths were set equal to one. A deviation of less than an order of magnitude from the macroscopic expectation is observed experimentally. Once more this fact demonstrates that the shell corrections vary smoothly and the α-halflives are affected only little by nuclear structure effects in this mass region. For a given element, α-halflives are getting shorter and shorter with decreasing neutron number, a trend well described by macroscopic models and observed for many elements beyond Z = 52. As isotopes approach the proton drip line, their halflives drop to the millisecond range. The short α-halflives observed are governed by this macroscopic trend, and nuclear structure is of minor importance only.

The contrary holds for the spontaneous fission lifetimes. To separate nuclear structure effects from macroscopic fission properties, we follow a slightly modified procedure [3] as originally introduced by Swiatecki [16]. The fission halflife is given by

$$T_{sf}(\text{in seconds}) = 3 \times 10^{-21}/P, \tag{1}$$

with P a barrier transmission factor, and the numerical factor the barrier knock-on time. Assuming a Hill-Wheeler-type transmission through the fission barrier with curvature $\hbar\omega_f$, we obtain

$$P = [1+\exp(2\pi B_f/\hbar\omega_f)]^{-1} \sim \exp(-2\pi B_f/\hbar\omega_f). \tag{2}$$

With $B_f = B^{macro} - \Delta E^{shell}$ and the shell corrections at the saddle point neglected, we obtain from Equations 1 and 2 an expression separating the nuclear structure effects from the macroscopic contribution:

$$\log(T_{sf/s}) = 2.73\ B^{macro}/\hbar\omega_f - 2.73\ E^{shell}/\hbar\omega_f - 20.5. \tag{3}$$

The experimental shell correction energies are obtained from mass excess values described in the previous section, whereas the macroscopic

fission barriers are calculated according to the semiempirical description given in [14]. For isotopes with small shell correction energies and
large values of B_f^{macro}, the barrier curvature parameter $\hbar\omega_f$ may be fitted
to the spontaneous fission halflives. With this value of $\hbar\omega_f$ kept
constant the macroscopic expectation may be calculated as a function of
the fissility parameter x. This macroscopic spontaneous fission halflife
is presented, together with the experimental halflives, in Fig. 6a. For
^{260}106 the macroscopic fission halflife is about 10^{-19} s, which compares
with 7 ms found experimentally. A stabilization by nuclear structure
effects of 17 orders of magnitude is observed.

Taking Equation 3 as a presentation with one adjustable parameter,
values for $\hbar\omega_f$ as a function of x may be obtained. Figure 6b shows $\hbar\omega_f$
for all even–even spontaneously fissioning nuclei. There is a smooth
trend of $\hbar\omega_f$ for all isotopes between U and Cf. For Fm and Z = 102 some
isotopes still follow the trend, whereas the isotopes of elements 104 and
108 show values of $\hbar\omega_f$ that are increased compared to the values for

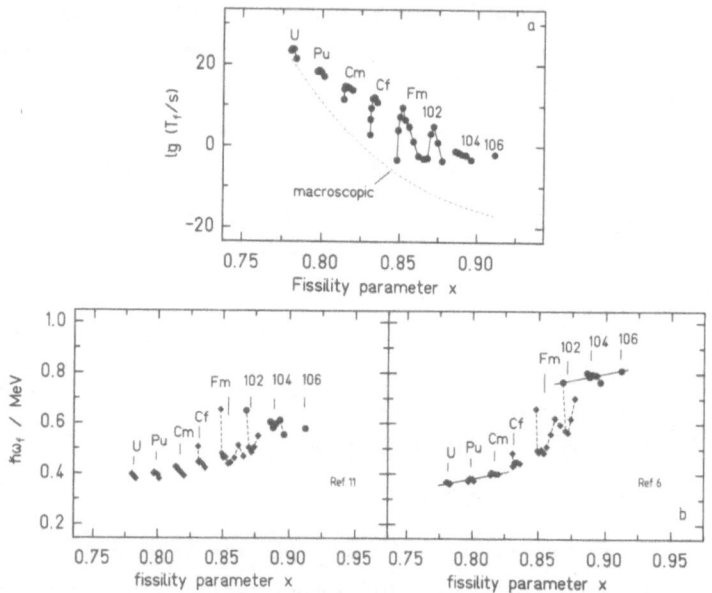

Fig. 6. Fission halflives of all known even–even isotopes together
with the halflives expected for nuclei with macroscopic
fission barriers [14] (a). The barrier curvature parameter is
fitted to ^{232}U and kept constant for all isotopes. Figure 6b
shows the curvature parameter $\hbar\omega_f$, obtained from fission
halflives, experimental shell correction energies, and the
macroscopic barriers of Fig. 5 [6,11]. Diamonds belong to nuclei
with a double-humped barrier and points are nuclei with a
predicted single-humped barrier. The lines are fits to the $\hbar\omega_f$
for the two groups.

uranium. The increased values of $\hbar\omega_f$ point to a narrowing of the fission barrier by factors of (1.5-2) or to a change in the inertia parameter by factors of (2-4). In Fig. 6b those isotopes for which a disappearance of the second fission barrier is predicted are indicated by points [17]. Their positions support the hypothesis [18-20] that the change of halflife systematics observed for Z = 104 may be caused by such a disappearance. Moreover, these isotopes are shell stabilized. Their macroscopic barriers are smaller than 1 MeV. The larger values of $\hbar\omega_f$ lead to a weaker dependence of the spontaneous fission halflife on the barrier height than that observed for lighter elements. The fission through the narrow single-humped barrier most probably gives a symmetric mass distribution of fission fragments. Until now only the mass distributions of two of these nuclei (^{260}104 and ^{258}102) have been measured and found to be symmetric [21]. In a calculation to explain the spontaneous fission halflives for nuclei around ^{258}Fm, Møller et al.[22] conclude that the inertia parameters for these nuclei have to be strongly reduced. A definite answer why $\hbar\omega_f$ is larger for the heaviest nuclei thus still is open.

The odd isotopes ^{255}104 and ^{257}105 have partial fission halflives of 2.7 and 8.2 s, respectively. Comparison to ^{256}104 allows us to determine the hindrance factors of spontaneous fission for odd protons and neutrons. Hindrance factors of 480 and 610 are obtained [4]. Odd-odd isotopes may be hindered by the product of the individual hindrance factors, that is by a factor of 2×10^5. The α-halflives of the Z = 107 and Z = 109 isotopes and of the odd isotopes of Z = 106 are all shorter than the fission halflives if we assume these hindrance factors and the fission halflives of 7.2 ms (^{260}106) [3] and > 0.4 ms (^{264}108) [2] for the neighbouring even-even isotopes as a reference.

Neither especially large hindrance factors nor nuclear structure effects in α-halflives are responsible for the absence of fission. The occurrence of α-decay for the isotopes of elements 106 to 109 is mainly a consequence of a strong groundstate shell stabilization giving rise to increased fission halflives.

Recent Theoretical Predictions of Shell Corrections and Halflives

An island of macroscopically unstable, but shell-stabilized, spherical nuclei around ^{298}114$_{184}$ was predicted as early as 1966 [23-26]. Many experiments sought to find these superheavy nuclei in nature or to produce them by nuclear reactions failed. Predictions of shell corrections for nuclei between the heaviest isotopes known and the 'superheavy' island are rare. Two calculations using the best macroscopic-microscopic models were published recently [12,13]. The shell corrections of the calculation of are shown in Fig. 7. Besides the strong shell effects at $^{298}114_{184}$ (-7.08 MeV) already known, another island at $^{272}109_{163}$ (-6.93 MeV) and $^{270}108_{162}$ (-7.97 MeV) has been predicted by Refs. 12 and 13, respectively. Nuclei not with N = 184 but with N = 178, 177 are found to have the largest shell corrections, e.g. -8.97 MeV for $^{291}114$ and -8.38 MeV for $^{294}116$. Following the successful path along N-Z = 48 leading to $^{266}109$ and further up to $^{276}114$, nearly constant shell corrections between -4.6 and -5.0 MeV are predicted. A continuous increase of the shell correction by another -4 MeV is expected going into the center of the island. For N = 163-166 a change from deformed nuclei to spherical

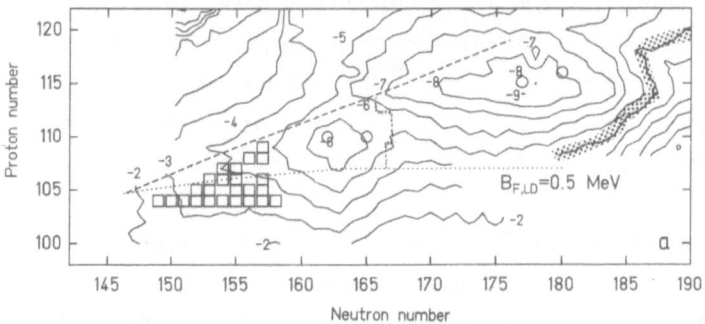

Fig. 7a Shell correction energies as calculated by Møller et al.[12]. The borderline at N ~ 165 between spherical and deformed nuclei (.....), the macroscopic fission barriers of B_f = (0.5) MeV[14] (⸱------⸱), and the region of shell stabilized superheavy nuclei defined by B_f > 4 MeV, B_p > 0 and B_{LD} ~ 0.5 MeV are given. The black squares show the known superheavy isotopes of elements 106-109.

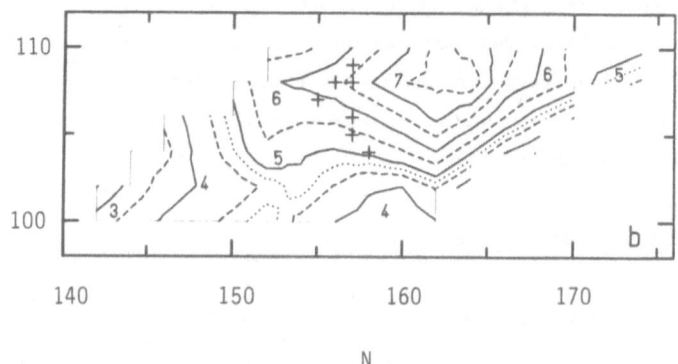

Fig. 7b Shell correction energies as calculated by Cwiok et al.[13] taken from Ref. 27. The heaviest isotopes of elements 104-109 are indicated.

nuclei is predicted [12] and is indicated in Fig. 7a. The N-Z = (48-49) paths lead to Z = 114 via a chain of deformed nuclei, whereas for N-Z > 50 spherical nuclei would be reached.

The new island around ^{272}109 consists of nuclei that are deformed (ε_2 = 0.21, ε_4 = 0.09)[12]. The most interesting feature of these nuclei are the large ε_4 deformations [12,13]. Positive values of ε_4 correspond to a sausage-like shape of the nucleus. These sausage-like nuclei were seen before around ^{180}Hf. Halflives have been calculated recently for the new region of stabilization. For the center of the region halflives of a few hours [27] and some ms [22] have been predicted, respectively.

What is it, a Superheavy Nucleus ?

In view of our experimental finding of high shell corrections for the elements discussed and the new calculations corroborating their stability, it seems appropriate to reconsider our definitions on superheavy nuclei. Nuclei stabilized by shell effects, but unstable as charged liquid drops towards fission are the species in a very general sense superheavy nuclei are supposed to be[23,24]. Their possible existence has been demanded after fission isomers have been observed [28] in a mass region between Z = 92 and Z = 97 where the liquid drop barrier and the shell corrections both are of importance. The shape dependent shell correction energies lead to the double humped barrier [24], the existence of which has been proven by many experiments in the years 1965-1975. For higher atomic numbers the liquid drop barriers become less and less important and all stabilization comes from shell effects. Starting with element 104 the fission barriers are predominantly due to shell correction energies. They may be - as discussed - narrow and single humped.

Superheavy nuclei exist due to shell correction energies only, and it is of minor importance how large their halflives actually become. The isotopes found for elements 108 and 109 with halflives in the few ms-range and until now undetected fission branches are compared to the fission time of unprotected liquid drops of 3 x 10^{-21} s stabilized by a factor 10^{18}. Nuclei with halfives approaching the age of the solar system would be stabilized just by another factor of 10^{18}, that is according to Equation 2 an increase by a factor of 2 in the height of the fission barrier. Not the height of the shell stabilized fission barrier makes a superheavy element, but the fact of pure shell stabilization in itself. To demonstrate the rise of the shell corrections mass measurements are decisive, and the mass measurements for ^{256}104, ^{260}106, and ^{264}108 are a first step in this direction. The finding of these mass measurements that fission barriers for macroscopically unstable systems as ^{266}109 are large, prove our hypothesis having made superheavy elements. In our case the nuclei are predicted to be deformed like sausages, positive ϵ_2 and ϵ_4 , but again this, as the height of the barrier or the halflife is an interesting, but minor detail. Shell correction energies are independent of the fissility parameter of the underlying liquid drop. The nearly constant values of the fission halflives for the even isotopes of elements 104 and 106 prove this independence and thus indirectly our classification that elements heavier than Z = 104, are superheavy in the general sense of their definition, is corroborated.

The kind of superheavy elements we have discovered has been foreseen by A. Bohr in 1974 [29], when he gave the following comment at the Nobel symposium on SHE:

'We have heard a great deal about the search for superheavy nuclei with a spherical shape. What about the possibility of SHE in other shapes stabilized by shell structure. The appropiate magic numbers would be different. Very likely the life-times would be much smaller than those estimated for spherical SHE, but perhaps these elements would be easier to produce'.

Figure 7a indicates the region of superheavy nuclei, nuclei with a predominantly shell stabilized fission barrier larger than 4 MeV giving rise to fission halflives of about 10^{-6} s, the actual halflife limit for unique isotope identification. With the discovery of elements 107 to 109 we entered along the ridge of largest shell corrections at N–Z $\sim$ 50 into the island of superheavy elements. Entering further is not a question of the groundstate stability of superheavy elements, a question we think settled principally by experiments and calculations, but a question of finding a way to overcome the blocking set to the fusion of heavy nuclei.

ON THE MAKING OF HEAVY ELEMENTS

Production Cross Sections

Regarding the possible combinations of elements to produce a wanted EVR out of a projectile and a target, we find a striking fact. Only about half of all possible combinations lead to EVR's. There is a limit beyond which nuclei do not fuse. For a symmetric collision system a limiting value of $Z^2/A \sim 38$, or x = 0.80 for the fissility parameter of the fused system is found.

In the experiments performed to produce highly fissionable isotopes, a three-fold limitation of the production process has been observed.

1. A thermal limitation in the exit channel by fission losses in the evaporation cascade, characterized by a survival probability $w(E^x)$. $w(E^x)$ is given by the evaporation cascade. These losses depend mainly on the number of evaporated particles and the size of Γ_n/Γ_f. For a long time this limitation was thought to be the only one to be overcome.

2. A structural limitation $S(E_D,E^x,A_i,Z_i)$ by the nuclear structure-dependent disappearance of the shell stabilization with excitation energy characterized by a shape dependent damping energy E_D of shell effects. For the latter limitation I refer to Refs. 1,30,31.

3. A dynamical limitation in the entrance channel due to the increasing repulsive Coulomb forces characterized by the fusion probability $p(E)$ [32-35]. The importance of the fusion probability as a limiting factor for fusion is discussed in the following sections.

These three limitations in mind the cross section at the barrier B_B may be presented as a factor formula:

$$\sigma^{EVR}(B_B) = \pi \lambdabar^2 \ell_{lim}^2 \cdot w(E^x) \cdot S(E_D,E^x,A_i,Z_i) \cdot p(B_B) \tag{5}$$

The first factors $\pi\lambdabar^2\ell_{lim}^2$ describe the formation of a fused system where the hindrances are not acting. With $\ell_{lim} \sim 15\ \hbar$ for highly fissionable nuclei[30,31] this cross section amounts to a few tens of millibarns. The other factors stand for the 3 limitations.

In Fig. 8 the production cross sections for heavy elements are presented. Figure 8 (a) compares the largest 4n cross sections ($E^x \sim 45$ MeV) using ^{249}Bk and ^{249}Cf targets[36] to produce isotopes of elements 102-106 with 1n cross sections ($E^x \sim 20$ MeV) using ^{208}Pb- and ^{209}Bi-targets, which have been used to produce isotopes of elements 100-109. Moreover, cross sections to produce isotopes of elements 102 and 104 using ^{238}U targets are given[37]. These cross sections are smaller by 1 to 2 orders of magnitude compared to the reactions using the heaviest actinide targets available. Beyond element 105 the cross section using Pb and Bi-based reactions are larger than for the most favourable actinides based reactions. Figure 8 (b) gives the cross sections as measured for reactions leading to isotopes of elements 102 to 109 via compound systems with N-Z = 50 and 49 for even and odd elements, respectively. All values are found in a listing in Ref. 1 except σ_{2n} = 190 nb and σ_{3n} = 7.3 nb obtained from the reaction ^{206}Pb(^{48}Ca,xn)^{38}. The slope of the cross sections for 3n and 2n reactions is larger than for 1n reactions. On the average the 1n cross section decreases by a factor of 3.5 for each element. For elements 104 and 105, the cross sections for 1n reactions become comparable to 2n cross sections and finally prevail for elements 107 to 109. The extrapolation of the production cross sections for the 3 groups of targets to element 110 give 3 pb for $^{62}Ni+^{208}Pb$, 0.1 pb for $^{23}Na+^{254}Es$ and 3 fb for $^{40}Ar+^{238}U$. These straightforward empirical estimates rely on the until now unbroken tendency of the cross sections to decrease with increasing element number. The crudeness of an extrapolation for the U-based reactions from element 104 to element 110 is evident.

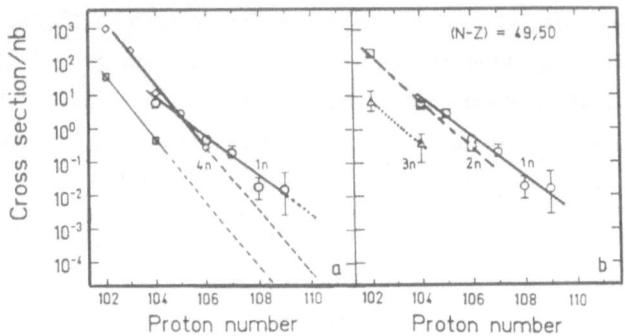

Fig. 8a. The 1n cross sections for cold fusion reactions compared to 4n cross sections for $^{249}Cf-$ and ^{249}Bk-based reactions [36], and the few values known for ^{238}U-based reactions[37]. The lines are fits to the data showing the extrapolation to element 110.

Fig. 8b. Experimental cross sections for different xn channels observed in cold fusion reactions leading to compound systems with N-Z = 49 and 50.

Evidence of Entrance Channel Limitation from Symmetric Fusion Reactions

Symmetric collision systems due to their larger mutual Coulomb repulsion will show the onset of entrance channel limitation already for production of elements below Z = 92 where fission barriers are still high. The EVR-cross section are in the μb-range, and systematic investigations became feasible[31,39-41].

The EVR cross sections measured for the systems $^{40}Ar+^{80}Hf$ and $^{124}Sn+^{96}Zr$ leading both to ^{220}Th are presented in Fig. 9. Coulomb barriers and $w(E^x)$ are given as a reference. The missing cross section for the 2n and 3n cross sections in the system $^{124}Sn+^{96}Zr$ indicates the entrance channel limitation, which is not seen in the $^{40}Ar+^{180}Hf$ system. Figure 10

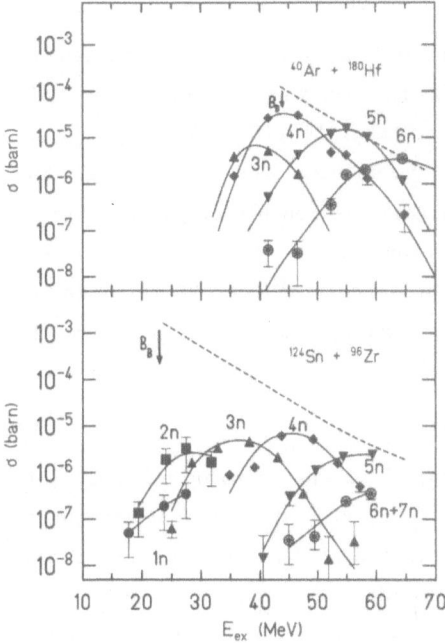

Fig. 9. The EVR cross sections for the reactions ^{40}Ar+^{180}Hf and ^{124}Sn+^{96}Zr 31. The hatched line indicates $w(E^x)$, the Bass barriers are given for orientation. The missing cross section for the ^{124}Sn-reaction in the barrier region is characteristic for entrance channel limitation.

Fig. 10. The EVR-cross sections for two nearly symmetric reactions, ^{90}Zr+^{90}Zr and ^{124}Sn+^{96}Zr 31,39. The onset of entrance channel limitation between Z^2/A = 35-37 is demonstrated.

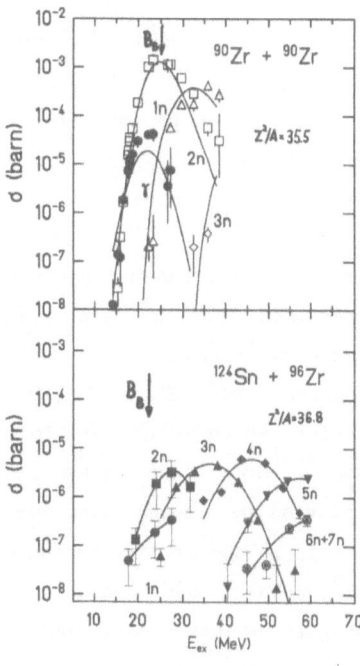

compares the cross sections of two nearly symmetric systems ^{90}Zr+^{90}Zr and again ^{124}Sn+^{96}Zr. The onset of entrance channel limitation between $Z^2/A = 35.5$ and 36.8 is clearly seen.

In order to discuss further the entrance channel limitation we introduce a fusion probability $p(E)$ for central collisions[31]. This quantity together with the survival probability $w(E+Q, \ell)$ allows us to separate numerically the phenomena in the entrance and exit channel. The procedure is especially adequate for highly fissionable compound systems, for which only in a small window near angular momentum zero EVR's surviving fission will be found. The cross section for EVR formation at the kinetic energy E in the center-of-mass system can be formulated by using the quantities $p(E, \ell)$ and $w(E+Q, \ell)$:

$$\sigma(E) = \pi \lambdabar^2 \sum_{\pi} (2\ell+1)p(E, \ell)w(E+Q, \ell) \tag{6}$$

Here, $E+Q$ is the excitation energy E^x in the compound nucleus, Q is the Q value of the reaction, and $\lambdabar$ is the de Broglie wavelength corresponding to the entrance channel. If the compound nucleus hypothesis is assumed to be valid, $w(E^x, \ell)$ does not depend on the entrance channel. We can now define an angular-momentum-weighted average of the fusion probability $p(E, \ell)$:

$$\langle p(E,\ell)\rangle = \pi \lambdabar^2 \sum_{\ell} (2\ell+1)p(E,\ell)w(E+Q,\ell)/\pi\lambda^2 \sum_{\ell} (2\ell+1)w(E+Q,\ell)$$

$$= \sigma(E)/\pi\lambda^2 \sum_{\ell} (2\ell+1)w(E+Q,\ell). \tag{7}$$

The fusion probability $\langle p(E,\ell)\rangle$ is obtained from the measured EVR cross section $\sigma(E)$, provided $w(E^x, \ell)$ is known. The function $w(E^x, \ell)$ can either be determined from the EVR cross section of a corresponding asymmetric reaction[31] which allows us to determine $w(E^x, \ell)$ in an energy region undisturbed by entrance channel effects, Fig. 9, or it can be determined from an evaporation cascade calculation[42], which has been tested to reproduce EVR cross sections adequately[39,40], Fig. 10. The dependence of $w(E^x, \ell)$ is determined by the ℓ-dependence of the fission barrier, which is assumed to behave as the barrier of a rotating liquid drop[43]. This ℓ-dependence restricts the ℓ-values contributing to EVR formation to a range much smaller than the broad distribution of angular momenta in the primary fusion process. Because the weighting function $w(E^x, \ell)(2\ell+1)$ used to obtain $p(E, \ell)$ is a narrow function peaking at all energies at small ℓ-values ($\sim 15\ \hbar$), $\langle p(E,\ell)\rangle$ becomes about $p(E,\ell = 0)$ and the term 'fusion probability for central collisions' $p(E)$ is justified.

In Fig. 11 we give a scheme for the general energy dependence of p(E). It is characterized by three energies, the adiabatic truncating barrier B_a, the Bass barrier B_B as a reference[44], and the dynamical barrier B. The logarithmic slope $(\hbar\omega)^{-1}$ of a one-dimensional tunnelling model describes p(E) at small energies. B_B-B_a and $p(B_a)$ characterize subbarrier fusion. $B-B_B$, the energy shift due to dissipation losses[34,35], and $p(B_B)$ fix the hindrance in the entrance channel. This six-parameter presentation describes both the subbarrier fusion effects and the entrance channel limitations. It replaces the two parameters B_B and $\hbar\omega$ of a one-dimensional tunnelling model.

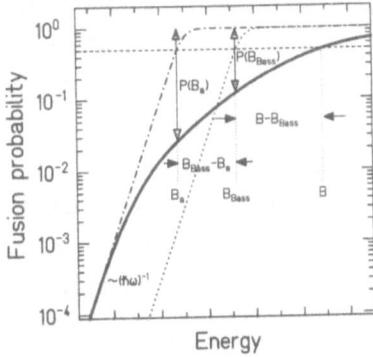

Fig. 11. Six parameters defining schematically the fusion probability. At low energies the slope $(\hbar\omega)^{-1}$ describes the Hill-Wheeler barrier penetration of the lowest possible adiabatic barrier B_a. Subbarrier fusion is parametrized by $p(B_a)$ and the subbarrier shift (B_B-B_a), the dynamical hindrance by $p(B_B)$ and the dynamical shift $(B-B_B)$ (extra-push).

For entrance channel limited fusion the shift $B-B_B$ and the fusion probability at the Bass barrier $p(B_B)$ mainly characterize p(E). For very large fluctuations and values of the barrier shift $B-B_B > 10$ MeV, as they are found for the systems of interest here, subbarrier fluctuations may be neglected and the Bass barrier B_B is chosen as truncating barrier. At an energy $E = B_B$ the system is assumed to have a single open barrier, the truncating barrier. Setting the transmission of this barrier equal to a Hill-Wheeler barrier a simple ansatz for $p(B_B)$, the fusion probability at $E = B_B$, is made

$$p(B_B) = 0.5 \exp{-(B-B_B)^2/2\sigma^2}. \tag{8}$$

This ansatz follows from an interpretation of the barrier shift phenomena along the following lines of reasoning. The measured barrier shifts $(B-B_B)$ are the intrinsic energy losses on the average trajectory leading to a fused system. This shift equals the extra-push energy as introduced in Refs. 34, 35. We define a probability $F(E')$ to pass the unshifted barrier B_B at a bombarding energy E' with an energy loss or extra-push of $E_x = (E'-B_B)$ which is smaller than the mean extra-push $\bar{E}_x = (B-B_B)$

$$F(E') = \exp-[(\bar{E}_x-E_x)^2/2\sigma^2] = \exp-[(B-E')^2/2\sigma^2].$$

At a bombarding energy $E' = B_B$ we obtain

$$F(B_B) = \exp-(\bar{E}_x^2/2\sigma^2) = \exp-[(B-B_B)^2/2\sigma^2].$$

For the fusion probability $p(B_B) = T(E = B_B) \times F(E' = B_B)$ follows Eq. (8), and $F(B_B)$ may be interpreted as the small chance of the system to avoid energy losses towards intrinsic degrees of freedom in the ascend of the barrier. The fluctuation parameter σ^2 has its origin in the statistically distributed small energy losses, which excite step by step the ensemble of nucleons. It should be proportional to the number of steps necessary to arrive at a barrier shift $\bar{E}_x = B-B_B$. This number equals the ratio of energy loss $\bar{E}_x$ and the mean energy loss ε for a single step. With $\varepsilon = const. \sigma^2$ increases with E_x. In case of decreasing values of ε at increasing values of $\bar{E}_x$, the dependence of σ^2 on $\bar{E}_x$ would be still stronger.

Figure 12 shows the fusion probability $p(E)$, for the system ^{96}Zr + ^{124}Sn of Fig. 9 as a function of bombarding energy. As a reference $p(E)$ for a hypothetical Hill-Wheeler barrier penetration of the Bass barrier is shown with dotted lines. The system $^{96}Zr+^{124}Sn$ shows a strong barrier shift $B-B_B = 27$ MeV and barrier fluctuation $\sigma = 9$ MeV. The values of the barriers and the corresponding fusion probabilities $p(B_B)$ are indicated, respectively.

Symmetric and nearly symmetric systems using beams of ^{86}Kr [40], ^{90}Zr [39], ^{124}Sn [31], and very recently ^{100}Mo [41] have been investigated systematically. Compound systems between $Z = 79$ and $Z = 92$ in the Z^2/A-range between 35 and 38 have been produced. An analysis of all the data except the ^{100}Mo results still in analysis, is compiled in Figs. 13-15. The barrier shifts $B-B_B$ as a function of the fissility of the system are shown in Fig. 13. Figure 14 shows the barrier fluctuations deduced from the analysis and in Fig. 15 the fusion probability at the Bass barrier is shown.

It is interesting to compare our data to the extra-push model of Swiatecki. The model assumes the dissipation losses to be due to the macroscopic motion of the walls and windows defining the potential of the collision system with respect to the Fermi motion of the nucleons

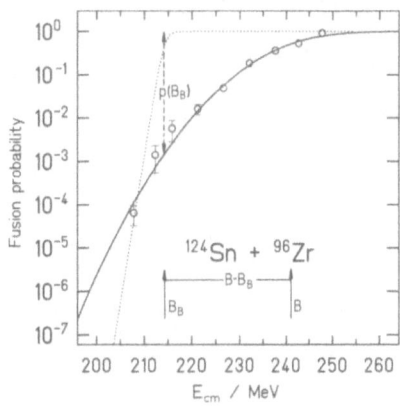

Fig. 12. Fusion probability $p(E)$ for the system $^{96}Zr+^{124}Sn$ as a function of bombarding energy. A strong barrier shift $B-B_B = 27$ MeV, and a large barrier fluctuation, $\sigma = 9$ MeV is demonstrated[31].

(wall–window friction). The model in its latest[34] as already in its first version[33] predicts an approximately parabolic dependence of the extra-push energy on the scaling parameter $(x-x_{thr})$.

$$B-B_B = \bar{E}_x = E_s \; [18(x-0.723) + 120(x-0.723)^2]^2 \qquad (9)$$

with

$$E_s = 7.6 \times 10^{-4}(z^2/A)^2_{crit} \; A_1^{1/3} \; A_2^{1/3} \; (A_1^{1/3} + A_2^{1/3})^2 \; / \; (A_1 + A_2),$$

$$x = (z^2/A) \; / \; (z^2/A)_{crit}$$

$$(z^2/A)_{crit} = 50.88 \; [1-1.78 \; (N-Z/N+Z)^2] \qquad (10)$$

Simulating symmetric collision systems the threshold value x_{thr} describing the onset of entrance channel limitation and the slope parameters in Eq. (9) are obtained from the model directly without any fits to open parameters. The extremely steep onset of extra-push heating prevents survival of EVR's for $x > 0.80$. Beyond $z^2/A \sim 38$ for the fused system a rapid disappearance of EVR's is predicted, as extra-push energies of hundreds of MeV experimentally cannot be verified.

The same threshold value of $x_{thr} = 0.72$ has been predicted by Sierk and Nix[32] already in a paper in 1974. In a more recent paper of the same authors[35] a slightly higher threshold value of $x_{thr} = 0.74$, and a slope value, as found in Ref. 34, have been published.

Figure 13 confronts the extra-push prediction with experimental barrier shifts of symmetric collision systems. The onset at $x_{thr} = 0.72$ is reproduced and the slope predicted is in agreement with experiments.

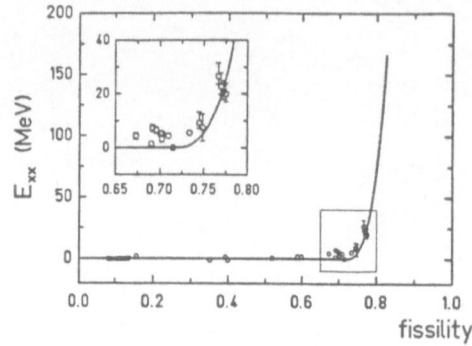

Fig. 13. Barrier shifts as a function of the fissility parameter x. The full curve is the prediction of the extra-push model [34]. Insert: Barrier shifts for symetric systems where EVR's have been detected and p(E) has been analyzed for EVR formation[31,39,40] in comparison with calculation.

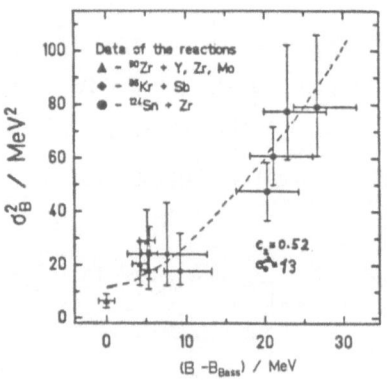

Fig. 14. The fusion barrier fluctuations for entrance channel limited reactions [31,39]. To the data a $E_x^{3/2}$ power law has been fitted with $\sigma_0^2 = 13$ and $c_2 = 0.52$.

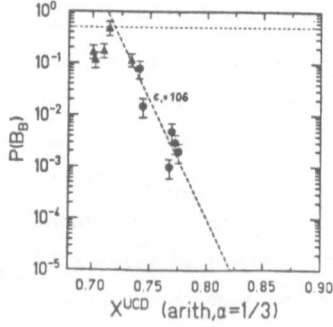

Fig. 15. The fusion probability $p(B_B)$ as a function of the fissility [31,39]. To the data points excluding the ^{90}Zr points (▲) a straight line has been fitted with a slope parameter $c_1 = 106$.

Actually only the data from Ref. 31 on $(Zr+{}^{124}Sn)$-systems fix the slope parameters. More data in the range of $0.74 < x < 0.78$ are needed[41] to definitely decide experimentally on the slope parameter. Trends in the isotopic dependences observed for the sequence of Zr-isotopes used in the investigations are not reproduced. The strongly shell stabilized collision systems ${}^{90}Zr+{}^{90}Zr$, and ${}^{90}Zr+{}^{124}Sn$ show comparatively small barrier shifts.

The increasing barrier fluctuations as presented in Fig. 14 and seen e.g. in Fig. 12 in the raw data for the system ${}^{96}Zr+{}^{124}Sn$ point to a dependence of σ on the barrier shift. $\sigma \sim (B-B_B)^n$ with $0 < n < 1$ follows on the one hand from the observed increase, $n > 0$, Fig. 14, and on the other hand from Eq. (8), which tells us for $\log p(B_B)$ to decrease with B_B, that n has to be smaller than 1. Assuming a linear relationship of $\log p(B_B)$ with the fissility as suggested by the data, Fig. 15, the fluctuation parameter σ^2 has in order Eq. (8) to be fulfilled to depend on $(B-B_B)$ with a power of $n = 3/2$. Figure 14 shows a fit of the data to a $(\sigma^2 - \sigma_o^2) = c_2 (B-B_B)^{3/2}$ power law. $\sigma_o^2 = 13$ takes into account a constant subbarrier fluctuation. With ${}^{90}Zr$-data omitted a value of $c_2 = 0.52$ is obtained. For this omission I refer to the arguments given in the following section concerning shell stabilized nuclei. The statistical interpretation of σ^2 demands a step size ε for the step-by-step energy loss process which decreases with $(\bar{E}_x)^{-1/2}$, in order to allow for $\sigma^2 \sim \bar{E}_x$ 3/2.

The barrier fluctuation is not contained in the models[34,35] which predict mean trajectories. In the model there is fusion above the barrier B and any kind of fusion below this barrier is neglected. As EVR's are amply found already below the extra-push barrier the model fails to make numerical predictions on the fusion probability at the energies which are of highest relevance for the production of heavy elements. To understand the fusion processes which lead to the heavy elements, we will see that just those processes have to be understood which the model assumes to be negligible. The extra-push losses are the sum of a great number of statistically independent single energy loss processes which microscopically are equivalent to nucleons changing orbits or collective excitations changing character. These losses are avoided in the fusion processes actually observed at the barrier B_B.

To the fusion proability at the Bass-barrier, $p(B_B)$, Fig. 15 an exponential, $\ln p(B_B) = -c_1 (x-x_{thr})$, is fitted. Omitting again the ${}^{90}Zr$ data points a slope parameter $c_1 = 106$ is obtained. The threshold at 0.72 is reproduced. From the same data points as used for the fit applying the relation Eq. (8) between the barrier shift and the fluctuation parameter

$p(B_B)$-values are calculated and a slope parameter $c_1 = 102$ is determined. This agreement in view of the crudeness of the assumptions made is remarkable.

Summarizing our finding on symmetric systems.

- From $x = 0.72$ on barrier shifts $B-B_B$, interpreted as the extra-push energy, are observed. The threshold value and the steep slope are in agreement with the models[34,35].

- The fluctuation parameter σ^2 has been determined. It increases with growing barrier shift. A $(\sigma^2 - \sigma_0^2) \sim (B-B_B)^{3/2}$ dependence has been deduced. A model including barrier fluctuations is needed to predict quantiatively the entrance channel limitation.

- The entrance channel limitation of fusion is described by the fusion probability at the Bass barrier. $p(B_B)$ strongly decreases with the fissility for fissilities above the threshold value 0.72. A linear decrease of $\log p(B_B)$ with $(x-0.72)$ is compatible with the data.

- Isotopic trends are found. These are not explained by the extra-push model. E.g., the doubly magic $^{90}Zr + ^{90}Zr$ system shows smaller barrier shift and small fluctuation values compared to the heavier Zr isotopes. The largest barrier shifts and fluctuations are found for the system using the soft ^{96}Zr. With the number of neutrons outside the $N = 50$ shell the barrier shift and fluctuation increase.

Entrance Channel Limitation of Asymmetric Collision Systems

The two-touching sphere entrance channel configuration of a mass asymmetric collision system is more compact than the corresponding configuration of a symmetric system. The distance to reach the Coulomb barrier is shorter for the asymmetric system than for the symmetric system and the dissipative losses in the fusion of the asymmetric system are expected to be smaller than for the symmetric system.

The ratio of disruptive Coulomb forces and attractive surface tension forces governs the amalgamation of two nuclei into one. For a monosystem this ratio is given by the fissility parameter x. For a two-touching sphere configuration, Bass[45] defined a corresponding parameter making use of the proximity force. Taking further into account that the proton and neutron ratio between the two partners is equilibrated very quickly (10^{-22} s), a modified fissility parameter x^{UCD} describing the ratio of Coulomb and nuclear forces for any two-touching sphere configuration follows. With $\kappa = (A_1/A_2)^{1/3}$ characterizing the asymmetry of the collision system the fissility parameter can be written:

$$x^{UCD}(\kappa) = x\ 4(\kappa^2 + \kappa + \kappa^{-1}\kappa + \kappa^{-2})^{-1} \tag{11}$$

For symmetric systems $f(\kappa) = 4(\kappa^2 + \kappa + \kappa^{-1}\kappa + \kappa^{-2})^{-1}$ approaches 1 and the fissility becomes equal to x with x defined as in fission, Eq. (10).

As the amalgamating nuclei at the Coulomb barrier for those systems of importance here are in closer contact than the two-touching sphere configuration, some average out of the fissilities of the mono- and binary system may present the appropriate asymmetry scaling.

$$x_{arith,\alpha} = x\ [(1-\alpha) + \alpha\ f(\kappa)] \quad \text{(arithmetic mean)}$$

$$\tag{12}$$

$$x_{geo,\alpha} = x\ f(\kappa)^{\alpha} \quad \text{(geometric mean)}$$

Here α is a parameter characterizing the overlap of the collision partners at the Coulomb barrier. Again for symmetric systems Eq. (12) gives the fissility x. The actual value of α has to be fitted to data or taken from a model. For small values of α the geometric and arithmetic mean become equal.

Simulating a great number of asymmetric collision systems, Blocki et al.[34] determined the theoretical value of the parameter. Figure 16 shows that for an arithmetic averaging a value of $\alpha = 1/3$ is obtained. All the asymmetric systems follow the slope given by the symmetric systems up to high extra-push energies. The same value $\alpha = 1/3$ has been found by Ref. 35 as well. The small value of α tells us that the influence of the two-touching sphere configuration on the effective fissility parameter is very weak. Asymmetric collision systems seem to profit less than we hoped from the smaller x-values of the two-touching sphere configuration[46].

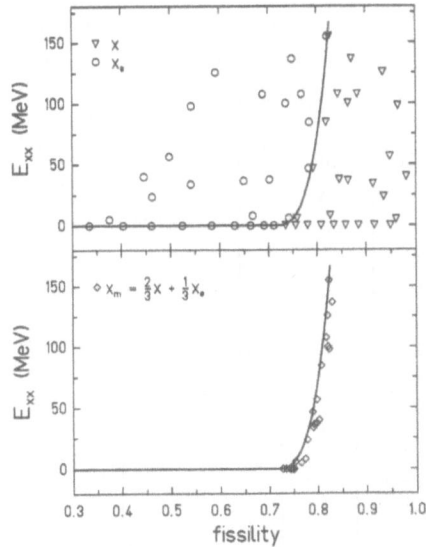

Fig. 16.

Model simulated extra-push energies for asymmetric collision systems for different fissility scalings [34]. For an x-scaling (▼) and the scaling Eq. (11) (O) the extra-push values are scattering over the plane (top). For an arithmetic mean scaling with $\alpha=1/3$ the asymmetric systems follow the extra-push curve as obtained for symmetric systems (bottom).

<u>Actinide-Based Systems</u>. According to the $\alpha = 1/3$ scaling already the actinide-based reaction $^{249}Cf(^{18}O,4n)^{48}$ leading to $^{263}106$ has a fissility of 0.73, which is beyond the threshold value. Actinide-based systems would be hindered for all elements beyond $Z = 106$ even using ^{254}Es as a target.

There is not much known experimentally on the fusion probability of actintide-based systems and a possible hindrance. A strong general hint to a hindrance is the world-wide failure to produce elements heavier than $Z = 106$ based on actinides as targets. A more direct result has been obtained in the various efforts to produce the compound system $^{264}104$. Table 3 compares three reactions leading to $^{260}104$ via 4n-channels. Measured[37,47-49] and calculated cross sections are compared. The small difference in excitation energy at the Bass barrier is of minor importance for the 4n-cross sections of the reactions considered. Referred to the $^{16}O+^{248}Cm$ reaction the hindrance of the more symmetric reaction seems to be evident. The numerical values would fit in the $p(B_B)$ systematics of Fig. 15, supporting the $\alpha = 1/3$ scaling.

For the production of heavy elements using actinide targets the limitation in the entrance channel most urgently has to be measured reliably. Beams of ^{22}Ne, ^{26}Mg, ^{30}Si, together with targets between ^{232}Th and ^{244}Pu should cover the critical range of x-values between 0.74 and 0.78. If a hindrance would be established, the asymmetry scaling of the extra-push models[34,35] would have been strongly supported. However, then the cross sections for a production of elements beyond $Z = 109$ using actinide targets would fall below the pb limit, as even for a reaction like $^{254}Es(^{23}Na,5n)^{272}110$ with $x = 0.78$ an appreciable hindrance in the entrance channel in addition to the large losses in the evaporation cascade has to be expected. Any search to synthesize heavy elements beyond $Z = 106$ using U- or Th-targets would be in vain.

Table 3. 4n-reactions to produce $^{260}104$.

Ref.	Reaction	Fissility ($\alpha = 1/3$)	E^x/MeV	σ/nb	$(\dfrac{\sigma/x^2}{\sigma_{16_O}/x^2})_{exp}$	$(\dfrac{\sigma/x^2}{\sigma_{16_O}/x^2})_{HIVAP}$	$p(B_B)$
48	$^{16}O+^{248}Cm$	0.72	44	6 ± 1	1	1	0.5
49	$^{22}Ne+^{242}Pu$	0.73	45	0.5	0.16	0.98	8×10^{-2}
37	$^{26}Mg+^{238}U$	0.745	47	0.14 ± 0.09	6×10^{-2}	0.77	4×10^{-2}

Table 4. Reactions leading to $Z = 110$ isotopes.

Reaction	Fissility ($\alpha = 1/3$)	E^x/MeV	$15^2\pi x^2$/mb	$w(E^x)$	$p(B_B)$	$S_{(E^x,E_p)}$	σ_{HIVAP}	$\sigma_\alpha = 1/3$
$^{254}Es(^{23}Na,5n)^{272}110$	0.78	51	61	10^{-8}	10^{-3}	1	0.6 nb	0.6 pb
$^{235}U(^{40}Ar,4n)^{271}110$	0.85	42	27	10^{-7}	5×10^{-7}	1	3 nb	1.3×10^{-3} pb
$^{208}Pb(^{64}Ni,1n)^{271}110$	0.88	15	15	5×10^{-5}	5×10^{-6}	1	750 nb	4 pb

Summarizing, Table 4 gives actinides-based reactions and the cross sections for the production of isotopes of element 110 with and without entrance channel limitation. As today the question of a hindrance for actinide-based systems cannot be answered definitely, the predictions of cross sections still cover order magnitudes.

Pb-Based Collision Systems. Using Pb and Bi targets and beams between Ar and Fe all heavy elements between 100 and 109 have been synthesized. The first experiments producing heavy elements by cold fusion were performed at Dubna in 1974 [50]. ^{40}Ar projectiles bombarded Pb targets and spontaneous fission events were observed, which could be assigned to Fm isotopes. The existing data are reproduced for all reaction channels by modern evaporation codes[38,46] and do not indicate any dynamical hindrance in the entrance channel, Fig. 17. The production of isotopes of element 102 by fusion of Ca and Pb isotopes was studied extensively[51-53]. Again an evaporation calculation reproduces the data without assuming any hindrance in the entrance channel[38]. More data on the weak channels would definitely help to determine whether or not a small hindrance might already be present. For elements 104 and 105 some excitation functions were measured using reactions with ^{50}Ti beams[4], Fig. 17. The 1n- and 2n-reaction channels have about equal yields. The small cross sections for the 2n channel and the observation of 1n channels are reproduced by the evaporation calculations only, if a dynamical hindrance at the barrier of about a factor of 25-30 is assumed. Such a hindrance is equivalent to a barrier shift of (21 ± 3) MeV and an increase of the barrier fluctuation to the large value of 12 MeV. It was shown that the same parameters in the evaporation calculation reproduce well the excitation function in the ^{249}Cf(^{12}C,4n)^{257}104 reaction without assuming any dynamical hindrance. The very sudden onset of a dynamical hindrance for the ^{50}Ti reaction and its absence in ^{48}Ca reactions may point to a nuclear structure effect caused by shell stabilized ^{48}Ca.

The excitation energy for all systems showing 1n cross sections between elements 104 and 109 decreases from 24 to 20 MeV, the fission barriers stay nearly constant, and still the cross section decreases by nearly three orders of magnitude, Fig. 18. If we assume for constant B_f approximately constant Γ_n/Γ_f values for all reactions, this decrease must be attributed to an increase of the entrance channel limitation. Figure 19 shows $p(B_B)$ as a function of the fissility for all systems investigated beyond $x = 0.71$. As fissility we have chosen the $\alpha = 1/3$ scaling of Ref. 34. In this presentation the cross section values known for heavy-element production beyond $Z = 104$ have been included. As all EVR's are found near the Bass barrier, the fusion probability at the

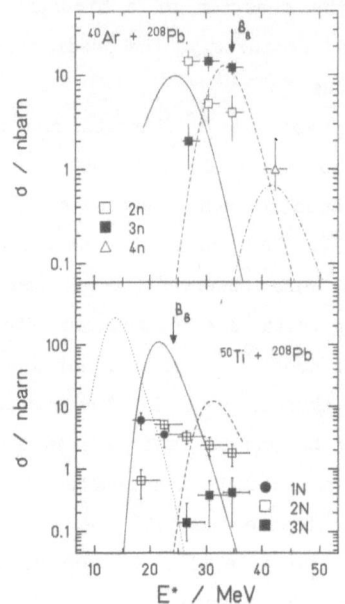

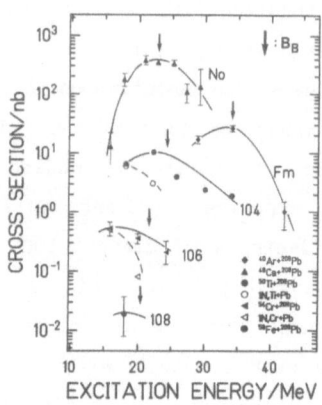

17

18

Fig. 17. EVR-cross sections for the reactions $^{40}Ar + ^{208}Pb$ and $^{50}Ti + ^{208}Pb$, 4. The ^{40}Ar reactions shows subbarrier fusion $\sigma = 5$ MeV and no entrance channel limitation, whereas the ^{50}Ti-reaction shows a barrier shift of 21 MeV and a large σ-value of 12 MeV.

Fig. 18. Total EVR-cross sections for production of the even elements $Z = 100$ to $Z = 108$, 54. The Bass-barriers are indicated. Hatched lines indicate the 1n-channel excitation functions.

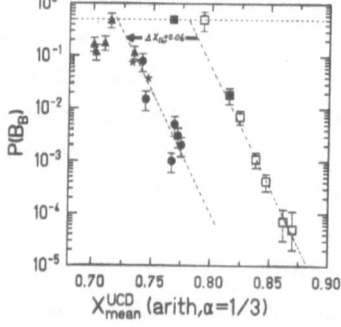

Fig. 19. The fusion probability $p(B_B)$ with $x_{arith,1/3}$ as scaling parameter 34. Data as in Fig. 15, actinide-based systems see Table 3, (✡), Pb-based systems from compilation Ref. 1. The arrow indicates the shift of $\Delta x_{thr} = 0.06$ to make the Pb-based systems ( ) coincide with the rest (▽ , ○ , ✡).

barrier $p(B_B)$ is obtained from the cross section values, if the $p(B_B)$ values are normalized to the $p(B_B)$ value obtained for the ^{208}Pb(^{40}Ti,xn) reaction and if we assume the decrease in cross section is caused by decrease of the fusion probability alone. The values of $p(B_B)$ thus obtained are lower values because Γ_n/Γ_f is assumed to stay constant between Z = 104 and 109. The value of $p(B_B)$ for production of ^{266}109 is 5 x 10^{-5}.

The slope of log $p(B_B)$, that is the dependence of the barrier fluctuation parameter σ^2 on the barrier shift (Fig. 14), seems to be unchanged compared to the symmetric systems, but we observe that the Pb- and Bi-based systems separate from the symmetric collision systems. This finding may be interpreted either as a failure of the α = 1/3 scaling or as a nuclear structure dependence of the threshold where the extra-push phenomenon sets in.

Following the latter interpretation we state: Instead of the predicted threshold value of x_{th} = 0.72 a shift to 0.78 for Pb- and Bi-targets is observed. For a given value of x such a shift would be equivalent to a gain factor in fusion probability of about 500 compared to another system of equal x. The x-values of ^{86}Kr+^{123}Sb and the actinide-based system ^{26}Mg+^{238}U are equal. With a shift of the threshold parameter by Δx_{th} = 0.06 for the system ^{50}Ti+^{208}Pb the barrier shifts and the fusion probabilities for all 3 systems become equal within a factor of two. $p(B_B)$ for the systems ^{86}Kr+^{123}Sb and ^{50}Ti+^{208}Pb indeed are measured to be equal and amount to 0.02. It would be interesting to verify by a new experiment the hindrance for ^{26}Mg+^{238}U estimated to be 0.04 in Table 3 [37].

Summarizing the finding on asymmetric collision systems

- The α = 1/3 scaling for actinide-based collision systems most probably is valid. A final test is possible using ^{232}Th and ^{238}U targets and ^{22}Ne, ^{26}Mg, and ^{30}Si beams to produce isotopes of elements 102 and 104 in the fissility range of 0.72 - 0.76.
- Pb- and Bi-based reactions show a shift of the extra-push threshold from x = 0.72 to x = 0.78. The delayed onset compared to the α = 1/3 scaling gives rise to a gain in fusion probability of about 500.
- The slope d ℓn $p(B_B)/d(x-x_{thr})$ seems to be independent from mass asymmetry, that is the barrier fluctuation parameter σ^2 should not depend on mass asymmetry. A change of 0.01 in the fissility parameter is equivalent to a decrease by a factor of 2.7 in fusion probability.

Fusion Using Pb-Like Targets

The high collective and intrinsic energies of the lowest excited states for N = 126 nuclei may prevent that these nuclei when ascending the fusion barrier loose radial energy in an early stage of approach. The mechanism of pumping radial energy into intrinsic energy may set in at higher threshold values x_{thr}, if the shell stabilized nuclei preserve their nuclear structures for deeper penetration, or for apparently larger values of α, than foreseen by the liquid drop dynamics. A value of α = 1/2 thus has been deduced by a former analysis from our data[1,46]. The retarded onset of extra-push limitation for the Pb- and Bi-based system means that these reactions are favoured twofold, first by the Q-values leading to 1n- and 2n-reactions instead of 4n- and 5n-reactions for actinide-based reactions, second by the delayed onset of the extra-push limitation.

The reaction $^{48}Ca+^{208}Pb$ may be regarded as the prototype of cold fusion reactions. For the production of Z = 102 isotopes compared to neighbouring systems as $^{54}Cr+^{198}Pt$ (E^x = 32 MeV) and $^{30}Si+^{226}Ra$ (E^x = 48 MeV) the cross section is expected to be increased by more than a factor 10^5. The latter reactions following the $p(B_B)$-values predicted by liquid drop dynamics, Fig. 19, would show EVR-production cross sections in the 10 pb-range.

The given factor of 10^5 for the Pb-based reactions refers to the 2n channel found in $^{208}Pb(^{48}Ca,xn)$ as the main channel. Going beyond Z = 106 the 1n-channel becomes dominating at energies where 2n-channels energetically still would be favoured. Probably the delayed onset of the entrance channel limitation for Pb-based systems is excitation-energy dependent and is seen only for very low excitation energies. One neutron emitted at an early stage during barrier passage could cool down the system to an energy below 15 MeV, an energy small enough to preserve shell structures even for spherical configurations. This cooling favours the chance to avoid decay by fission. The odd proton in ^{209}Bi seems to affect the fusion probability only little. Pair breaking does not prevent the system from fusion.

Other Rearrangement Processes Involving Spherical Nuclei

The few reactions known which yield spherical closed shell nuclei in large scale rearrangement processes show striking similarities. In the following different cases are compiled, but an adequate discussion of the links between the observations must be subject of some future lecture.

a)   The ^{208}Pb-cluster. ^{208}Pb-like clusters with an unbroken N = 126 neutron shell are observed in spontaneous fission like decays of heavy nuclei. The emission of ^{14}C from ^{223}Ra [55] or ^{24}Ne from ^{232}U [56] together with all other known cases can be understood as a rearrangement process into a ^{208}Pb-like cluster and a medium weight rest[57]. There are cases known with broken Z = 82 shell and odd nucleon numbers, but the N = 126 neutron shell has always been found intact. On the dependence of this new radioactivity on excitation energy of the emitting system nothing is known. The two-touching sphere configuration is more compact than the fusion barrier of the system, or the fissility of the system is smaller than the critical value x = 0.72. Wherever spherical ^{208}Pb-like clusters are produced by radioactive decay of heavy nuclei the configuration of the two separating nuclei is, as in α-decay, protected by a barrier, the tunnelling through which governs the decay probability[58].

This is different in rearrangement processes involving heavy transient systems formed in heavy ion collisions and decaying into spherical end products near ^{208}Pb, as ^{110}Pd + ^{238}U → ^{212}Po + ^{136}Xe [59], or ^{48}Ca + ^{248}Cm → ^{216}Rn + ^{80}Zn [60]. Here the compact exit configuration is on the other side of the fusion barrier. The heavier the intermediate system becomes the more excited the most compact scission configuration will be. These reactions are especially suited to investigate the stability of closed shell configurations to intrinsic excitation energy and deformation.

b)   The ^{132}Sn-cluster. The doubly closed shell configuration ^{132}Sn manifests itself clearly in fission. Spontaneous fission of ^{258}Fm shows total kinetic energies of fission products exhausting the Q-value of the reaction almost completely[61]. It fissions into two nuclei of ^{129}Sn. This phenomenon has been predicted as early as 1964 by Faisner and Wildermuth [62], but theoretially treated only very recently[22,63]. The high-kinetic energy fission is known also for ^{260}Md, an odd-odd isotope[21] In the case of thermal neutron fission of ^{257}Fm at an intrinsic excitation energy of only a few MeV's it has nearly disappeared[64].

In thermal neutron fission of ^{233}U, ^{235}U, and ^{239}Pu a rare disintegration mode with no neutrons emitted called cold fission[65,66] has been observed. Again, the Q-value of the reaction is found to be transferred completely into kinetic energy of the fission fragments. A compact scission shape with no deformation energy of the fragments is possible, if one of the fragments has a closed shell configuration as ^{130}Sn, ^{132}Sb, ^{134}Te. The cold fragmentations show odd and even isotopes in the exit channel. Pair breaking does not prevent the process. The preferential yields of the ^{132}Sn-like nuclei disappear, if the intrinsic excitation energy of the scission configuration is raised to a value of about 10 MeV[67]. Berger et al.[68] have shown that for configurations with a mass split containing spherical ^{134}Te nuclei a transition from the fission valley to the fusion valley becomes possible at the small distances of separation corresponding to a two touching sphere configuration of the fission products in their ground states. Such a transition between the two valleys has been proposed also to explain the ^{258}Fm spontaneous fission by Møller et al. in their latest paper[22]. The formation of spherical strongly shell stabilized nuclei in the fission of systems with fissilities in the range 0.78 - 0.80 points to the existence of a fast necking-in mode, which operates between closed shell nuclei in the range of the critical fissility[67,69]. This mode breaks nucleon pairs, as odd and even isotopes have been observed in the fissioning systems as well as in the primary fragments. Already a minor intrinsic excitation energy makes the phenomenon disappears.

c) The ^{78}Ni-cluster. Besides the neutron rich double magic nuclei ^{208}Pb, ^{132}Sn a third strongly shell stabilized neutron rich nucleus, ^{78}Ni, may exist. It could be possibly produced like the two havier configurations, in cold rearrangement reactions of systems in the critical fissility range x = 0.72-0.80, either in thermal fission or in fusion-fission reactions, as ^{48}Ca+^{254}Es.

The "Soft Way to Element 109"

Figure 20 qualitatively projects the finding in fission to fusion of heavy nuclei. All collision systems enter via the fusion valley and have to pass to the fission valley in order to fuse. However, systems with spherical partners make this crossing at smaller distances of approach using the fast necking-in mode for transition into a configuration with a compactness not far from the compactness of the system at the Coulomb barrier. The final passage over the barrier is achieved with a minimum transfer of radial energy into intrinsic excitation energy.

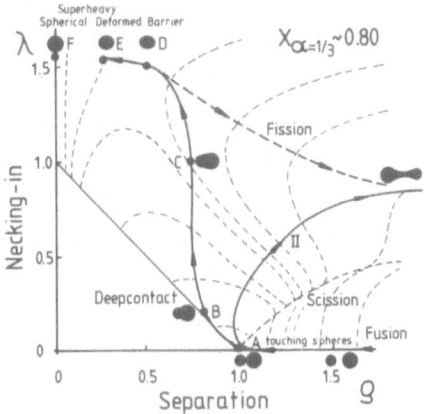

Fig. 20.

Schematic presentation of the "soft way" using Pb-like targets to produce superheavy deformed nuclei in a two-dimensional necking-in versus separation plot[33]. The collision system reaches the deep contact point B without intrinsic energy losses. It opens the neck via a fast necking-in mode, $\overline{BC}$, approaching a spheriodal $2:1$ configuration, again with only minor intrinsic energy losses. For the short passage over the saddle $\overline{CD}$, the new dynamics[34] may act with the fusion probability $p(B_B)$ giving the chanche to pass without further intrinsic energy losses. After the remaining excitation energy has been taken off by a γ-cascade the ground state E of a deformed superheavy nucleus is reached. The pitfalls of a spherical end product F and a new dynamic trajectory (II) where extra-push heating is acting are circumvented.

The production of superheavy elements using the "soft way" minimizes the losses of evaporation residues by the 3 limitations of Eq. (5). Making heavy elements by cold fusion $E^x = (20\text{--}24)$ MeV, avoids the high excitation energies introduced into the compound system using actinides as a target. This advantage of increasing the survivable probability in the evaporation cascade was well recognized as early as 1974 by Y. Oganessian et al.[18]. But two other pit-falls on our way to superheavy elements were not known at that time.

The first pit-fall was recognized in 1979[30], when we tried to make evaporation residues, which were supposed to be spherical and highly fissionable, e.g. ^{216}Th in simulation of the superheavy doubly closed shell nucleus ^{298}114. It was shown that the ground state shell correction to the fission barrier was helping to stabilize evaporation residues only at excitation energies smaller than 10 MeV. Superheavy nuclei with such a low excitation energy cannot be made using actinide targets even in the most favourable cases as ^{248}Cm(^{48}Ca, 3n)^{293}116 [70]. On the other hand it is known since long that the shell correction energies have to be taken into account in order to understand the production of the known isotopes of the heaviest elements which are deformed, see e.g. the discussion of this point in Ref. 1. Superheavy elements made via fusion should have a deformed nucleus, as it withstands excitation energy better than a spherical nucleus.

The second pit-fall, the hindrance in fusion proabability was supposed to be overcome by brute force, by extra-push[32]. This recommendation was good for the production of nuclei still stabilized by a liquid drop fission barrier, as was shown in Fig. 9 for Th-isotopes[31]. However, all reactions leading to the heaviest elements do not show any surviving nuclei at the shifted barrier, Fig. 18. Here the extra-push recommendation failed completely. The remaining evaporation residues were found at the unshifted barrier. The problem of heavy element production is not to succeed using extra-push energy, but how to avoid the transfer of radial energy into intrinsic excitation energy, that is how to circumvent extra-push heating. Here the strongly shell stabilized Pb-like target nuclei seem to have a definite advantage compared to actinide-based systems. They fuse without extra-push under conditions, where extra-push should have been expected.

The production of the superheavy elements as described in these lectures circumvents the pit-falls, the N-Z = 48 isotopes produced are deformed nuclei, and using spherical Pb-like target nuclei they are produced cold and in a way which shows a delayed onset of extra-push heating.

To demonstrate the manyfold action of shell effects the "soft way" to the formation of ^{266}109 is described[71].

- ^{209}Bi, a highly shell stabilized stable nucleus meets its partner ^{58}Fe, the strongest bound stable nucleus we know (B/A = 8.792 MeV/A).

- The high ground-state shell correction of the collision system (-14 MeV) together with a favourable mass ratio of only 3.6 allows to form a compound system with 20 MeV excitation energy only, (cold fusion).

- The strong binding of its nucleons and its high vibrational energy allow the closed shell nucleus ^{209}Bi to approach ^{58}Fe and to form a compact configuration without loosing its shell stabilization, that is without being intrinsically excited, $\overline{AB}$ in Fig. 20.

- The opening of the neck of the compact configuration creates only little excitation energy. This rapid process compared to mass asymmetry equilibration may lead to a breaking of a few neutron pairs. The system, now not far from the Coulomb barrier, is still cold, $\overline{BC}$ in Fig. 20.

- The system passes the barrier. Its excitation energy may be reduced by an emission of one neutron at this early state of amalgamation, $\overline{CD}$ in Fig. 20.

- The system with an excitation energy of (15-20) MeV is cold enough to profit from the shell stabilization of the deformed superheavy system ^{266}109. A γ-cascade finally cools the system to reach the ground state of ^{266}109, $\overline{DE}$ in Fig. 20.

- A superheavy nucleus in its ground state has been made. The high ground-state shell correction of the deformed superheavy nucleus protects it against spontaenous fission long enough to allow its disintegration by α-emission.

The formation of heavy elements by Pb-based reactions is a symphony of shell effects modifying the liquid drop behaviour in all aspects, shell effects in the collision partners, shell effects in the dynamics, and finally shell effects in the superheavy end product. We have been taught how to profit from highly organized nucleonic systems, how to transform this order into the new order of a superpheavy element, and we learnt that even its order is different from what we formerly thought a superheavy nucleus would be. The making of element 107 to 109 showed how to overcome the restrictions set by our early models of nuclei as liquid drops. It is a late triumph of the shell model. The idea of superheavy elements was challenging, but the complex interplay of shell effects in their realization showed a beauty nobody imagined in 1966, when the concept was borne.

Outlook to Make Still Heavier Elements

Actinide-based reactions produce hot compound systems, (40-50) MeV. Compared to cold fusion reactions their chance to survive fission is smaller by factors of 2×10^{-4} and 2×10^{-3} for a 5n- and 4n-channel, respectively. This larger chance to disintegrate during deexcitation may be compensated by a larger fusion probability compared to Pb-based reactions. e.g., the hindrance in the entrance channel reduces the production of ^{266}109 by a factor 10^4. The actinide-based reaction only has a chance to compete, if no entrance channel limitation is acting. The naive hope for unhindered fusion has not been strengthened by the latest theoretical results, Fig. 16, and the analysis given in the previous section. To go beyond Z = 109 using actinide-based reactions stays a gamble which may still be justified as long as the question of hindrance for these reactions is not finally solved.

The prospects to produce element 110 by a ^{208}Pb(^{64}Ni,n) reaction are easier to foresee. p(B$_B$) will decrease further. Cross sections of around a pb can be expected. An experiment performed by our group last year gave a lower limit of 5 pb^{72}. New separators to push cross-section limits to the pb-limit are needed. With cold fusion we may be rewarded by the extraordinary properties of ^{208}Pb once more.

Finally, we have to accept that the way to the discovery of further elements is blocked. We are going to understand why we arrived at our limits, but until now we do not know whether we may overcome these limits. We may have to accept a situation similar as for neutron rich nuclei: WE have made some of them, but we know we cannot make all of them. Our "Opus Magnum" is one atom of ^{266}109. It unites all what shell effects may achieve. Maybe the message is to accept the victory of the "soft way" gratefully and acknowledge the limitations set to the game.

REFERENCES

1. P. Armbruster, Ann. Rev. Nucl. Part. Sci. 35 (1985), 135-94
2. G. Münzenberg et al., Z. Phys. A324 (1986) 489
3. G. Münzenberg et al., Z. Phys. A322 (1985) 227
4. F.P. Heßberger et al., Z. Phys. A321 (1985) 317
5. A.H. Wapstra, G. Audi, Nucl. Phys. A432 (1985) 1
6. P. Møller, R.J. Nix, At. Data Nucl. Data Tables 26 (1981) 165
7. W.D. Myers, Droplet Model of Atomic Nuclei, New York, IFI/Plenum (1983)
8. H. von Groote et al., At. Data Nucl. Data Tables 17 (1976) 418
9. P.A. Seeger, W.M. Howard, At. Data Nucl. Data Tables 17 (1976) 428
10. S. Liran, N. Zeldes, At. Data Nucl. Data Tables 17 (1976) 431
11. P. Møller, private communication (1986)
12. P. Møller et al., Z. Phys. A323 (1986) 41
13. S. Cwiok et al., Nucl. Phys. A410 (1983) 254
14. M. Dahlinger et al., Nucl. Phys. A376 (1982) 94
15. J.O. Rasmussen, Phys. Rev. 113 (1959) 1593
16. W.J. Swiatecki, Phys. Rev. 100 (1955) 937
17. J. Randrup et al., Phys. Rev. C13 (1976) 229
18. Yu.Ts. Oganessian, Lect. Notes Phys. 33 (1974) 221
19. A. Baran et al., Nucl. Phys. A361 (1981) 83
20. J. Randrup, Nucl. Phys. A217 (1973) 221
21. E.K. Hulet et al., Phys. Rev. Lett. 56 (1986) 313
22. P. Møller et al., to be published Nucl. Phys.
23. W.D. Myers, W.J. Swiatecki, Nucl. Phys. 81 (1966) 1
24. V.M. Strutinsky, Nucl. Phys. A95 (1967) 420
25. H. Meldner, Proc. Int. Conf. "Nuclides far off the Stability Line (1966), Ark. Fys. 36 (1967) 593
26. T. Sikkeland, Proc. Int. Conf. "Nuclides far off the Stability Line" (1966), Ark. Fys. 36 (1967) 539
27. A. Sobiczewski et al., Z. Phys. A326 (1986) to be published
28. S. Polikanov et al., JETP42 (1962) 1464
29. A. Bohr, Physica Scripta 10A (1974) 52
30. K.-H. Schmidt et al., Proc. Symp. Phys. Chem. of Fission, Jülich 1979; Vienna IAEA 1 (1980) 409

31. C.C. Sahm et al., Nucl. Phys. A44 (1985) 316
32. A.J. Sierk and J.R. Nix, Proc. Symp. Phys. Chem. of Fission, Rochester (1973), Vol. II, p. 273, IAEA Vienna 1974
33. W.J. Swiatecki, Phys. Scr. 24 (1981) 113
34. J.P. Blocki et al., Nucl. Phys. A459 (1986) 145
35. K.T.R. Davies, A.J. Sierk and J.R. Nix, Phys. Rev. C28 (1983) 679
36. A. Ghiorso, in Actinides and Perspectives, ed. by M. Edelstein, Pacific Grove, Calif. (1982) 23
37. V.M. Vasko et al., Report P7-81-863 (1981) Dubna
38. F.P. Heßberger, Thesis, TH-Darmstadt (1984)
39. J.G. Keller et al., Nucl. Phys. A452 (1986) 173
40. W. Reisdorf et al., Nucl. Phys. A44 (1985) 154
41. B. Quint et al., GSI 86-1 (1986) 62
42. W. Reisdorf, Z. Phys. A300 (1981) 227
43. S. Cohen et al., Ann. Phys. 82 (1974) 557
44. R. Bass, Proc. Symp. Deep Inelastic and Fusion Reactions with Heavy Ions, Lect. Notes Phys. 117 (1980) 281
45. R. Bass, Nucl. Phys. A231 (1974) 45
46. H. Gäggeler et al., Z. Phys. A316 (1984) 291
47. A. Ghiorso et al., Phys. Rev. Lett. 33 (1974) 1490
48. P. Somerville et al., Phys. Rev. C31 (1985) 1801
49. Yu.Ts. Oganessian et al., Sov. Journ. At. Energy 28 (1970) 502
50. Yu.Ts. Oganessian, Nucl. Phys. A239 (1975) 353
51. J.M. Nitschke et al., Nucl. Phys. A313 (1979) 235
52. O.A. Orlova et al., JINR P 7-12061-Dubna (1978)
53. H. Gäggeler et al., to be published
54. G. Münzenberg, GSI-Report 86-19 (1986) 222
55. M.J. Rose and G.A. Jones, Nature 307 (1984) 245
56. P.B. Price and S.W. Barwick, to be published
57. A. Sandulescu et al., Sov. J. Part. Nucl. 11/6 (1980) 528
58. Yi-Jin Shi and W.J. Swiatecki, Phys. Rev. Lett. 54 (1985) 300
59. W. Mayer et al., Phys. Lett. B152 (1985) 162
60. P. Armbruster et al., GSI 84-1 (1984) 81;
    H. Gäggeler et al. GSI 85-1 (1985) 92
61. D.C. Hoffman, Phys. Rev. C21 (1980) 972
62. H. Faissner and K. Wildermuth, Nucl. Phys. 58 (1964) 177
63. U. Brosa et al., Z. Phys. A325 (1986) 241
64. W. John et al., Phys. Rev. Lett. 27 (1971) 45
65. C. Signarbieux et al., J. Physique Lett. 42 (1981) L437
66. H.-G. Clerc et al., Nucl. Phys. A452 (1986) 277
67. P. Armbruster, Lecture Notes in Physics 158 (1982) 1
68. J.F. Berger, Nucl. Phys. A428 (1984) 389
69. P. Armbruster et al., CERN 81-09 (1981) 675
70. P. Armbruster et al., Phys. Rev. Lett. 54 (1985) 406
71. G. Münzenberg et al., Z. Phys. A309 (1982) 89;
    G. Münzenberg et al., Z. Phys. A315 (1984) 145
72. G. Münzenberg et al., GSI 86-1 (1986) 29

QUASI-ELASTIC TRANSFER AND SUB-BARRIER FUSION

IN THE SYSTEMS ^{32,36}S+^{58,64}Ni

Alberto M. Stefanini

I.N.F.N. - Laboratori Nazionali di Legnaro
35020 Legnaro (Padova, Italy)

ABSTRACT

   Quasi-elastic transfer reactions between ^{32,36}S and ^{58,64}Ni have been
experimentally studied, measuring cross sections and Q-value distributions
for the various exit channels. The influence of such transfer channels on
the sub-barrier fusion mechanism is discussed. Proton stripping channels
are pointed out to be quite relevant in the case of ^{32}S+^{64}Ni.

INTRODUCTION

   A number of experimental studies is being currently devoted to quasi-
-elastic transfer reactions between heavy ions around the Coulomb barrier.
This is a field where the available experimental data are still quite
scarce, preventing thereby a clarification of the role of the transfer
channels in the enhancements[1] of the sub-barrier fusion cross sections,
although there is a general "theoretical" agreement on the significance of
that role. It is appearing clear[2] that a large fraction of the reaction
cross section in heavy systems is actually due to neutron transfer channels,
and this implies substancial modifications of the ion-ion potential at the
corresponding interaction distances[3]. Here I will present the measurements
of transfer cross sections recently performed[4,5] at Legnaro in the medium-
-heavy systems ^{32,36}S+^{58,64}Ni at three energies around the Coulomb barrier.
These experiments were triggered by the previous studies[6] on sub-barrier
fusion in the same systems, where large enhancements were found, as well as
large isotopic effects which were tentatively related[7,8] to the fact that
neutron transfer channels have positive ground state Q-values in some cases.

EXPERIMENTAL METHOD AND RESULTS

   The ^{32,36}S beams were delivered by the XTU Tandem accelerator of Legnaro National Laboratory at energies ranging from 94 MeV up to 112 Mev.
Self-supporting Ni targets were used, whose thicknesses were 235(265)μg/cm^2,

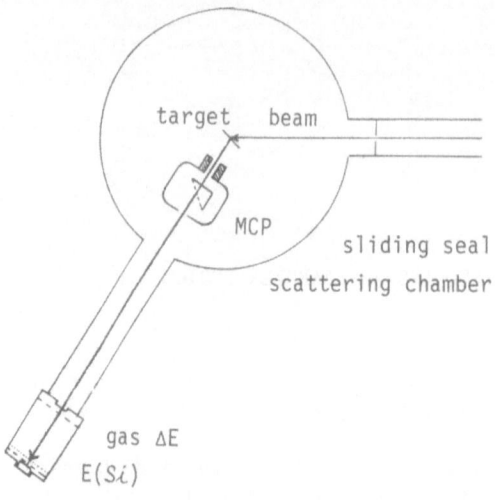

target   beam

MCP

sliding seal
scattering chamber

gas ΔE
E(Si)

Fig. 1. Experimental set-up.

with isotopic enrichments of 99.8 (96.7)% for $^{58(64)}$Ni. The experimental
set-up is sketched in fig. 1. It consisted of a sliding seal scattering
chamber, two monitor detectors symmetrically mounted at ±26° with respect
to the beam direction and a OF-ΔE/E counter telescope, detecting the
beam-like particles in the angular range $34° \leqslant \vartheta_{lab} \leqslant 111°$ by a suitable rotation of the chamber. The first element of the counter telescope was a
micro-channel plate time-zero detector, followed by an ionization chamber
for the ΔE signal and by a Silicon surface barrier detector giving the time
and the residual energy.

The mass and Z resolutions were such that the yields in the individual
transfer channels could easily be extracted. Fig. 2 shows two mass spectra
from the ^{32}S+^{64}Ni reaction, where one can see the stripping of up to 4
nucleons and the pick-up of 1, 2 and possibily 3 particles. The thickness
of the targets prevented us from observing the population of final quantal
states in the Q-value spectra of the various channels. Some examples of
Q-value spectra for one- and two-particle transfer channels, which will be
relevant in the discussion of the data, are reported in fig. 3. We note
that only a negligible part of the transfer cross section populates the
Q>0 ground states of the 2n-pick-up and of the 2n-stripping channel available in the ^{32}S+^{64}Ni and in ^{36}S+^{58}Ni respectively.

The angular distributions of the transfer channels are systematically
bell-shaped (we will see some example): from integrating those distributions
and the Q-value spectra we get the total transfer cross sections into the
different channels plotted in fig. 4. The sum of all channels at each
energy is also reported (marked with "tot"). The plotted errors are only
statistical; another ±15% is estimated to come as an average from angle
integration and a comparable uncertainty is connected with the absolute
normalization. The arrows mark the Coulomb barriers, as deduced from the
systematics of ref.[9], and the lines are only visual guides.

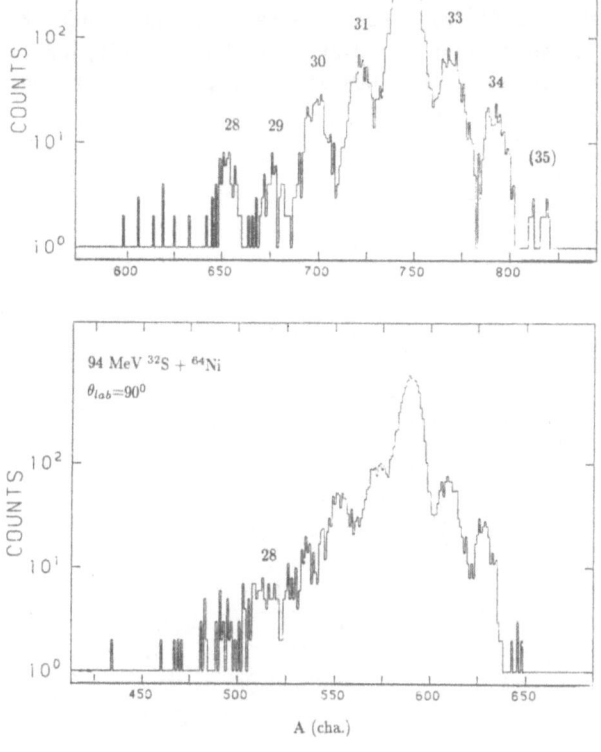

Fig. 2. Mass spectra of Sulphur-like reactions
products (logarithmic count scale).

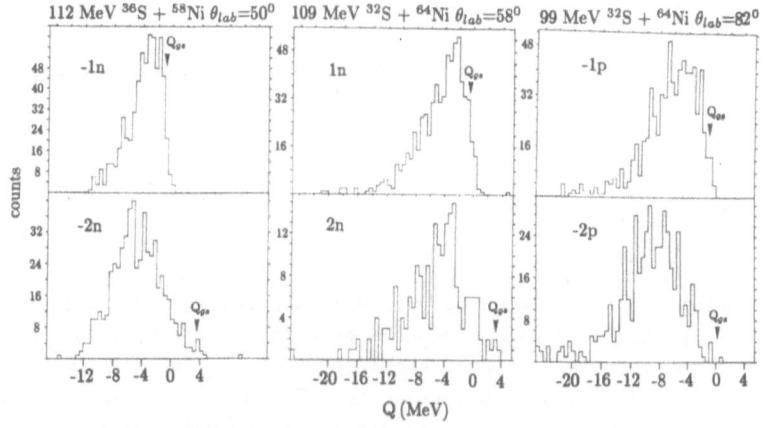

Fig. 3. Q-value distributions of one- and two-nucleon transfer channels.
The arrows mark the ground state Q-values. Here and in the following figures, -1n indicates the stripping of one neutron on
the Ni target and 1n is the corresponding pick-up; the notation
is similar for the other channels.

There is a basic and important difference in the transfer strength
distribution between the ^{32}S- and the ^{36}S-induced reactions: in these
latter cases neutron transfer dominates, summing up to 84% of the total
as an average, whereas strong charged particle stripping channels are
observed in both ^{32}S+^{58,64}Ni systems where they constitute 72% of the
total transfer cross section as an average. The α-stripping is the second
strongest channel in ^{32}S+^{58}Ni. Qualitatively, we observe that in these

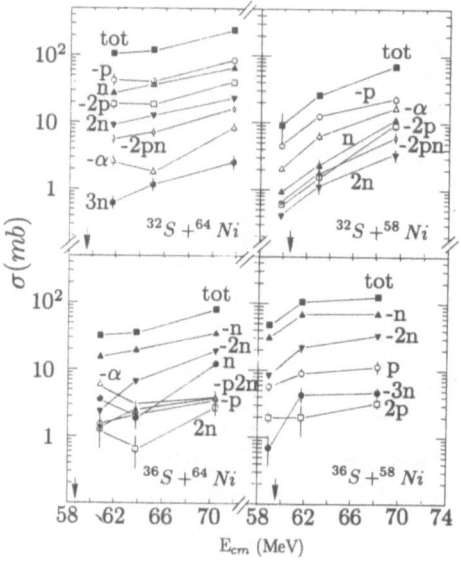

Fig. 4. - Angle- and Q-integrated transfer cross sections
measured in the present work.

systems which are not so much heavy, Coulomb effects do not deplete at all
the charged particle transfer channels even in the vicinity of the barrier.
In view of the observed strong isotopic differences, we can predict that
the paths leading to fusion may be quite different in the various systems;
the existence of the proton (and α) stripping channels has to be taken into
account, not forgetting that the transfer of a charged particle from the
light to heavy collision partner automatically reduces the Coulomb barrier.

OPTICAL MODEL ANALYSIS

Calculations were done within the optical model, using the code
Ptolemy[10] where the usual Woods-Saxon parametrization of the ion-ion
potential is made. The (unresolved) elastic and inelastic scattering
angular distributions were calculated, <u>not</u> fitted, in the coupled-channel
approach. The lowest 2^+ and 3^- states of both colliding nuclei were individually coupled to the corresponding ground states with strengths taken
from the experimental $B(E\lambda)$'s. The potential parameters were taken from
ref.[11] where elastic scattering measurements for $^{28}Si+^{58,62}Ni$ around
20 MeV above the Coulomb barrier are reported; those parameters are
V=50 MeV, W=10 MeV, $a_R=a_I$=0.65 fm and $r_R=r_I$=1.22 fm (=1.20 fm at the
lowest energy for each system). The experimental data and the calculations

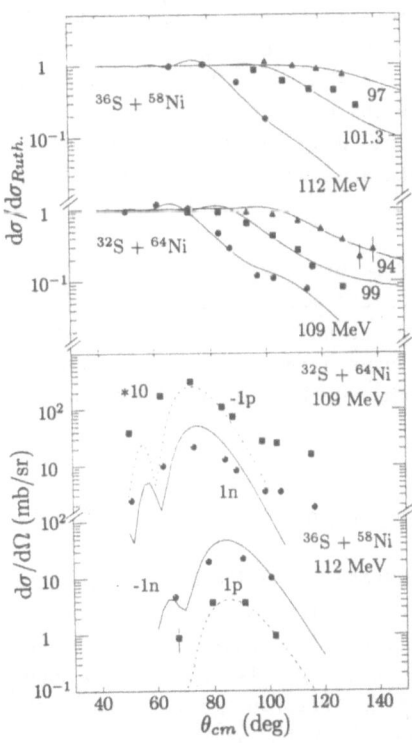

Fig. 5. Elastic plus inelastic (upper part) and
one-nucleon transfer (lower part) angular
distributions. Experimental errors are
only statistical. The lines are the optical
model calculations described in the text.

are shown in fig. 5 (upper part) for two systems and a reasonable agreement is seen; the need of better elastic scattering data is also clear.

The same potential was then used to calculated DWBA one-nucleon transfer cross sections (some selected cases are shown in fig. 5, lower part)). For each reaction, the spectroscopic factors were derived from the corresponding light particle induced transfers[12] and a sum over a number of final states was done covering the measured Q-value distribution. Also, states populated with less than 5% of the full shell model strength in the light ion-induced reaction, were excluded from the sum. Where spectroscopic factors are not available (^{35}S, ^{37}Cl), the full shell model strength was assigned to the lowest states of a particular spin and parity.

As the potential was chosen "a priori", rather than by any fit to the elastic and inelastic scattering data, I believe that any comparison of the DWBA cross sections with the measured ones cannot exceed the qualitative level. Anyway, the situation is such that the angle-integrated one-neutron transfer data are systematically overpredicted by 20% or more and the largest discrepancies (up to factors 2.5) are found at the lowest energies, except for ^{36}S+^{64}Ni where the data are underpredicted by 20% (a factor 2 at the lowest energy). On the contrary, the one-proton transfer DWBA cross sections are within ±20% in agreement with the experiment, except for ^{32}S+^{58}Ni where the data are underpredicted by factors around 2 at all energies.

The inelastic and reaction cross sections, calculated in the coupled--channel approach described above, are reported in Table 1, together with the experimental total transfer cross sections and the fusion cross sections previously measured for the same systems. The influence of the transfer channels on the sub-barrier fusion mechanisms is the last topic I am going to cover.

TRASNFER AND SUB-BARRIER FUSION

Let us consider ^{32}S+^{64}Ni and ^{36}S+^{58}Ni, i.e. the two systems where two--neutron transfer channels with positive Q-values (+3.75 MeV and + 3.35 MeV respectively) are available, and whose different sub-barrier fusion cross sections are not explained[6] by only taking into account the couplings to the low-lying inelastic excitations (i.e. by accounting for the smaller "softness " of ^{36}S and ^{58}Ni with respect to the other two nuclei). Speci-fically, the fusion cross sections of ^{32}S+^{64}Ni are almost one order of magnitude larger than those of ^{36}S+^{58}Ni when plotted vs. scaled energies: this is shown in fig. 6. Alternatively, one can say that the actual barrier of ^{32}S+^{64}Ni is 1.2-1.5 MeV lower than in the other system.

On the other hand, very similar total transfer cross sections are measured in the two cases, so that a simple correlation between total transfer yield and probability of sub-barrier fusion does not hold there. It does, by the way, comparing with ^{32}S+^{58}Ni which is also shown in fig. 6.

Table 1. Total transfer cross-sections measured in this work, fusion cross
sections[6], coupled-channel inelastic cross sections (including
the lowest 2^+ and 3^- states of the two colliding nuclei); σ_r are
the sums of the three preceding quantitites and σ_r^{cc} are the
reaction cross sections calculated with the same coupled-channel
procedure indicated in the text.

		$^{32}S+^{64}Ni$			$^{32}S+^{58}Ni$		
E_{cm} (MeV)		71.7	65.0	61.7	69.5	63.0	59.9
σ_{tr} (mb)		241±52	121±26	106±25	70±16	26±6	9±5
σ_{fus} (mb)		469±92	237±41	140±29	307±76	106±28	26±6
σ_{in} (mb)		257	214	180	236	179	143
σ_r (mb)		956±106	571±49	426±38	613±78	311±29	178±8
σ_r^{cc} (mb)		997	675	437	780	445	250

		$^{36}S+^{58}Ni$			$^{36}S+^{64}Ni$		
E_{cm} (MeV)		68.2	61.7	58.9	70.4	63.8	60.8
σ_{tr} (mb)		127±27	106±25	48±10	80±18	36±9	32±7
σ_{fus} (mb)		256±52	107±21	30±7	449±91	291±53	153±30
σ_{in} (mb)		153	122	99	172	149	130
σ_r (mb)		539±59	335±33	177±12	701±93	476±54	315±31
σ_r^{cc} (mb)		769	420	226	910	567	397

A more quantitative analysis of the correlation was therefore tried.
Representative coupling strengths were extracted from the measured transfer
cross sections and Q-value distributions for the various channels, assuming that they do not change qualitatively at low energies where the
transfer data are missing, and neglecting any mismatching of fusion and
transfer in the angular momentum space (a feature which should be absent
at low energies).

Based on the analogy between the pairing vibration and the collective
surface vibration, a compact formulation of the two-nucleon transfer formfactor was derived by C.H. Dasso et al.[13]. In the present work it was used
both for two- and for one-nucleon transfer channels, although being formally incorrect in the latter cases where less steep radial slopes of the
formfactors are expected. Indeed, preliminary results from experiments at
ORNL involving heavier systems[14] show quite similar radial dependences of
one- and two-neutron formfactors, possibly due to the pairing interaction.
The present data give a hint that such a situation holds as well in the
S+Ni systems[15], even if more detailed measurements are needed here.

All the one- and two-nucleon transfer channels were thus assimilated[16] to inelastic excitations (with multipolarities 0^+). The centroids of the measured Q-value distributions defined their energies and the DWBA calculations were performed using the same potential parameters listed above, and no Coulomb couplings. The strength or "deformation" parameters β was varied to get agreement between the calculated cross sections and the experiment. The resulting coupling strengths were found to depend weakly on the bombarding energy, and thus averaged over that parameter. As the flux going into the available Q>0 channels is very small (see fig. 3 again), very small coupling strengths result for those particular states, whose influence on the fusion process is then predicted to be negligible.

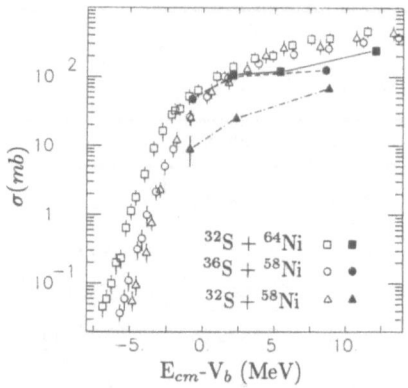

Fig. 6. Fusion[6] and total transfer cross sections vs. the energy distance from the barrier[9]. Lines are only visual guides.

The fusion cross sections were calculated in the constant coupling approximation[7,8] using the proximity potential[17] and the WKB transmission coefficients below the barrier. The nuclear radius parameter was adjusted to best fit the experimental cross sections above the barriers. The lowest 2^+ and 3^- levels were chosen to represent the contributions of collective surface excitations to the fusion, with coupling strengths derived from the collective model formulation. The deformation length were extracted[6,18] from the experimental B(Eλ) to the ground state.

Fig. 7 shows the results of this analysis for two of the systems (full lines). The measured fusion and total transfer cross sections are also reported, as well as the results of coupling the inelastic excitations only (dashed lines) and the no-coupling limits (dotted lines, one-dimensional

barrier penetration). The good agreement with the data receives a substantial contribution from the transfer channels in both systems, but meaningful
differences are there. In fact, whereas proton pick-up effects are negligible in ^{36}S+^{58}Ni where neutron transfer accounts for about 50% of the
asymptotic barrier reduction, proton stripping is very important in
^{32}S+^{64}Ni. Also, inelastic excitations have a comparatively more dominant

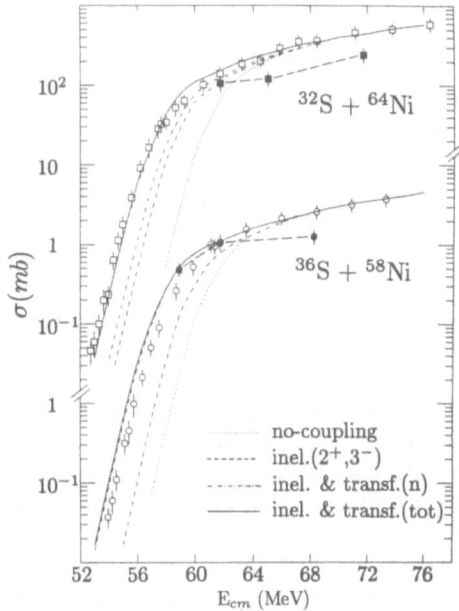

Fig. 7. Experimental (open symbols) and calculated fusion cross
sections, and the total transfer cross sections (full
symbols connected by long-dashed lines). Coupling only
one- and two-neutron transfers, besides the inelastic
channels, yields the dash-dotted lines.

role in this latter case, whose relative sub-barrier fusion enhancement
(fig. 6) seems therefore to be built on surface collective modes and charged
particle stripping channels. It is also to be noted as a systematic feature
that the calculated effects of one-particle transfers are smaller than those
of the corresponding two-particle channels, althoug these have lower cross
sections. This feature is a consequence of the different Q-value distributions which are in all cases centered at less negative values for one-
-nucleon transfer channels.

ACKNOWLEDGMENTS

This work results from the collaboration of many colleagues, whose names I am pleased to list here: G. Montagnoli, G. Fortuna, R. Menegazzo, S. Beghini, C. Signorini, A. De Rosa, G. Inglima, M. Sandoli, F. Rizzo, G. Pappalardo, G. Cardella. Several discussions with W. Henning, C.H. Dasso, S. Landowne and R.A. Broglia were quite fruitful. Grateful thanks are also due to the Legnaro Tandem staff and to all personnel of LNL.

REFERENCES

1. M. Beckerman, Phys. Rev. 129:145 (1985).
2. K. E. Rehm, Proc. Symp. on The Many Facets of Heavy-Ion Fusion Reactions, March 24-26, 1986, Argonne National Laboratory (USA), Rep. ANL-PHY-86-1, p. 27.
3. K. E. Rehm, F. L. H. Wolfs, A. M. Van den Berg and W. Henning, Phys. Rev. Lett. 55:280 (1985).
4. G. Montagnoli, A. M. Stefanini, G. Fortuna, R. Menegazzo, S. Beghini, C. Signorini, A. De Rosa, G. Inglima, M. Sandoli, F. Rizzo, G. Pappalardo, Proc. Symp. on The Many Facets of Heavy-Ion Fusion Reactions, March 24-26, 1986, Argonne National Laboratory (USA) Rep. ANL-PHY-86-1, p. 555.
5. A. M. Stefanini, G. Montagnoli, G. Fortuna, R. Menegazzo, S. Beghini, C. Signorini, A. De Rosa, G. Inglima, M. Sandoli, F. Rizzo, G. Pappalardo, G. Cardella, Phys. Lett. B, in press.
6. A. M. Stefanini, G. Fortuna, R. Pengo, W. Meczynski, G. Montagnoli, L. Corradi, A. Tivelli, S. Beghini, C. Signorini, S. Lunardi, M. Morando and F. Soramel, Nucl. Phys. A456:509 (1986).
7. R. A. Broglia, C. H. Dasso, S. Landowne and G. Pollarolo, Phys. Lett. 133B:34 (1983).
8. C. H. Dasso, S. Landowne and A. Winther, Nucl. Phys. A405:381 (1983); Nucl. Phys. A407:221 (1983).
9. L. C. Vaz, J. M. Alexander and G. R. Satchler, Phys. Rep. 69:373 (1981).
10. M. H. Macfarlane and S. C. Pieper, Argonne National Laboratory Report No. ANL-76-11 (Rev. 1), 1978.
11. Y. Sugiyama, Y. Tomita, H. Ikezoe, K. Ideno, N. Shikazono, N. Kato, H. Fujita, T. Sugimitsu, S. Kubono, Phys. Lett. B176:302 (1986).
12. P. M. Endt, Nucl. Data Sheets 19:23 (1977); ibid. 39:641 (1983); ibid. 28:559 (1979); T. W. Burrows and M. R. Bhat, ibid. 47:1 (1986); N. J. Ward and J. K. Tuli, ibid. 47:135 (1986).
13. C. H. Dasso and G. Pollarolo, Phys. Lett. B155:223 (1985); R. A. Broglia, C. H. Dasso and S. Landowne, Phys. Rev. C32:1426 (1985).
14. S. Juutinen, X. T. Liu, S. Sørensen, B. Cox, R. W. Kincaid, C. R. Bingham, M. W. Guidry, W. J. Kernan, C. Y. Wu, E. Vogt, D. Cline, M. L. Halbert, I. Y. Lee, C. Baktash, preprint, September 1986.

15. A. M. Stefanini et al., to be published.
16. S. C. Pieper, M. J. Rhoades-Brown and S. Landowne, Phys. Lett. B162:43 (1985).
17. J. Blocki, J. Randrup, W. J. Swiatecki, C. F. Tsang, Ann. of Phys. 105:427 (1977).
18. R. A. Broglia, and F. Barranco, Proc. Int. Conf. on Fusion Reactions below the Coulomb Barrier, June 13-15, 1984, Cambridge, MA (USA).

TOWARDS A MICROSCOPIC DESCRIPTION OF

HEAVY ION DYNAMICS

M. Baldo and A. Rapisarda

INFN Sez. Catania
Corso Italia 57
95129 Catania, Italy

INTRODUCTION

The enormous effort in interpreting and understanding the experimental data on heavy ion reactions at low energy has produced in the past years a large variety of theoretical models. They range from the most microscopic ones, like the Time Dependent Hartree Fock (TDHF), to the most phenomenological ones, like those based on classical calculations with phenomenological frictional forces. Despite this effort many questions appear to be still open, both because the experimental data seem not to be able to sharply distinguish among different models and because a simple physical picture of the relevant processes involved has not yet fully emerged.

At low energy one can expect that the evolution of the system is dominated by the mean field, because the average velocity (essentially the Fermi velocity) of the nucleons is much larger then the two ions relative velocity. Based on this consideration, the TDHF model[1] has been introduced and applied to the description of few fusion and deep-inelastic reactions. Because of the assumed dominance of the mean field, in the TDHF model the wave function of the entire system is assumed to be a single Slater determinant, whose orbitals are obtained by solving numerically the corresponding (self-consistent) TDHF equations, starting from static Hartree-Fock solutions, boosted in order to get the correct initial conditions. The model appears to be able to describe the gross features of the considered reactions in a few respects, like the total mean energy loss, the average scattering angles, the trend in the fusion cross sections etc. However, besides the fact that it has not been systematically applied to the analysis of experimental data, it does not shed light on the main mechanisms underlying the reaction processes. In particular it is not clear in this model how the reaction mechanisms evolve from the quasi-elastic regime, for which the theoretical methods are well established and a great amount of experimental data are avaliable, into the deep-inelastic one. Furthermore, because of the large computing time needed, in practical calculations one is obliged to use an oversimplified description of the two ions nuclear structure, like the level distributions and the nucleon-nucleon forces. Finally it is not yet completely clear to which extent the nucleon-nucleon collisions can affect the results and in particular can explain the width of many physical quantities (energy loss, mass distribution, etc.), essentially absent in the TDHF calculations.

A simpler model[2], which tries to isolate the main physical processes relevant for heavy ion collisions, has been proposed by the Copenhagen group. In this model the semiclassical macroscopic description of inelastic excitation in the quasi-elastic region is extended to the deep-inelastic case by solving a large set of classical coupled equations of motion. The energies and strenghts of each excitation is taken from microscopic R.P.A. calculations. The identity of the two nuclei is preserved throughout the reaction process. More details on this model can be found in the lectures by G.Pollarolo at this school, where success and failures of the model can also be found. In this treatment the transfer processes are considered in a completely classical way, following the so-called "proximity model".

The proximity friction, arising from the exchange of particles, is the basis of a simple classical model[3] for heavy ion reactions. It has been developed a few years ago and is based on the assumption that exchange of particles is the main mechanism for energy and angular momentum dissipation. This dissipation is essentially of the one-body type. The motion of a particle which is transferred from one nucleus to the other is rapidly randomized by the mean field, which is assumed to be irregular enough for this kind of randomization to occur. The proximity treatment has been systematically applied, with fairly good success, for calculations on fusion cross sections of light-medium nuclei[4]. The most successful application of the model seems to be the explanation of the correlation between the mass or angular momentum distribution width and the energy loss in deep-inelastic collisions[5]. It is not yet clear, however, if it is able to reproduce also the other features of the collisions. Furthermore the theoretical foundations of this model has not been explicitly studied within a full quantal microscopic treatment.

2. THE GENERAL SCHEME

Our general aim is to develop a model which should be able to describe in a unified microscopic way heavy ion collisions from the quasi-elastic region down to the deep-inelastic regime. As mentioned in the introduction, at low energies the mean field is expected to dominate the time evolution of the system, whereas the collisions should affect the fluctuations of different quantities around the average values.

For peripheral collisions the mean field of each colliding nucleus is only slightly distorted during the reaction. So in this case one can have a simple picture of the collision by considering two undistorted potential wells moving along a classical trajectory, according to a suitable ion-ion potential. The justification of the classical approximation is of course the extremely short wave lenght of the two ions in the relative motion. In this picture, the particles move independently in the time-dependent undistorted potential produced by the sum of the two potential wells. Thus one is led to solve the quantal evolution of the single-particle wave functions describing the internal structure of the two nuclei. This Time Dependent Single-Particle Model (TDSPM) has already been introduced in Ref.6, where it has been shown to be, in principle, an approximation to TDHF. However the TDSPM, although assuming an undistorted mean field, can make use of realistic single-particle nuclear levels and formfactors. In order to extend this model from the quasi-elastic region to the deep-inelastic one, it is necessary to include the mean field distortions in the description of the particle motion, as well as, possibly, the collision processes and self-consistency for the relative motion. From these considerations the following approximation scheme emerges, distinguishing three steps:

i) Solve the single-particle dynamics for the single-particle Hamiltonian ($\hbar = 1$)

$$H = -\frac{\nabla^2}{2m} + V_1(t) + V_2(t) \quad ,$$

where V_1, V_2 are the two potential wells moving along the classical trajectory. Of course, the introduction of self-consistency in the trajectory is trivial.

ii) Add the distortions of the mean field, i.e. consider the residual interaction among the particles. This can be done, for instance, within the R.P.A. scheme.

iii) Add collision terms in the evolution. The simplest way to do it is the relaxation time approximation, as described in the sequel.

We shall focus this presentation on step i), which allows a close comparison with the classical proximity treatment. This restrict the discussion to peripheral collisions. We shall present a few results also for the part iii), while for step ii) only the formal development is introduced in the Appendix.

3. THE SINGLE-PARTICLE EVOLUTION

In this section we develop the formalism for the TDSPM in the framework of a discrete basis formulation, where only the bound states associated with each potential well are introduced. At low energies this is fully justified, as discussed in Ref.7. Our formalism is closely related to the semiclassical approach to heavy ion collisions of Ref.8 and to the second-quantized formalism of Ref.9.

Let us consider two potential wells moving along a classical relative trajectory. Then the Hamiltonian of the system can be written

$$H(t) = -\frac{\nabla^2}{2m} + V_1(\vec{r} - \vec{R}_1(t)) + V_2(\vec{r} - \vec{R}_2(t)) \quad , \quad (1)$$

where R_1 and R_2 are the positions of the centers of potential 1 and 2 respectively. In Ref.6 the time-dependent Schroedinger equation corresponding to the Hamiltonian (1) has been solved numerically in coordinate space. In our approach we shall use a discrete basis set of single particle wave functions to describe a particle moving in the combined potential of the two wells. As it will be shown, for the collisions we are going to consider this is an enough large basis. Furthermore it allows a better separation between transfer and inelastic processes. The most convenient basis set is provided by boosting, with a generalized Galilei transformation, the proper bound state wave functions of each well. Thus, if we call $\{\Phi_{b_1}(\vec{r})\}$ and $\{\Phi_{b_2}(\vec{r})\}$ the set of bound eigenstates of potential 1 and 2

$$\left[-\frac{\nabla^2}{2m} + V_1(\vec{r})\right] \Phi_{b_1}(\vec{r}) = E_{b_1} \Phi_{b_1}(\vec{r}) \quad ,$$

(2)

$$\left[-\frac{\nabla^2}{2m} + V_2(\vec{r})\right] \Phi_{b_2}(\vec{r}) = E_{b_2} \Phi_{b_2}(\vec{r}) \quad ,$$

the corresponding orbitals in the "moving frame" are obtained according to

$$\Psi_{b_1}(\vec{r},t) = \exp\left[i(m\dot{\vec{R}}_1\vec{r} - \gamma_1(t) - E_{b_1}t)\right] \Phi_{b_1}(\vec{r}-\vec{R}_1(t)) \quad ,$$

(3)

$$\Psi_{b_2}(\vec{r},t) = \exp\left[i(m\dot{\vec{R}}_2\vec{r} - \gamma_2(t) - E_{b_2}t)\right] \Phi_{b_2}(\vec{r}-\vec{R}_2(t)) \quad ,$$

where

$$\gamma_1(t) = \int^t\left[\tfrac{1}{2}m\dot{\vec{R}}_1^2(t) + m\ddot{\vec{R}}_1(t')\vec{R}_1(t')\right] dt' \quad , \quad \gamma_2(t) = \int^t\left[\tfrac{1}{2}m\dot{\vec{R}}_2^2(t') + m\ddot{\vec{R}}_2(t')\vec{R}_2(t')\right] dt' \quad (4)$$

being m the nucleon mass. These orbitals satisfy the time dependent Schroedinger equations ($\hbar = 1$)[(*)]

$$i\frac{\partial}{\partial t}\Psi_{b_1}(\vec{r},t) = \left[-\frac{\nabla^2}{2m} + V_1(\vec{r}-\vec{R}_1(t)) - m\ddot{\vec{R}}_1(\vec{r}-\vec{R}_1(t))\right]\Psi_{b_1}(\vec{r},t) \quad ,$$

$$i\frac{\partial}{\partial t}\Psi_{b_2}(\vec{r},t) = \left[-\frac{\nabla^2}{2m} + V_2(\vec{r}-\vec{R}_2(t)) - m\ddot{\vec{R}}_2(\vec{r}-\vec{R}_2(t))\right]\Psi_{b_2}(\vec{r},t) \quad .$$

(5)

The acceleration terms appearing in Eqs. (5) take into account the non-inertial force coming from the non-uniformity of the motion. For a constant velocity motion Eq. (3) amounts to a proper Galilei transformation.

Let us then consider a particle moving in the two potential wells. Since we are planning to consider transfer processes, its wave function $\Psi(t)$ has a component in the potential 1 and a component in the potential 2, and thus it can be expanded according to

---

(*) Throughout tha paper we shall use the units corresponding to $\hbar = c = 1$, MeV.

$$\Psi(t) = \sum_{b_1} c_{b_1}(t)\, \Psi_{b_1}(t) + \sum_{b_2} c_{b_2}(t)\, \Psi_{b_2}(t) \tag{6}$$

and has to satisfy the time dependent Schroedinger equation

$$i\frac{\partial}{\partial t}\Psi(t) = \left[-\frac{\nabla^2}{2m} + V_1(\vec{r}-\vec{R}_1(t)) + V_2(\vec{r}-\vec{R}_2(t))\right]\Psi(t) \tag{7}$$

Inserting (6) into (7) and by using Eqs. (5), one gets

$$i\left(\sum_{b_1}\dot{c}_{b_1}\Psi_{b_1} + \sum_{b_2}\dot{c}_{b_2}\Psi_{b_2}\right) = \sum_{b_1}\left[V_2 + m\ddot{\vec{R}}_1(\vec{r}-\vec{R}_1)\right]\Psi_{b_1} + \sum_{b_2}\left[V_1 + m\ddot{\vec{R}}_2(\vec{r}-\vec{R}_2)\right]\Psi_{b_2} \tag{8}$$

We then take the scalar products of Eqs. (9) with the basis set $\{\Psi_{b_1}\}$ and $\{\Psi_{b_2}\}$, in order to get a set of coupled differential equations for the amplitudes c_{b_1} and c_{b_2} . In doing that we have to keep in mind that the basis functions satisfy

$$(\Psi_{b_1}, \Psi_{b_1'}) = \delta_{b_1 b_1'}, \quad (\Psi_{b_2}, \Psi_{b_2'}) = \delta_{b_2 b_2'}, \quad (\Psi_{b_1}, \Psi_{b_2}) = \mathcal{E}_{b_1 b_2}, \tag{9}$$

where we have denoted by $\mathcal{E}_{b_1 b_2}$ the overlaps between functions of the well 1 and of the well 2. In general, in fact, the functions Ψ_{b_1} and Ψ_{b_2} are not orthogonal, and their overlaps vanish only asymptotically at large distance. We then get

$$i\dot{c}_{b_1} + i\sum_{b_2}\dot{c}_{b_2}\mathcal{E}_{b_1 b_2} = \sum_{b_1'}\left[(b_1, V_{b_2} b_1') + m\ddot{\vec{R}}_1(b_1,(\vec{r}-\vec{R}_1) b_1')\right]c_{b_1'} +$$
$$+ \sum_{b_2}\left[(b_1, V_{b_1} b_2) + m\ddot{\vec{R}}_2(b_1,(\vec{r}-\vec{R}_2) b_2)\right]c_{b_2}$$

$$i\dot{c}_{b_2} + i\sum_{b_1}\dot{c}_{b_1}\mathcal{E}_{b_1 b_2} = \sum_{b_2'}\left[(b_2, V_1 b_2') + m\ddot{\vec{R}}_2(b_2,(\vec{r}-\vec{R}_2) b_2')\right]c_{b_2'} + \tag{10}$$
$$+ \sum_{b_1}\left[(b_2, V_2 b_1) + m\ddot{\vec{R}}_1(b_2,(\vec{r}-\vec{R}_1) b_1)\right]c_{b_1}$$

This set of equations can be put in a more compact form by introducing the overlap matrix

$$G(t) = \begin{pmatrix} \mathbb{1}_1 & \mathcal{E} \\ \mathcal{E}^+ & \mathbb{1}_2 \end{pmatrix} \tag{11}$$

and the vector $C = \begin{pmatrix} c_{b_1} \\ c_{b_2} \end{pmatrix}$. Then Eqs. (10) read

$$i\,G(t)\,\dot{C} = \left[\mathcal{V}(t) + A(t)\right]C \tag{12}$$

In Eq. (11), $\mathbb{1}_1$, $\mathbb{1}_2$ denote the unit matrices in the subspaces of $\{\Psi_{b_1}\}$ and $\{\Psi_{b_2}\}$ respectively. Furthermore

$$\mathcal{V} = \begin{pmatrix} W_1 & Z \\ \tilde{Z} & W_2 \end{pmatrix} \quad , \quad A = \begin{pmatrix} a_1 & 0 \\ 0 & a_2 \end{pmatrix}$$

$$(W_1)_{b_1 b_1'} = (b_1, V_2\, b_1') \quad , \quad Z_{b_1 b_2} = (b_1, V_1 b_2) + m\ddot{\vec{R}}_2\,(b_1, (\vec{r}-\vec{R}_2)\,b_2) \quad , \text{(13)}$$

$$(W_2)_{b_2 b_2'} = (b_2, V_1\, b_2') \quad , \quad \tilde{Z}_{b_2 b_1} = (b_2, V_2 b_1) + m\ddot{\vec{R}}_1\,(b_2, (\vec{r}-\vec{R}_1)\,b_1)$$

and

$$(a_1)_{b_1 b_1'} = m\ddot{\vec{R}}_1\,(b_1, (\vec{r}-\vec{R}_1)\,b_1') \quad , \quad (a_2)_{b_2 b_2'} = m\ddot{\vec{R}}_2\,(b_2, (\vec{r}-\vec{R}_2)\,b_2') \tag{14}$$

are the acceleration terms. For convenience, we have included also the acceleration term in Z and $\tilde{Z}$.

The matrices W_1 , W_2 describe processes in which the particle is scattered inside each potential well, while Z , $\tilde{Z}$ describe processes in which the particle jumps from one potential well to the other. In the present model they are readily identified with the inelastic and transfer formfactors respectively.

Having assumed the linear independence of the basis functions $\{\psi_{b_1}\}$, $\{\psi_{b_2}\}$, we can rewrite Eq. (12) in the form

$$i\,\dot{c} = G^{-1}\left[\mathcal{V}(t) + A(t)\right]c \quad , \tag{15}$$

where G^{-1} can be easily found to be

$$G^{-1} = \begin{pmatrix} (1-\varepsilon\varepsilon^+)^{-1} & -(1-\varepsilon\varepsilon^+)\varepsilon \\ -(1-\varepsilon^+\varepsilon)^{-1}\varepsilon^+ & (1-\varepsilon^+\varepsilon)^{-1} \end{pmatrix} . \tag{16}$$

In the second-quantized formalism of Ref. 9 one introduces a dual basis in order to treat the problem of non-orthogonality. In this way the annihilation operators are not the hermitean conjugate of the creation operators. We shall follow another procedure and consider a "symmetrized representation" which will be useful in the applications. This is also suggested by the fact that the matrix $\mathcal{V}$ is non-hermitean, because in general $Z \neq \tilde{Z}^+$. Actually, by using Eqs. (5), one can easily verify that the matrix elements appearing in Z and $\tilde{Z}$ are related by

$$Z_{b_1 b_2} - (\tilde{Z}^+)_{b_1 b_2} = -i\,\dot{\varepsilon}_{b_1 b_2} \quad . \tag{17}$$

A similar relationship was derived in Ref. 8. Thus only asymptotically, where the overlap is vanishing small the potential $\mathcal{V}$ is hermitean.

Let us introduce the vector $d = G^{1/2}c$, which is well defined because G is a positive definite matrix. Then by using (17), Eq. (15) can be rewritten in terms of d according to

$$i\,\dot{d} = \tfrac{1}{2}\,i\left[\dot{G}^{1/2},\,G^{-1/2}\right]d + G^{-1/2}\left[\tfrac{1}{2}(\mathcal{V}+\mathcal{V}^+) + A\right]G^{-1/2}d \quad . \tag{18}$$

In this symmetrized form the interaction matrix is clearly hermitean.

We shall consider in this paper only peripheral collisions, where the overlaps are at most of the order 0.1. A fairly good approximation can be then to neglect in Eq. (18) all the terms of order $\mathcal{E}^2$ and $\mathcal{E}\dot{\mathcal{E}}$. With the help of (16) one can easily verify that to this order

$$\left[\dot{G}^{1/2}, G^{-1/2}\right] \simeq 0 \quad ,$$

(19)

$$G^{-1/2}\tfrac{1}{2}(\mathcal{V}+\mathcal{V}^+)G^{-1/2} \simeq \begin{pmatrix} W_1 & V_t + \tfrac{1}{4}(W_1\mathcal{E}+\mathcal{E}W_2) \\ V_t^+ + \tfrac{1}{4}(\mathcal{E}^+W_1+W_2\mathcal{E}^+) & W_2 \end{pmatrix} \equiv \mathcal{H} \quad , \quad V_t = \tfrac{1}{2}(Z+\tilde{Z}^+).$$

Eq. (18) then reduces to

$$i\dot{d} = \left(\mathcal{H} + A\right)d \quad ,$$

(20)

where we have considered also the acceleration terms to be of order $\mathcal{E}$.

If one considers only transfer processes, as in most of our applications, Eq. (20) further simplifies as

$$i\dot{d} = \mathcal{V}_t\, d$$

$$\mathcal{V}_t = \begin{pmatrix} 0 & V_t \\ V_t^+ & 0 \end{pmatrix} \quad .$$

(21)

Till now we have considered the evolution of the wave function of a single particle.

The many-body case, in the independent-particle picture we are considering, can be, in the most simple way, treated by considering the single-particle density matrix ρ , which we define by:

$$\rho_{bb'}(t) = \langle \Psi(t)|a_{b'}^+ a_b|\Psi(t)\rangle \quad ,$$

(22)

where $a^+ a$ are the creation and annihilation operators for the basis states $\{\Psi_b\}$.

According to Thouless theorem, if we assume the original total wave function to be a Slater determinant, it will be a Slater determinant all the time and each time-dependent orbital has components satisfying Eq. (15). The equation of motion of ρ can then be obtained accordingly. It is however again convenient to use the symmetrized representation and introduce the density matrix

$$\bar{\rho} = G^{-1/2}\, \rho\, G^{-1/2} \quad .$$

(23)

Then $\bar{\rho}$ can be expressed as

$$(\bar{\rho})_{bb'} = \sum_{b''b'''}(G^{-1/2})_{bb''}\sum_{\beta}(\Psi_{b''},\Psi_\beta)(\Psi_\beta,\Psi_{b'''})(G^{-1/2})_{b'''b'} =$$

$$= \sum_{b''b'''}(G^{1/2})_{bb''}\sum_{\beta} c_{b''}^\beta c_{b'''}^{\beta*}(G^{1/2})_{b'''b'} = \sum_\beta d_b^\beta(t)\, d_{b'}^{\beta*}(t) \quad ,$$

(24)

where $\{c^\beta\}$ and $\{d^\beta\}$ are the component vector of the orbital Ψ_β in the original and symmetrized representation respectively. Thus the index β indicates the initial condition for Ψ i.e , $\Psi_\beta(-\infty)=\Psi_\beta(-\infty)$ and $c_b^\beta(-\infty) = d_b^\beta(-\infty) \simeq \delta(\beta,b)$.

In deriving Eq. (24) use has been made of the relation

$$(\Psi_b, \Psi_\beta) = \sum_{b'} (G)_{bb'} \, c_{b'}^\beta .$$ (25)

From the expression (24) and the equation of motion for d, Eq. (18), one immediately gets the equation of motion

$$i\dot{\bar{\rho}} = \frac{i}{2}\left[\left[\dot{G}^{1/2}, G^{-1/2}\right], \bar{\rho}(t)\right] + \left[G^{-1/2}(\tfrac{1}{2}(\mathcal{V}+\mathcal{V}^\dagger)+A)\,G^{-1/2}\,\bar{\rho}\right] .$$ (18 bis)

Once the one-body density matrix is known, the average value of any one-body quantity Q can be calculated. In fact in second quantization formalism Q can be written

$$Q = \sum_{bb'b''b'''} G_{bb'}^{-1} \, Q_{b'b''} \, G_{b''b'''}^{-1} \, a_b^\dagger d_{b'''} ,$$

$$Q_{b'b''} = (\Psi_b, Q\,\Psi_{b''}) .$$ (26)

Consequentely

$$\langle Q \rangle = \mathrm{tr}\,(G^{-1}Q\,G^{-1}\,\rho^T) = \mathrm{tr}(Q\,G^{-1}\bar{\rho}^T G^{-1}) =$$
$$= \mathrm{tr}\,(Q\,G^{-1/2}\,\bar{\rho}\,G^{-1/2}) = \mathrm{tr}\,(\bar{Q}\,\bar{\rho}) .$$ (27)

The last expression is the mean value of Q in the symmetric representation.

According to the preceding discussion one can neglect the terms of order ε^2 and $\varepsilon\dot{\varepsilon}$ to get, assuming only transfer

$$i\dot{\bar{\rho}} = [\mathcal{V}_t , \bar{\rho}] ,$$ (28)

which correspond to Eq. (21).

It has to be noticed that the matrix $\bar{\rho}$ satisfies $\bar{\rho}^2 = \bar{\rho}$ all the time as a consequence of the hermiticity of $\mathcal{V}_t$. Thus, despite the fact that the basis is non-orthogonal, we can use it, in the symmetrized representation we are adopting, as if it were orthogonal. In particular, the mean value of the potential $\mathcal{V}_t$ is given by

$$\langle \mathcal{V}_t \rangle = \mathrm{tr}\,(\mathcal{V}_t \, \bar{\rho}) ,$$ (29)

with $\bar{\rho}$ solution of Eq. (28). For all the other quantities, like the internal angular momentum, there can be ambiguities in the definition, when the two nuclei overlap. It seems a rather natural choice, in the symmetrized representation and in the approximation of small overlap we are here considering, to adopt the general expression

$$\langle Q \rangle = \mathrm{tr}\,(Q\,\bar{\rho}) ,$$ (30)

where Q is the matrix $Q_{bb'}$ of Eq. (26).

It is not particularly difficult to include the overlap to all orders, according to Eqs. (15) or (18). However for our considerations the above described approximation scheme is a fairly good one.

4. THE FRICTIONAL FORCES

a) The Proximity Treatment

In a classical picture the exchange of particles produces a frictional force between the two ions, because the particle momentum distribution in the two nuclei are shifted against each other due to the relative motion. Each transferred particle brings, in fact, a mismatch of momentum which generates a dissipative force, in great similarity to the mechanism which produces the shear viscosity in a classical gas.

Following these classical considerations, the so-called "proximity friction" has been proposed to describe damped heavy ion collisions. This model provides simple expressions for the flux of particles and the corresponding frictional forces, which depend only on geometrical quantities and universal functions. In the presentation we shall follow closely the original paper by J.Randrup[3].

The first assumption of the model is that the single-particle degrees of freedom relax in a time much shorter than the typical collision time and consequentely one can choose a time interval $\Delta t \gg \frac{1}{\Delta \varepsilon}$, being $\Delta \varepsilon$ the typical single-particle level spacing, during which the formfactors can be considered approximately constant, i.e.

$$\frac{1}{\Delta \varepsilon} \ll \Delta t \ll \tau \quad , \tag{31}$$

where τ is the characteristic time of the formfactors variation.

If the interaction is weak enough, in the interval Δt one can use perturbation theory to evaluate the transfer rate. According to Eq.(28), assuming ($Z = \tilde{Z}^+$ for simplicity, in agreement with Ref.3)

$$Z_{b_1 b_2} \simeq V_{b_1 b_2} e^{i(\varepsilon_{b_1} - \varepsilon_{b_2})t} \tag{32}$$

where $V_{b_1 b_2}$ is time-independent, one can get the variation of the density matrix up to second order in $V_{b_1 b_2}$ easily. Because of the assumption of the randomization of the single-particle degrees of freedom, the initial density matrix is diagonal and consequentely the first order variation for any diagonal operator is zero if only transfer processes are present. One then easily gets

$$\Delta N = \sum_{b_2} \rho_{b_2 b_2}^{B} (\sum_{b_1} |Z_{b_1 b_2}|^2) - \sum_{b_1} \rho_{b_1 b_1}^{A} (\sum_{b_2} |Z_{b_1 b_2}|^2) \tag{33}$$

for the change in particle number in nucleus A. Naturally for nucleus B the variation is $-\Delta N$. The $\rho^{A,B}$ are initial density matrices for nucleus A and B respectively.

Because of the assumption (31) one can consider the limit $\Delta t \to \infty$ in evaluating (33). This leads directly to an expression of the Golden-rule type

$$\lim_{\Delta t \to \infty} |z_{b_1 b_2}(t)|^2 \simeq 2\pi \, \delta(\mathcal{E}_{b_1} - \mathcal{E}_{b_2}) |V_{b_1 b_2}|^2 \Delta t \quad , \tag{34}$$

i.e. the number of transferred particles is proportional to the Δt and the transfer occurs essentially on-shell. So the transfer rate is given, according to Eq.(33), by the difference of two fluxes, which can be interpreted as the one-way fluxes from nucleus A to nucleus B and vice-versa.

These results remind of the classical picture in which particles move from one nucleus to the other through a window, see Fig.1. When the two nuclei overlap, in fact, the nuclear surface tends to disappear from the contact region and the nucleons should be essentially free to move from one nucleus to the other.

This analogy with the classical picture can be used to evaluate the frictional forces produced by the transfer. In fact, the momentum deposited $\Delta \vec{p}$ by the particles which move from one well to the other can be calculated in a fullyclassical way if one approximates the single-particle levels by two continuous Fermi distributions

$$\vec{F} = \int j(\vec{s}) \Delta \vec{p}(\vec{s}) \, d^3 \vec{s} \quad , \tag{35}$$

where $j(\vec{s})$ is the classical one-way current and $\vec{s}$ the particle velocity. Here $\Delta \vec{p} = m(\vec{s} - \vec{v})$ and $\vec{v}$ is the relative velocity of the two wells. This leads to the so-called "window formula", which connects the frictional forces to the one-way flux

$$\vec{F} = n_0 \Delta \sigma (2 \vec{v}_\| + \vec{v}_\perp) \quad , \tag{36}$$

where $\vec{v}_\|$ and $\vec{v}_\perp$ are the parallel and orthogonal components of the relative velocity with respect to the normal through the window, $\Delta \sigma$ is the window area and n_0 the bulk flux in one direction. Eq.(36) gives the force exerted by one system to the other. Thus the radial and tangential frictional coefficients c_r and c_t are given by

$$c_r = 2 c_t = f^t = 2 n_0 \Delta \sigma \quad , \tag{37}$$

where f^t is the sum of the two one-way fluxes, i.e. the total number of exchanged particles per unit time. If one assumes that the relationship (36) is still approximately valid also in the more microscopic treatment of Eqs.(33),(34), then the frictional coefficients can be ralated to the formfactors $V_{b_1 b_2}$, that is

$$c_r = 2 c_t = 2\pi m \int d\mathcal{E} \, (f^B(\mathcal{E}) + f^A(\mathcal{E})) \, \rho^A(\mathcal{E}) \rho^B(\mathcal{E}) |V(\mathcal{E})|^2 \sim$$
$$\sim 4\pi m \int_0^{\mathcal{E}_F} \rho^A(\mathcal{E}) \, \rho^B(\mathcal{E}) |V(\mathcal{E})|^2 d\mathcal{E} \quad , \tag{38}$$

having assumed two continuous Fermi distributions for the single-particle levels with level densities ρ^A and ρ^B. Analogously the net flux is obtained from Eqs.(33),(34) as

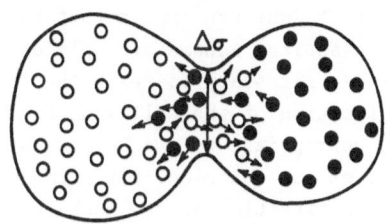

Fig. 1. Imagining the two nuclei as Fermi gases, when they approach
each other, there is a transfer of particles from one to the
other through the window $\Delta\sigma$. In the proximity model the momentum deposited by the particles generates a frictional force
very similar to the one which produces the shear viscosity in
a classical gas.

$$\frac{\Delta N}{\Delta t} \simeq 2\pi \int d\mathcal{E} \, (f^B(\mathcal{E}) - f^A(\mathcal{E})) \, \rho^A(\mathcal{E}) \, \rho^B(\mathcal{E}) |V(\mathcal{E})|^2 \simeq 2\pi F_A \, \rho^A(\mathcal{E}_F) \, \rho^B(\mathcal{E}_F) |V(\mathcal{E}_F)|^2 \simeq$$
$$\simeq \tfrac{1}{2} m F_A \frac{\partial c_r}{\partial \mathcal{E}_F} \quad , \tag{39}$$

where it has been assumed $f^B(\mathcal{E}) - f^A(\mathcal{E}) \simeq F_A \, \delta(\mathcal{E} - \mathcal{E}_F)$, being F_A the difference
between the two Fermi energies. In fact the difference is sharply peaked
around the average Fermi energy $\mathcal{E}_F$ ($|F_A| \ll \mathcal{E}_F$).

According to the "proximity procedure", one then calculates the one-
way flux along the same line followed by the calculation of the proximity
potential[10] and one can use the Eqs.(37),(38) for extracting the frictional
coefficients and the net flux respectively. This procedure leads to simple
formulae for different quantities. In particular the total frictional force
is

$$\vec{F} = 2\pi n_0 \bar{R} \, b \, \Psi(\zeta)(2\vec{v}_{\parallel} + \vec{v}_{\perp}) \quad , \tag{40}$$

where $\bar{R}$ is the so-called reduced radius of the two nuclei, b is the surface
diffusivity parameter (usually $b \simeq 1$ fm and $n_0 = 0.263$ MeV 10^{-22} s fm^{-4}) and
Ψ is a universal function which depends only from the distance between
the two surfaces ζ . According to Ref.3 the formfactor Ψ can be parametrized as follows

$$\Psi(\zeta) = \begin{cases} 1.4 - \zeta & \zeta \leq -.04 \\ 1.6 - .5\zeta - \frac{1.8}{\pi} \sin\left(\frac{\zeta + .4}{3.6}\pi\right) & \\ 0 & \zeta \geq 3.2 \end{cases} \quad .$$

Comparing (40) and (36) one can see that the proximity frictional force
can be considered a refinement of the window formula in that it gives a
more realistic estimate of the window area $\Delta\sigma$ as a function of the
distance ζ .

It has to be stressed again that the frictional coefficients c_r, c_t
corresponding to the frictional force, Eq.(40), are functions only of the
distance between the two nuclei.

b) The Microscopic Treatment

The general scheme described in Sections 2 and 3 can be used to develop
a quantal treatment for the frictional forces arising from particle exchange
and check the limits of validity of the classical treatment shown in the
preceding subsection.

On general grounds, in fact, one can expect that the basic hypothesis outlined above for the classical scheme are only of limited validity. In particular the delay time, necessary to reach the long time regime, is usually a substantial fraction of the collision time and is doubtful that this long time regime, appropriate to the Golden rule (34), can be really approached at any stage of the reaction. Furthermore, quantal effects, like tunnelling and particle diffraction at the neck between the two nuclei, could change appreciably the classical picture. In this second subsection we discuss how to extract from a quantal treatment the proper frictional forces to compare with the classical result of Eq.(40).

One can ask, first of all, if in the quantal treatment we are here considering there is any mechanism similar to the one of mismatch of momentum, that in the classical scheme is the origin of frictional forces. This can be readily identified with the optimal Q-value effect, which is endowed in the single-particle formfactors V_t and which will tend to favour particular orbitals in the transfer process producing a definite pattern in the energy and angular momentum losses.

One can then try to extract from the evolution of the energy and angular momentum also the evolution in time of the frictional forces by properly identifying the dissipative part of the interaction. Unfortunately, in a microscopic treatment, this is not a trivial task, because it is always possible to add (or subtract) a conservative force to a given frictional force without altering the final asymptotical value of the excitation energy, which is obviously an observable quantity.

For instance the transfer process can give rise to a polarization potential[11] which acts during the collision and is clearly a conservative one. However, it has to be noticed that the loss of angular momentum can be unambiguosly related to a tangential friction (see Eq.(42) below).

A natural choice would be to identify the dissipative energy with the total intrinsic excitation E^* of the two ions. In this case the frictional force F can be extracted from the definition

$$\dot{\vec{r}} \cdot \vec{F} = - \frac{dE^*}{dt} \quad , \tag{41}$$

where $\dot{\vec{r}}$ is the relative velocity. Eq.(41) can be further split into a radial and a tangential component making use of the equation

$$\vec{r} \times \vec{F} = - \frac{d\vec{L}}{dt} \quad , \tag{42}$$

where $\vec{L}$ is the intrinsic angular momentum. if the motion is planar, Eqs. (41),(42) give the radial friction F_r and the tangential friction F_φ uniquely.

The essential requirements for the force F to be accepted as a frictional force are:

(i) it must have "direction opposite to the velocity", i.e. it has to be "negative";

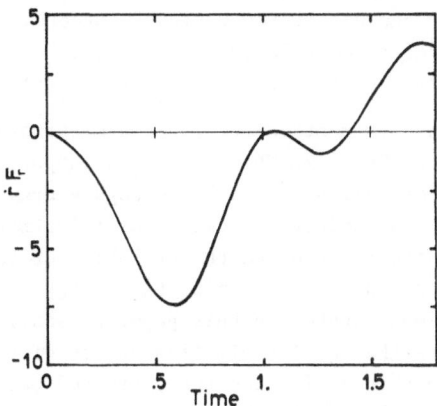

Fig. 2. The work per unit time as a function of time in the case of
^{86}Kr + ^{208}Pb at Ecm=308 MeV for a central collision. As one
can see, $\dot{r} F_r$ is not always negative nor symmetric with respect to the instant of closest approach ($t \simeq 1$). Notice that
in our units ($\hbar = c = 1$, MeV) $t = 0.6 \, 10^{-21}$ s.

(ii) for low velocities, it has to be proportional to the velocity
itself.

The choice of Eq. (41) seems not to satisfy these requirements. This
is illustrated in Fig. 2, where is reported the result of a calculation
for the system ^{86}Kr + ^{208}Pb for a central elastic trajectory. In the
figure is reported the work rate per unit time of the radial force F_r

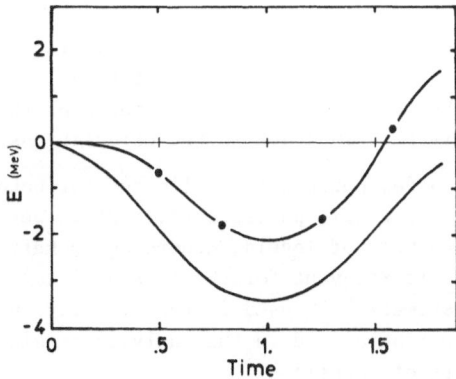

Fig. 3. Comparison between the adiabatic energy (full drawn curve)
and the mean total energy (dash-dotted curve) for the same
case of Fig. 2. It is evident a tendency towards the adiabatic regime.

extracted from the average excitation energy E^* according to Eq. (41). For a "bona fide" frictional force this work rate should be always negative and symmetrical with respect to the distance of closest approach.

We have adopted a different choice for the dissipative energy, following the suggestion of Ref. 12. This choice is supported by the observation that for peripheral collisions at low enough energy the evolution of the system is close to the adiabatic regime, as illustrated in Fig. 3 for the same trajectory. Here is reported the evolution of the total energy of the system in comparison with the adiabatic energy, defined as the energy of the local ground state. In this paper we shall then use Eqs. (41), (42) with E^* defined as the total excitation energy with respect to the adiabatic energy. With this definition the extracted energy dissipation $\dot{r}$ F_r has the behaviour of Fig. 4, which appears more satisfactory. In fact, it looks like if any energy loss above the adiabatic energy is nearly an irreversible one. More details on these calculations will be given in the next section, where we report further applications and studies in relation to a proper definition of the frictional forces.

A closer insight into the dissipation mechanism can be obtained by looking at the adiabatic energy levels as a function of distance, as reported in Fig. 5. One can see that no level crossing or quasi-crossing occurs up to distances sligthly smaller than the sum of the radii, which is the range of distances we shall consider in all our applications. Under these circumstances the dissipation is produced by this "moving away" from the adiabatic limit (corresponding to infinitely slow motion).

A more detailed discussion on the frictional forces can be found in Ref. 13.

5.APPLICATIONS

In most of the applications we used elastic trajectories calculated by means of the proximity potential[10] for the nuclear part of the ion-ion interaction. The energy and impact parameter have been chosen in such a way that the distance of closest approach was nearly the sum of the two nuclear radii. The restriction to elastic trajectories is not relevant for our purposes, because we want to study the frictional forces due to the relative motion of the two heavy ions. The effects of the forces on the trajectory can be trivially included in the calculations.

We have studied a few peripheral collisions for both symmetrical and asymmetrical systems.The single-particle formfactors were calculated microscopically for each pair of levels, according to Ref. 14. Only the tails of these formfactors are relevant for the present discussion. Such formfactors have been systematically used in the calculation of the imaginary part of the ion-ion potential and in the analysis of single-particle transfer data in a variety of reactions[15] .

a) Symmetrical Systems

We considered the collision ^{40}Ca + ^{40}Ca at Ecm = 83 MeV and impact parameter b = 5 fm. The single-particle levels included in the calculation are reported in table 1 - only neutrons have been considered for simplicity.

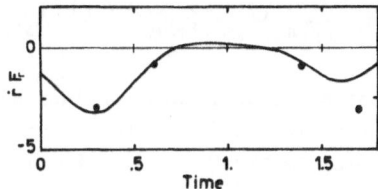

Fig.4.The work per unit time after subtraction of the adiabatic
energy for the same case of Figs. 2, 3. Now ṙ F_r is almost
always negative, but not completely symmetric. However,
there is a tendency towards symmetry with respect to the
instant of closest approach (t≈1). This is a memory effect,
i.e. friction depends on the trajectory. The reported dots
are the result of a parametrization, see Sect. 6 The agreement is good in the approaching phase, but not in the outgoing one.

The density matrix ρ at time t can be expressed in terms of the
initial density matrix at time t = 0 by means of a unitary transformation
U(t)

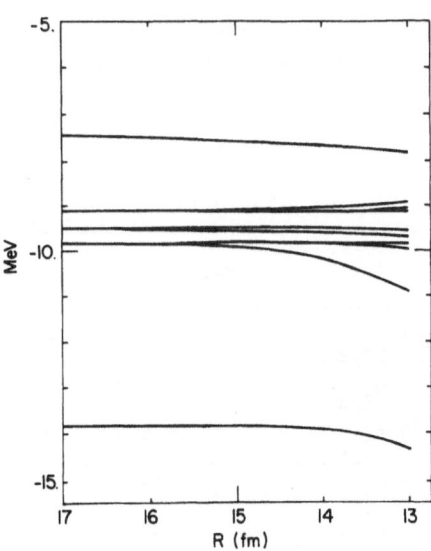

Fig.5.Adiabatic occupied levels as a function of the relative
distance for the system ^{86}Kr + ^{208}Pb.

$$\rho_{bb'}(t) = \sum_{b''b'''} U_{bb''} \, \rho_{b''b'''}(0) \, (U^+)_{b'''b'} \quad . \tag{43}$$

At t = 0 the two potential wells are considered at a distance much larger than the sum of the two radii, so that we can assume for $\rho_{bb'}(0)$ a diagonal form corresponding to two Fermi distributions, i.e.

$$\rho = \begin{pmatrix} F & 0 \\ 0 & F \end{pmatrix} \quad , \tag{44}$$

where F is the diagonal matrix with diagonal elements equal to 1 for occupied levels and 0 for non-occupied ones. Thus, denoting by subscripts 1, 2 the levels pertaining to wells 1, 2 respectively, Eq. (43) can be rewritten

$$\rho_{bb'}(t) = \sum_{b_1}^{occ.} U_{bb_1}(t)\,(U^+)_{b_1 b'} + \sum_{b_2}^{occ.} U_{bb_2}(U^+)_{b_2 b'} = \rho_{bb'}^{(1)}(t) + \rho_{bb'}^{(2)}(t) \quad . \tag{45}$$

Therefore it is possible to split the density matrix $\rho(t)$ in two parts, one which receives contribution only from the levels originally belonging to well 1 and the other only from the levels originally belonging to well 2. In particular the total particle density at the point $\vec{r}$ is given by

$$\rho(\vec{r},t) = \sum_{bb'} \Psi_b^*(\vec{r},t)\, \rho_{bb'}(t)\, \Psi_b(\vec{r},t) =$$
$$= \sum_{bb'} \Psi_b^*(\vec{r},t)\left[\rho_{bb'}^{(1)} + \rho_{bb'}^{(2)} \right] \Psi_b(\vec{r},t) = \tag{46}$$
$$= \rho^{(1)}(\vec{r},t) + \rho^{(2)}(\vec{r},t) \quad ,$$

where $\Psi_b(\vec{r},t)$ is the wave function of the orbital b according to the definition of Section 2. It has to be noted that

Table 1. Single-particle levels for ^{40}Ca.

	^{40}Ca
(1 $p_{1/2}$	- 4.3)
1 $p_{3/2}$	- 6.2
0 $f_{7/2}$	- 8.3
0 $d_{3/2}$	- 15.6
1 $s_{1/2}$	- 18.1
0 $d_{5/2}$	- 21.6
(0 $p_{1/2}$	- 27.6)

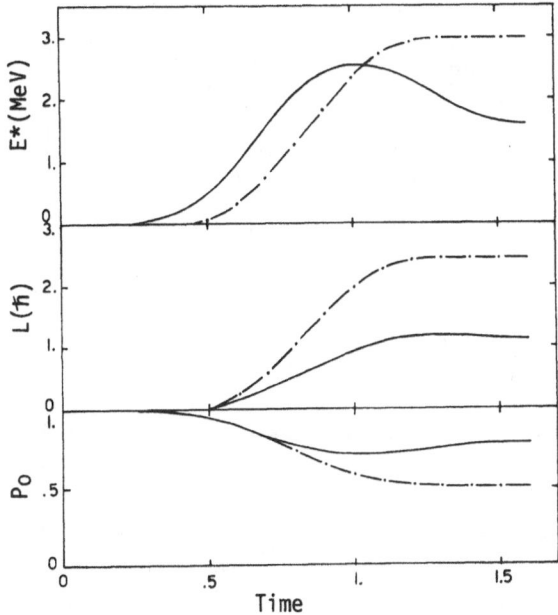

Fig.6. In the upper part of the figure and in the middle, a
comparison between the predictions of the proximity
model (dash-dotted curve) and the microscopic approach
calculations (full drawn curve) for the system ^{40}Ca $+^{40}$Ca
at Ecm = 83 MeV and impact parametr b = 5 fm. The
proximity predictions have been reduced by a factor N/A
having considered only neutrons. In the bottom, the
probability P_o for the system to remain in the ground
state as a function of time. The quantal re ult (full
curve) is shown in comparison with the predictions of
Ref. 14 (dash-dotted curve). The distance of closest
approach is reached at t = 0.8.

$$\int d^3\vec{r} \; \rho^{(1)}(\vec{r},t) = N_1 \qquad , \qquad \int d^3\vec{r} \; \rho^{(2)}(\vec{r},t) = N_2 \qquad (47)$$

N_1, N_2 being the number of particle originally in the well 1 and 2 respectively. Despite the indistinguishability of the particles the density
$\rho^{(1)}$ can give an idea of how the particles, initially in well 1, have been
redistributed between the two wells because of the transfer process.
Analogous considerations holds for $\rho^{(2)}$. According to Eq. (46), in spite
of the fact that no net flux of particles can occur between the two identical nuclei, there is a mixing between the orbitals of the two wells. As
a consequence the density $\rho^{(1)}(\vec{r},t)$, for instance, gets contributions also
from orbitals of the well 2 and thus it will be appreciably different from
zero also at the location of nucleus 2. This mixing can be considered as
the analogous of the classical (one-way) flux of particles from one well
to the other, which in the classical picture is the origin of frictional
forces, as discussed in Section 4a.

217

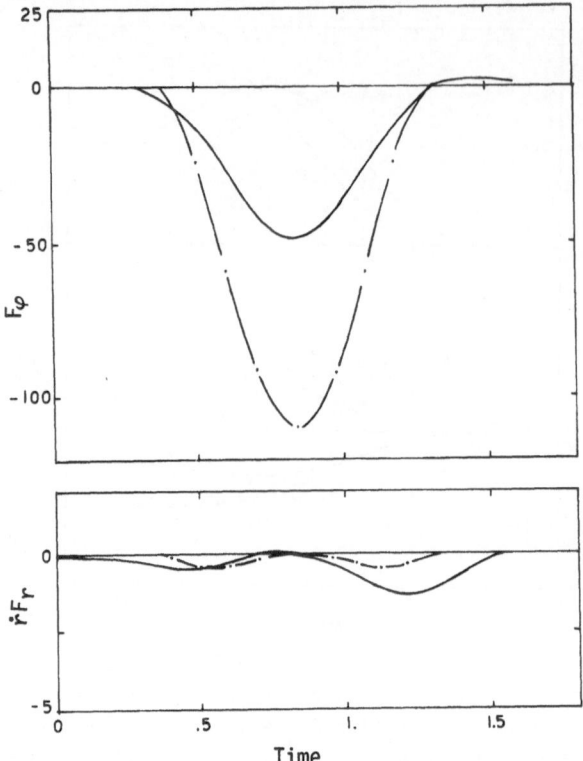

Fig.7.Comparison between the quantal calculations (full drawn
curve) and the resul of the proximity model (dash-dotted
curve) for the same case of Fig. 6. The units used correspond to $\hbar = c = 1$, MeV.

For the same trajectory, we report in Fig. 6 the total excitation
energy and intrinsic angular momentum as a function of time, in comparison
with the predictions of the proximity model[3] . From these quantities one
can extract the tangential and radial friction according to the discussion
of the previous section. The results are reported in Fig. 7, again in comparison with the classical model.

The first striking difference is that the quantal treatment predicts
a smaller angular momentum loss than the proximity model by about a factor
2.

The two radial frictions look more similar. However the quantal one
is asymmetrical with respect to the distance of closest approach. This can
be a consequence either of a conservative force component or of some memory
effects, as suggested by the discrease of the excitation energy in the
latest stage of the collision, see Fig. 6.

This feed-back effect is also evident in the probability P_0 for the
system to remain in the ground state, as depicted in the lower part of
Fig. 6. This probability is defined by $P_0 = |\langle \Psi(o) | \Psi(t) \rangle|^2$ where
$\Psi(o)$ is the initial ground state of the system and $\Psi(t)$ the
quantal state at time t. The prediction for P_0 according to our quantal
treatment is compared to the one obtained with the treatment of Ref. 14,

where the different transfer processes are considered as indipendent and incoherent. The same formfactors, single-particle basis and trajectory have been used in both calculations.

We have checked that all these results are essentially unaffected by enlarging the single-particle level basis. In Fig. 8 is reported the effect of adding to the basis the two levels indicated in parenthesis in Table 1. The stability of the results give confidence on the validity of the comparison of our calculations with the classical ones.

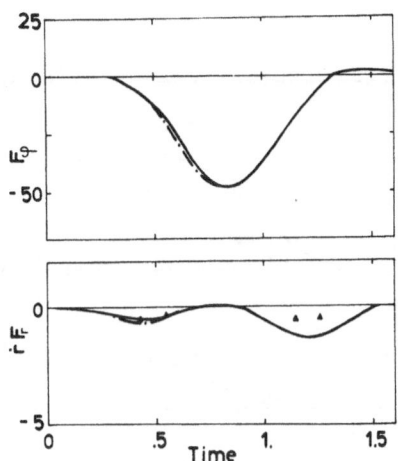

Fig.8.Quantal calculations for the frictional forces with 5 levels (full drawn curve) and with 7 levels (dash-dotted curve) for the same case of Figs. 6, 7. The triangles refers to the result of a parametrization, see Sect. 6.

b) Asymmetrical Systems

Table 2. Single-particle levels for the system ^{86}Kr + ^{208}Pb.

^{86}Kr		^{208}Pb	
0 $g_{7/2}$	- 3.9	0 $i_{11/2}$	- 4.1
2 $s_{1/2}$	- 4.6	1 $g_{9/2}$	- 6.2
1 $d_{5/2}$	- 7.4	2 $p_{1/2}$	- 7.2
0 $g_{9/2}$	- 9.1	2 $p_{3/2}$	- 9.5
1 $p_{1/2}$	- 13.8	1 $f_{5/2}$	- 9.8

As a representative example for asymmetric systems we considered the reaction ^{86}Kr + ^{208}Pb at Ecm = 350 MeV and impact parameter b = 4.5 fm. The single-particle levels included in the calculations are reported in Table 2 - again, for simplicity, only neutrons have been considered.

A detailed comparison with the proximity model for different quantities as a function of time is reported in Figs. 9,10. As for ^{40}Ca + ^{40}Ca, the quantal treatment seems to give again a smaller angular momentum dissipation, but now the energy dissipation is bigger. Furthermore, the radial friction as a function of time looks more asymmetrical with respect to the distance of closest approach (at t = 0.8 in this case) and displays a positive part, which seems to indicate some departure from irreversibility - in the sense discussed in Section 2.

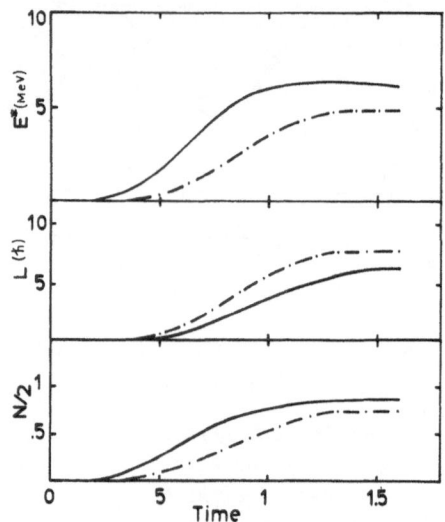

Fig.9. Comparison between the quantal calculations (full drawn curve) and the proximity predictions reduced by N/A (dash-dotted curve) for the asymmetric system ^{86}Kr + ^{208}Pb at Ecm = 350 MeV and impact parameter b = 4.5 fm. The quantity N/2 is the net number of transferred particles.

However, the final results of the quantal treatment for E^*, L and the number of transferred particles are, in this case, closer to the classical ones. This indicates that the definition of the frictional forces in the proximity model cannot be easily related to the one we have adopted.

Analogously to the case of symmetrical systems, we have calculated the probability P_0 to remain in the ground state as a function of time, see Fig.11 (b). In this case P_0 is much smaller and follows the trend obtained with the method of Ref.14. The smaller value of this probability is in agreement with the fact that no strong feed-back is present in this

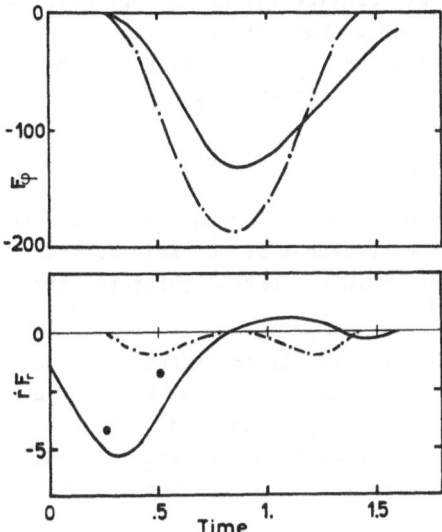

Fig. 10.The frictional force evolution in the quantal treatment
(full drawn curve) and in the proximity model (dash-dotted
curve) for the same case of Fig.9. The reported dots are
the result of a parametrization, see Section 6.

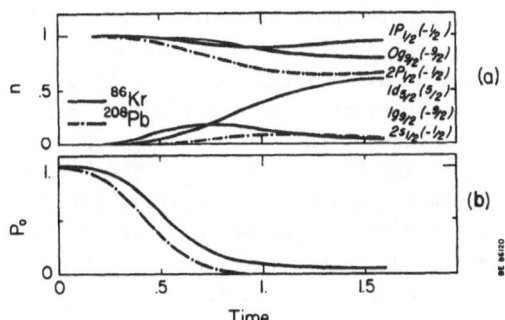

Fig. 11.For the system ^{86}Kr + ^{208}Pb at Ecm =350 MeV and impact
parameter b = 4.5 fm: a) evolution of the occupation
numbers of the levels most affected by the transfer
process. In brakets is shown the magnetic quantum number;
b) the probability P_0 to remain in the ground state in
the quantal treatment (full drawn curve) and in the
approach of Ref.14 (dash-dotted curve).

calculation for the different quantities reported in Fig.9, although the occupation numbers of the different levels, see Fig.11(a), show some oscillations as a function of time. In fig.11(a) only those levels close to the Fermi energy are considered, all the others, in fact, display occupation numbers with much weaker variations in time. The latter result supports the assumption that, for the trajectory considered, the size of the single-particle basis is large enough.

6. GENERAL PROPERTIES OF FRICTION

In this section we study in more details the properties of friction and we try to see to what extent it is trajectory-dependent. This is an essential requirement to infer general rules for its estimate like in the proximity treatment.[3]

In order to study separately the tangential and radial friction one can let the two wells move along particular prescribed trajectories, for which only one of the two components is different from zero. Furthermore, one can move the two wells with uniform velocity and check the possible proportionality of the extracted forces with respect to the velocity itself.

To isolate the tangential friction, the only possible choice is moving the two wells along a relative circular trajectory, for which the radial velocity vanishes so that no radial frictional force is present.

We have then considered diferrent circular trajectories of different radius and, starting from the ground state, we have let the system evolve along each trajectory for different uniform velocities. Typical results for the system $^{40}Ca + {}^{40}Ca$ are reported in Fig.12, for the frictional tangential force F_φ , extracted according to Eq.(42). One can see that, after a time delay, F_φ reaches a maximum and then it starts oscillating around some average value which is about one half of the first maximum. We have taken, for each distance and velocity, the value of F_φ at the first maximum, which occurs essentially at the same time in all cases. In Fig.13 we reported those values as function of velocity for different radii of the trajectory. One can notice a remarkable linear dependence of F_φ , expecially at lower velocities and larger distances. The linearity is closely followed up to velocities which are characteristic for the collisions we have taken into account. From the slopes of these curves at the origin (indicated by the full straight lines in Fig.13), one can extract the tangential frictional coefficient, which is shown in Fig.14 as a function of distance. Because we have taken the value of the frictional force at the first maximum, according to Fig.12, in Fig.14 we have divided this force by a factor of 2. Such a choice, once used to calculate the tangential frictional force along the more realistic trajectory of Fig.8, reproduces quite closely the results for F_φ obtained solving the equation of motion along the same trajectory. In the same figure, the proximity result is also reported.

The separation of the radial frictional force can be obtained by considering central trajectories followed at different uniform velocities. Of course, in this case the procedure to adopt is different from the one used in the tangential case, because the relative distance changes necessarily. We have then followed the suggestion of Ref.12, i.e. we have let the

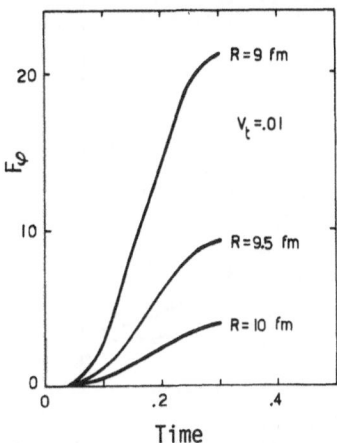

Fig.12.Absolute value of the tangential frictional force as a
       function of time for a circular trajectory followed with
       a velocity v_t = 0.01 ($\hbar$ = c = 1, MeV) for the   system
       ^{40}Ca + ^{40}Ca. Each curve correspond to a different radius R.

two wells approach each other along a central trajectory, starting from an
asymptotical distance, with uniform velocity, recording friction at different distances – several velocities have been used.

     The results for the radial frictional force F_r are reported in Fig.15.
Again one can notice a linear dependence on the relative velocity, from
which one can extract the frictional coefficient as a function of distance
(see the full straight lines) reported in Fig.16. If one uses this frictional coefficient to calculate the work per unit time along the trajectory
of Fig.8, one gets the values (indicated by triangles) reported in the
figure. It seems that, at least for the approaching phase of the collision,

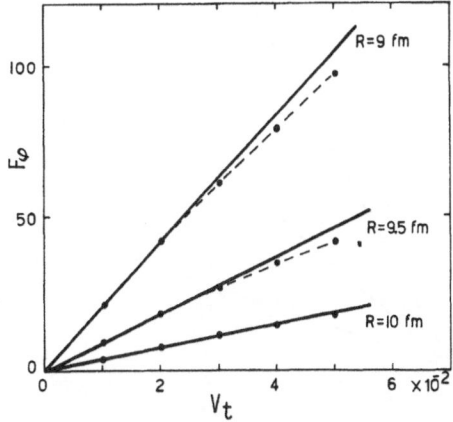

Fig.13.Absolute value of the tangential frictional force (full
       dots) as a function of tangential velocity v_t for the
       same trajectories of fig.12. The full lines correspond
       to the slopes near the origin. The dashed lines are only
       to guide the eye and show the deviation from linearity.

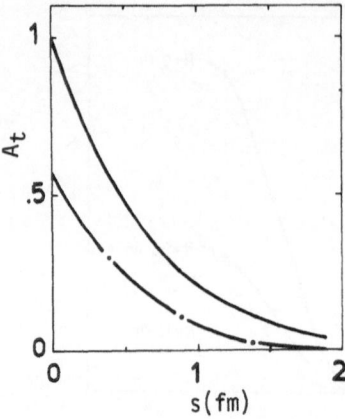

Fig.14.The tangential frictional coefficient A_t extracted from
Fig.13 (dash-dotted curve) in comparison with the proximity estimate (full drawn curve). In the abscissa is reported the distance between the surface of the two nuclei.

the adopted parametrization of the radial frictional force is reasonably
in agreement with the actual calculations performed along the realistic
trajectory.

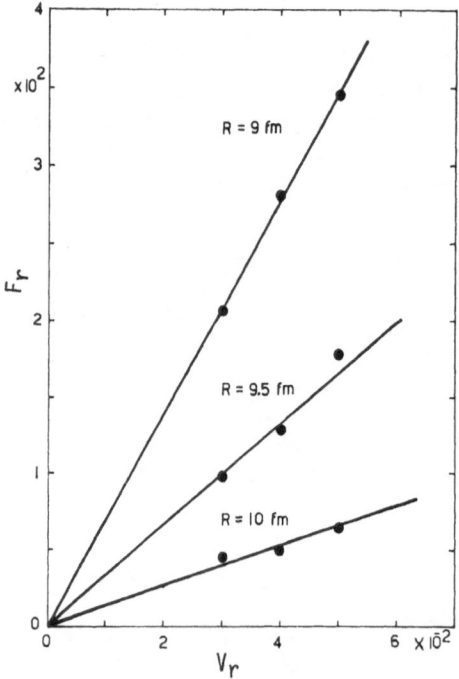

Fig.15.The absolute value of the radial frictional force, taken
at different distances R, as a function of the radial velocity v_r (full dots) for the system $^{40}Ca + ^{40}Ca$ along
the central trajectory. The full straight lines give the
average slope.

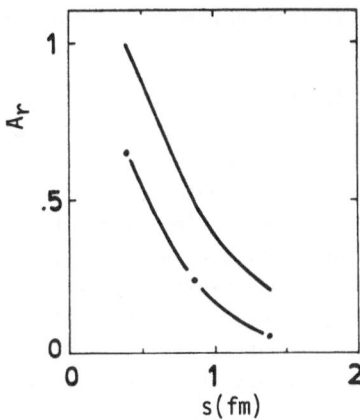

Fig.16.The radial frictional coefficient A_r extracted from
Fig.15, (full drawn curve) as a function of the distance
between the surface of the two nuclei. The proximity
prediction is also reported for comparison (dash-dotted
curve).

Similar calculations for the radial friction were presented in Ref.12
for the same system $^{40}Ca + {}^{40}Ca$. In these calculations the Schroedinger
equation in coordinate space has been solved for each single-particle
orbital along radial trajectories. In our formalism this amounts to include
both transfer and inelastic processes. The radial frictional force reported
in Fig.15 turns out to be roughly a factor of two smaller than the one
reported in Ref.12. As we shall see the inclusion of inelastic processes
brings the total radial frictional force in close agreement with the results
by Pal and Gross, as shown in Fig.20. This comparison gives further evidence
that our single-particle basis is large enough for collisions at these
energies. This is in agreement with the results of the study reported in
Ref.7.

Following the same procedure used for the system $^{40}Ca + {}^{40}Ca$, we have
calculated the frictional coefficients in the case of the asymmetric system
$^{86}Kr + {}^{208}Pb$, considering only transfer. By means of the radial frictional
one, one gets the points reported in Figs.4,10. The agreement is again
reasonable, expecially for the central trajectory and in the approaching
phase. Analogous results hold for the tangential friction.

From all these results and the above discussion one can conclude that
it is not possible, in general, to parametrize in a simple way the frictional forces as a function of distance and velocity only. No simple prescription in fact can be extracted for calculating the friction along the whole
trajectory[13].

7.INCLUDING INELASTIC

Until now we have considered only transfer processes in order to
compare with the classical picture given by the proximity model. In this
section we study the effect of including also the inelastic processes in
the description of the collision, according to the general scheme of Sect.2.

The single-particle inelastic formfactors have been calculated according to Ref.16.

In Fig.17 we show, as functions of time, the total internal excitation energy, the angular momentum and the net number of transferred particles for the system ^{86}Kr + ^{208}Pb, along the same elastic trajectory of section 5. The three curves correspond to calculations where only inelastic, only transfer processes or both of them are included. These comparison show how much the transfer and the inelastic phenomena give contribution and to which extent they are coupled. The result for excitation energy seems to indicate that the inelastic processes contribute at least as much as the transfer ones. Furthermore, the curves look like as if the two degrees of freedom couple in a constructive way in increasing the total excitation

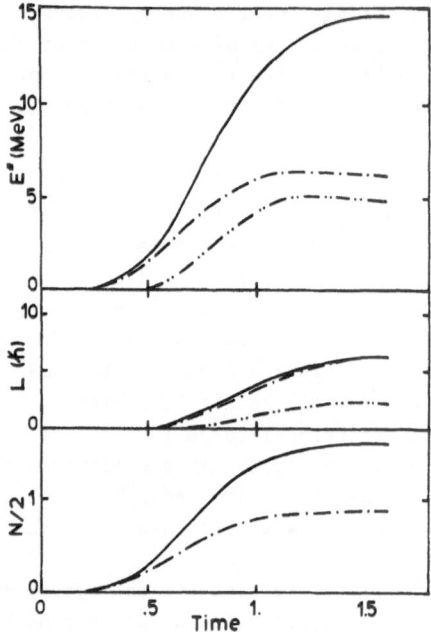

Fig.17.Quantal calculations, including inelastic excitation, as a function of time for the system ^{86}Kr + ^{208}Pb at Ecm = 350 MeV and impact parameter b = 4.5 fm. The dash-dotted curve indicates the result with transfer only and it is the same of Fig.9. The dashed-double-dotted curve indicates the result obtained with the inelastic only, while the full drawn curve is the result one gets including both transfer and inelastic processes.

energy. On the contrary, the inelastic processes seem to be less effective in increasing the intrinsic angular momentum and the coupling with transfer looks destructive. Finally, the net number of transferred particles N/2 seems to be appreciably influenced by the presence of the inelastic form-factors. A possible explanation of this effect could be that the higher orbitals populated by the inelastic processes have transfer formfactors with longer tails because they are in general less bound levels.

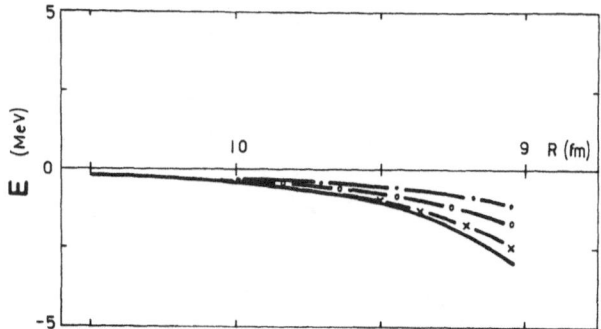

Fig.18. Comparison between the adiabatic energy (full drawn
curve) and the mean total energy for the system
$^{40}Ca + ^{40}Ca$ along the central trajectory, including
inelastic excitations. The trajectory has been followed
with different velocities, that is $v_r=0.06$ ($-\cdot-$),
$v_r=0.04$ ($-o-$) and $v_r=0.02$ ($-x-$).

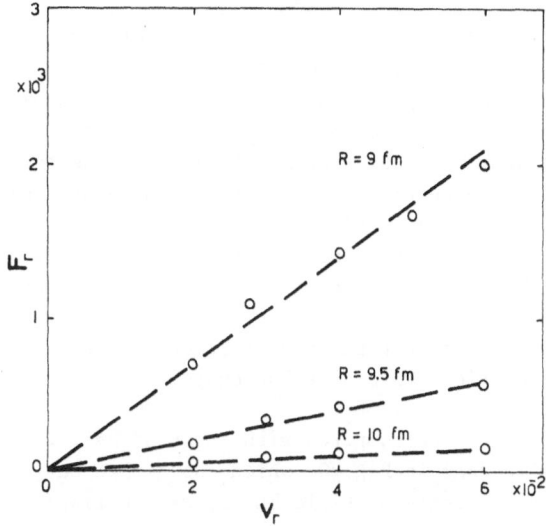

Fig.19. The absolute value of the radial frictional force as a
function of radial velocity for the system $^{40}Ca + ^{40}Ca$
including inelastic, cf. Fig.15. The result of our calculation has been multiplied by a factor of 2 to take
into account also protons and compare with the results
of Ref.12, cf. Fig.20. The dashed line indicate the
average slope.

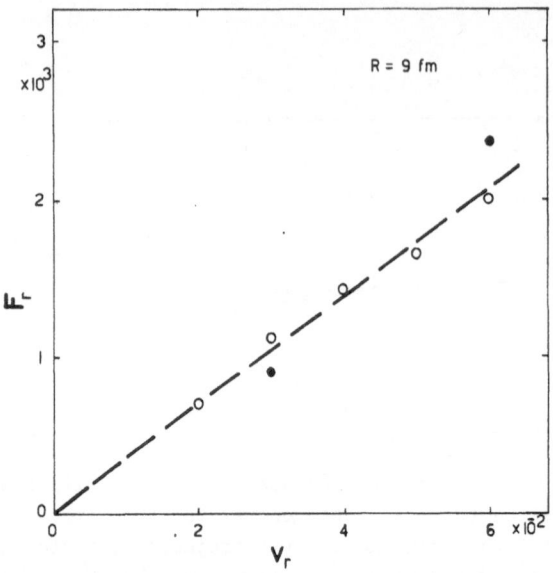

Fig.20.The same results of Fig.19 compared with the calculations
of Ref.12 (full dots).

Of course, all these effects need a more sistematic study before any
general rule can be inferred.

At this point, one can ask if, with the inclusion of the inelastic
processes, it is still possible to introduce frictional forces to describe
the loss of energy and angular momentum. So we start again by considering
the total average energy in comparison with the adiabatic energy. This is
done in Fig.18, where for the system ^{40}Ca + ^{40}Ca, the total average energy
is reported along the central trajectory, followed with different velocities
in comparison with the adiabatic energy. Again the shift from this adiabatic energy increases smoothly and uniformly with increasing velocities.
Following the same procedure used in the case of transfer only, one can
extract the radial frictional force as a function of velocity at different
distances. Fig.19 shows that the radial frictional force is remarkably
linear up to velocities typical of low energy collisions.

These results can be compared with those of Ref.12 as anticipated in
Section 6. This is done in Fig.20, which shows that our scheme, which makes
use of the discrete single-particle basis, as described in Section 2, is
appropriatein this low energy regime.

With the inclusion of the inelastic processes, the model can be considered realistic enough to try a comparison with the experimental data. A
direct comparison is of course impossible, because experimentally only
asymptotic values can be measured. The energy and angular momentum losses
can give however rather clear indications about the dissipative mechanisms
which occur in the reaction. The angular momentum loss can be deduced by
multiplicity measures . In the inelastic region we are considering here,
the reactions are strongly focussed in scattering angle and the experimental
angular range usually encompasses most of the distribution. Under these

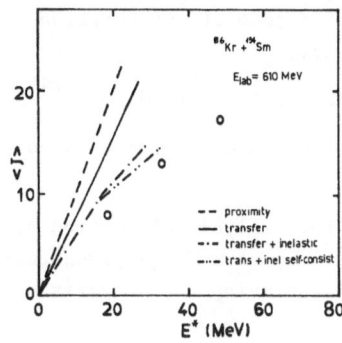

Fig.21.Comparison among different theoretical predictions and the experimental data of Ref.5 for ^{86}Kr + ^{154}Sm at E_{lab} = 610 MeV.

conditions the experiments can be considered to give the results integrated over the whole angular range.

The only reaction which has been systematically studied along these lines is the reaction ^{86}Kr + ^{154}Sm. In ref.17 the gamma multiplicity as a function of the total kinetic energy loss has been measured for two bombarding energies E_{lab} = 610 MeV and E_{lab} = 490 MeV. The data have been further analysed in Ref.5. In Fig.21 we report these experimental results, in the regime of small energy losses, in comparison with

i) the proximity prediction as reported in Ref.5,

ii) the results of the microscopic calculations which include only transfer processes,

iii) the results one gets by including also the inelastic processes,

Table 3. Single-particle levels for ^{154}Sm.

	^{154}Sm
0 $i_{13/2}$	- 4.
2 $p_{3/2}$	- 4.7
0 $h_{9/2}$	- 5.8
1 $f_{7/2}$	- 6.4
0 $h_{11/2}$	- 10.4

229

iv) the predictions of a complete calculation including the self-consistency of the trajectory.

The single-particle levels considered for ^{154}Sm are shown in Table 3.

The main conclusions one can draw from this comparison are the following:

I) in agreement with the analysis of Section 5, the microscopic calculation gives less angular momentum for a given energy loss with respect to the classical estimate, improving the comparison with the experimental data;

II) the inelastic processes appear to be essential to get a closer agreement with experiments;

III) the inclusion of self-consistency in the trajectory seems to improve further this agreement, but it is not essential for obtaining the general trend.

From this analysis one can then conclude that we are beginning to understand the dissipative phenomena involved in peripheral heavy ion collisions, at least for not too high excitation energies. It appears in this respect very desirable a systematic work, both theoretical and experimental, along these lines, on different systems and energies, in order to further elucidate the dissipative mechanisms not only in this energy regime, but also down to the deep-inelastic one.

8. BEYOND THE INDEPENDENT-PARTICLE APPROXIMATION

As mentioned in the introduction, there are essentially two physical processes, not included in the independent-particle description, which become important for closer collisions and larger excitation energies. First of all, if the two nuclei start overlapping and to be highly excited, one can expect the mean field to be distorted. This distortion, with respect to the static situation, modifies the interaction potential between the two ions as well as the motion of the particles. The latter effect in particular is responsible for the possible collective character of the excitation. To first order this distortion can be described in the Random Phase Approximation (R.P.A.), which is known to give a fairly good description of the low lying collective excitation in spherical nuclei. In the Appendix we show how to incorporate the R.P.A. correlations in the formalism developed in Section 3.

On the other hand, when the excitation is large enough, the particle mean free path becomes shorter, because the Pauli principle becomes less effective in hindering the collision between particles[18]. Particle collisions are the source of internal dissipation and damping of all collective motions. One can expect that the collisions, besides introducing the two-body dissipation, not included in a mean field description, give rise to a loss of memory in the evolution of many physical quantities. In particular the collisions tend to relax the one-body density matrix towards the local equilibrium. Unfortunately a complete microscopic description of the particle collisions is very difficult to include in a realistic model, because in a quantal treatment particle collisions are essentially off-shell and till now they have been considered only in schematic models[19] or in a semiclassical picture for higher energy ion-ion collisions[20].

In this section we want to study, in a very phenomenological way, the effect of particle collisions. We shall consider only one-body physical quantities, so that we can expect they will be not directly affected by particle collisions. However, as mentioned above, the one-bodydensity will tend towards local equilibrium, so that also the evolution of a one-body quantity has to be influenced by collisions processes, at least in this indirect way.

The simplest way to introduce the collisions is to implement the equation of motion for the one-body density matrix, Eq.(28) with the general Hamiltonian $\mathcal{H}$, by a so-called "relaxation time term", which describes in a phenomenological way its relaxation towards equilibrium, that is

$$\dot{\rho} = -i\left[\mathcal{H},\rho\right] - \frac{1}{\tau}\left(\rho - \rho_{eq}\right) \ , \tag{48}$$

where $\rho_{eq}(t)$ is the density matrix at the local equilibrium, constrained by the total average energy E, the total average angular momentum L and the total number of particle N

$$\rho_{eq}^{(t)} = \rho_{eq}(E,L,N) \ . \tag{49}$$

In Eq.(48) τ is the characteristic time with which the density matrix tends to equilibrium. In fact, in the absence of dynamical evolution, $\mathcal{H} = 0$, the solution of Eq.(48) is

$$\rho(t) = \rho_{eq} + \left(\rho(0) - \rho_{eq}\right) \exp(-t/\tau) \ , \tag{50}$$

i.e. ρ tends exponentially to ρ_{eq} in a time τ . In Eq.(48) the evolution of ρ will be influenced by the interplay between τ and the characteristic times of $\mathcal{H}$. In general, τ depends on the excitation energy - an estimate for it can be found in Ref.21. However, since we are interested only in looking at the possible effects of particle collisions, we shall consider τ as a parameter to be varied.

In this approach there is anyway a difficulty. When, in fact, the two nuclei are not overlapping particle collisions occur only within the orbitals of each nucleus, i.e. the matrix elements $\rho_{bb'}$, with b and b' belonging to different wells, should not relax to zero. On the contrary, when there is an overlap, i.e. a neck, between the two nuclei, then also collisions between orbitals belonging to different wells can take place, however at a different rate than in each separate nucleus. One should then consider two different relaxation times, depending on the overlap of the two nuclei. Since, however, we are considering only peripheral collisions, we shall neglect these particle collisions across the two nuclei. We have to keep in mind that, in this way, near the distance of closest approach, we are missing some particle collisions which are occurring around the neck.

Let us consider, for simplicity, again the central collision for the system ^{86}Kr + ^{208}Pb of Section 4 at Ecm = 308 MeV, including only transfer. We have calculated for different values of τ , the total excitation energy

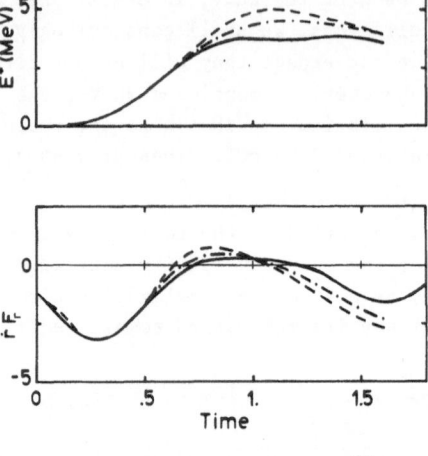

Fig.22. Quantal results for the system ^{86}Kr + ^{208}Pb   at
Ecm = 308 MeV and b = 0 including the relaxation
time τ . The full drawn curve corresponds to
$\tau = \infty$, while the dash-dotted one to τ = 0.25
and the dashed one to τ = 0.05.

and the work per unit time along the trajectory, see Fig.22. As expected the
excitation energy and the frictional force are very little affected even
for very short relaxation times τ . However, the latter clearly shows a
tendency to become more "symmetrical" with respect to the instant of closest
approach. This suggests that the density matrix relaxation tends to reduce
the memory of the radial frictional force,in the sense that, in the outgoing
phase, it behaves very similarly to the approaching phase, as it should
exactly be if no memory were present. Notice anyway the slight increase of
the positive part around the instant of closest approach, which can be attributed to the above discussed lacking of "collisions in the neck".

     In any case, to get a considerable effect, we need a very short relaxa-
tion time, much shorter than the collision time and the estimate of Ref.21[*].
Of course, a complete loss of memory cannot be reached because there is no
damping in the transfer process and therefore, during the collision, the
particles are redistributed among the levels. Hence in the outgoing phase
the physical situation is in any case different than in the approaching one.

     In conclusion, particle collisions are expected to introduce little
effects on the considered (one-body) frictional forces and other one-body
quantities. Knowing the forces acting between the two nuclei at a   given
position is thus necessary to follow the whole trajectory and the evolution
of the system.

---

[*] This estimate is around 0.9 in this case.

9.CONCLUSIONS AND PROSPECTS

We have presented a general microscopic scheme to describe peripheral heavy ion collisions at low energies. In the simplest approximation the model is an independent-particle description, in which one solves the single-particle dynamics in the undistorted mean field of the two colliding nuclei.

By restricting the treatment to transfer processes only, we showed a detailed comparison with the classical picture. The results show a very definite discrepancy with the classical treatment. The microscopic quantal calculations give systematically a smaller ratio between tangential and radial frictional forces. The calculations depend on the shell structure of the nuclei and on the followed trajectory. Consequentely, they can be used only for qualitative corrections to the classical predictions[13]. It seems not possible to extract a simple prescription to calculate the frictional forces, despite the fact that the forces show a remarkable linear dependence on the trajectory in the approaching phase of the collision.

However, the microscopically extracted forces can be straightforwardly used for a self-consistent calculation of the trajectory and of the full reaction.

The inclusion of the inelastic processes affects considerably the energy and angular momentum dissipation rate. In the case of $^{40}Ca + ^{40}Ca$ the transfer and the inelastic processes contribute approximately in equal part. This is true also for the asymmetric system $^{86}Kr + ^{208}Pb$, where the transfer rate either seems to be affected by the presence of the inelastic processes in a considerable way. More systematic work is needed, anyway, to draw general quantitative conclusions. However, the predictions of the model seems to give the right trend in comparison with the few available experimental data. This gives confidence on the possibility of understanding the main mechanisms underlying the dissipation phenomena in peripheral heavy ion collisions as well as the possible extension of this model to more inelastic reactions. To this purpose we have discussed how to include the distortions of the mean field and the inplementation of particle collisions. With these extensions the model should be able to give a microscopic description of deep-inelastic reactions, wherethe collective character of the inelastic excitation is usually essential. Moreover this approach is flexibleenough such as to allow calculations of cross sections for different kinds of collision phenomena, like fusion or for other specific outgoing channels.

In conclusion, more theoretical work in this direction has to be done and more experimental data at these low energies are quite desirable. So that one can check the validity of the physical ideas underlying the different existing models , hence allowing a definite discrimination among them.

The authors would like to thank the Niels Bohr Institute for the warm hospitality and the financial support during the several visits in these years. This work is part of a long term project in collaboration with R.A. Broglia and A. Winther.

APPENDIX

In this appendix we show how to incorporate in the general scheme of Section 3 the effects of the distortions of the mean field, due to the residual interaction, which can give rise to collective excitation of the nuclei, in the particle dynamics. To this purpose it is simpler to use a second quantized formalism. Let us then introduce creation and annihilation operators for the states of the bound state basis set introduced in Section 3. For later convenience it is however useful to adopt the Schroedinger picture, i.e. the states of Eq.(3) are replaced by

$$\chi_b(\vec{r},t) = \exp\left[i(m\vec{R}\cdot\vec{r} - \gamma(t)\right] \Phi_b(\vec{r} - \vec{R}(t)) \quad , \tag{A1}$$

Φ_b being still defined by Eq.(2). The basis states χ_b satisfy

$$\left[i\frac{\partial}{\partial t} + \frac{\nabla^2}{2m} - V(\vec{r}-\vec{R}(t)) + m\ddot{\vec{R}}\cdot(\vec{r}-\vec{R}(t)) - E_b\right]\chi_b(\vec{r},t) = 0 \tag{A2}$$

and we shall call a_b^+, a_b the corresponding creation and annihilation operators. The single-particle Hamiltonian (1) can be written

$$H(t) = \sum_{bb'}(G^{-1}h\,G^{-1})_{bb'}\,a_b^+(t)\,a_{b'}(t) \quad ,$$

$$h_{bb'} = (b, H(t)b') = (G'H_0 + \gamma(t) + A)_{bb'} \quad , \tag{A3}$$

where γ, A are defined in Eq.(13) and $G'_{bb'} = \exp(i(E_b - E_{b'})t)G_{bb'}$ with G defined in Eq.(11). The matrix H_0 is diagonal with diagonal elements equal to the single-particle energies. The operators a^+, a satisfy the commutation relations

$$\left[a_b^+(t), a_{b'}(t)\right] = (G'(t))_{bb'} \quad . \tag{A4}$$

The equation of motion for the creation operators crealy reads

$$i\,\dot{a}_b^+(t) = \sum_{b'}\left[H_0 + (G')^{-1}(\gamma + A)\right]_{bb'}\,a_{b'}^+(t) \equiv \sum_{b'}\mathcal{L}_{b'b}\,a_{b'}^+(t) \quad , \tag{A5}$$

which corresponds to Eq.(15). The residual interaction $v^{(1)}$ pertaining to nucleus 1 can be written

$$v^{(1)} = \frac{1}{4}\sum_{b_1 b_1' b_1'' b_1'''}(b_1 b_1'|v_A^{(1)}|b_1'' b_1''')\,a_{b_1}^+ a_{b_1'}^+ a_{b_1'''} a_{b_1''} \quad , \tag{A6}$$

where

$$(b_1 b_1'|v_A^{(1)}|b_1'' b_1''') = \int d^3\vec{r}_1 d^3\vec{r}_2\,\chi_{b_1}^*(\vec{r}_1,t)\,\chi_{b_1'}^*(\vec{r}_2,t)\,v^{(1)}(\vec{r}_1-\vec{r}_2) \times$$

$$\times\left[\chi_{b_1''}(\vec{r}_1,t)\chi_{b_1'''}(\vec{r}_2,t) - \chi_{b_1'''}(\vec{r}_1,t)\chi_{b_1''}(\vec{r}_2,t)\right] \quad . \tag{A7}$$

At this point, we notice that the matrix elements (A7), because of translational and Galilei invariance, coincide with the usual static two-body matrix elements of the residual interaction

$$(b_1 b_i' | v_A^{(1)} | b_1'' b_1''') = \int d^3\vec{r}_1 d^3\vec{r}_2 \, \Phi^*_{b_1}(\vec{r}_1) \, \Phi^*_{b_i'}(\vec{r}_2) \, v^{(1)}(\vec{r}_1 - \vec{r}_2) \times$$

$$\times \left[\Phi_{b_1''}(\vec{r}_1) \Phi_{b_1'''}(\vec{r}_2) - \Phi_{b_1'''}(\vec{r}_1) \Phi_{b_1''}(\vec{r}_2) \right] \quad , \tag{A8}$$

so that they are time-independent. In presence of the residual interaction the equation of motion (A5) will be modified accordingly. as in Section 3 we shall consider the equation of motion for the one-body density matrix, that is for the product $a^\dagger a$

$$i\frac{\partial}{\partial t}(a^\dagger_b(t) \, a_{b'}(t)) = i \, \dot{a}^\dagger_b a_{b'} + i \, a^\dagger_b \dot{a}_{b'} =$$

$$= \sum_{b''} \mathcal{L}_{b''b} \, a^\dagger_{b''} a_{b'} - \sum_{b''} \mathcal{L}^*_{b''b'} a^\dagger_b a_{b''} + [v^{(1)} + v^{(2)}, \, a^\dagger_b a_{b'}] \quad . \tag{A9}$$

The first two terms at the r.h.s. of Eq.(A9) correspond to the independent-particle dynamics developed in Section 3, while the last term gives the contribution of the residual interaction in nucleus 1 and 2 respectively. In general these equations cannot be easily solved, however they can be linearized according to the standard procedurefollowed in deriving the R.P.A. approximation, see for instance Ref.22. This affects only the matrix elements of the density matrix corresponding to particle-hole. For instance, in nucleus 1, for the matrix elements $\rho_{h_1 p_1}$ of the density matrix correspon-ding to the particle state p_1 and the hole state h_1 , one gets

$$i\frac{\partial}{\partial t} \rho_{h_1 p_1}(t) = \sum_b \mathcal{L}_{b \, p_1} \rho_{h_1 b} - \sum_b \mathcal{L}_{h_1 b} \rho_{b p_1} + \sum_{p_1' h_1'} \{ [(\mathcal{E}_{p_1} - \mathcal{E}_{h_1}) \delta_{p_1 p_1'} \delta_{h_1 h_1'} +$$

$$+ (h_1 p_1' | v_A^{(1)} | p_1 h_1')] \rho_{h_1' p_1'} + (h_1 h_1' | v_A^{(1)} | p_1 p_1') \rho_{p_1' h_1'} \} \quad . \tag{A10}$$

An analogous equation is obtained for $\rho_{p_1 h_1}$ and for the nucleus 2. The matrix appearing inside the curly brakets is just the usual R.P.A. matrix. Eq.(A10) is still a linear equation and can be solved by the same method used in the case of the independent-particle approximation of Section 3.

REFERENCES

1. For a review see: J. W. Negele, Rev. Mod. Phys., 54 (1982) 913.
2. R. A. Broglia, C. H. Dasso and A. Winther, Proceedings of the Interna-tional School of Physycs "E. Fermi", Course LXXVII, Nuclear Structure and Heavy Ion Collisions, Eds. R. A. Broglia and R. A. Ricci, North Holland (1981), p.327.
3. J. Randrup, Nucl. Phys., A307 (1978) 319.
   J. Randrup, Ann. Phys., 112 (1978) 356.
4. J. R. Birkelund, L. E. Tubbs, J.R. Huizenga, J. N. De and D.Sperber, Phys. Rep., 56 (1979) 107.
5. T. Døssing and J. Randrup, Nucl. Phys., A433 (1985) 280 and references therein.
6. G. F. Bertsch and R. Schaeffer, Nucl. Phys., A277 (1977) 509.
   M. Baldo, F. Catara, E. G. Lanza, U. Lombardo and L. Lo Monaco, Nucl. Phys., A391 (1982) 249.
7. H. Esbensen, R. A. Broglia and A. Winther, Ann. Phys., 146 (1983) 149.
8. R. A. Broglia and A. Winther, Phys. Rep., 4C (1972) 153.
9. K. Dietrich and K. Hara, Nucl. Phys., A211 (1973) 349.
10. J. P. Blocki, J. Randrup, C. F. Tsang and W. J. Swiatecki, Ann. Phys., 105 (1977) 427.
11. C. H. Dasso, S. Landowne, G. Pollarolo and A. Winther, Nucl. Phys.,

A459 (1986) 134.

12. S. Pal and D. H. E. Gross, to be published.

13. M. Baldo, A. Rapisarda, R. A. Broglia and A. Winther, to be published.

14. R. A. Broglia, G. Pollarolo and A. Winther, Nucl. Phys., A361 (1981) 307.
    R. A. Broglia, R. Liotta, B. S. Nilsson and A. Winther, Phys. Rep., 29C (1977) 291.

15. G. Pollarolo, R. A. Broglia and A. Winther, Nucl. Phys., A406 (1983) 369.
    J. M. Quesada, Ph. D. Thesis, University of Sevilla, unpublished.
    J. M. Quesada, R. A. Broglia, V. Bragin and G. Pollarolo, Nucl. Phys., A428 (1984) 305C.
    G. Pollarolo and R. A. Broglia, Nuovo Cimento, 81A (1984) 278.
    J. M. Quesada, G. Pollarolo, R. A. Broglia and A. Winther, Nucl. Phys., A442 (1985) 381.

16. R. A. Broglia, C. H. Dasso, G. Pollarolo and A. Winther, Phys. Rep., 48C (1978) 351.

17. P. R. Christensen, F. Folkmann, O. Hansen, O. Nathan, N. Trautner, F. Videbaek, S. Y. van der Werf, H. C. Britt, R. P. Chestnut, H. Freisleben and F. Puhlhofer, Nucl. Phys., A349 (1980) 217.
    P. R. Christensen, O. Hansen, O. Nathan, F. Videbaek, H. Freisleben, H. C. Britt and S. Y. van der Werf, Nucl. Phys., A390 (1982) 336.

18. M. T. Collins and J. J. Griffin, Nucl. Phys., A348 (1980) 63.

19. P. Danielewicz, Phase Space Approach to Nuclear Dynamics, p.37, Eds. M. Di Toro, W. Noremberg, M. Rosina, S. Stringari, World Scientific (1986).

20. See for instance: G. Bertsch, H. Kruse and S. Das Gupta, Phys. Rev., C29 (1984) 673.

21. G. Bertsch, Zeit. Phys., A289 (1978) 103.

22. P. Ring and P. Schuck, The Nuclear Many body Problem, (1980) Springer.

A SIMPLE MODEL FOR DEEP-INELASTIC REACTIONS

G. Pollarolo

Dipartimento di Fisica Teorica, Università di Torino

and I.N.F.N Sezione di Torino, Torino, Italy

1. INTRODUCTION

In these lectures I would like to discuss a model[1] for deep-inelastic heavy-ion reactions that takes explicitly into account the surface degrees of freedom of the two impinging ions.

It is unnecessary to go into the details of a discussion about deep inelastic reactions. In fact there is a monumental monography on this subject by W. U. Schröder and J. R. Huizenga[2] that discuses everything that has been done in this field. So let me just stress the main features of these reactions that are relevant for the justification of the model I am going to present. In order to elucidate them I will use the data[3-6] for the reaction $^{136}Xe + ^{209}Bi$ at the bombarding energy $E_{lab} = 1130\ MeV$ shown in Fig. 1.

i) The angular distribution of the reaction products (see Fig. 1b) integrated over all the masses, charges and energies, shows a pronounced peak at an angle close to the grazing (the actual value of the maximun of this cross section is not very relevant, depending on the adopted Q-value cut-off) . This angular distribution, typical of grazing collision, indicates that the collision time is very short, of the order of 10^{-22} sec.

ii) The excitation function (see Fig. 1c) shows a sharp peak at small energy loss corresponding to the quasi-elastic reactions and a broad bump at larger energy loss corresponding to the deep-inelastic events. It is interesting to point out that many fragments emerge with an energy that is smaller than the Coulomb energy of a pair of touching spheres, indicating that during the interaction the two ions develop large deformations.

iii) The charge distribution (see Fig. 1d) of the emerging fragments presents two peaks centered at the initial values of projectile and target. As a function of the Q-values they have a width that is proportional to the energy loss. This suggests that the transfer of particles between the two ions should provide the dominant mechanism in the evolution of the reaction.

Many model have been developed for the study of these reactions (cf. ref. [2]) in which the particle transfer plays the central role as a dissipating mechanism for energy and angular momentum. This is treated as a stochastic process i.e. the transfer of a nucleon does not influence the transfer of the next. In all these models one mimics the large deformations involved by introducing an adiabatic potential or appropiate neck degrees of freedom.

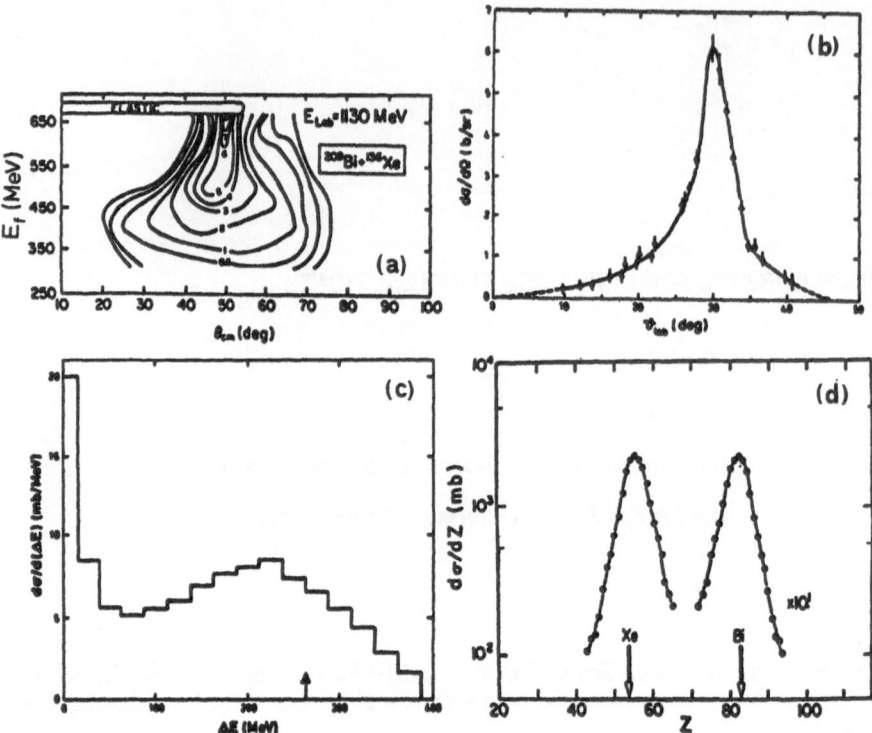

Fig. 1. *Experimental data[3-6] for the reaction $^{136}Xe + ^{209}Bi$ at $E^{lab} = 1130 MeV$. The Wilczyn- ski plot is shown in (a) while the distributions, in angles, in excitation energy and in the charge of the fragments are displayed in (b), (c) and (d) respectively. The arrow in (c) indicates the position of the Coulomb barrier for two touching spheres.*

It is well known that the nucleus displays surface degrees of freedom. The collective low- lying modes and the giant isoscalar resonances can , in fact, be identified as surface oscillations. It is thus tempting to develop a model in which these degrees of freedom are taken into account explicitly and thus have a consistent description of the deformation. Such a model should be able to describe the full range of heavy ion reactions from the quasi-elastic regime to the fusion.

2. SHAPE DEGREES OF FREEDOM AND NUCLEAR RESPONSE FUNCTION

In the study of a system with a large number of particles the concept of elementary modes of excitation proves to be extremely useful in the description of this system. In particular the spectrum of the nuclei around the closed shell can be described in terms of surface vibrations, pairing vibrations and single particle excitations. The properties of these elementary modes is quite well established. Large enhancements in the inelastic and two-particle transfer reactions leading to the corresponding collective states has been clearly seen in heavy-ions. It is just natural to devise a model in which these degrees of freedom are treated explicitly. Up to now we have only been able to develop a model that treats consistently the surface vibrations. The particle transfer degrees of freedom must be incorporated through transport equations in the one-body approximation.

I introduce the shape's degrees of freedom by generalising the liquid drop model with the usual multipole expansion of the surface:

$$R(\hat{r}) = R_0 \left[1 + \sum_{n,\lambda\mu} \alpha_{n,\lambda\mu} Y^*_{\lambda\mu}(\hat{r}) \right] \tag{1}$$

where R_0 is the equilibrium radius and the $\alpha_{n,\lambda\mu}$ are the parameters describing the displacement of the nuclear surface from its spherical shape. In the hydrodynamical model we have only one mode for each multipolarity. Due to the shell structure this mode is actually split into several ones. This is the reason for the extra label n in eq. (1).

In the harmonic approximation the $\alpha_{n,\lambda\mu}$ degrees of freedom are described by the Hamiltonian:

$$H_{vib} = \sum_{n,\lambda\mu} \left\{ \frac{|\Pi_{n,\lambda\mu}|^2}{2D_{n,\lambda}} + \frac{1}{2} C_{n,\lambda} |\alpha_{n,\lambda\mu}|^2 \right\} \tag{2}$$

where $\Pi_{n,\lambda\mu}$ are the conjugate momenta to the $\alpha_{n,\lambda\mu}$

$$\Pi_{n,\lambda\mu} = D_{n,\lambda} \dot{\alpha}^*_{n,\lambda\mu} \tag{3}$$

while $D_{n,\lambda}$ and $C_{n,\lambda}$ are the mass and the restoring force parameters, respectively. These two parameters are related to the energy $E_{n,\lambda} = \hbar\omega_\lambda$ of the mode

$$\hbar\omega_\lambda = \hbar \sqrt{\frac{C_{n,\lambda}}{D_{n,\lambda}}} \tag{4}$$

and to the transition amplitude from the ground state to the one phonon state

$$< 1_{n,\lambda} \| M(\lambda) \| 0 > \propto \sqrt{\frac{\hbar\omega_\lambda}{D_{n,\lambda}}} \tag{5}$$

where $M(\lambda)$ is the mass multipole operator defined by:

$$M(\lambda\mu) = \int r^\lambda \rho(r) Y^*_{\lambda\mu}(\hat{r}) d^3 r \tag{6}$$

Among the many states that define the complexity of a nuclear spectra only few can be ascribed to surface vibrations, one has thus to learn how to recognize them. First of all we must remember that we want to descibe the shape of a nucleus. We are thus interested in vibrations in which neutrons and protons oscillate in phase with the same amplitude, i.e. the isoscalar mode (T=0).

In order to get a shape deformation many nucleons have to be involved in a coherent way to form the corresponding excited state, only collective states can thus be ascribed to surface oscillations. A convenient measure of the collectivity of a state is given by its strength defined as:

$$S_n(\lambda) = \hbar\omega_{n,\lambda} | < 1_{n,\lambda} \| M(\lambda) \| 0 > |^2 \tag{7}$$

$$= \hbar\omega_{n,\lambda} B(\lambda, 0 \rightarrow 1_{n,\lambda})$$

where $B(\lambda, 0 \to 1_{n,\lambda})$ represents the probability for the transition from the ground state to the one phonon state. The usefulness of this quantity relies on the fact that we can define, for any given multipolarity, an energy weigthed sum rule (EWSR)

$$S(\lambda) = \sum_n S_n(\lambda) = \frac{3\lambda(\lambda+1)}{4\pi} \frac{\hbar^2 A}{2m_0} R_0^{2\lambda-2} \tag{8}$$

whose values can be estimated in a model independent way. The ratio,

$$f_{n,\lambda} = \frac{S_n(\lambda)}{S(\lambda)} \tag{9}$$

gives the fraction of the EWSR exhausted by the given state. The plot of this fraction as a function of the excitation energy (called response function) allow us to identify the energies for which one has an appreciable concentration of strength. The reason this concentration of strength can be interpreted as surface vibration will be clear in the following. For instance if one makes a plot for the quadrupole states $\lambda = 2^+$ one sees, indeed, a concentration of strength in two regions of excitation energy. One at few MeV, corresponding to the low-lying state and one at an energy around $60A^{-1/3}$ corresponding to the giant state.

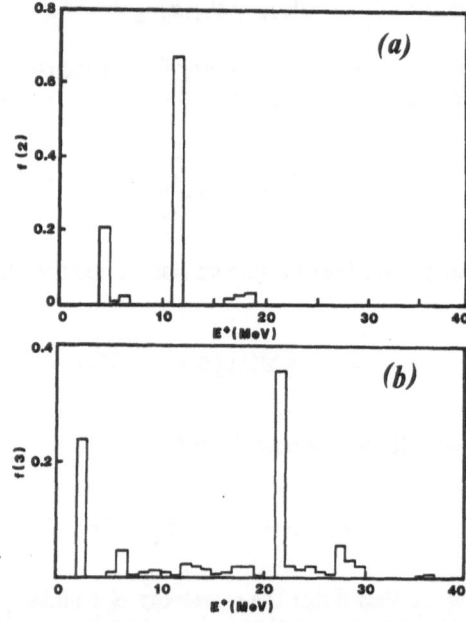

Fig. 2. RPA responce function for ^{208}Pb. In (a) the responce function for the quadrupole states while for the exadecapole states is given in (b).

The existence of this concentration of strength at high energy is very well established for the quadrupole states. Evidence is still lacking for the higher multipolarities so our knowledge of the high energy part of the response function must depend on theoretical calculations. These are usually performed in the random-phase-approximation (RPA). A calculation of this sort proceeds as follows: starting from a collection of single particle levels constructed from a shell-model potential (Hartree-Fock or phenomenological) one builds, for a given multipolarity, all possible particle-hole (p-h) states. In this (p-h) base the residual interaction, i.e. the part of the nucleon-nucleon interaction not taken into account by the average field, is diagonalized.

In Fig. 2 we show, for the case of ^{208}Pb, the result of such a calculation for $\lambda = 2$ and $\lambda = 3$. These calculations were obtained by diagonalizing a separable interaction in a (p-h) base constructed with Hartree-Fock single particle levels generated by a Skyrme III two body force. It is apparent from this figure that, at least for the low multipolarities, the EWSR is essentially exhausted by two states, one at low energy with 10-20 % of the EWSR and one at high energy exhausting the remaining fraction of the sum rule.

A considerable part of our knowledge about the low lying states comes from the analysis of inelastic scattering in grazing collisions. These analyses are performed in the formalism of the DWBA using a form factor derived from the macroscopic model. Their results are in good agreement with the one obtained from the Coulomb excitation experiments giving support to the hypothesis that neutron and proton oscillate in phase. The form factors for the excitations can also be calculated microscopically using the RPA wave functions[8]. They are in good agreement (at least in the tail region) with the macroscopic one. This agreement seems also to extend to the high lying states. We can therefore interpret these giant states as surface vibrations.

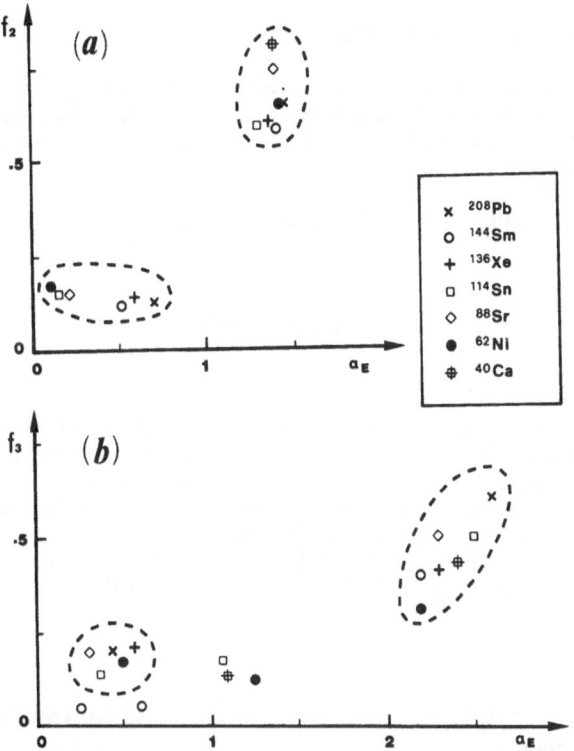

Fig. 3. *The fraction of the EWSR exasted by the RPA collective roots for the quadrupole (a) and exadecapole (b) states are shown as a function of the parameter α_E intruduced in (12). The centroinds of the circled areas define the universal response function.*

Summarizing the discussion, the $C_{n,\lambda}$ and the $D_{n,\lambda}$ parameters of the vibrational hamiltonian (2) are determined from the energy and the strength of the modes:

$$E_{n,\lambda} = \hbar \sqrt{\frac{C_{n,\lambda}}{D_{n,\lambda}}} \qquad (10)$$

$$S_{n,\lambda} = \left(\frac{3AR_0^\lambda}{4\pi}\right)^2 (2\lambda + 1)\frac{\hbar^2}{2D_{n,\lambda}} \qquad (11)$$

The values of these two quantities are deduced either from experiments or from RPA calculations of the nuclear response.

For the low multipolarities the RPA strength is usually concentrated in a few isolated roots (cfr. Fig.2.). Around the root at high excitation energy there exist other kinds of excitations like (2p-2h) that couple to the RPA state, giving rise to a broadening of the strength distribution. In our description besides the energy and the strength of the mode we must also give a third quantity, the width $\Gamma_{a,\lambda}$ of the state. For multipolarities larger than 3 the strength distribution is usually fragmented over several RPA roots. These are also coupled to the background of (2p-2h) states giving rise to additional width.

How the spreading width and the fragmentation of strength is incorporated in our model will be discussed later on.

2.1 Universal response function

During the collision, exchange of particles takes place between the two ions. The response function we must use should not be that of target and projectile but it should be therefore an "average" responce function describing the concentration of collective strength also of the neighbouring nuclei.

We have seen that the energy of the states scales like $A^{-1/3}$,

$$E_{a,\lambda} = \alpha_E A^{-1/3} \tag{12}$$

Now, plotting for each multipolarity λ the fraction $f_{a,\lambda}$ of the EWSR exhausted by each state as a function of the parameter α_E, we see that, for all nuclei, the RPA collective roots concentrate in two regions (cfr. Fig. 3). It is natural to take the centroids of these two regions as representative of the states. Following this procedure for all multipolarities, we obtain the "universal responce function" of Table 1. The values of the widths are given as indicative since very little is known about them. The large numbers used for $\lambda = 4$ and $\lambda = 5$ reflect the fact that for this multipolarity the RPA response is fragmented over several RPA roots.

TABLE 1. - *Parameters for the modes corresponding to the universal response function as discussed in the text.*

λ	α_E	$\Gamma(MeV)$	%EWSR
2+	.41	–	20
	1.40	2	80
3–	.40	–	25
	2.36	4	50
4+	1.43	–	25
	3.20	6	60
5–	0.65	–	12
	2.25	8	50

2.2 Adiabatic cut-off

So far we have discussed the response function, we must now discuss those states which are important in the time evolution of the reaction.

Let us define the collision time τ_{coll} as the ratio between the sum of the two nuclear radii and the relative velocity at the top of the Coulomb barrier

$$\tau_{coll} = \frac{R_a + R_A}{\sqrt{\frac{2}{m_{aA}}(E_{CM} - E_B)}} \tag{13}$$

where by a and A we label the radius of projectile and target, by m_{aA} the reduced mass and with CM and B we have labelled the center of mass bombarding energy and Coulomb barrier respectively.

In the range of bombarding energies between 5 - 20 MeV/N we get:

$$\frac{\hbar}{\tau_{coll}} \simeq 1 - 2 \ MeV \tag{14}$$

This would imply that only states with an energy of about 1 MeV are excited during the collision. However it is through the form factor that a given state can be excited. This form factor has a tail with a decaying length $a \simeq 0.6 fm$, i.e. in $1 fm$ its value drops one-order of magnitude. A better estimation of the time scale for a given excitation is thus the time needed to cover this distance. We call this charactheristic time τ_{ch}. For the bombarding energies mentioned above one gets:

$$\frac{\hbar}{\tau_{ch}} \simeq 10 - 20 \ MeV \tag{15}$$

so also the giant states can be excited in a heavy-ion collision.

3. THE INTERACTION

In the previous section we have discussed at length our choice of variables for the description of the intrinsic states of the two nuclei. From the analysis of the reaction products we see that the two ions maintain their identity throughout the reaction so, for the description of the relative motion, we are entitled to use the vector $\vec{r}$ between the centers of mass of the two impinging particles. For the mass transfer degrees of freedom we use the mass and charge of the projectile.

TABLE 2. *List of the collective variables used in the model in the case of planar trajectories (cfr. Fig. 4).*

relative motion	intrinsic motion	mass transfer
r, ϕ	$\alpha_{n,\lambda\mu}(i) \ \ i = (a, A)$	A_p, Z_p

To describe the evolution of the reaction we have now to introduce the interaction. Since the number of degrees of freedom we have chosen are less than the actual number, our interaction must have a dissipative component as well as a conservatice one. This describes the coupling among the degrees of freedom treated explicitly and the those that are ignored. Let me discuss these two components separately.

3.1 Conservative interaction

First of all the two ions interact via the Coulomb force. This interaction U^C consists partly of the monopole term and partly of terms arising from the interaction of all the multipoles of A with the full charge of a and viceversa. The multipole-multipole terms are neglected. The expression we have used for U^C is :

$$U^C = \frac{Z_a Z_A e^2}{r} + Z_a e \sum_{\lambda\mu}^{(A)} \frac{3 Z_A e (R_A)^\lambda}{2\lambda+1} \frac{Y^*_{\lambda\mu}(\hat{r})}{r^{\lambda+1}} \alpha_{\lambda\mu}(A)$$
$$+ Z_A e \sum_{\lambda\mu}^{(a)} \frac{3 Z_a e (R_a)^\lambda}{2\lambda+1} \frac{Y^*_{\lambda\mu}(-\hat{r})}{r^{\lambda+1}} \alpha_{\lambda\mu}(a) \quad (16)$$

where the multipole moments have been evaluated in the linear approximation in the deformation parameters.

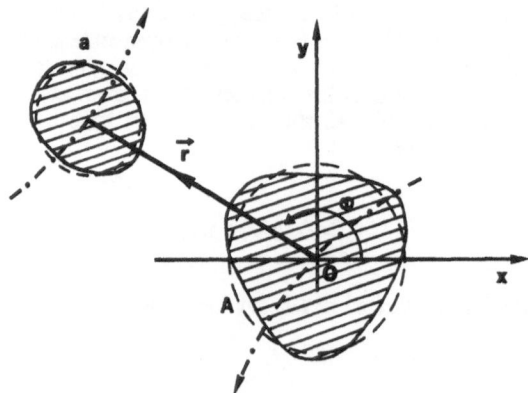

Fig. 4. *The coordinates used to describe the relative motion.*

When the two ions come close enough so that the two nuclear densities start to overlap, an additional interaction takes place. Information about this interaction (al least to its tail) can be obtained through the study of elastic scattering data in the formalism of the optical model. A straightforward way to calculate this interaction is to evaluate the double-folding integral

$$U^N(\vec{r}) = \int \rho_1(\vec{r}_1)\rho_2(\vec{r}_2) V_{12}(|\vec{r} - \vec{r}_1 - \vec{r}_2|) d^3 r_1 d^3 r_2 \quad (17)$$

where ρ_i are the mass densities of projectile and target and V_{12} is the effective nucleon-nucleon interaction.

Another approach to calculate $U^N(\vec{r})$ is the proximity approximation[3] .In this approximation one calculates the interaction between two gently curved surfaces. One gets:

$$U^N(r) = 4\pi\gamma \bar{R} b \Phi(s/b) \quad (18)$$

where γ ($\simeq 1\ MeV\ fm^{-2}$) is the surface tension and b ($\simeq 1\ fm$) is the surface thickness. The actual shape of the surfaces enters through the parameter $\tilde{R}$ that depends on the curvatures of the target and projectile at the point of closest approach.

$$\tilde{R} = \left[(C_\parallel^t + C_\parallel^p)(C_\bullet^\perp + C_\perp^p) \right]^{-1/2} \qquad (19)$$

The function Φ has been parametrized as follows:

$$\Phi(\xi) = \begin{cases} -3.437 \; exp\left(-\frac{\xi}{0.75}\right) & \xi \geq 1.5 \\ -0.5(\xi - 2.54)^2 - 0.085(\xi - 2.54)^3 & \xi \leq 1.5 \end{cases} \qquad (20)$$

The quantity s represents the distance of closest approach between the two facing surfaces. For a spherical configuration the tail of this potential is in good agreement with the elastic scattering data.

The universal function $\Phi(s/b)$, beside the attractive component at large distances, presents a repulsion when the two surfaces start to overlap. This repulsion reflects the enormous energy required to create regions where the density exceeds the normal nuclear matter value.

3.2 Non-conservative interaction. Particle transfer

At short distances the two ions interact not only through the residual nucleon-nucleon interaction. Nucleons can be exchanged between the two ions, this exchange resulting in an additional interaction. This is not conservative and it will be described with a dissipative force. The effect of the particle-transfer process can be incorporated in our model treating it as a diffusion process. Following J. Randrup[9] the particle transfer gives rise to a frictional force $\vec{F}$ that, in the proximity approximation, can be written as:

$$\vec{F} = -4\pi n_0 \tilde{R} b \left(\vec{U}_n + \frac{1}{2}\vec{U}_t \right) \Psi(s/b) \qquad (21)$$

where n_0 represents the flux in the bulk of the nuclear matter. The two quantities $\vec{U}_n$ and $\vec{U}_t$ represent the relative velocities of the two ions in the direction perpendicular and parallel to the window respectively. The function Ψ can be parametrized as follows:

$$\Psi(\xi) = \begin{cases} 1.4 - \xi & \xi \leq -0.4 \\ 1.6 - 0.6\xi - 0.57sin\left(\frac{\xi+0.4}{1.14}\right) & -0.4 \leq \xi \geq 3.2 \\ 0 & \xi \geq 3.2 \end{cases} \qquad (22)$$

The relative velocity $\vec{U}$ can be evaluated at each instant of time as a function of the time derivatives of the collective variables defining the intrinsic states on the two nuclei.

4. EQUATIONS OF MOTION

In the discussion above we stressed that the dynamics of our system is determined both by conservative and non conservative forces. The Hamilton equations of motion must be modified as follows:

$$\frac{dp_s}{dt} = -\frac{\partial H}{\partial q_s} - \frac{\partial \mathcal{F}}{\partial \dot{q}_s} \qquad (23)$$

$$\frac{dq_s}{dt} = \frac{\partial H}{\partial p_s}$$

where with $\{q_s\}$ and $\{p_s\}$ we have indicated, in short hand, the coordinates and the corresponding conjugate momenta of our system.

In these equations by H we mean the Hamiltonian describing the conservative part of our system, given by

$$H = T(r,\phi,p_r,p_\phi) + H_{vib}(a) + H_{vib}(A) + U^C(r,\alpha_{n,\lambda\mu}(i)) + U^N(r,\alpha_{n,\lambda\mu}(i)) \qquad (24)$$

where T is the kinetic energy part for the relative motion. For planar trajectories we have:

$$T = \frac{p_r^2}{2m_{sA}} + \frac{p_\phi^2}{2m_{sA}r^2} \qquad (25)$$

We denote the Rayleigh function by $\mathcal{F}$. This function must be introduced in order to take into account the dissipative forces. It is defined in such a way that twice its value gives the rate at which the energy is dissipated:

$$\frac{dE}{dt} = -2\mathcal{F} \qquad (26)$$

This function can only be defined if the dissipative force is proportional to the velocity. The force (21), arizing from the transfer of particles, is proportional to the relative velocity at the window, so we have:

$$\mathcal{F}_{MT} = \vec{F}_{MT} \cdot \vec{U} \qquad (27)$$

In the discussion of the response function we have seen that the high lying modes are characterized by a dumping (or spreading) widths. These widths are responsible for the dumping of energy from deformation to the thermal bath (uncorrelated particle motion) of the two ions. This effect can be incorporated in our description remembering that the equation:

$$D_\lambda \ddot{\alpha}_{\lambda\mu} + \gamma_\lambda \dot{\alpha}_{\lambda\mu} + C_\lambda \alpha_{\lambda\mu} = 0 \qquad (28)$$

describes a dumped harmonic oscillator. The dumping coefficient γ is related to the width of the mode by the relation:

$$\gamma_\lambda = \frac{2\Gamma_\lambda D_\lambda}{\hbar} \qquad (29)$$

The corresponding Rayleigh function can be written as:

$$\mathcal{F}_{vib} = \sum_{n,\lambda\mu} \frac{2\Gamma_{n,\lambda}D_{n,\lambda}}{\hbar} |\dot{\alpha}_{n,\lambda\mu}|^2 \qquad (30)$$

In order to write explicitly the equations of motion (23) I must still specify the distance of closest approach s between the two surfaces and the relative velocity $\vec{U}$ at the window of the two nuclear matter pieces. These two quantities enter in the expression of the proximity interaction (18) and of the proximity frictional force (21) respectively.

The distance s between the two nuclear surfaces is calculated along the vector $\vec{r}$ joining the center of mass of the two ions. In terms of the deformation parameters $\alpha_{n,\lambda\mu}(i)$ it is written as:

$$s = r - R_0^a\left[1 + \sum_{n,\lambda\mu}^{(a)} \alpha_{n,\lambda\mu}(a)Y_{\lambda\mu}^*(-\hat{r})\right] - R_0^A\left[1 + \sum_{n,\lambda\mu}^{(A)} \alpha_{n,\lambda\mu}(A)Y_{\lambda\mu}^*(\hat{r})\right] \tag{31}$$

The proximity potential is a function of the deformation parameters $\alpha_{n,\lambda\mu}(i)$ also through the reduced radius $\tilde{R}$. This is calculated using (4.31) and (4.32) of ref. [1] .

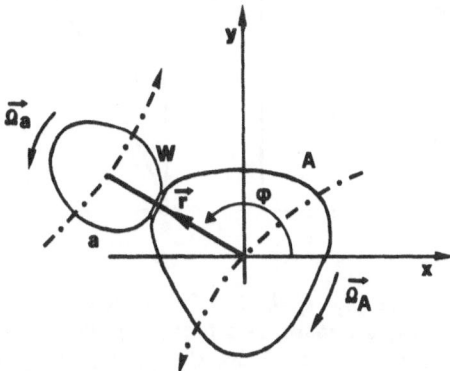

Fig. 5. *The same as Fig. 4. With W we denote the position of the window and with Ω_i the intrinsic angular velociteis of the two ions.*

The velocity $\vec{U}$ defining the frictional force due to the mass transfer should be calculated at the window (W), as for the distance s we suppose that the window opens along the vector $\vec{r}$. Following the notation of Fig. 5 this is defined as follows:

$$\vec{U} = \vec{v}_a(W) - \vec{v}_A(W) \tag{32}$$

$$= \vec{r} + \vec{\Omega}_a \times \vec{R}_a(W) - \vec{\Omega}_A \times \vec{R}_A(W) + \vec{v}^{irr}(W)$$

with

$$\vec{v}^{irr}(W) = \vec{v}_a^{irr}(W) - \vec{v}_A^{irr}(W) \tag{33}$$

The angular velocity $\vec{\Omega}_i$ is calculated from the intrinsic angular momentum l_i using the rigid moment of inertia of nucleus i. With $\vec{v}_i^{irr}(W)$ I indicated the velocity of the nuclear surface of nucleus i due to the vibrational degrees of freedom. Using the irrotational flow approximation[9] we have:

$$\vec{v}_i^{irr}(W) = \vec{\nabla}\left[\sum_{n,\lambda\mu} \frac{R_0^i}{\lambda}\left(\frac{R_W^i}{R_0^i}\right)^\lambda \dot{\alpha}_{n,\lambda\mu}(i)Y_{\lambda\mu}^*(\theta_W,\phi_W)\right] \tag{34}$$

Althougth the equations (23) look quite complicated, they can be easily solved by computer [10] as they are coupled linear equations. The physical picture is also quite simple since it describes the collision between two classical bodies which interact in the surface, and subsequently deform. The deformation follows dump harmonic oscillations. In addition there is a frictional force, arising from particle transfer, that dissipates energy from the relative motion to the intrinsic excitation of the two ions.

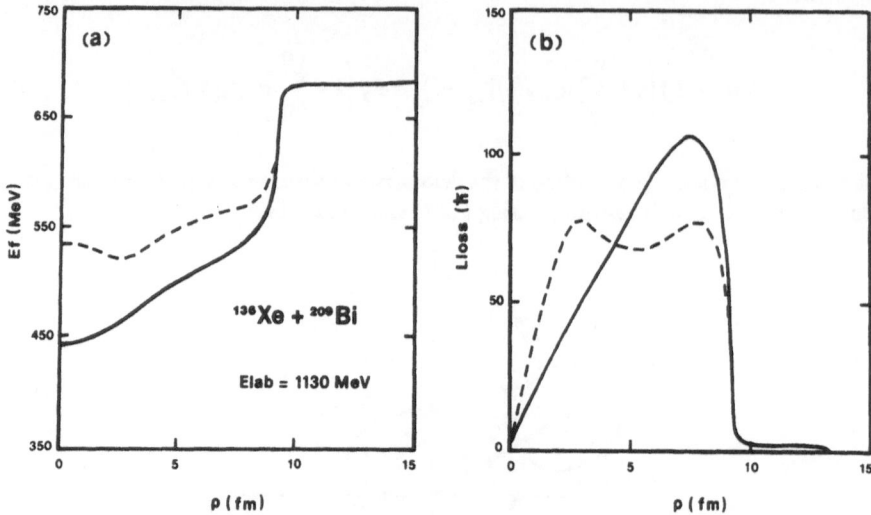

Fig. 6 *The total kinetic energy (E_f) of the two fragments after the collision and the total angular momentum loss (L_{loss}) are shown as a function of the impact parameters (ρ). The dashed lines indicate the result of the calculation when for the velocy $\vec{U}$) we have used the full expression (32). The continous line indicates the result when the irrotational component is neglected.*

From the solution of this system of coupled equations one can extract rather detailed information on the reaction. At the moment let me concentrate on a few entities, namely the final energy in the relative motion and the total angular momentum loss which we follow as a function of the impact parameter ρ (initial angular momentum). In Fig. 6, we display these two quantities for the reaction $^{136}Xe + ^{209}Bi$ at $E_{lab} = 1130MeV$. The dashed curves correspond to a calculation in with we have utilized the full expression (32) for the velocity $\vec{U}$, while the full curves are the result of the same calculation where we have neglected the irrotational velocity flow. As we see the influence of this term is relevant both for the energy and the angular momentum loss. From the analysis of the time evolution of the reaction one sees that this term plays a dominant role in the way-out of the trajectory where the two ions display the maximum deformation. The velocity flow correspsonding to the collective modes of a nucleus has been evaluated for several RPA roots[11]. From these calculations it seems that the irrotational approximation is justified for the high lying modes, relevant in the approaching phase of the trajectory, while it is very poor for the low lying states. These are the relevant degrees of freedom for the description of the deformation and play the main role in the way-out of the trajectory.

Since it is very difficult to incorporate in our model the results of the microscopic calculations for the irrotational velocity flow, we have decided to drop this term from the velocity $\vec{U}$. Clearly further studies of this subject are necessary.

4.1 Energy and angular momentum balance

The equations of motion (23) do not conserve energy and angular momentum because of the presence of dissipative terms. The angular momentum balance (since we consider only planar trajectories, only the z component is relevant) has to be obtained as follows. In addition to the angular momentum in the relative motion L_{rel} the deformation degrees of freedom carry angular momentum. The z component of the angular momentum carried by the mode n, λ is given by[9] :

$$L_{\mathfrak{a},\lambda}(i) = \sum_{\mu} \mu \alpha_{\mathfrak{a},\lambda\mu}(i) \tau_{\mathfrak{a},\lambda\mu}(i) \tag{35}$$

The total angular momentum carried by the vibrations is thus, for the target:

$$L_{vib}(A) = \sum_{\mathfrak{a},\lambda} L_{\mathfrak{a},\lambda}(A) \tag{36}$$

A similar expression old for the projectile angular momentum.

The angular momentum dissipated in either of the two nuclei through the dumping term (29) can be calculated by integrating, along the trajectory, the rate of angular momentum loss, i.e.

$$L_{vib,dis}(i) = \int_{-\infty}^{+\infty} dt \sum_{\mathfrak{a},\lambda} \frac{\gamma_{\mathfrak{a},\lambda}(i)}{D_{\mathfrak{a},\lambda}(i)} L_{\mathfrak{a},\lambda}(i) \tag{37}$$

Similarly one may calculate the loss of angular momentum from the relative motion due to the mass transfer, by the integral:

$$L_{trans,dis} = \int_{-\infty}^{+\infty} dt (\vec{r} \times \vec{F}_{MT})_z \tag{38}$$

Since the frictional force (21) is supposed to act at the point of contact between the two nuclei we can calculate how the angular momentum (38) is shared between them. As we have furthermore assumed that the point of contact lies on the line connecting the two centers, we get:

$$L_{trans,dis} = L_{trans,dis}(a) + L_{trans,dis}(A) \tag{39}$$

with

$$\dot{L}_{trans,dis}(i) = \frac{R_i(t)}{R_a(t) + R_A(t)} \dot{L}_{trans,dis}(t) \tag{40}$$

where the R_i are given by (1).

If we sum all the angular momenta:

$$L_{tot} = L_{rel} + L(a) + L(A) \tag{41}$$

where

$$L(i) = L_{vib}(i) + L_{vib,dis}(i) + L_{trans,dis}(i) \tag{42}$$

we find the overall angular momentum conservation.

The total angular momentum that is dissipated in each nucleus is assumed to contribute to an overall rotation around the z-axis. Assuming that the moment of inertia of the nucleus i is J_i, we find that the nucleus i rotates with angular frequency

$$(\vec{\Omega}_i)_z = \frac{L_{vib,dis}(i) + L_{trans,dis}(i)}{J_i} \qquad (43)$$

In actual calculations we have used the rigid moment of inertia for a sphere.

The energy which is initially present as kinetic energy of relative motion is gradually transformed into excitation energy of the fragments. In analogy with (41) we may write the energy balance in the form:

$$E_{tot} = E_{rel} + E(a) + E(A) + U^N + U^C \qquad (44)$$

where

$$E_{rel} = \frac{p_r^2}{2m_{aA}} + \frac{p_\phi^2}{2m_{aA}r^2} \qquad (45)$$

The excitation energy of the fragments is partly present in the vibrational modes and partly in disordered motion due to the work of the dissipative forces.

$$E(i) = E_{vib}(i) + E_{trans,dis}(i) + E_{vib,dis}(i) \qquad (46)$$

The vibrational energy is:

$$E_{vib}(i) = \sum_{n,\lambda} E_{n,\lambda}(i) \qquad (47)$$

$$E_{n,\lambda}(i) = \sum_{\mu} \frac{1}{2D_{n,\lambda}(i)} |\Pi_{n,\lambda\mu}(i)|^2 + \frac{1}{2}C_{n,\lambda}|\alpha_{n,\lambda\mu}(i)|^2 \qquad (48)$$

The energy dissipated through the damping term (29) is calculated by integrating the rate of energy loss along the trajectory ,i.e.:

$$E_{vib,dis}(i) = \sum_{n,\lambda\mu} \int_{-\infty}^{+\infty} dt \frac{\gamma_{n,\lambda}(i)}{D_{n,\lambda}(i)^2} |\Pi_{n,\lambda\mu}(i)|^2 \qquad (49)$$

whereas the total energy dissipated through the mass transfer is given by the time integral:

$$E_{trans,dis} = E_{trans,dis}(a) + E_{trans,dis}(A) = \int_{-\infty}^{+\infty} dt(\vec{r}\cdot\vec{F}_{MT}) \qquad (50)$$

This energy in divided between the two nuclei in the ratio of their masses.

4.2 Fluctuations and Zero Point Motion

The description of the collision process in terms of the average trajectory, as with the system of coupled equations (23), would be complete if the fluctuations about the expectation values of the dynamical variables were small. As seen from the data in Fig. 1) this is not the case. Large fluctuations are indeed present in all the measured quantities like the energy loss, the scattering angle, the mass and charge of the fragments.

Inspired by the results of the linear coupling model (cf. Ref [1] and ref. therein) The fluctuations in the energy loss E_{loss} (excitation energy) can be incorporated in a naive way. Making use of the results of the linear coupling model the dispersion around the average energy loss due to the excitation of the surface modes can be written as:

$$(\Delta E_{loss})^2_{vib} = \sum_{n,\lambda}\left[<N_{n,\lambda}(a)> (\hbar\omega_{n,\lambda}(a))^2 + <N_{n,\lambda}(A)> (\hbar\omega_{n,\lambda}(A))^2 \right] \tag{51}$$

where $<N_{n,\lambda}(i)>$ is the average number of phonons present in the mode n,λ of nucleus i, and $\hbar\omega_{n,\lambda}$ indicates the corresponding energy of the mode.

Besides the excitation of the surface modes energy is also dissipated through the mass transfer. This is a statistical process, and we can estimate its contribution to the dispersion in energy as proportional to the portion of energy loss due to the mass transfer $E^{trans,dis}$. The total dispersion in the energy loss can be written:

$$(\Delta E)_{loss} = \sqrt{(\Delta E_{loss})^2_{vib} + E^2_{trans,dis}} \tag{52}$$

Clearly this procedure can only give an idea regarding the fluctuations present in the excitation energy of the fragments. The translation in the spread of the kinetic energy of relative motion may not be adequate. In this framework we do not know how to feed this information about the fluctuations back to the relative motion and thus to the scattering angle. To overcome this difficulty one must have a treatment of the fluctuations that can be extended beyond the linear coupling limit. This treatment may be provided by the Wigner transform. For a general discussion see ref. [12]. Here I limit myself to few details of how the problem of the fluctuations can be incorporated in a semiclassical description.

The ground-state wave function of a harmonic system ascribes both to the coordinate and the conjugate momentum a Gaussian distribution of probability with zero-expectation value and with standard deviations

$$(\Delta x)^2 = \frac{\hbar\omega}{2C} \tag{53}$$

$$(\Delta p)^2 = \frac{\hbar\omega D}{2} \tag{54}$$

One can thus think of the ground state of an oscillator as classically represented by an ensamble of point in the phase-space (x,p) which are spread around the origin with Gaussian-like density.

In the linear coupling limit this distribution of points would exactly follows the displacement of the centroid. The spread of the excitation energy is thus given by (51). For the general case of non linear coupling one can find approximatly the final phase-space distribution. One solves the system of coupled equations (23) with random initial conditions for the coordinate and momenta associated with the surface modes, in accordance with their initial phase-space distribution. Later on I will show some calculations performed with random initial conditions for the harmonic variables.

5. APPPLICATIONS.

In this section I will discuss some solutions of the classical equations of motion (23) pertinent to actual cases. We do not expect that the simple model outlined above is in condition

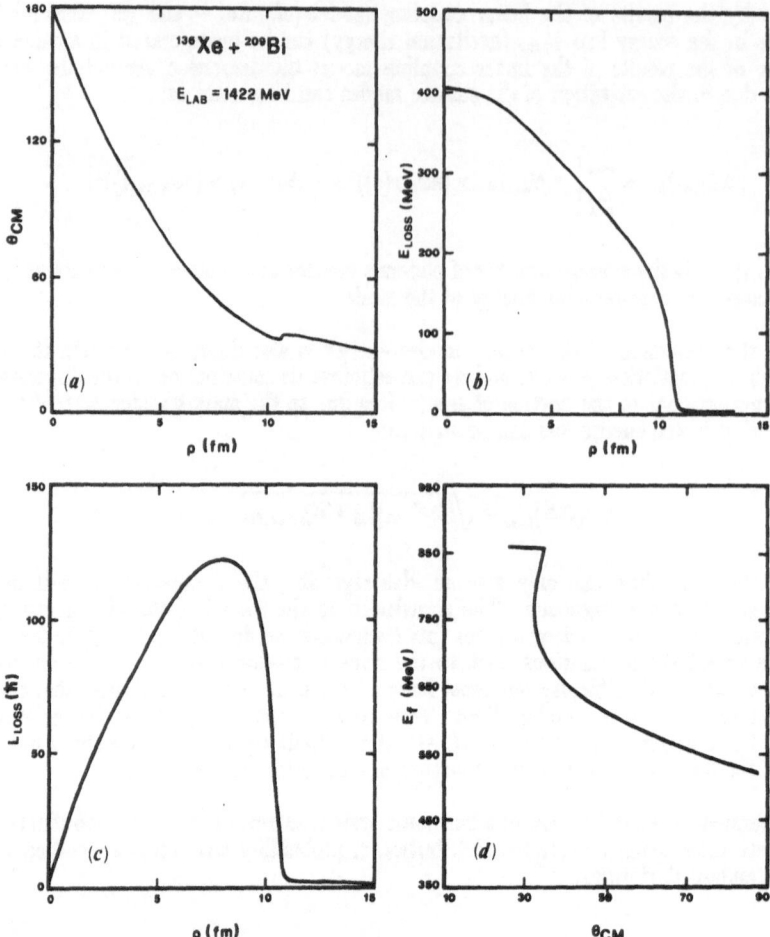

Fig. 7. *The deflection function (a), the energy loss (b) and the angular momentum loss (c) as a function of the impact parameter (ρ) for the reaction $^{136}Xe + ^{209}Bi$ at 1422MeV. The correlation between the scattering angle and the total kinetic energy (E_f) is shown in (d).*

to describe in full details the many facets of the experimental data. On the other end, we expect, that it is able to describe the evolution shown by these data respect to different combinations of target and projectile and at different bombarding energies. With our choice of collective variables we should, in fact, be in condition to describe the full range on heavy-ion reactions, from the quasi elastic regime to the deep-inelastic and even fusion reactions.

This section will be organized as follows. First we discuss the reaction $^{136}Xe + ^{209}Bi$ in order to illustrate the different quantities we can extract from our model. In this way we also illustrate the output of TORINO code[11] that as been developed for the numerical solution of the system of coupled equations (23). Then we will show some specific applications for the angular momentum transfer, the energy sharing and some examples of fusion reactions. To the study of the quasi-elastic reactions will be devoted the last section.

5.1 The $^{136}Xe + ^{209}Bi$ reactions

As an example I use the reaction $^{136}Xe + ^{209}Bi$. This has been measured in great details [2-6] at several bombarding energies. Let me start looking at the quantities pertinent to the relative motion. These are shown in Fig. 7 and Fig. 8. Figure 7 displays an overview of the results at

$1422MeV$. In this figure, as a function of the impact parameter ρ are displayed the scattering angle θ, the energy loss E_{loss} and the angular momentum loss L_{loss}. The correlation between the scatteing angle θ and the final kinetic energy in the relative motion E_f (Wilczynski plot) is shown if Fig. 7d.

For large impact parameters the deflection function (Fig.7a) follows the Rutherford law (pure Coulomb fiels):

$$\theta_C = 2artg\frac{\rho}{a_0} \tag{55}$$

where:

$$a_0 = \frac{Z_a Z_A e^2}{m_{aA} v^2} \tag{56}$$

The final energy (in Fig. 7b is shown the energy loss E_{loss}) remains nearly constant and equal to the bombarding energy. In fact only Coulomb excitations take place in this region.

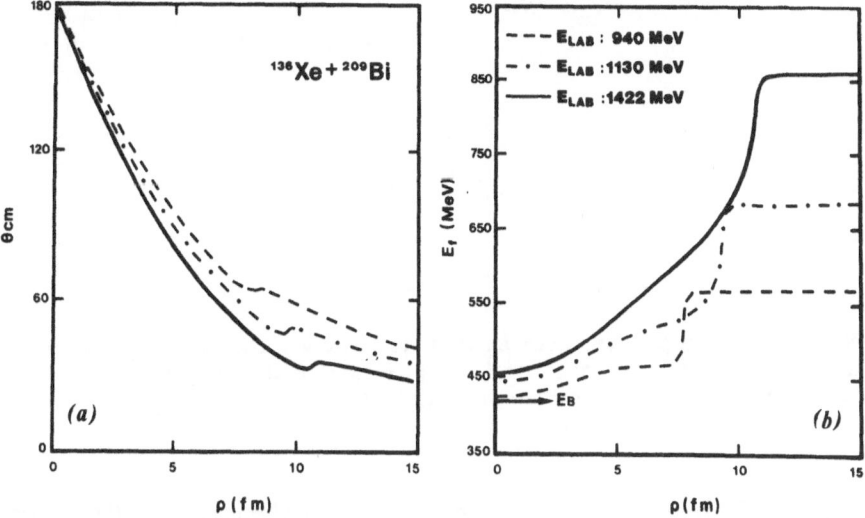

Fig. 8. *Deflection functions (a) and total kinetic energy E_f (b) as a function of the impact parameter (ρ) for the indicated bombarding energies. In (b), with the arrow is indicated the position of the Coulomb barrier.*

For smaller impact parameters the surface interaction causes the deflection function to bend in the forward direction so to display a maximum, the rainbow angle (θ_r). This is moving towards smaller angles increasing the bombarding energy (cfr. Fig. 8a). At the corresponding impact parameters one starts to have an appreciable energy and angular momentum loss (cfr. Fig. 7b and 7c). This is the regime of grazing collisions where we can check the responce function used in the calculation.

For even smaller impact parameters all the available energy and angular momentum is dissipated in intrinsic excitation. The two ions emerge from the reaction with a final energy that is independent from the bombarding energy and close to the Coulomb barrier of two touching spheres (cfr. Fig. 8b where with an arrow we have indicated the Coulomb barrier E_B). This range of impact parameters is thus identified with the deep-inelastic events. Note that in this regime the deformation plays an important role in that it tends to reduce the nuclear attraction and thus allowes the two ions to come apart.

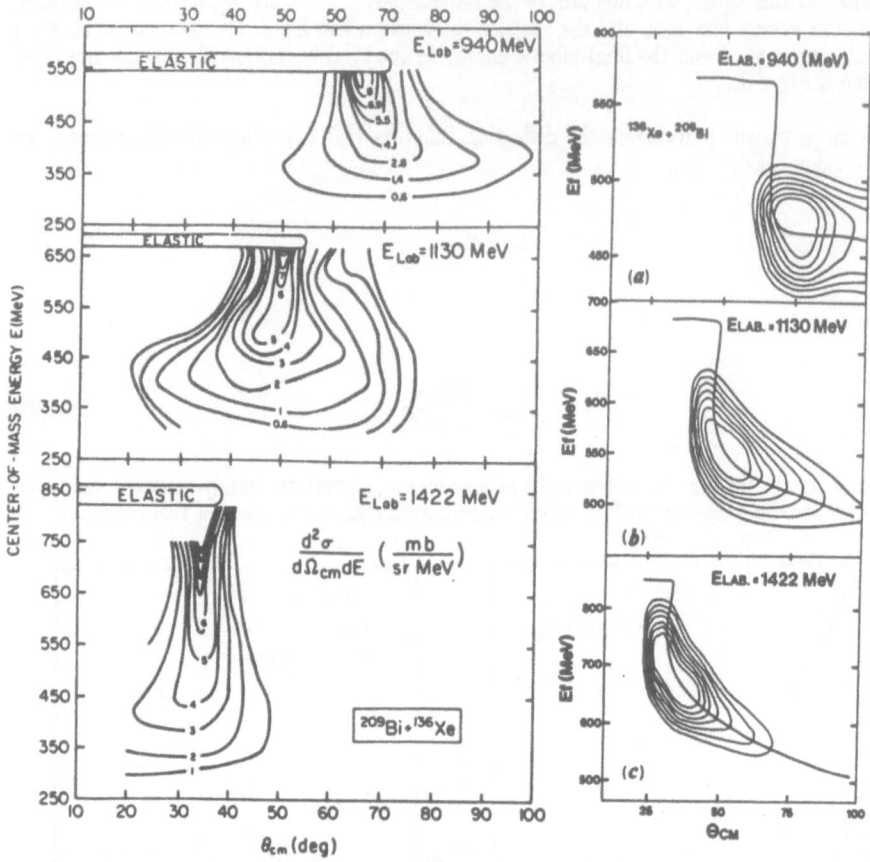

Fig. 9. *The Wilczynski plots are shown for the three bombarding energies in comparison with the experimental data[2-6] (left-hand side). Only events with an energy-loss larger then 50 MeV have been spread. The contour line are in linear scale.*

The plot in Fig. 7d shows the correlation between the total kinetic energy of the emerging fragments E_f and the scattering angle. This has to be compared with the ridge of the experimental Wilczynski plot of Fig. 9. The calculation reproduces quite nicely the focalization at the rainbow but it seems to display also cross section at large angle.

In order to allow a better comparison with the experimental data, these display large fluctuations around the ridge, one can try to spread the cross section both in energy and scattering angle. The cross section corresponding to a given energy loss and at a given scattering angle is spread in accordance to a Gaussian distribution, both in energy and scattering angle, with a standard deviation given by (52) for the energy and a standard deviation of 8 degrees for the scattering angle as it has been suggested by evaporation calculations[13]. The results for the three bombarding energies are shown in Fig. 9 in comparison with the experimental data. For all the energies the calculations do not show, for the large energy loss, any drift toward smaller angles. This component should, indeed, correspond to fast-fission events.

In Fig. 10 we show the same Wilczynski plot for the reaction $^{208}Pb + ^{208}Pb$ at $7.57 MeV/N$[15]. In this case the bombarding energy is very close to the Coulomb barrier and the cross section for deep inelastic events shows a drift toward larger angle as it is well riproduced by the calculation (note that we have spread only the cross section corresponding to Q-value $\leq -50 Mev$).

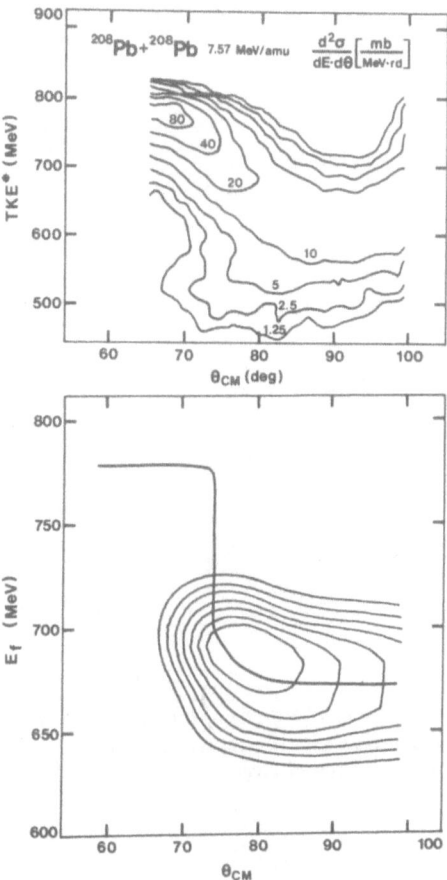

Fig. 10. *The same as Fig. 9 but for the reaction* $^{208}Pb + ^{208}Pb$[15].

The inclusive angular distributions and the excitation functions for the deep-inelastic events can be obtained by partial integration of the Wilczynski plots discussed above. The angular distributions are shown in Fig. 11. in comparison with the experimental data. Fig. 12, display instead the excitation functions. The broadening of the deep-inelastic bump, with the increas of the bombarding energy, is quite well riproduced by our calculation but our energy distribution are centered at somewhat larger values of the final energy (smaller energy loss). The actual value of the maximum in the angular distribution is not very indicative depending on the adopted Q-value cut-off.

Many models have ben developped to study these reactions. I decided to compare our results with the time-dipendent Hartree-Fock (TDHF) calculations [16]. This model contains both our mechanism, particle transfer and the excitation of the surface degrees of freedom, treated in a self-consistent way. The correlation between the average scattering angle and the average final energy are compared if Fig. 13. for the 940 and 1130 bombarding energies. Both models agree quite well in the regime of small energy loss but the TDHF calculation predicts larger energy loss and, as our model, it desplays cross section in the backward direction. A much more complete camparison with the TDHF calculation has been performed for an head-on collision of $^{208}Pb + ^{208}Pb$ at $1535 MeV$ (cfr. ref. [1]).

As I have discussed in section 4.1, from our model we can also get informations about the intrinsic states of the two fragments. For all the bombarding energies I display in Fig. 14 the excitation energy and the intrinsic angular momentum of the light and heavy products of the reaction. For the lower bombarding energy the excitation energy is shared equally between

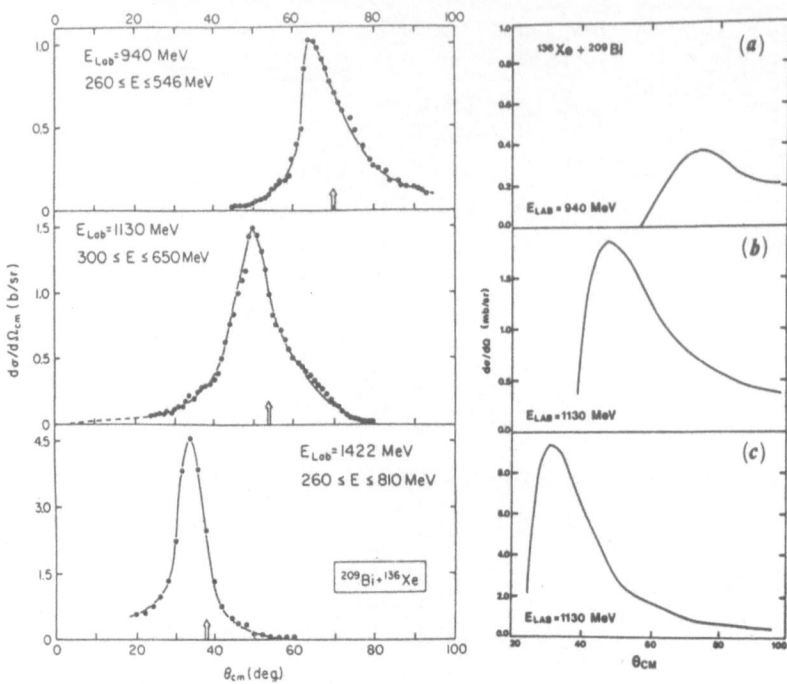

Fig. 11. *Center-of-mass angular distributions of the deep-inelastic events with an energy-loss larger that 50 MeV in comparison with the experimental data (left-hand side). Note that the model overestimates the cross section at large angles.*

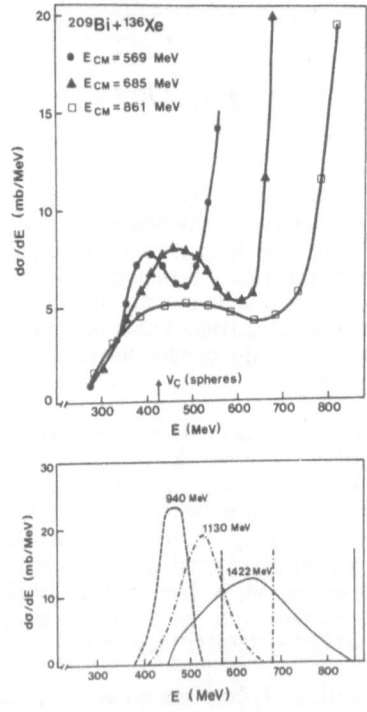

Fig. 12. *Total kinetic energy distribution of the final fragments in the reaction $^{136}Xe + ^{209}Bi$ at the indicated bombarding energies in comparison with the experimental data (top).*

the fragments while at heigher energies the heavy fragment seems to get more excitation. The angular momentum instead is shared more equally and does not display any energy dependence. In Fig. 14c, for the heavy fragment and for all the bombarding energies, we show the ratio between the energy dissipated through the mass transfer to the total excitation energy. For the lower bombarding energy and for the full range of impact parameters the two mechanisms contribute equally to the energy dissipation. Increasing the bombarding energy the mass transfer tends to dominate at large energy loss.

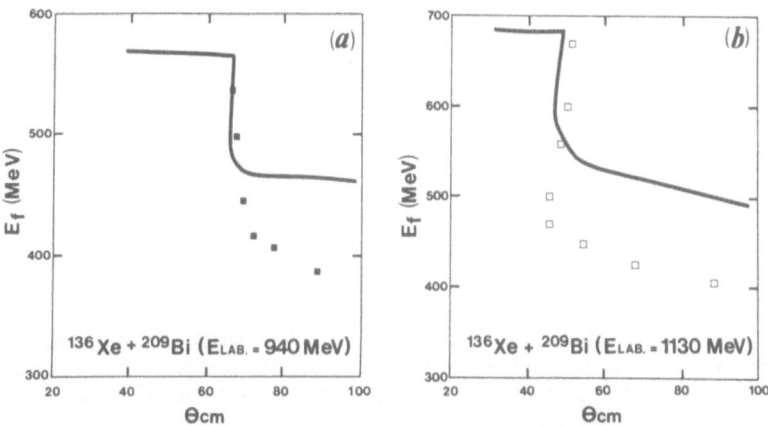

Fig. 13. *The correlation between scattering angle and final energy are compared with the TDHF calculations of ref.[16] (boxes) for the two indicated bombarding energies.*

A ward of warning about the previous results. In fact, these may be modified by changing the prescriptions for sharing the energy and the angular momentum dissipated through mass-transfer among the fragments. We recall that this part of the dissipated energy has been divided in accordance to the masses (thermal equilibrium) while the angular momentum has been divided in accordance to the position of the window (cfr. eq. (40)) relative to the center of mass of the two fragments.

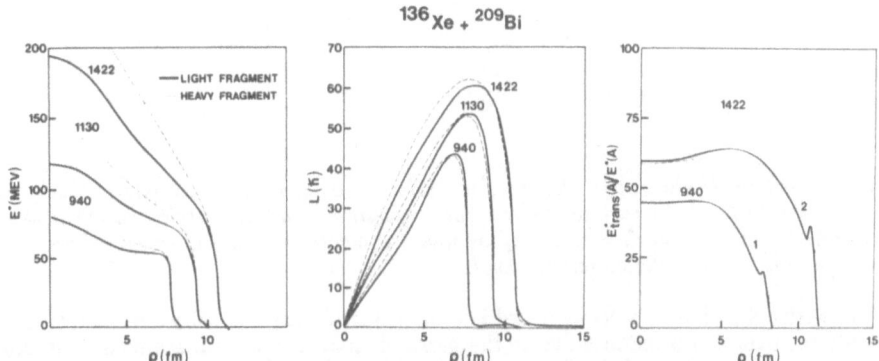

Fig. 14. *Excitation energy and spin of the ligth (full line) and heavy (broken line) fragments of the reaction $^{136}Xe + ^{209}Bi$ as a function of the impact parameter at the three bombarding energies. In (c), as a function of the impact parameter is shown, for the target, the ratio among the energy dissipated through mass-transfer with the total excitation energy.*

The results about the width and the drift of the charge (mass) of the target are displayed in Fig. 15 in comparison with the experimental charge distribution. These where obtained by integrating along the trajectory the equations proposed by J. Randrup[9]. The drift toward simmetry seems to be well reproduced but the width are under-estimated by a factor of 4. This last result is mainly due to our treatment of the deformation that reduces the flux of particle.

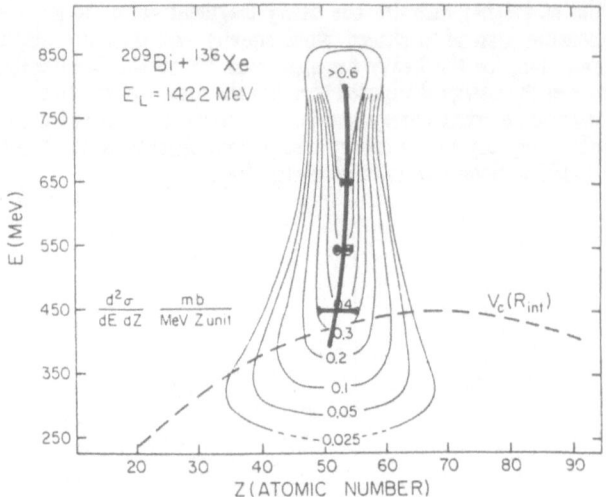

Fig. 15. *Double differential cross section $d^2\sigma/dEdZ$ of projectile-like fragments from the reaction $^{136}Xe + ^{209}Bi$ at 1422 MeV in comparison with the calculated mean value and the corresponding full-width-half maximum. The dashed curve indicates the Coulomb repulsion of two spheres touching at their interaction radius.*

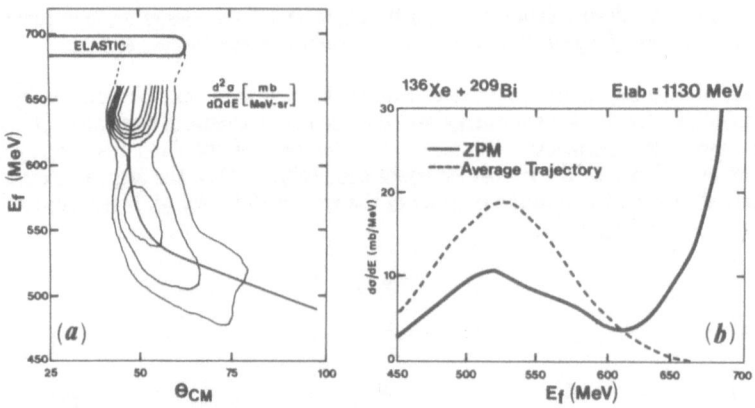

Fig. 16. *Wilczinski plot (a) for the reaction $^{136}Xe + ^{209}Bi$ at 1130 MeV calculated taking into account the ZPM of the surface modes. The contours are in linear scale. In (b) is shown the corresponding energy distribution of the final fragments (full line) in comparison with the average trajectory calculation (broken line).*

In section 4.2 I have breafly indicated how our classical model can be generalized to incorporate the quantal fluctuations due to the harmonic nature of the surface modes. In order to get a nice description of the full cross section I have decided to performe 600 trajectories calculations. These are distributed over the full range of relevant impact parameters in such a way each trajectory carries the same cross section. For each of these calculations I have choosen random values for the deformation parameters $\alpha_{n,\lambda\mu}$ and the corresponding conjugate momenta $\pi_{n,\lambda\mu}$ in agreement with the Gaussian distribution of the ground state wave function. The resulting Wilczinski plot is shown if Fig. 16a. Superimposed is also shown the result of the average trajectory calculations (in this case the initial values of the $\alpha_{n,\lambda\mu}$ and the $\pi_{n,\lambda\mu}$ are set to zero). The corresponding energy distribution is obtained by integrating, over the full angular range, this double differential cross section. This is displayed in Fig. 16b. For comparison we have also reproduced the energy distribution obtained with the approximate procedure outlined above. This calculation reproduces nicely the experimental minimum at

610 MeV corresponding to the transition from the quasi-elastic regime to the deep-inelastic one. Also the values of the cross section at the maximum of the deep inelastic bump (this is shifted at to higher energy) is in good agreement with the experiment.

5.2 Excitation energy and angular momemtum sharing between heavy and light fragments in a deep-inelastic reaction

We have seen that in a deep-inelastic reaction a large fraction of the energy and angular momentum available in the relative motion of the nuclei is converted into intrinsic excitation and intrinsic spin of the reaction fragments. The question of the division of the total kinetic energy loss (TKEL) and the total angular momentum loss among the fragments has attracted considerable interest. This was motivated by the hope that the understanding of these divisions would help to identify the different mechanism responsable for the large observed energy and angular momentum transfer.

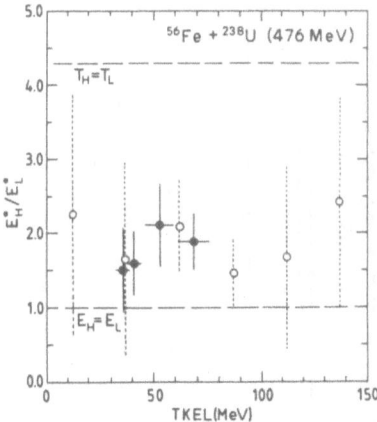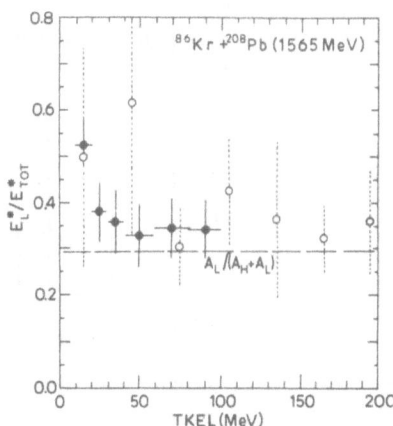

Fig. 17. *Ratio of the excitation energy distributed between the heavy and light fragments as a function for the energy loss for the reaction* $^{56}Fe+^{238}U$ *at476MeV[17] (a) and the ratio of the light fragment excitation energy and the total one as a function of the energy loss for the reaction* $^{86}Kr+^{208}Pb$ *at1565MeV[18] (b). The value expected in the limits of equal division of the excitation energy and division leading to thermal equilibrium are shown by dashed horizontal lines. Full circle are the experimental data and the empty circle are the results of the calculations. The estimated fluctuations are shown by vertical dotted lines.*

Empirical evidence for the energy sharing is provided by measurments of the energy distributions of neutrons emitted by the highly excited reaction products while the nuclear spins of primary reaction fragments can be inferred from the measurments of the multiplicity of the emitted γ rays or from the measurments of the angular distibution of the induced fission fragments.

Advocates of the statistical pictures have naturally tendet to interpret the early results on neutron evaporationspectra as consistent with the division of excitation energy between the fragments given by the same ratio as that of their masses. This is needed in order to ascribe equal temperature to projectile and target (thermal equilibrium). Recent measurements on $^{56}Fe+^{238}U$[17] and $^{86}Kr+^{208}Pb$[18] have shown a transition from the small energy loss where the energy is shared equally between the fragments to the high energy loss where a thermal equilibrium seems to be reached. This behaviour is quite similar to the one displayed in Fig. 14 where is seen a transition from a situation where the inelastic processes dominate for small energy loss and for energy not to high over the Coulomb barrier to a condition in which particle transfer takes over for large energy loss and very high bombarding energies.

To substantiate these simple order of magnitude expectations we have performed a full calculation[19] with the model discussed above. These have been performed by using for all the nuclei the universal responce function of section 2.1. The simple procedure would be to calculate the average trajectories as a function of the impact parameter (as it has been done in Fig. 14) and than to unfold the ratio of excitation energies as a function of the total kinetit energy loss (TKEL). However, since that ratio E_H^*/E_L^* is not expected to be in one-to-one correspondence with the partial wave number, we have performed the calculations in such a way as to obtain the magnitude of the fluctuations which originate in the coupling to the surface modes. In Fig. 17a the results fo the reaction $^{56}Fe + ^{238}U$ at $476 MeV$ are shown in comparison with the experimental data of ref. [17]. The estimates of the spread around the average results (empty circles) are indicated by the dotted lines. We have also performed calculations for the same system at higher bombarding energies which show a gradual increase of the value for the E_L^*/E_L^* as expected from simple arguments.

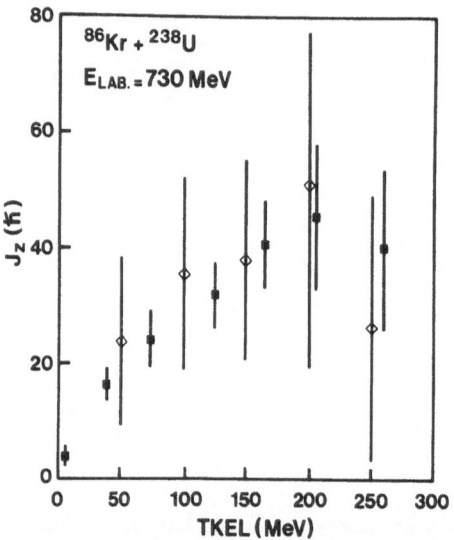

Fig. 18. *Aligned component of the spin transferred to the U in the reaction $^{86}Kr + ^{238}U$ as a function of the total kinetic energy loss. The full square represent the data of ref.* [21] *while the open one are the calculated point with the corresponding standard deviations.*

In Fig. 17b I compare the results of a similar calculation for the reaction $^{86}Kr + ^{208}Pb$ at $1565 MeV$ with the data extracted from the recent measurments of ref. [18]. Both theoretical and experimental resuls are in this case closer to the thermal ratio. We note that in both reactions the estimated size of the fluctuations is larger than the reported experimental errors. It would thus be interesting to have measurments for the second moments of these quantities. Since the expected width from statistical origins are relatively small, evidence for large fluctuations may comfirm the importance of quantal effects in the process.

In our model the orbital angular momentum is transformed in intrinsic spin of the two fragments due to the tangential component of the one-body frictional force (21) and due to the dumping term (29) in the equations for the surface modes (cfr. eqs. 37 and 38). Since I consider only planar trajectories the predictions of the model may only be compared with the aligned component of the fragment spins. To be able to have predictions on the fragment spin projections in the reaction plane I have to generalize the model to allow out of plane scattering. This generalization has been done and the obtained results, for the reaction $^{86}Kr + ^{208}Pb$ have been discussed in ref. [20]. Here I limit myself to show the results concerning the reactions $^{86}Kr + ^{238}U$ (Fig. 18) and $^{86}Kr + ^{154}Sm$ (Fig. 19). In the first case the angular momentum transferred to the U-like fragments has been inferred from sequential fission measurments[21] while in the second case the angular momentum of the Sm-like fragments has been obtained from the measurment of the multiplicity of the γ-rays emitted by the higly excited fragments[22]. In order to have an estimation of the quantal fluctuation I have performed the calculations taking into account the zero-point motion of the surface modes. For the description of these surface modes I have utilized the universal responce function of section 2.1.

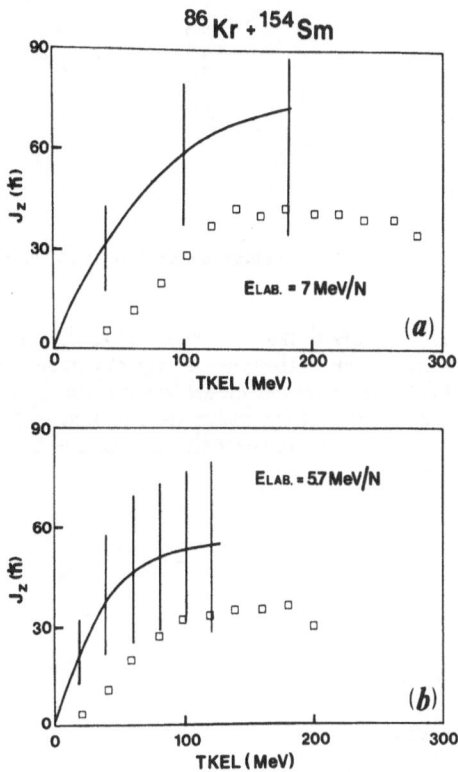

Fig. 19. *Same as Fig. 18 but for the reaction $^{85}Kr + ^{154}Sm$ at the indicated bombarding energies.*

5.3 Energy dependence of the fusion reactions

The results of Figs. 7 and 8 are typical of heavy projectile and target combination. For light ions $(A \leq 40)$ and at not to high bombarding energy (respect to the Coulomb barrier) no emerging trajectories are present for impact parameters smaller than the grazing. For these impact parameters enough energy is dissipated from the relative motion so that the two nuclei cannot overcome the surface-surface actractive interactions and remain in contact for a long time. The question whether the outward motion is stopped, and the two ions are caught in the attractive field is decided by the competition between the rate of energy dissipation and the rate at which the barrier dwindles due to the deformation and the angular momentum transfer (this reduces the centrifugal barrier). A pictorial example where this situation is met is shown in Fig. 20 for the reaction $\alpha + ^{208}Pb$ at $65 MeV$ of bombarding energy. In this figure is shown the evolution of two trajectories corresponding to impact parameters smaller than the grazing.

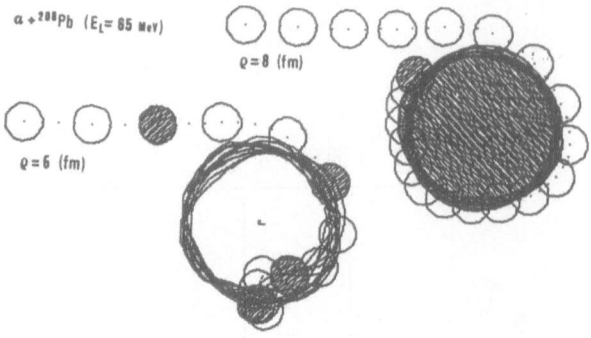

Fig. 20. *Pictorial representation of two captured trajectories in the reaction $\alpha + ^{208}Pb$ at 65 MeV.*

The composite system, thus formed, may be considered as the precursor to the compound-nucleus formation. The identification of the range of impact parameters for which no emerging trajectories are present allows us to use our model to calculate the fusion cross-section. The accurate description of the variation of the fusion cross section with energy becomes an important test of the ability of the model to cover the full range of heavy-ion reactions.

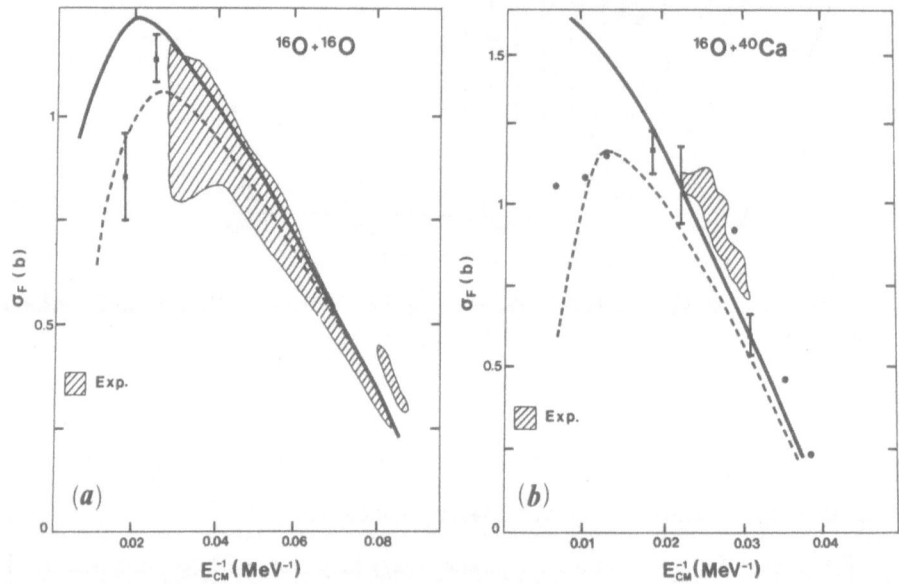

Fig. 21. *With the full line is indicated the fusion cross section for the reaction $^{16}O + ^{16}O$ (a) and $^{16}O + ^{40}Ca$ (b) as a function of the bombarding energy. With the dashed line I have reproduced the calculations of ref.* [25]. *The shaded area represents the data.*

In Fig. 21, as function of the inverse of the center-of-mass bombarding energy $(E_{CM}^{-1},$ is shown the calculated fusion cross section (full drown curve) for the reactions $^{16}O + ^{16}O$ and $^{16}O + ^{40}Ca$ in comparison with the experimental data[23-24] (shaded area). With the broken line we have, also, reported the calculation of ref. [25] where only the mass transfer has been included as a dissipative mechanism and where the two ions are kept spherical throughout the collision. In mine calculations, for the description of the deformation, I have utilized the universal response function of section (2.1). The calculations reproduce quite well the data for the light ion combination while they overestimate, at high energy, the fusion cross section.

The extension of this model to heavier system is not straightforward. The large deformations and the large angular momentum loss tend to reduce the barrier in the way-out of the trajectory so that the two ions can escape. This may indicate that either the nuclei are to soft, and then the two ions obtain large deformation without loosing enough energy, or the tagential component of the friction is too strong. It should be interesting to see if a modification of the ratio between tangential and radial friction force can give rise to a better description of the fusion cross section for medium-heavy nuclei and at the same time account for the angular momentum loss that it is overestimated (cf. Fig. 19).

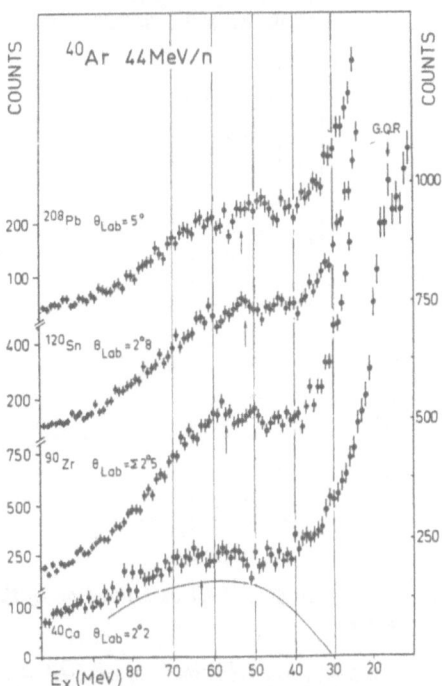

Fig. 22. *Energy spectra of the inelastically scattered* ^{40}Ar *on the indicated targets at a bombarding energies of* $44MeV/N$. *For the reaction on the* ^{40}Ca *is indicated the estimated contribution from the transfer plus evaporation channels.*

6. INTERMEDIATE ENERGY LOSSES IN QUASI-ELASTIC COLLISIONS.

Many experiments testify the existence of structures, ranging up to very high excitation energy, in the energy spectra of some fragments emitted in several heavy-ion collisions[26]. These observations were carried out with several combinations of target and projectile and at several bombarding energies from $10 MeV/N$ to $44 MeV/N$. In Fig. 22 we show some of the ^{40}Ar inclusive spectra in the collision at $44MeV/N$ on several targets. The features characterizing the observed structures have been summarized as follows[27]:

a) the structures extend up to very high excitation energy ($\sim 100, 150 MeV$);

b) their position is independent from the bombarding energy and, for a given target, is independent from the projectile;

c) their width increase with the excitation energy;

d) the structures at $E_s \le 50MeV$ are peaked at angles close to the grazing while the structures at higher excitation energy are more forward peaked, this seems to suggest that their excitation is achieved by a direct process;

f) these structures are also present in few nucleon transfer channels.

The fact that the observed structures show up at the same excitation energy for different incidence energies and for different projectiles has been taken as strong evidence that they represent target excitations. It has been also suggested that these structures are due to the multiple excitation of giant states. It is thus very tempting to use the model discussed above to investigate this suggestion. As a test case we will concentrate on the reaction $^{36}Ar + ^{208}Pb$ at $11MeV/N$[28]).

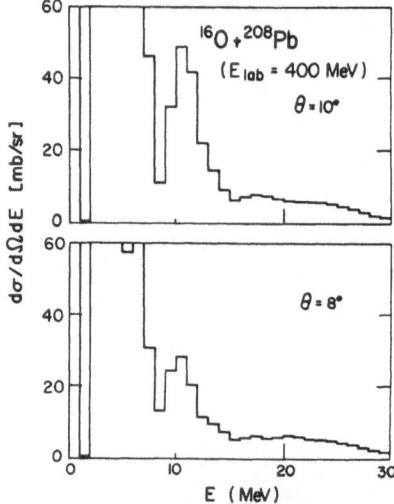

Fig. 23. *Double differential cross section $d^2\sigma/dEd\Omega$ as a function of the excitation energy for the collision $^{16}O + ^{208}Pb$ at 400 MeV.*

Most of the cross-section in a heavy-ion collision is associated with deep-inelastic events with a broad energy distribution (cf Fig.16b). This is determined by the energy loss partly due to the excitation of the surface mode and partly due to the transfer of particles. The energy spectrum of the detected projectile-like fragments, thus, receives contribution from the spectra of several heavier scattered particles, which due to the particle evaporation, decay into the observed mass partition before they reach the detector. For certain target and projectile combination these processes (trasfer followed by evaporation) can give rise to bump in the energy spectra, their position depends on the bombarding energy and they can thus be easily identified[29]. In other cases these processes ,because of the many channels involved, should give rise to a rather smooth background in the energy distribution [30]. If I disregard this background, I may treat the transfer reactions as a depopulation mechanism of the initial mass partition through an imaginary potential. In this way I may use the model to estimate the absolute cross section in the inelastic channel. In this approach, which I follow throughout this section, particle transfer does not affect the trajectory of relative motion.

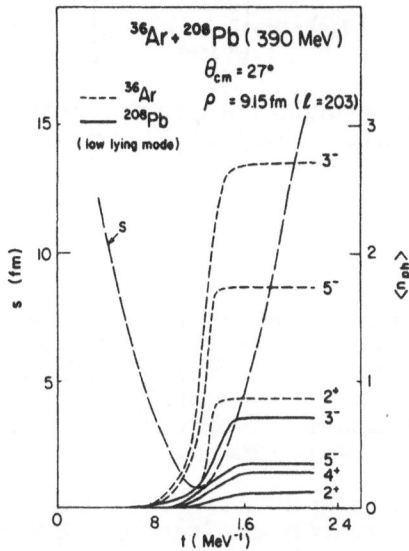

Fig. 24. *Average number* $< n_{ph} >$ *of low-lying modes excited in the collision* $^{36}Ar + ^{208}Pb$ *as a function of time. Also displayed is the distance s between the surfaces of the two ions.*

6.1 Energy spectra.

The average occupation number $< N_i(\rho) >$ for each mode $i \equiv (n, \lambda)$ excited in a collision with impact parameter ρ is given by

$$< N_i(\rho) >= \frac{E_i(\rho)}{\hbar \omega_i} \tag{57}$$

where $E_i(\rho)$ is the excitation energy associated with the mode of energy $\hbar \omega_i$. The probability distribution for having a number n_i of phonons is approximatly

$$P_{n_i} = \frac{< N_i(\rho) >^{n_i}}{n_i!} exp\big(- < N_i(\rho) > \big) \tag{58}$$

and thr probability for an enegy loss E at that impact parameter is

$$P(\rho, E) = \sum_{\{n_i\}} \delta\big(E - \sum_i n_i \hbar \omega_i\big) \prod_i P_{n_i}(\rho) \tag{59}$$

The double-differential cross section can now be written as

$$\frac{d^2\sigma}{dEd\omega} = \sum_r ho \frac{\rho |d\rho/d\theta|}{2\pi sin\theta dE} P(\rho, E) T(\rho) \tag{60}$$

where the sum extends over all impact parameters that feed the choosen scattering angle $\theta(\rho)$, and where

$$T(\rho) = exp\left[\frac{2}{\hbar}\int_{-\infty}^{\infty} W_{tr}(r)dt\right] \qquad (61)$$

is a coefficient describing the probability that the system has to remain in the initial mass partition. The function $W_{tr}(r)$ is the imaginary part of the ion-ion potential due to mass transfer.

6.2 The $^{36}Ar + ^{208}Pb$ reaction.

It has been suggested that the structures observed in the reaction $^{36}Ar + ^{208}Pb$ are associated with the excitation of states of ^{208}Pb. In order to assess the validity of this interpretation I have to use, in the analysis of the data, a good description of the lead spectrum. This is shown in the left-hand-side of Table 3; it was constracted by making use of empirical information and of microscopic calculations based on the RPA formalism. For the ^{36}Ar[31] I have decided to use the universal responce funtion of section 2.1.

Recently, the inelastic scattering of ^{16}O on ^{208}Pb has been studied[29] at a bombarding energy of 400 MeV in the laboratory system. At grazing angle ($\theta \leq 12°$) the low-lying states and the some of the giant resonances have been clearly identified. The analysis of these data thus provide a test to check the responce function chosen for ^{208}Pb. The calculated spectra for angles close to the grazing (cf. Fig. 23) reproduce quite well the experimental features. The structures found experimentally between ~ 20 and $\sim 45MeV$ of excitation energy has been attributed to pich up processes followed by evaporation[29].

TALBE 3. *Spectra of* ^{208}PB. *With Copenhagen I have indicated the spectra, used in ref.*[31] *constracted from experimental data and RPA calculations. With Paris the spectra used by the Orsay group (see ref.*[34]*).*

	^{208}Pb (Copenhagen)			^{208}Pb (Paris)		
λ	$E(MeV)$	$\Gamma(MeV)$	$\%EWSR$	$E(MeV)$	$\Gamma(MeV)$	$\%EWSR$
0+	13.6	2.0	100	14.2	1.0	45
2+	4.1	–	16			
	10.8	2.7	62	12.0	2.0	80
3−	2.6	–	17			
	17.0	5.0	80	13.0	0.6	3
				21.5	0.5	55
4+	4.3	–	6			
	10.9	2.5	23	12.1	–	15
	24.0	7.0	70	30.0	5.0	30
5−	3.3	–	5			
	20.0	9.0	40	16.0	0.5	7
				22.0	2.5	21
6+				14.5	1.0	10

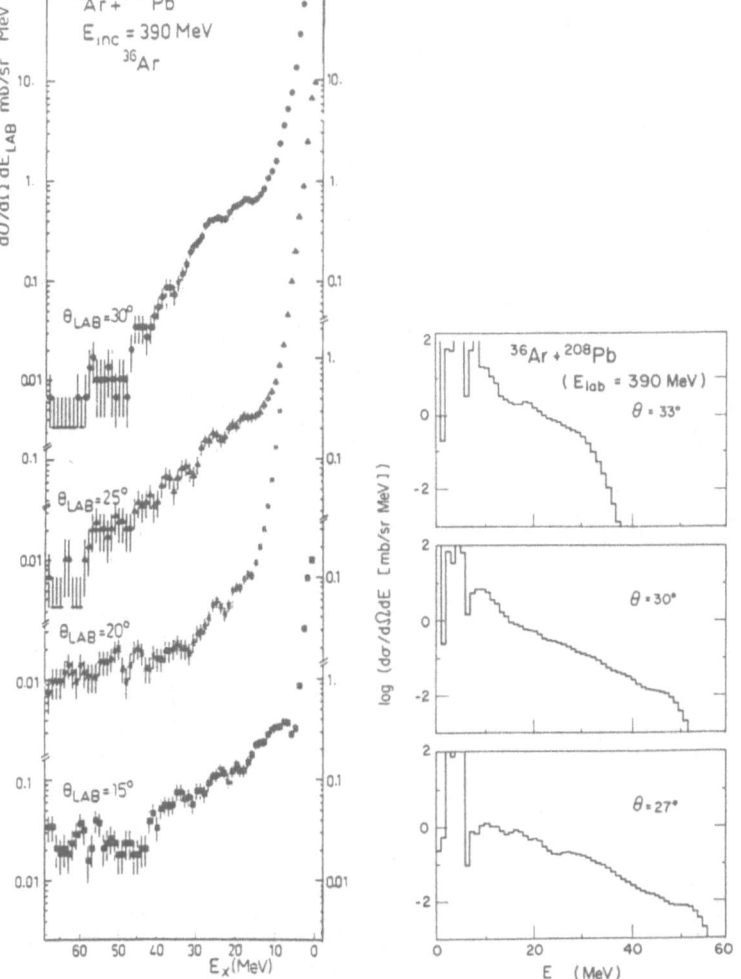

Fig. 25. *Double differential cross section as a function of the excitation energy for the reaction* $^{36}Ar + ^{208}Pb$ *and for three angles smaller than the grazing angle. In these histograms, obtained with the responce function 'Copenhagen', events leading to projectile excitation larger then 12 MeV have been excluded. To the right an experimental fig. corresponding to similar scattering angles is adapted from ref.* [28].

For an impact parameter close to the grazing ($\rho \equiv 9 fm$) and for the reaction $^{36}Ar + ^{208}Pb$ at $390MeV$, the average number of phonons associated with the low-lying and giant resonances are displayed in Fig. 24 as a function of time. As expected the low-lying modes are most strongly excited. This a consequence of the large collectivity and relative low energy of these modes. The probability of multiple excitation of giant resonances is small. In fact, typical values of $< \langle N \rangle >$ for these high-lying states are $0.1 - 0.3$.

Trajectories with smaller values of the distance of closest approach could, in principle, have a better chance of exciting the giant resonances. However the probability $T(\rho)$ for these trajectories to remain in the inelastic channel are vanishingly small. We note that not even in deep-inelastic processes do any of the modes normally acquire a large number of phonons. If this were the case, rather extreme deformations may arise, putting into question the validity of the model.

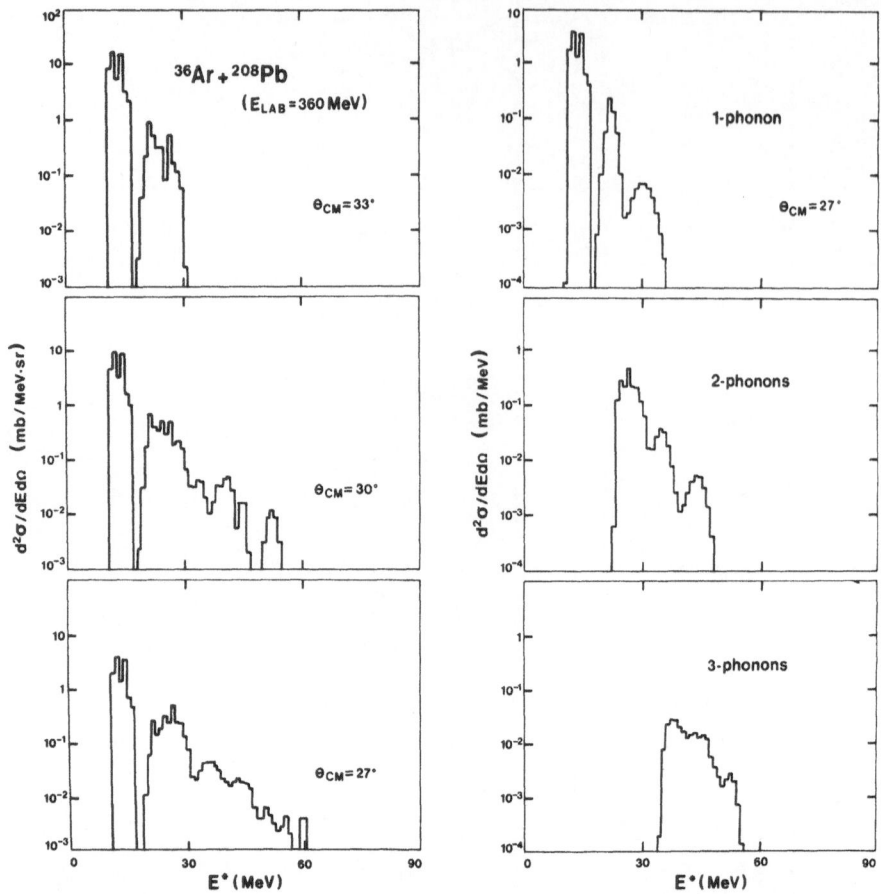

Fig. 26. *Double differential cross section $d^2\sigma/dEd\Omega$ for the reaction $^{36}Ar + ^{208}Pb$ and for three angle close to the grazing angle. These histograms have been calculated using the 'Paris' response function of Table 3. One should notice the nice structures present in these spectra that extend up to high energies. To demonstrate that these spectra are dominated by the multiple excitation of the giant quadrupole I display to the right-hand-side the multiphonon decomposition of the spectra at $\theta = 27°$. The Ar is kept spherical.*

The inelastic spectra are compared with the data in Fig. 25 for angles close to the grazing. A conspicuous feature of these results is that most of the cross section is associated with events where a variety of different modes is excited. One should note that energy loss of the order of magnitude, although somewhat smaller, than those observed experimentally are obtained, and that the calculated cross section have the right order of magnitude.

In orde to have a good estimation of the imaginary potential W_{tr} I used the microscopic formalism of ref. [32]. In these calculations all one-particle transfer processes with Q–value larger than $-20MeV$ have been included. The energy of the single-particle transitions entering into the calculation are obtained making use of a Saxon-Wood potential with standard parameters (for more details cf. ref. [32]).

Other approaches have been developed to calculate the excitation of giant resonance in heavy-ion collision. In particular the adiabatic time- dependent Hartree-Fock (ATDHF) approximation[33] and the linear responce or randoom-phase approximation (RPA)[34]. In this last model the projectile is described as a moving Saxon-Wood potential with a fixed shape following a classical trajectory. The evolution of the target states is calculated with a

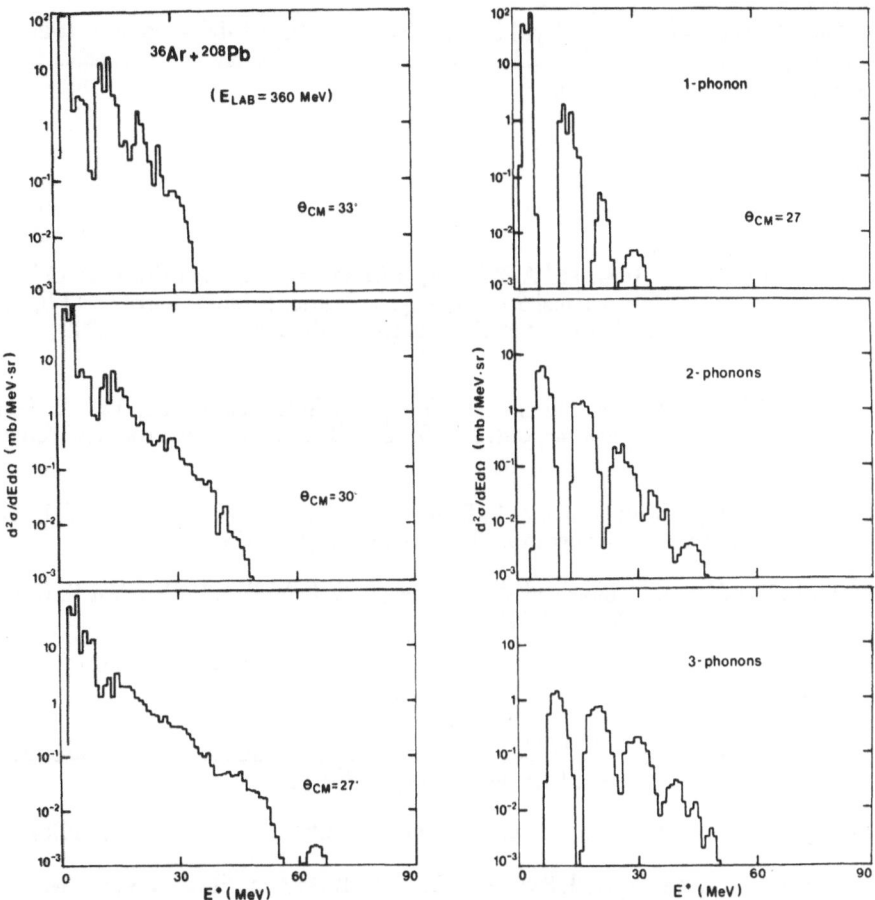

Fig. 27. *Same as Fig. 26. Here the excitation functions are calculated implementing the 'Paris' response function with the empirical informations for the low-lying modes of* ^{208}Pb. *As it is apparent from the multiphonon decomposition (right-hand-side) of the spectra at* $\theta = 27°$ *these are dominated by the multiple excitation od the low-lying states and they display structures only for angles close to the grazing.*

time-dependent Schrödinger equation in a non interacting quasi-boson approximation. The strength function and the formfactor for a single phonon excitation are calculated in the RPA. In these calculations a strong probability to excite multiphons, built for a large part on the giant quadrupole resonance, is found for small distance off closest approach while for a grazing distance of closest approach ($d = 13.1 fm$) a small number on phonons are excited.

Since the model of ref. [34] is a linear approximation to the model discussed in these lectures but with the formfactors calculated microscopically (from the RPA wave functions) it is interesting to compare the two results. Using for the lead the responce function of ref.[34] and reported on the right-hand side of Table 3, I obtain the excitation functions of Fig. 26. In order to stress that these spectra are dominated by the multiple excitation of the giant quadruple I show the multiphonons decomposition of the excitation function at $\theta = 27°$.

Both from these spectra and from the responce function of Table 3 it is clear that, in this description, are missing the excitation of the low-lying states of ^{208}Pb. In order to see the influence of these states on the excitation functions I performed a similar calculation where for the lead I added the informations for the low-lying states as contained in the right-hand-side of Table 3. The results are displayed in Fig. 27. Due to the low-lying states no structures is now apparent from the calculated excitation function. This are now dominated by the excitation of the low-lying states.

REFERENCES

[1] - R. A. Broglia, C. H. Dasso and A. Winther, in *Nuclear Structure and Heavy Ion Collision, Peoceedings of the International School of Physics "Enrico Fermi", Course LXXVII*,edited by R. A. Broglia, C. H. Dasso and R. Ricci (North-Holland, Amsterdam, 1981), p. 327 and ref. therein.

[2] - W. U. Schöder and J. R. Huizenga, in *Treatise on Heavy-Ion Science, Vol. II, Fusion and Quasi-Fission Phenomena*, edited by D. Allan Bromley (Plenum Press,New York and London, 1985), p. 115.

[3] - W, W, Wilcke, J. R. Birkelund, A. D. Hoover, J. R. Huizenga, W. U. Schröder, V. E. Viola Jr., K. L.Wolf and A. C. Mignerey, Phys. Rev. C **22**,(1980)128.

[4] - W. U. Schröder, J. R. Birkelund, J. R. Huizenga, K. L. Wolf, J. P. Unik and V. E. Viola Jr., Phys. Rev. Lett. **36**,(1976)514.

[5] - W. U. Schröder, J. R. Birkelund, J. R. Huizenga, K. L. Wolf, and V. E. Viola Jr., Phys. Rep. **45**,(1978)301.

[6] - H. J. Wollersheim, W. W. Wilcke, J. R. Birkelund, J. R. Huizenga, W. U. Schröder, H. Freieslaben and D. Hilsher, Phis. Rev. C **24**,(1981)2114.

[7] - R. A. Broglia, C. H. Dasso, G. Pollarolo and A. Winther, Phys. Rep. C, **48**,(1978)351.

[8] - J. Blocki, J. Randrup, W. J. Swiatecki and T. C. Tang; Ann. Phys. **105**,(1977)427.

[9] - J. Randrup, Nucl. Phys. **307A**,(1978)319;
J.Randrup, Ann. of Phys. **112**,(1978)356;
J.Randrup, Nucl. Phys. **327A**,(1979)490;
J.Randrup, Nucl. Phys. **383A**,(1982)468;
J.Randrup, Nucl. Phys. **452A**,(1986)105;
J.Randrup, Ann. of Phys. **171**,(1986)28.

[10] - A. Bohr and B. Mottelson, *Nuclear Structure Vol. II*, (Benjamin, New York, 1975).

[11] - C.H. Dasso and G. Pollarolo, Computer code TORINO. To be pubblished in Comm. Com. Phys.

[12] - F. E. Serr, T. S. Dumitrescu, T. Suzuky and C. H. Dasso, Nucl. Phys. **A404** (1983)359.

[13] - H. Esbensen, in *Nuclear Structure and Heavy Ion Collision, Peoceedings of the International School of Physics "Enrico Fermi", Course LXXVII*,edited by R. A. Broglia, C. H. Dasso and R. Ricci (North-Holland, Amsterdam, 1981), p. 572 and ref. therein.

[14] - M. Baldo, private comunication.

[15] - T. Tanabe, R. Bock, M. Dakowski, A. Gobbi, H. Sann, H. Stelzer, U. Lynen, a. Olmi and D. Pelte; Nucl. Phys. **A342**,(1980)194.

[16] - A. K. Dhar, B. S. Nilsson, K. T. R. Davies and S. E. Koonin; Nucl. phys. **A364**,(1981)105.

[17] - R. Vanden bosch, A. Lazzarini, D. Leach, D.K. Lock, A. Ray and A. Seamster; Phys. Rev. Lett. **52**(1984)1964.

[18] - H. Solhbach, H. Freiesleben, P. Braun-Munzinger,W. F. W. Schneider, D. Schull, B. Kohlmeyer and F. Puhlhofer; Phys. Lett. **153B**(1985)386.

[19] - C. H. Dasso, M. Lozano and G. Pollarolo; Phys. Rev. **C32**(1985)2195.

[20] - R.A. Broglia, G. Pollarolo, C.H. Dasso and T. Dossing; Phys. Rev. Lett. **43**(1979)1649.

[21] - R. Vandenbosch; Phys. Rev. **C20**(1979)171;
R.J. Puigh, P.Dyer, R. Vandembosch, T.D. Thomas, L. Nunnelly and M.S. Zisman, Phys. Lett. **86B**(1979)24.

[22] - P.R. Christensen, F. Folkmann, O. Hansen, O. Nathan, N. Trautner, F. Videbaek, S.Y. Vander Werf, H.C. Britt, R.P. Chestnut, H. Freisleben, and F. Pühlhofer; Nucl. Phys. **A349**(1980)217.

[23] - A. Weidinger, F. Busch,G. Gaul, W. Trautmann and W. Zipper; Nucl. Phys. **A263**(1976) 511;
B. Fernandez, C. Gaarde, J. S. Larsen, S. Pontoppidan and F. Videbaek; Nucl. Phys. **A306**(1978)259;
H. Spinka and H. Winkler; Nucl.Phys. **A233**(1974) 456.
J.J Kolata, R.C. Fuller, R.M. Freeman, F. Haas, B. Heush and A. Gallman; Phys. Rev. **C16**(1977)891.

[24] - D.F. Geesaman, C.N. Davids, W. Henning, D.G. Kovar, K.E. Rhem, J.P. Schiffer, S.L. Tabor and F.W. Prosser; Phys. Rev. **C18**(1978) 284;
S.E. Vigdor, ANL/PHY 76-2(1976)95.

[25] - J.R. Birkelund, L.E. Tubbs, J.R. Huizenga, J.N. De and D. Sperber; Phys. Rep. **56**(1979) 107.

[26] - N. Frascaria. Proceeding of the *XXIV International Winther Meeting on Nuclear Physics, Bormio (Italy), 20-25 January 1986*;and ref. Therein.

[27] - N. Van Giai, Proceedings of the *Winther College on Fundamental Nuclear Physics, Trieste (Italy), 7 February-30 March 1984.*

[28] - P. Chomaz, Y. Blumenfeld, N. Frascaria, J. P. Garron, J.C. Jacmart, J.C. Roynette, W.Bohne, A. Gamp, W. vom Oertzen, N. Van Giai and D. Vautherin; Z. Phys. **A319** (1984) 167.

[29] - T.P. Sjoreen, F.E. Bertrand, R.L. Auble, E.E. Gross, D.J. Horen, D.S. Shapira and D.O. Wright; Phys. Rev. **C29**(1984) 1370.

[30] - Y. Blumenfeld, J.C. Roynette, P. Chomaz, N. Frascaria, J.P. Garron and J.C. Jacmart; Nucl. Phys. **A445**(1985) 151.

[31] - G. Pollarolo, R.A. Broglia and C.H. Dasso; Nucl. Phys. **A451**(1986) 122.

[32] - R.A. Broglia, G. Pollarolo and A. Winther; Nucl. Phys. **A361**(1981)307.
G. Pollarolo, R.A. Broglia and A. Winther; Nucl. Phys. **A406**(1983) 369.

[33] - H. Tricoire, C. Marty and D. Vautherin; Phys. Lett. **B100**(1981) 109.

[34] - P. Chomaz, thése 3éme cycle, Insitut de Physique Nucléaire Orsay, IPNO-84-01.
P. Chomaz and D. Vautherin Phys. Lett. **B139**(1984) 244.

EXCITATION FUNCTION FLUCTUATIONS FOR DISSIPATIVE HEAVY ION

COLLISIONS

G.Pappalardo'[+],G.Cardella[+],F.Rizzo'[+],A.De Rosa[0],
G.Inglima[0],V.Russo[0],M.Sandoli[0],G.Fortuna[^],G.Monta-
gnoli[^],C.Signorini" and A.M.Stefanini[^]

'Dipartimento di Fisica Università di Catania;[0]Di-
partimento di Fisica Università di Napoli;"Dipartimento di Fisica Università di Padova;'INFN Sezione
di Catania;[^]INFN Laboratori Nazionali di Legnaro,
Italy

The nucleus-nucleus collisions, in the energy region of
about 10 MeV/A, are characterized, as it is well known, by
quasi-elastic and dissipative mechanisms. The former are generally associated with the presence of strongly populated peaks
in the high energy side of the emitted fragment spectra; the
second ones are revealed by large cross sections for transitions
corresponding to low or intermediate outgoing energies.
Excitation functions for elastic and quasi-elastic processes have shown, when measured with good incident energy resolution (typically $\Delta E \simeq 100$ KeV), more or less pronounced structures
which have been interpreted as due to:

  i) isolated resonances corresponding to the intermediate sys-
     tem single levels with a defined J^{π} value;
 ii) few overlapping resonances;
iii) fluctuations of statistical origin, when $\Gamma \gg D$;
 iv) dynamical effects due to the orbital angular momentum co-
     herence.

In many cases, the experimental results have been interpreted
in terms of a "dinucleus" formation in the early stage of the
collision giving rise to quasi-molecular resonances[1]. The lifetime τ of such a system has been deduced from the width Γ of
the single resonances or, in the statistical case (iii), from
the "coherence energy" Γ, extracted from the experimental points
after a suitable analysis, through the relationship $\tau \simeq \hbar/\Gamma$. In
some cases the life-time of the quasi-molecular states ($\tau \simeq 10^{-21}$
sec) has been experimentally distinguished from that of the more complex compound nucleus states ($\tau \simeq 10^{-20}$ sec), within a doorway picture of the heavy ion collisions[2]. Confining our attention to the statistical $\Gamma \gg D$ case we stress here the important
point that the visibility of the cross section fluctuations is
related to the number N of incoherently contributing channels;
this is expressed (see T.Ericson[3]) by:

1) $$C(0) \simeq \frac{1}{N}$$

where

2)
$$C(0) \simeq \frac{<\sigma^2>-<\sigma>^2}{<\sigma>^2}$$

is the relative autocorrelation function for $\varepsilon=0$, and N is the effective number of channels each being characterized by $\alpha,l,s,$ J,π,α',l',s' quantum numbers.

Recently[3], for the first time, excitation functions have been measured with good incident energy resolution, also for dissipative processes i.e. taking into account the low and intermediate energy region of the residual fragment spectra. We discuss here the results obtained studying the reaction $^{28}Si+^{64}Ni$ at incident energy from 120 MeV to 126.750 MeV in 250 KeV steps with an overall energy resolution $\Delta E=400$ KeV in the laboratory frame. For experimental details see ref.4 .

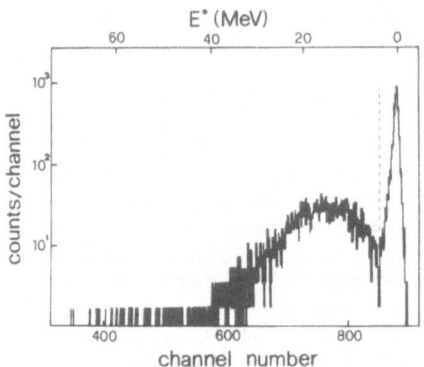

Fig.1- Energy spectrum corresponding to Z=12
fragment atomic number for the $^{28}Si+^{64}Ni$
reaction at E(lab)=124 MeV. The dashed
line separates the quasi-elastic peak
from the damped part of the spectrum,the
only one included in the excitation function.

In Fig.1 the energy spectrum of the Z=12 fragment is shown and in Fig.2 the energy spectra for Z=12,13 fragments at various incident energies, are displayed. As can be seen, in this experiment there is a rather sharp separation between quasi-elastic and dissipative processes, the latter being characterized by the broad "bump" centered at about 15 MeV of excitation energy; the excitation functions, obtained by integrating over this dissipative part of the spectra, show a clear oscillating behaviour (Fig.3) for all the considered transitions. The data have been analyzed with statistical methods. In particular the coherence energies Γ have been extracted from these curves by constructing the correlation functions. The corresponding times ($\tau \simeq \hbar/\Gamma$) are

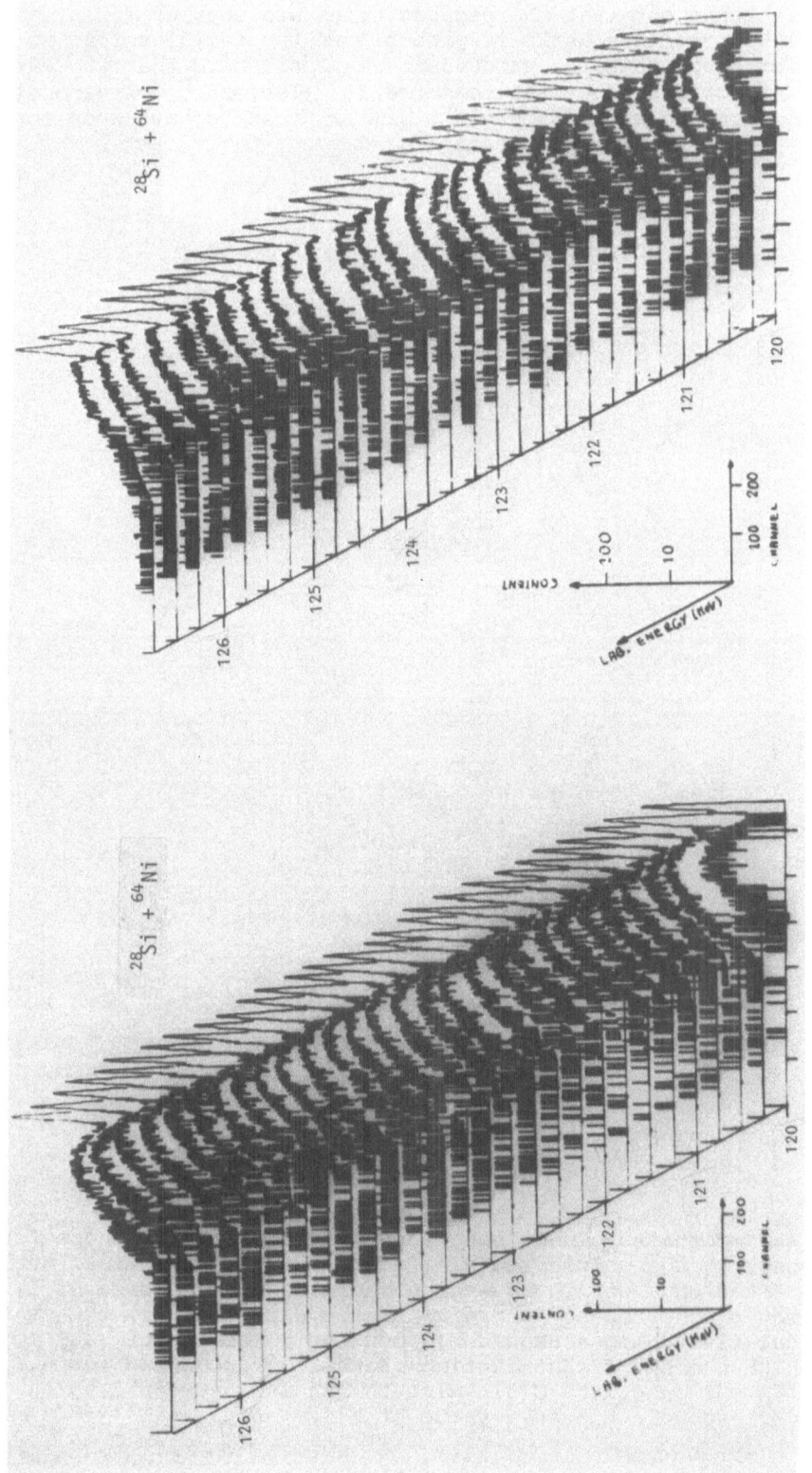

Fig.2- Energy spectra from the $^{28}Si+^{64}Ni$ reaction relative to Z=12 and Z=13 atomic numbers.

reported in Fig.4 for different Z fragments.

It turns out that the deduced times are shorter for projectile-like fragments as it is expected in the semiclassical models for deep inelastic processes (we recall here that the excitation functions have been measured at $\Theta(lab)=25^0$, corresponding to the focusing emission angle). Similar results have been obtained in more recent experiments[5,6].

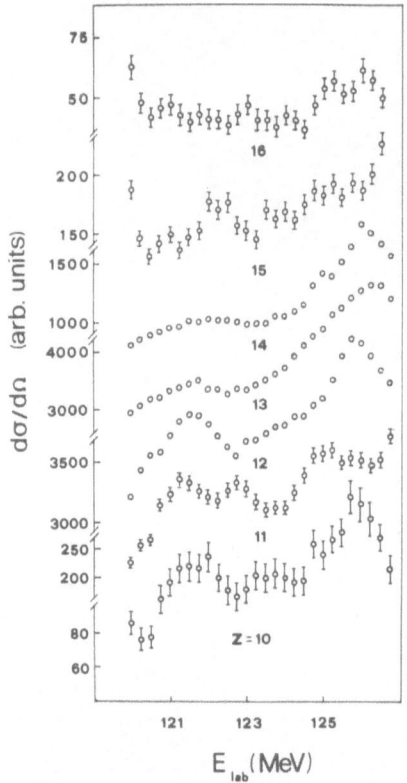

Fig.3- Excitation functions of the $^{28}Si+^{64}Ni$ reaction for different fragment atomic numbers.

An important comment concerns the visibility of these fluctuations. At the considered excitation energies the final nuclei are excited in a continuum region where several hundred of levels would contribute. According to the expression 1) the statistical fluctuations should be completely washed out, due to this high number of contributing channels. However we found that, in our case, the C(0) value is ranging between 0.15 to 0.25 corresponding to an N value of a few units.

A stimulating interpretation of this apparent contradiction could be given in terms of dinuclear system formation, in deep inelastic dissipative processes, <u>acting as a doorway state which decays into a very few selectively excited final states having similar configurations</u>. As an example we could suppose a dinuclear system formed by two nuclei excited into quasi-giant degrees of freedom; after a time of the order of $\tau \simeq 10^{-21}$ sec this system would decay to G.R. levels of the residual nuclei. This hypothesis has to be tested by measuring excitation functions for high energy γ in coincidence with the emitted fragments.

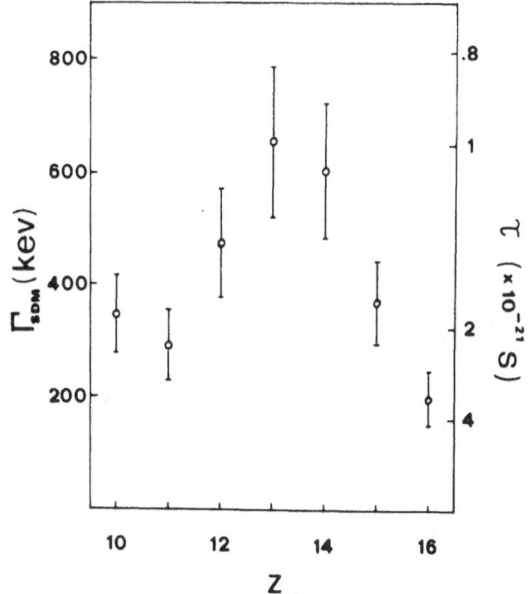

Fig.4- Coherence energies extracted by means of the SDM analysis from the excitation functions of Fig.3. The Γ(SDM) widths are reported versus fragment atomic number Z. On the right side is shown the time scale related to the energy one by the uncertainty principle.

A quantitative theoretical description of the above phenomena is not yet established. However:
1) it has been shown that all the observed features are, in a general framework, consistent with a statistical theory, as the Ericsons's one, whenever some correlations on the width amplitudes $\gamma_{\alpha i}$ and on the orbital angular momenta matrix elements $S_{\alpha l s}$ are retained[7];

2) more recently[8], R.Bonetti established a relationship between λ_{corr} (the " correlation length " i.e. the number of inter-fering partial waves) and Γ_{corr} (the correlation energy) through the expression:

$$3) \qquad \lambda_{corr} = (\frac{d^2}{\hbar^2 c^2 \delta}) \; \Gamma_{corr}$$

obtained by A.Y.Abul-Magd and M.H.Simbel in 1979[9].

In both cases the oscillations of the cross section allow to extract an energy width Γ which gives, through $\tau \simeq \hbar/\Gamma$, the life-time of the dinuclear system formed at the early stage of the reaction. It would be worthwhile, in our opinion, to analy-ze to what extent this dinuclear model effect can be quantita-tively understood within a theory based on the statistical door-way states concept.

REFERENCES

1) Nuclear Molecular Phenomena, Hvar, Yugoslavia, 1977, edited by N.Cindro (North-Holland, Amsterdam, 1978); Resonant Beha-viour of Heavy ion Systems, Aegeon Sea, Greece, 1980, edited by G.Vourvopoulos (National Printing office, Athens, 1981) and references therein.
2) A.De Rosa,G.Inglima,V.Russo,M.Sandoli:Phys.Rev.C27(1983)2688.
3) T.Ericson: Ann.Phys. 23(1963)390.
4) A.De Rosa,G.Inglima,V.Russo,M.Sandoli,G.Fortuna,G.Montagnoli, C.Signorini,A.M.Stefanini,G.Cardella,G.Pappalardo,F.Rizzo: Phys.Lett. 160B(1985)239.
5) A.Glaesner,W.Dunnweber,W.Hering,D.Konnrth,R.Ritzka,R.Singh: Phys.Lett. 169B(1986)153.
6) R.Lucas,B.Berthier,M.C.Mermaz,J.Suomijarvi,J.P.Coffin, G.Guillame,B.Heusch,F.Jundt,F.Rami: Proceedings of the Inter-national Nuclear Physics Conference, Harrogate U.K., 1986, Vol.1 p.287.
7) A.De Rosa,G.Inglima,V.Russo,M.Sandoli,G.Fortuna,G.Montagnoli, C.Signorini,G.Cardella,G.Pappalardo,F.Rizzo: Proceedings of the 4th International Conference on Nuclear reaction Mecha-nisms edited by E.Gadioli, Varenna, Italy, 1985, p.276.
8) R.Bonetti,M.S.Hussein: Phys.Rev.Lett. 57(1986)194.
9) A.Y.Abul-Magd,M.H.Simbel: Phys.Lett. 83B(1979)27.

STATISTICAL MULTIFRAGMENTATION OF NUCLEI

by J. P. Bondorf

The Niels Bohr Institute, University of Copenhagen

Blegdamsvej 17, DK 2100 Copenhagen Ø, Denmark

ABSTRACT: After an introduction to the main subject: break-up of nuclear matter, some elementary concepts from heavy ion dynamics are introduced. We discuss the collision process within a three step scenario, including fragmentation into many pieces of the interacting nuclei. The partition of the multiparticle system into clusters is discussed. The various components of the energy of a fragment assembly at the break up stage are introduced. They are generalizations of the Weissäcker-energies. We consider the thermodynamics of a gaseous mixture of fragments in equilibrium. The thermodynamic probability determining the weight of a given partition is calculated. Some results of calculations are discussed.

FRAGMENTATION OF MATTER

In daily life we often experience the formation of fragmented matter. Let me remind of the crushing into powder of a solid or the splashing into small drops of a liquid. In both cases the matter has been influenced by violent external stress which has been spread through the volume so that a variety of rupture zones were formed. The physical conditions for rupture are twofold. First the microscopic structure of the matter varies from one position to another. Thus the threshold for breaking the matter, defined by the rupture strength, may depend on the degrees of freedom which are involved (in a quantitative description these degrees of freedom have to be defined). Secondly, the matter under stress is influenced by dynamical external forces which stir up a pattern of oscillations giving rise to varying strain among the dynamical variables. The rupture in a local macroscopic or microscopic dynamical variable happens whenever the "strain" σ exceeds the "rupture strength" σr of that variable, i. e. when

$$\sigma > \sigma r \qquad (1)$$

By "strain" we mean force or force density in the Newtonian sense. Often σ and σr are not independent. It can easily happen that onset of rupture by fluctuations at some spots influences the dynamical evolution of the system at other spots. Then the ruptures proceed in a pattern which depends on the first small fluctuations. One can imagine that in a solid a network of rupture surfaces and zones are formed throughout the volume. These rupture zones are the beginning of the separation between the fragments. In the breakup of a liquid, important processes are probably formation of cavities followed by droplet formation. The ruptures in liquids happen at the most weak spots in the necks just as in nuclear fission. An important thing in these processes for both solids and liquids is that a large volume is affected within the same time interval.

All these processes result in the formation of many separated fragments. We call the processes <u>multifragmentation</u>. They should be considered in contrast to evaporation, where the excited piece of matter looses mass fragments and energy from the surface. It would be of great interest to construct microscopic models for fragmentation processes by using eq. (1). In this paper we shall, however, approach the phenomenon with a much lower level of ambition and just study the phase space for possible outcomes of the processes.

In multifragmentation an almost trivial but also the most profound physical feature is the change of the system from being one coherent body to consist of many independent bodies. A consequence of this is the dramatic increase of the total surface area of the matter. If the fragments are all of similar size this area scales like $M^{(1/3)}$ where M is the number of fragments, also called the multiplicity. Thus we have an average internal surface area after break up

$$S(M) \approx S(1) \cdot M^{(1/3)} \qquad\qquad (2)$$

where S(1) is the surface area of the original body. We shall later look more systematically into consequences of these fundamental changes of the system when multifragmentation occurs.

So far we considered fragment formation where M increases with time. Fragments can be formed in a quite different way, namely by condensation processes where the initial state of the matter is a gas the constituents of which join into bigger aggregates. In this process M decreases with time. In nuclear physics the formation of light reaction products in a high energy heavy ion collision has been viewed as a result of such a process called coalescence /1/. If one could keep a system in the same condition for a long time, by a hypothetical container, for example, it is obvious that both fragmentation and coalescence would happen side by side in order that equilibrium could be achieved.

In the statistical model which we shall consider below we calculate the properties of a nuclear system under the hypothesis of such a thermal equilibrium at a time towards the end of the expansion just when the fragments separate. We then assume that the nuclear interaction between the fragments vanishes suddenly so that the distribution of the primary fragments is frozen. From then on the Coulomb force between the fragments will accelerate the charged fragments further during the expansion. The assumption of thermal equilibrium is clearly not quite correct because the system is in dynamical evolution during the whole expansion process. By making the assumption of equilibrium we however obtain some advantages. One is that the calculations become simpler. Another and more important one is that in this way the properties of the nuclear binding and the nuclear energy spectra on the fragmentation can be handled in a transparent way.

The statistical thermodynamical theory for breakup of nuclei was introduced by Randrup and Koonin /2/ and later together with Fai /3/. Other statistical model descriptions have been made by Mekijan /4/, who treats mainly fragmentation at higher energies. In this paper we shall in particular discuss the model of Bondorf et al. /6/-/8/ the basic physical principles of which were introduced in /5/. The role of the long range Coulomb force for the fragmentation was emphasized by Gross et al. /9/. Aichelin and Hüfner /10/ consider the fragmentation in a statistical way with no thermodynamic requirements. The treatment of fragmentation in a dynamic, collective description has been done by several authors, with Hartree-Fock by Knoll et al. /11/, Bonche et al. /12/, progressing breakup by Biro et al. /13/, cascade by Cugnon, Boal et al. /14/, Vlasov equation by Remaud et al. /15/, Uehling-Uhlenbeck (Boltzmann) model by Bertsch et al. /16/.

The fragmentation relates to the liquid gas phase transition for nuclear matter, Mosel et al. /17/, Lamb et al. /18/. Fragmentation of ordinary condensed matter has been studied within the so called percolation theory which was taken up for nuclei by Campi and Desbois /19/, Bauer et al. /20/ and Biro et al. /21/ for nuclear problems. The connection between the percolation and the other fragmentation models were so far not well understood but some start to attack this problem was recently suggested /22/. Experimental studies are still scarce. Therefore the most comprehensive studies were so far made by means of emulsions, Jakobsson et al. /23/, Waddington et al. /23a/, analyzed by Aichelin et al. /23b/. More indirect studies were the semi-exclusive experiments with correlations between a few detected fragments (see experimental results /26/,/27/ mainly from BEVALAC, CERN-SC, MSU, SARA and GANIL). Also studies of non nucleonic emission from the fragmenting systems have been made theoretically and experimentally, with the theoretical works of Bertsch /24/,Greiner et al /24a / , and experiments /23/ on pion emission. Such particle studies give an independent account of properties of the hot matter, and they play an increasing role. For more comprehensive outlines of recent theoretical and experimental developments I refer to various conferences , f. ex. the Visby meeting 1985 /26/ or the Caen meeting 1986 /27/. For a review see Csernai et. al. /28/.

The paper is organized as follows. First we discuss ways to create a hot nucleus ready for nuclear multifragmentation. We then discuss some mathematical features of the splitting of a finite particle system into pieces. Thereafter first the energetics of fragmentation and then some relevant thermodynamic concepts are introduced. Finally we calculate and discuss some characteristic results.

CREATION OF A HOT SOURCE

The aim of the fragmentation theory is to describe the break up into many fragments of a highly excited extended piece of nuclear matter. We shall mention some examples of how to inject energy into the nucleus prior to the fragmentation.

In a heavy ion collision there is a good chance to have a uniform distribution of the energy in extended nuclear volumes prior to fragmentation. In peripheral collisions, fig.1, the two nuclei collide only partially. Two colder matter pieces (spectators) are formed together with the hot overlap zone (participants). In central collisions the energy is more uniformly distributed over the whole colliding system. All the hot zones can undergo multifragmentation if the energy is sufficiently high. In the so called "reaction phase diagram", fig.1, we indicate how various characteristic reaction mechanisms depend on the impact parameter b and the bombarding kinetic energy per nucleon εlab. The regions denoted by PS and TE are of special interest to us. In an inelastic collision between two pieces 1 and 2 of nuclear matter the excitation energy per nucleon is

$$\varepsilon^* = (\sqrt{(M1+M2)^2 \cdot c^4 + 2 \cdot M2 \cdot c^2 \cdot a1 \cdot \varepsilon lab - (M1+M2) \cdot c^2 + Q)}/(a1+a2) \qquad (1)$$

The non-relativistic limit gives the following expression

$$\varepsilon^*/\varepsilon_{lab}=a1\cdot a2/(a1+a2)^2\approx1/4 \text{ for } a1\approx a2 \tag{2}$$

In these expressions a1 and a2 are the nucleon numbers of the interacting projectile and target parts respectively, and M1 and M2 their rest masses. In the clean fireball picture the interacting nuclear volumes are determined by the geometrical overlap. However, the fireball is not quite independent of the spectator parts. Some energy is transferred to excitation of the spectators. It means that if this energy is high enough, also the spectators can disintegrate. We have indicated this with the curve ----- in the figure. The transport of energy to the spectators depends in a complicated way on the motion and the transport properties of the nuclear matter. In this paper we shall just point out that the spectator parts can under some circumstances recieve so much energy that they can perform multifragmentation.

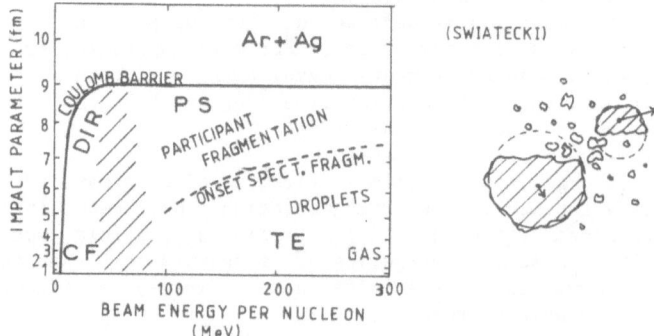

Fig.1 Left: Schematic reaction "phase diagram" for the collision between Ar and Ag (A1=40, A2=107). The shaded zone separates the low and the intermediate energy regions. The separation curve between PS (participant-spectator) and TE (total explosion) is calculated by combining the model of /32/ with the theory of multifragmentation. Right: Artists wiew of a PS reaction.

There are other ways to excite the nuclei. We shall now mention some of them. The excitation of nuclei by high energy protons has been known for a long time. In such reactions the injection of energy into the nuclear volume will always be rather local. Even in the case that a nucleus is struck centrally by a relativistic proton, some of the energy is presumably first deposited along a cylinder or cone. From there it spreads to the rest of the nucleus.

Annihilation of an antiproton when it hits a nucleus is another way to excite the nucleus locally. Dependent on the incident momentum of the antiproton, an energy of the order 2GeV will be deposited at the surface or more deep in the nucleus. The typical decay of a slow antiproton leads to emission of some pions (maybe 5-6) followed by the deexcitation of the residual nucleus. Evaporation and fission are typical decay modes (Bocquet et al. /29/), but also multifragmentation is a possibility.

Another way is the injection of one delta-resonance after a charge exchange reaction. This gives an excitation energy of the target nucleus in the range 250-350 MeV. In this case there is a considerable chance that fast nucleons dominate the deexcitation in a non-equilibrium reaction, where the energy is only spread to a few degrees of freedom. A reaction with a more uniform spread of the energy over the nuclear volume is, however, also possible.

A highly excited nucleus will be compressed at the first collision stages. Calculations of the central density in a heavy ion collision as function of time during a collision event have been made in several models, see f. ex. ref. /28/. From such calculations one finds that first the central density increases rather quickly to a maximum. Then the composite compressed nucleus expands, in the beginning as one coherent body, but around some later time the density becomes so low, that the matter becomes unstable. Low density zones may develop, the matter in between will condense, and fragments are formed. This process is very chaotic, and fluctuations and complicated interaction patterns make a detailed description rather impossible. Instead we have to turn to statistical methods. This is our starting point of the statistical theory.

In the fragmentation theory which is described below we shall in all cases consider the various reactions as just means to inject energy into the nuclear volume. The main input into the theory will be the baryon and charge numbers and the excitation energy of an initially coherent piece of highly excited nuclear matter.

As the total nuclear binding energy is of the order of 7-8 MeV per nucleon it is clear that at an excitation energy per nucleon of this order of magnitude the system must disintegrate. It is also evident that if the fragments are not just nucleons but nuclei, then not all of the binding energy is used for the fragmentation. This means that the fragmentation can happen at a somewhat lower excitation energy. Off hand it is not obvious that multifragmentation is at all a preferred decay mode. Successive or multiple fast evaporation could also carry excess energy away so that a multifragmentation mechanism would be surpassed. However, the experimental evidence shows that in heavy ion collisions at intermediate energy, events with a number of not too small fragments are frequent. This gives sufficient motivation for development of a multifragmentation description.

PARTITIONS

The most basic concept of multifragmentation is the partition /6/. For a body of Ao constituent identical particles we define a partition by the so called partition vector N(A) which fulfills the condition

$$Ao = \sum_{A=1}^{Ao} N(A) \cdot A \tag{3}$$

Thus N(A) is the number of particles of mass A. Another important quantity related to the partition vector is the multiplicity

$$M = \sum_{A=1}^{Ao} N(A) \tag{4}$$

In table 1 we show for Ao = 4 the 5 possible partitions together with their multiplicity.

Table 1. The partitions for a system with mass number 4

partition vector	multiplicity
(4,0,0,0)	4
(2,1,0,0)	3
(1,0,1,0)	2
(0,2,0,0)	2
(0,0,0,1)	1

Table 2. Values of the function $P(A,M)$ for $A \leq 15$, and values of $P(A)$ as defined by (7).

A \ M	1	2	3	4	5	6	7	8	9	10	11	12	13	14	15	P(A)
1	1															1
2	1	1														2
3	1	1	1													3
4	1	2	1	1												5
5	1	2	2	1	1											7
6	1	3	3	2	1	1										11
7	1	3	4	3	2	1	1									15
8	1	4	5	5	3	2	1	1								22
9	1	4	7	6	5	3	2	1	1							30
10	1	5	8	9	7	5	3	2	1	1						42
11	1	5	10	11	10	7	5	3	2	1	1					56
12	1	6	12	15	13	11	7	5	3	2	1	1				77
13	1	6	14	18	18	14	11	7	5	3	2	1	1			101
14	1	7	16	23	23	20	15	11	7	5	3	2	1	1		135
15	1	7	19	27	30	26	21	15	11	7	5	3	2	1	1	176
20																627
50																204 226
100																190 562 992

The multiplicity has a clear physical significance. This can intuitively be inferred from everydays experience. It is well known that violent processes in a piece of matter usually result in the splitting of the matter into a number of pieces. This number increases vith the degree of violence. In order to study the multiplicity one must know in what ways a piece of matter can be split into several pieces. We shall start by investigating in how many different ways $P(A,M)$ a given piece of matter of A equal constituents can be subdivided into M pieces.

One can easily verify that the following recurrence relation holds:

$$P(A,M) = P(A-1,M-1) + P(A-M,M) \qquad (5)$$

Here $P(A-1,M-1)$ is the number of partitions which have at least one fragment of mass 1, and $P(A-M,M)$ is the number of all the other partitions having no fragments of mass 1. By iterative use of (5) we get

$$P(A,M) = \Sigma\ P(A-kM-1,M-1) \qquad (6)$$
$$k=0..A/M-1$$

where the starting values for the iterative procedure are $P(1,0) = 0$ and $P(1,1) = 1$. Eq. (6) is then used for an iterative calculation of $P(A,M)$. We show some results of a calculation in table 2, and a calculated M-distribution for A=100 is shown in fig. 3. It is seen that the distribution is broad and peaked at some value between M=0 and M=A/2. The total number of different partitions for a given A is

$$P(A) = \sum_M P(A,M) \tag{7}$$

This number is strongly increasing with A. We show some values in table 2.

Since nuclei has both protons and neutrons a realistic partition vector has two indices, Z and A, and besides the baryon conservation the partition vector also fulfills charge conservation,

$$Zo = \sum_{Z=1}^{Zo} N(A,Z) \cdot Z \tag{8}$$

In the following treatments we will simplify the problem to only using (3) and then assume an analytic connection Z=Z(A). From a physical point of view the partition vector, as defined in this section, is not at all sufficient for defining the complete physical configuration of a fragmented system. This requires a complete specification of all the quantum numbers of the fragments, i. e. isospin, energy, parity, momentum, spin and angular momentum. However, the most basic and visible concept for the fragmentation is the partition. In the following theory we shall treat only the partitioning in the most elaborate way and apply approximate methods when dealing with the other quantities of the fragmentation.

ENERGY RELATIONS FOR A MIXTURE OF DROPLETS

We simplify the composite system to be a spherical nucleus at rest with mass number Ao, charge number Zo and with an excitation energy pr. nucleon $e^* = E^*/Ao$ like in a compound nucleus theory. The main idea is now to assume that the system at the density of breakup consists of a mixture of nuclear droplets of varying size. We assume that the energy of this mixture is thoroughly spread over the degrees of freedom of the system so that it is close to thermal and chemical equilibrium. We shall assume that just beyond this density the droplets don't interact any longer. Thus the expansion flow energy which must be present, is replaced by thermal energy in this first version of the theory. The concept of partition applies to this composite nucleus. Each possible partition (i) (defined in connection with eq. (3)) gives rise to one droplet composition, the thermodynamical probability of which can be calculated as

$$W(i) = \exp(S(i)) \tag{9}$$

where S(i) is the entropy of partition (i). We have to determine this entropy for the mixture of fragments in the nuclear break up volume.

Consider for partition (i) the mixture of M fragments in a fixed volume Vb. We assume that the conservation laws (3) and (8) hold, and that the fragments move as in an ideal gas. The total energy of the system can be expressed as

$$\begin{aligned} Etot &= (3/5) \cdot (Zo \cdot e)^2/Rb + \sum N(A,Z) \cdot E(A,Z) \\ &= Eground + E^* = Eground + e^* \cdot Ao \end{aligned} \tag{10}$$

where Rb is the radius of the volume Vb. Note that an important part of the Coulomb energy is taken into account by the first term in (10). The energy E^* is the excitation energy of the nucleus. The additional energy for each fragment of mass number A and charge number Z is:

$$E(A,Z) = Etransl + Ebulk + Esurf$$
$$+ Esym + Ecoul \qquad (11)$$

with
$$Etransl = 3/2 \cdot T \text{ (independent of A)} \qquad \varepsilon o \approx 16 MeV$$
$$Ebulk = (-B(A,Z) + T^2/\varepsilon o) \cdot A \quad \text{for } A>4 \text{ ; } B(A,Z) \approx 16 MeV$$
$$\qquad -B(A,Z) \cdot A \qquad \text{for } 1<A \leq 4$$
$$\qquad 0 \qquad \text{for nucleons}$$
$$Esurf = (\beta - T \cdot d\beta/dT) \cdot A^{(2/3)} \text{ ;}$$
$$\beta = \beta(0) \cdot [(To^2-T^2)/(To^2+T^2)]^{(5/4)}$$
$$\beta(0) \approx 18 \text{ MeV ; } To \approx 15-20 \text{ MeV}$$
$$Esym = \gamma \cdot (A - 2 \cdot Z)^2/A \text{ ; } \qquad \gamma \approx 25 \text{ MeV}$$
$$Ecoul = (3/5) \cdot (Z \cdot e)^2/R \cdot [1 - R/Rcell]$$

The radius of fragment A: $R = ro \cdot A^{(1/3)}$; $ro \approx 1.2 fm$

The terms in (11) are just generalized Weisssäcker mass energies. The surface energy has been given a temperature dependence (see /18/ and /6/) so that it vanishes at the critical temperature To like in a van der Waals gas. The bulk intrinsic excitation energy of the fragments is contained in the term $T^2/\varepsilon o$ and in the temperature dependent part of the surface energy. A very important property of the model is that we determine the free volume in which the fragments move just at breakup. This is done by demanding that the average distance 2d between the fragments at this stage is of the order of the maximum range of the nuclear force between fragments. In the calculations below we have put $2d \approx 2.8 fm$. This gives the breakup volume Vb:

$$Vb = Vo + Vf \qquad (12)$$
where
$$Vo = 4/3 \cdot \pi \cdot Ro^3$$
$$Ro = ro \cdot Ao^{(1/3)}$$

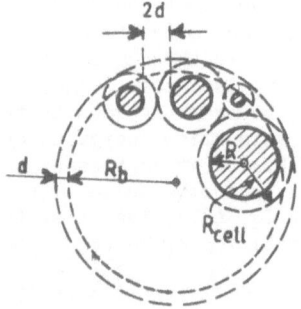

Fig.2 Schematic 2-dimensional analog illustration of the 3-dimensional packing of fragments used in the theory.

The volume Vf in which the fragments are assumed to move freely is determined in the following way. We approximate the shapes of the fragments by spheres which are packed in a compact way (see fig. 2). We then have the following approximate relation between the total breakup radius Rb and the fragment radii Ri:

$$(Rb + d)^3 = \Sigma_i (Ri + d)^3 \qquad (13)$$

Calculation of the total volume of the nucleus at normal density gives the following constraint on the fragment radii

$$Ro^3 = \sum_i Ri^3 \equiv M \cdot \langle Ri \rangle^3 \qquad (14)$$

The free volume is

$$Vf = Vb - Vo = Vo((Rb/Ro)^3 - 1) \qquad (15)$$

In a simplified description which we shall apply in the following, the fragment radii Ri are all put equal to an average value $\langle Ri \rangle$ defined in (14). We then get from (13) and (14)

$$Rb = Ro + d \cdot (M^{(1/3)} - 1) \qquad (16)$$

where M is the total number of fragments (the multiplicity).

The Coulomb energies in E(A,Z) in (11) are just additional energies to those of the uniformly charged sphere of radius Rb. They are obtained by noticing that the charge of a fragment is contained within a sphere of radius R and not spread over the whole cell of radius Rcell. The two radii are related as

$$R/Rcell = Ro/Rb \qquad (17)$$

TRERMODYNAMICS OF A GAS OF FRAGMENTS

Before calculating the entropy which should be used in (9), let us first look at some relevant thermodynamical properties of the system. Each partition (i) has a temperature T and a thermodynamic probability W(i) which depend on the partition i. The temperature T also depends on (i) and will fluctuate in a narrow interval around some average value. This is a consequence of the finiteness of the system with energy conservation. It is worth noticing that in our ansatz for the fragment energies we use the average energy E(A,Z) for each (A,Z) component rather than the individual energy for each fragment. This means that the model in the version described here is neither strictly microcanonical nor strictly canonical.

In order to determine the entropy we now use the general relation for the total energy

$$E = F + T \cdot S \qquad (18)$$

where F is the free energy (Gibbs), T the temperature and S the entropy. By inspecting the nature of the energy components in eq's (11), we see that all components are additive with exception of the overall Coulomb energy term in (10), which depends on the partition only through the breakup radius. We shall, in close analogy with chemical terminology call the (A,Z) components "chemical" components. The set of energy terms: transl, bulk, surface, symmetry and residual Coulomb, we call "physical" components of the system. Since (18) is linear in F and S, also these two quantities are additive with respect to both the chemical and the physical components.

For a system with fixed E we have from (18):

$$S = -\delta F/\delta T \big|_{Vb} \qquad (19)$$

which is also valid for for each (chemical,physical) component. Thus we have besides the energy relation (10) also the relation for the free energy

$$F = \Sigma\ N(A,Z)\cdot F(A,Z) \tag{20}$$

and the entropy

$$S = \Sigma\ N(A,Z)\cdot S(A,Z,) \tag{21}$$

with
$$S(A,Z) = -\delta F(A,Z)/\delta T\Big|_{Vb} \tag{22}$$

In order to use (22) to determine S we have to find the free energy components. Most of them can most easily be guessed directly from the energy terms in (11). For the translational motion we calculated the free energy directly from the definition

$$F = -T\cdot \ln\Sigma_{n}\ \exp(-En/T) \tag{23}$$

where the last sum is the "sum over states" in the Gibbs ensemble. All the chemical and physical components of energy, free energy and entropy are shown in table 3. Note the term $-\ln N(A,Z)!/N(A,Z)$ in the translational entropy. It is added in order to reduce the probability due to $N(A,Z)$ identical particles in each chemical component of the system.

Table 3
Energy, free energy and entropy for one fragment (A,Z)

	$E_{A,Z}(T)$	$F_{A,Z}(T)$	$S_{A,Z}(T) = -\dfrac{\partial F_{A,Z}}{\partial T}\Big|_{V_b}$
transl.	$\tfrac{3}{2}T$	$-T\ln\!\big(g_{A,Z}\tfrac{V_f A^{3/2}}{\Lambda^3}\big)+\dfrac{T}{N_{A,Z}}\ln N_{A,Z}!$	$\tfrac{3}{2}+\ln\!\big(g_{A,Z}\tfrac{V_f A^{3/2}}{\Lambda^3}\big)-\dfrac{1}{N_{A,Z}}\ln N_{A,Z}!$
bulk	$\big(-B(A,Z)+\tfrac{T^2}{\varepsilon_o}\big)A$	$\big(-B(A,Z)-\tfrac{T^2}{\varepsilon_o}\big)A$	$\dfrac{2T}{\varepsilon_o}A$
surf.	$\big(\beta(T)-T\tfrac{\partial\beta(T)}{\partial T}\big)A^{2/3}$	$\beta(T)A^{2/3}$	$-\dfrac{\partial\beta}{\partial T}A^{2/3}$
sym.	$\gamma\dfrac{(A-2Z)^2}{A}$	$\leftarrow same$	0
Coul.	$\tfrac{3}{5}\dfrac{(Ze)^2}{R_{A,Z}}\big(1-\dfrac{R_{A,Z}}{R(A,Z)_{cell}}\big)$	$\leftarrow same$	0

$$\Lambda = \big(\tfrac{2\pi\hbar^2}{m_N T}\big)^{1/2}$$

$$\beta = \beta(0)\big(\tfrac{T_o^2-T^2}{T_o^2+T^2}\big)^{5/4}\qquad \beta(0) \approx 18\ MeV \qquad T_o = 15-20\ MeV$$

$$\gamma = 25\ MeV$$

$$\varepsilon_o = 16\ MeV$$

The final expression for the thermodynamical probability (9) is now /6/:

$$W(i)= \prod_{A,Z}[g(A,Z)\cdot Vf\cdot A^{(3/2)}\cdot \exp(3/2+Sint(A,Z))/\ (T)^3]^{N(A,Z)}/N(A,Z)! \tag{24}$$

The intrinsic entropy of the fragments is $Sint(A,Z) = (2 \cdot T/\epsilon_0) \cdot A - (d\beta/dT) \cdot A^{\wedge}(2/3)$ for A>4 as seen from table 3. The statistical factor $g(A,Z)$ is specified individually for the fragments with A≤4 and otherwise put equal to 1. The quantity $\Lambda(T)$ in (24) is the thermal wave length for a nucleon with temperature T. It is also given in table 3.

CALCULATION AND DISCUSSION

In order to calculate the entropy $S(i)$ to be used in (4) we now in principle run through all the possible partitions. For each partition (i) with given $N(A,Z)$ a temperature $T(i)$ is determined from (10) and (15) and then $W(i)$ is determined from (24). In practice it is very difficult to run through all the partitions for a big nucleus because of their large number. Therefore we have used a Monte Carlo method. We show some characteristic calculated results in figures 3, 4, 5, and 6 . The mass ·number is $A_o = 100$ and the charge. $Z_o = 50$.

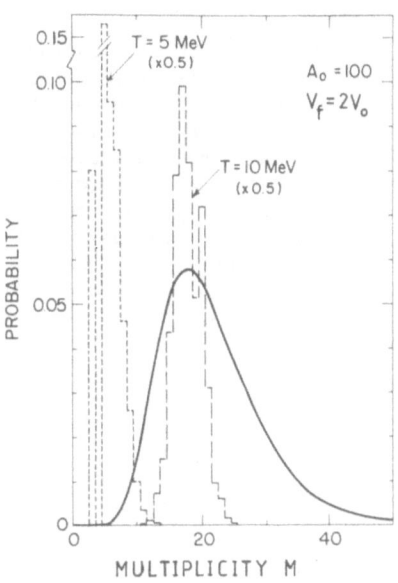

Fig 3. Distributions of calculated multiplicities normalised to 1. Solid curve: full partition space. Histograms: thermodynamic calculations at two different temperatures. The calculation was made with fixed temperature, and the free volume was kept fixed as Vf=2·Vo.

One sees from fig. 3 how narrow the thermodynamic multiplicity distribution is compared to the unbiased one, eq. (6). This means, that for a Monte Carlo simulation one can very quickly learn the relevant M-range which governs the fragmentation, and then bias the sampling accordingly. Our investigations show that this method leads to some improvement of the Monte Carlo simulation economy. In fig. 4 we show how the average multiplicity M varies with excitation energy. At the onset of fragmentation, called the crack-threshold or the crack-point, M rises smoothly from 1. When energy is added to the system just above the crack-point one finds that the energy is largely used for the creation of new surface (2) somewhat modified by the M-dependent Coulomb energy. At higher energy the bulk translational energy plays an increasing role in the energy balance. As the system approaches the critical temperature for bulk matter To, the vanishing of the surface energy will furthermore boost the multiplicity.

In fig. 5 we show how the temperature varies with excitation energy.

The onset of fragmentation happens at a characteristic temperature, <u>the crack temperature</u>, which is of the order of 5-6 MeV. This is also known from other investigations, see f.ex. /12/. It is fairly constant with mass Ao, with a tendency to rising towards the small masses Ao. The figure shows how the temperature remains almost constant over a large range of excitation energy. This constancy of the temperature is connected with the creation of new surface and volume above the crack point and is one of the most striking features of the theory. Fig. 6 shows a few calculated mass distributions for different excitation energies. One sees that for the energy just above the crack point the mass distribution has a modified U-shape. It mostly reveals partitions with one big fragment.

12

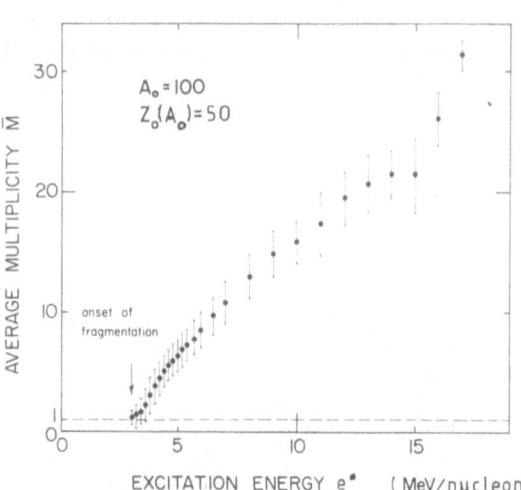

Fig 4. Calculated average multiplicity as function of the excitation energy per nucleon. The fluctuations do not indicate calculational errors but are fluctuations from the use of the many different partitions.

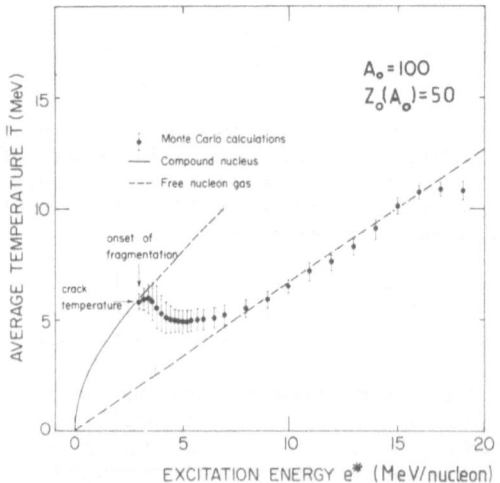

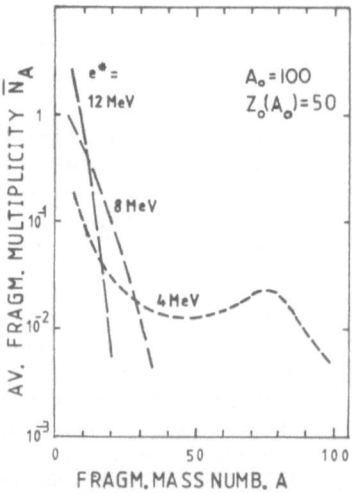

Fig 5. Calculated average temperature as a function of the excitation energy per nucleon. The fluctuations have the same meaning an in fig.4.

Fig 6. Calculated primary mass distributions for various excitation energies.

In the calculation only the mass numbers have been considered for the partitions. However, the theory is formulated so that charge numbers enter in all relevant quantities. A method for forming partitions with both A and Z has been suggested by Sneppen /33/. Because of the enormous size and complexity of the phase space, more advanced methods to get around to all relevant parts of the phase space are desirable. Recently two groups /34/,/35/ have applied the Metropolis statistical method in order to overcome this problem.

CONCLUDING REMARKS

The statistical model of multifragmentation is one of several theories which can be applied to understand the "total explosion" of a highly excited nucleus.

The theory is a phase space approach which uses the assumption that the reaction time is so long that there is established an approximate thermal equilibrium in the expanding nuclear matter just before the freeze out of the fragment composition. The theory enables a prediction of the crack temperature around 5-6 MeV, above which the nuclear matter breaks up into several pieces. This temperature is lower than the liquid gas phase transition temperature by a factor of the order 2-3. The crack temperature remains almost constant over a large interval of excitation energy per nucleon. For an excitation energy above 10-15 MeV per nucleon the number of bigger fragments (droplets) decreases at the expense of light fragments. The theory has some possibilities of non equilibrium simulation. As an example one can change the intrinsic heat capacity of the fragments by the level density parameter and in this way simulate both hot and cold fragmentation. The theory takes full account of the fluctuations in the partition space, and is in this way similar to the nuclear percolation theory.

In order to compare the theory with experiments it is nessesary to combine it with realistic dynamic models of the collision process. This includes decay by evaporation of the primary fragments. Both the composition and the spectra of the primary products will be modified by this secondary decay.

ACKNOWLEDGEMENTS

The author wants to thank H. W. Barz, R. Donangelo, B. Jakobsson, I. N. Mishustin, C. J. Pethick, K. Sneppen and H. Schulz for collaboration and stimulating discussions.

REFERENCES

/1/ Randrup, J. and Koonin, S. E., Nucl. Phys. A356 (1981) 223.
/2/ Fai, G. and Randrup, J., Nucl. Phys. A281 (1982) 557.
/3/ Fai, G. and Randrup, J., Nucl. Phys. A404 (1983) 551.
/4/ Mekjian, A., Phys. Rev. C17 (1978) 1051.
/5/ Bondorf, J. P., Proc.Nucl.Phys.Workshop I.C.T.P.Trieste
     Oct.5-30,1981, ed.C.H.Dasso, North Holland (1982)765 and
     Nucl. Phys. A387 (1982) 25c.

/6/ Bondorf, J., Mishustin, I. N. and Pethick C., Proc. Int. School
    seminar on heavy ion physics, ed. Yu. Ts. Oganessian et al.
    Dubna (1983) 354.
    Bondorf, J. P., Donangelo, R., Mishustin, I. N., Pethick, C. and
    Sneppen, K., Phys. Lett. 150B (1985) 57.
    Bondorf, J. P., Donangelo, R., Mishustin, I. N., Pethick, C.,
    Schulz, H. and Sneppen, K. Nucl. Phys. A443 (1985) 321.
    Bondorf, J., Donangelo, R., Mishustin, I. N. and Schulz, H.
    Nucl. Phys. A444 (1985) 460.
    Bondorf, J. P., Donangelo, R., Schulz, H. and Sneppen K.
    Phys. Lett. 162B (1985) 30.
/7/ Mishustin, I., Nucl. Phys. A447 (1985) 67c.
/8/ Barz, H. W., Bondorf, J. P., Donangelo, R., Mishustin, I. N. and
    Schulz, H., Nucl. Phys. A448 (1986) 753.
/9/ Gross, D. H. E., Phys. Scr. T5 (1983) 213.
    Sa Ban-Hao and Gross, D. H .E., Nuclear Physics 437 (1985) 643.
/10/ Aichelin, J. and Hüfner, J., Phys. Lett. 136B (1984) 15.
/11/ Knoll, J. and Strack, B., Phys. Lett. 149B (1984) 45.
/12/ Bonche, P., Levit, S. and Vautherin, D.,
    Nucl. Phys. A436 (1985) 265.
/13/ Biro, T. S. and Knoll, J., GSI preprint 85-29.
/14/ Cugnon, J., Phys. Rev. C22 (1981) 2094.
    see also
    Boal, D. H. and Goodman, A. L., (subm. Phys. Rev. C) (1985).
/15/ Remaud, B., Sebille, F., Gregoire, C., Vinet, L. and Raffray, Y.,
    Nucl. Phys. A447 (1985) 555c.
    see also
    GANIL preprints 85.12 and 86.01
/16/ Bertsch, G., Kruse, H. and Das Gupta, S., Phys. Rev. C29
    (1984) 673. Bertsch, G. et al. Phys. Rev. C29 (1984) 675.
/17/ Mosel, U., Zint, P. G. and Passler, K. H., Nucl. Phys. A236
    (1974) 252.
/18/ Lamb, D. Q., Lattimer, J. M., Pethick, C. and Ravenhall, D. G.,
    Phys. Rev. Lett. 41 (1978) 1623. Nucl. Phys. A360 (1981) 459,
    Univ. of Illinois at Urbana preprint P/84/3/34.
/19/ Campi, X. and Desbois, J., Contrib. XIII Int. Meet. Bormio
    (1985) 497.
/20/ Bauer, W., Post, U., Dean, D. R. and Mosel, U., preprint
    Giessen (1985).
/21/ Biro, T. S., Knoll, J. and Richert, J., preprint
    GSI-85-67.
/22/ Barz, H. W., Bondorf, J. P., Donangelo, R. and Schulz, H.,
    Phys. Lett. 169B (1986) 318.
/23/ Jakobsson, B., Jönsson, G., Lindkvist, B. and Oskarsson, A.,
    Z. f. Phys. A307 (1982) 293.
/23a/ Waddington, C. J. and Freier, P. S., Phys. Rev. C31 (1985) 888.
/23b/ Aichelin, J. and Campi, X., MPI, Heidelberg preprint (1986)
    HD-TVP-86-6.
/24/ Bertsch, G., Phys. Rev. C15 (1977) 713.
/25/ Benenson, W. et al., Phys. Rev. Lett. 43 (1979) 683.
    Johansson, T. et al., Phys. Rev. Lett. 48 (1982) 732.
    Grosse, E., Proc. VI H.E.H.I. study, Berkeley (1983).
/26/ Proc. of the Conf. on Nucleus Nucleus collisions II, Visby,
    ed. H. Å. Gustafsson et al., Nucl. Phys. A447 (1986).
/27/ Proceedings HICOFED86, Journal de Physique 47,
    Colloque C4, Suppl. au no 8.
/28/ Csernai, L. P. and Kapusta, J. I., Phys. Rep. 131
    (1986) 223.
/29/ Bocquet, J. P. et al., preprint CERN-EP/87-11.
/30/ Aichelin, J., Nucl. Phys. A447 (1985) 569c.

/31/ Barz, H. W., Bondorf, J. P., Guet, C., Lopez J. and Schulz, H.,
     under preparation.
/32/ Bondorf, J. P., De, J. N., Fai, G. and Karvinen, A. O. T.,
     Nucl. Phys. A430 (1984) 445.
/33/ Sneppen, K., Niels Bohr Institute preprint NBI86-44
/34/ Zhang Xiao-Ze., Gross, D. H. E., Xu Shu-Yan. and Zheng Yu-Ming
     Nucl. Phys. A461 (1987) 641,668.
/35/ Koonin, S. and Randrup, J., preprint LBL-21165 (1987)

DYNAMICAL AND STATISTICAL ASPECTS OF INTERMEDIATE ENERGY HEAVY ION
COLLISIONS: LECTURES ON THREE SELECTED TOPICS

Jörn Knoll
GSI, Postfach 110541, D-6100 Darmstadt RFA

1. Preface

The lectures presented deal with three different topics relevant for the discussion of nuclear collisions at medium to high energies. The first lecture concerns a subject of general interest, the description of statistical systems and their dynamics by the concept of missing information. It presents an excellent scope to formulate statistical theories in such a way that they carefully keep track of the known (relevant) information while maximizing the ignorance about the irrelevant, unknown information. The last two lectures deal with quite actual questions of intermediate energy heavy-ion collisions. These are the multi-fragmentation dynamics of highly excited nuclear systems, and the so called subthreshold particle production. All three subjects are self-contained, and can be read without the knowledge about the other ones.

I like to thank all the collegues who helped me on various occasions. For the subjects presented I enjoyed to work together with T. Biro, S. Bohrmann, J. Richert, R. Shyam.and J. Wu. Interesting discussions with J. Bondorf P. Braun-Munzinger, W. Cassing, J. Cugnon, H. Feldmeier, E. Grosse, C. Guet, J. Hüfner, B. Jakobsson, U. Mosel, B. Schürmann, and J. Zimanyi are gratefully acknowledged.
It is a wonderful occasion to dedicate these lectueres to Friedrich Beck. Over the last decade he followed the developments of this field and my contributions to it with more than just interest. I like to thank him for his continuous support and encouragements.

2. Missing Information

A statistical system is characterized by the fact that its actual 'state' is not precisely known. Rather, one has only some limited information upon it, expressed by the fact that each of its micro-states occurs with a certain probability. A quantum system is described by an incoherent mixture of micro-states. In thermodynamics even a microscopic system attains this property due to its coupling to a heat-bath. For isolated systems, however, the concept of a heat-bath is not available and one has to face the question in which sense such systems do behave statistically, that is to say, one has to know which information is known ($\rightarrow$ relevant information) and which one is actually lacking ($\rightarrow$ irrelevant information).

The problem is to find a description of the system by such statistical operator **D** which respects the fact that some relevant information is known while all other information is unknown. This can be achieved by the concept of *missing information*. Developed about 40 years ago this concept can be found in only a few text books (e.g. [1]). Also it has not yet found the corresponding appreciation in the nuclear physics community, although it is needly adaptable for their problems (see however the recent considerations in [2]).

2.1. Basic definitions of missing information

About any physical system Σ which one likes to describe, already quite some information is known: this is the Hilbert-space of all the possible states the system can take, the Hamiltonian which governs its dynamics. This one may consider as the preassumed information. Furtheron due to certain preparation of the system one may know some information about certain observables. Else, nothing more may be known. To quantify the 'unknown' we like to assign to any such system a real number $I(\Sigma) \geq 0$. Intuitively it should somehow reflect the number of decision steps which are still neccessary to precise the system in its actual state.

For simplicity let us start with a system which has a choice of n apriory equally possible states, like the six faces of a die, or the 37 choices of a roulette. Since all states are apriori equally probable it is clear that the missing information about the system depends only on the number of possibilities n. The follwing properties are selfevident:

$$I(n) = 0 \longleftrightarrow n = 1,$$
$$I(n) > I(m) \quad \text{if} \quad n > m \tag{2.1}$$

expressing the fact that no information is missing if the system has only one choice, and the monotony of the missing information with an increasing number of choices. Futher on, if a statistical system Σ is composed of two statistically independent systems, i.e. $\Sigma = \Sigma_1 \times \Sigma_2$, one likes to demand additivity in the missing information

$$I(\Sigma) = I(\Sigma_1) + I(\Sigma_2), \qquad \text{if } \Sigma = \Sigma_1 \times \Sigma_2. \tag{2.2}$$

If there are m and n apriori equal choices for the two systems, this implies

$$I(mn) = I(m) + I(n). \tag{2.3}$$

Straight forward contemplation upon the above axioms yields besides an irrelevant normalisation factor

$$I(n) = \ln n. \tag{2.4}$$

That is, the missing information is just the logarithm of the number of equal choices. If I would be defined on the logarithm of base 2 the missing information is identical to the minimum number of binary decisions which are necessary to precise the actual state.

In general one considers a system which again consists of n a priori equally probable states, however, due to special circumstances or preparation of the system each of the different states i occurs with a certain probability P_i. Thus, we may ask how much information are we lacking about a system prepared in such a way?

The following consideration shows that the missing information $I(\{P\})$ can be deduced from the preceeding simple case. One just has to do same as measuring the probability distribution P. That is, one takes N (N→∞) equally prepared systems (Gibb's ensemble). Each of these systems happens to be in one of the states i=1...n. Yet, we do not know in which state any of the considered systems is. All what we know is, that the frequency to encounter a given states i follows from its probability and therefore is about $N_i = N P_i$. Evidently there $N!/(N_1! \; N_2! \; .... \; N_n!)$ equally possible realisations for the N prepared systems. Therefore the missing information per system results to

$$I = 1/N \ln\{N!/ \; \Pi \; N_i!\} \tag{2.5}$$

or employing Stirling's formulas for N → ∞

$$I(\{P\}) = - \sum_{i=1}^{n} P_i \ln P_i. \tag{2.6}$$

This is a well known expression. For constant $P_i = 1/n$ it reduces to (2.4). Further on, $I(\{P\}) = 0$,if and only if P is a sharp distribution.

The generalisation to quantum systems is straight forward. There a statistical system is decribed by the density operator (statistical operator) **D** with Tr **D** = 1. Given **D**, its missing information can be found to be

$$I(\mathbf{D}) = - \text{Tr}\{ \mathbf{D} \ln \mathbf{D} \} \tag{2.7}$$

which is von Neumann's entropy expression.

2.2. Principle of Max I

For the decription of a statistical system one employs the following postulates:
- apriori equal probabilities to any state of an orthonormal basis of quantum states.
- the statistical operator takes such a form which maximizes the missing information $I(\mathbf{D})$ while respecting as constraints all known (relevant) information (Jaynes' principle [3]).

For simplicity we call a statistical system that can be described in such a way in *equilibrium with respect to the constraints given*. Essentially two types of constraints can be formulated:

i) *generalized canonical constraints*: the expectation values of certain observables are known, and

ii) *micro constraints*: the eigenvalues of certain observables (e.g. energy) are only permitted in a given interval.

2.3. Generalized canonical ensembles

Let us first discuss the canonical case. There, for a certain set of relevant observables R_μ $\mu = 1...k$ the mean values

$$< R_\mu > \ = \ \text{Tr} \{ \ R_\mu \ D \ \} = r_\mu \qquad \mu = 1...k$$

(2.8)

are known. Thus, we are looking for the particular density operator D_0 which decribes the system in equilibrium given the constraints (2.8). Therefore D_0 maximises $I(D)$ given these constraints and the normalisation condition $\text{Tr}\{D\} = 1$, i.e. it follows from the variational problem

$$\delta \{ \ I(D) - \lambda_0 \ \text{Tr} \ D - \sum_{\mu=1}^{k} \lambda_\mu \text{Tr}\{ \ D \ R_\mu \} \ \} = 0,$$

(2.9)

where the λ_μ are the Lagrange parameters corresponding to each of the constraints (2.8). Although the derivation is not straight forward in the quantum case, the solution of this problem takes up the expected and quite familiar exponential form

$$D_0 \ = \ D_0(r_1,... ,r_k) \ = \ 1/Z \ \exp\{ \ - \sum_{\mu=1}^{k} \lambda_\mu \ R_\mu \ \},$$

(2.10)

where the partition sum

$$Z(\lambda_1,....\lambda_k) \ = \ \text{Tr} \{ \ \exp\{ \ - \sum_{\mu=1}^{k} \lambda_\mu \ R_\mu \ \}\}$$

(2.11)

cares about the absolute normalisation. The Lagrange parameters $\lambda_1,.... \lambda_k$ have to be chosen such that D_0 reproduces the desired mean values, i.e.

$$r_\mu \ = \ < R_\mu > \ = \ - \frac{\partial}{\partial \lambda_\mu} \ \ln \ Z(\lambda_1....\lambda_k). \qquad \mu = 1...k$$

(2.12)

The missing information in equilibrium takes the value

$$I(D_0) \ = \ \ln \ Z(\lambda_1....\lambda_k) + \sum_{\mu=1}^{k} \lambda_\mu \ r_\mu.$$

(2.13)

Particular examples are, if one choses the *energy* , or the *energy* and the *particle number* to be given on the mean. Then D_0 takes the form of the *canonical* or *grand canonical* ensemble of thermodynamics, respectively. The formulation presented here, however, gives a new view point to it, in as much that one knows precisely which information is accounted for in the density operator and which is not. Further on, the information theory formulation allows to fix any possible set of observables by its mean values, even in case that they do not commute. For example, one can fix the mean position and the mean angular momentum of a system. In a case like this the interpretation is such that the prepared system, if measured by one of the observables R_μ (without measuring the other ones at this time), gives its mean value r_μ on the ensemble average.

2.4. Generalized micro ensembles

While the cononical ensembles consider the mean values of certain observables as the relevant information, micro ensembles request a sharp value of such observables R_μ. Such a treatment is not only advisable if one treats systems with a small number of particles but also in cases which sensitively depend on the overall balance of conservation laws. As an example we may mention the subthreshold particle production discussed in sect. 4. In the quantum case the micro description implies that all states of the statistical ensemble are eigenstates of the constraining operators R_μ with the prescribed eigenvalues r_μ. It is evident that in the case of more than one observable such ensembles are only meaningful if all operators cummute with one another. This of course restricts the application of micro ensembles to particular cases. If no further constraints are given a micro ensemble leads to an equal distribution among all possible states which are in line with the prescribed eigenvalues. Thus the statistical operator takes up the form

$$D(r_1,\ldots r_k) = \prod_{\mu=1}^{k} \delta(R_\mu - r_\mu).$$

(2.14)

This permits the definition of the *level density* D with respect to this set of eigenvalues

$$D(r_1,\ldots r_k) = \mathrm{Tr}\left\{ D(r_1,\ldots r_k) \right\} = \mathrm{Tr}\left\{ \prod_{\mu=1}^{k} \delta(R_\mu - r_\mu) \right\}$$

(2.15)

The formal link between canonical and the corresponding micro ensembles is given by an inverse Laplace transformation over the corresponding Lagrange parameters λ_μ. Since in many cases canonical ensemble are much easier to be formulated and calculated, the inverse Laplace transform evaluated in saddle point approximation opens the possibility to estimate level densities. The level density of a non-interacting gas of Fermions is a famous example for such a procedure.

In most of theoretically investigated cases the considered observables R_μ are such that they are additive with respect to a separation of the system into two parts, i.e.

$$R_\mu = R_\mu(1) + R_\mu(2).$$

(2.16)

Under these conditions it is straight forward to express the probability P of finding a given subsystem 1 with the eigenvalues $r_1(1)$ to $r_k(1)$ in a system where the total values are r_1 to r_k. This is just given by the number of states of the system given the fact that the parts 1 and 2 take the corresponding eigenvalues relative to the total number of states the system can take

$$P = \frac{D^1(r_1(1),\ldots r_k(1)) \, D^2(r_1(2),\ldots r_k(2))}{D(r_1,\ldots r_k)}, \qquad r_\mu = r_\mu(1) + r_\mu(2).$$

(2.17)

This relation is frequently used to calculate single particle spectra of micro ensembles. The precribed observables R_μ are normally all those which are related to conservation laws, as energy, momentum, charge, and baryon number for example. In turn the integrated probability over $r_1(1)$ to $r_k(1)$ adds up to unity. Therefore eq.(2.17) gives rize to a recursive relation for the level densities

$$D(r_1...r_k) = \sum_{r_1(1),...r_k(1)} D^1(r_1(1),... r_k(1)) \; D^2(r_1 - r_1(1),...r_k - r_k(1)) \tag{2.18}$$

This form is frequently used to establish level densities of non-interacting many body systems, like relativistic phase-space integrals [6,7].

2.5. Semi-classical determination of level densities.

In many cases it is advantage to calculate the level density in the semi-classical limit. This amounts to replace all operators by their classical analog in definition (2.15) and the trace by the integration over the classical phase-space $\int d^3r \, d^3p \, / \, (2\pi\hbar \,)^3$ of each particle, devided by the combinatorical factor (factorials) for all identical particles.

Let us discuss the specific case of a non-interacting system of A particles in a given volume V^* where the total energy and the total momentum are sharply given. Then the repective level density D(P,E) is related to the so called phase-space integrals Φ

$$\Phi = \int d^3P_1... \; d^3P_A \; \delta(P - \sum_{a=1}^{A} P_a) \; \delta(E - \sum_{a=1}^{A} \varepsilon_a) \; \text{by} \tag{2.19a}$$

$$D(P,E) = F \, \Phi(P,E), \text{ where} \tag{2.19b}$$

$$F = \{ \frac{V}{(2\pi\hbar \,)^3} \}^A / \{ \prod_v M_v! \}. \tag{2.19c}$$

Here P_a and ε_a are the momenta and energies of each particle. The system may be composed of different types of identical particles. For each type v it contains M_v of them in number. At this level the formula applies both for Bosons and for Fermions and is only valid in the so called non degenerate limit, i.e. at low occupation rates for each quantum state. The derivation of level densities in the degenerate case of high occupations has to start from the corresponding grand canonical partition sum employing the inverse Laplace transform. We will not do this here and rather confine the discussion to the simple phase-space integrals (2.19b). Simple scaling arguments show that they can be obtained in closed form in two limits, (a) in the ultra relativistic limit, where $\varepsilon_a = P_a$ and the non-relativistic limit, where $\varepsilon_a = m_a + P_a^2 /2m_a$. The results are

$$\Phi(P,E,N) = (\frac{\pi}{2})^{N-1} \frac{(4N-4)!(2N-2)}{\{(2N-1)!\}^2(3n-4)!N!} \; \{ E^2 - P^2 \}^{(3N-4)/2}, \tag{2.20a}$$

in the relativistic case of N massless particles, and

$$\Phi(P,E,A) = \frac{Q^{2/3}}{M^{2/3}} \frac{(2\pi)^{3(A-1)/2}}{\Gamma(3/2(A-1))} \; \{ E - M - p^2/2 \, M \}^{(3A-5)/2} \tag{2.20b}$$

in the non-relativistic case on A massive particles where $\quad M = \sum_{a=1}^{A} m_a \; \text{and} \; Q = \prod_{a=1}^{A} m_a.$

_____

* In the relativistic case this is taken to be the volume in the system's rest-frame, where P=0.

These are the type of equations which were used to estimate the available phase space in the case of pion and energetic gamma production at so-called subthreshold conditions(sect. 4).

2.6. Transport theories and missing information

Sofar we discussed time-independent statistical systems which are in equilibrium under the given constraints. By far more interesting is the case of time-dependent processes, which are to be described keeping track of only some of the dynamical variables (the relevant ones) instead of the exact and time-reversal equations of motion. This leads to a description in terms of transport equations. Robertson [4] has shown a way how to formulate such a transport theory from the the concept of missing information.

Any physical experiment can of course be described by solving the exact equations of motion, say von Neumann's equation for the exact density operator **D** starting from its initial value at $t = -\infty$. Then one has to reduce the complete information contained in $\mathbf{D}(t = \infty)$ to the one observed in the experiment (say the cross sections). Such a procedure would imply no loss of information through the dynamical equations of motion as they are reversible. The loss of information is introduced through the measurement which is normally limited to a few of observables. This procedure is not only unnecessarily complicated, it is in fact untractable for the type of experiments we are interested in. Thus, there is a need to limit the description to a certain set of relevant variables $r_\mu(t) = \mathrm{Tr}\{\mathbf{R}_\mu \mathbf{D}(t)\}$. The aim is to select such a set for which the corresponding minimum information density operator $\mathbf{D}_0(r_1(t0,...,r_k(t))$ as given by eq.(2.10) to (2.12) gives identically the same result for the observed data as the exact $\mathbf{D}(t)$. The corresponding equations of motion for $\mathbf{D}_0(t)$ are then clearly irrevesible and correspondingly imply a loss of information during the time evolution. Two things are worth to be méntioned (a) the set of relevant degrees of freedom depends on the type of data observed, e.g. for a correct description of a reaction cross section one certainly needs less relevant variables, than for a detailed differential cross section; (b) any set of observed data can be described by different sets of assumed relevant observables (e.g. to a given set one is always entitled to add a few irrelevant ones without spoiling the results!). The last statement implies that depending on the theory chosen one achieves a different amount of information loss or entropy gain. Thus, apriori there is no *unique entropy* that one can attribute to a certain *experiment* (the notion experiment comprising the the type of reaction studied, together with the type of data observed). Still, one may consider the existence of an optimally chosen minimum set of relevant variables leading to the maximum of missing information compatible with the observed data. This one may consider as the entropy attributed to this experiment. Any other theoretical model should reach an entropy less than this one [2].

The procedure to determine the relevant observables is normally very complicated and there is no apriori recipe to find them. Rather, one proceeds pragmatically, assuming certain relevant variables, like the one-body density [5] (e.g. Boltzmann, Boltmannn Ühling Uhlenbeck, Vlasov, TDHF, or hydrodynamic equations), and tries to justify the assumptions by comparison to measured data. In this respect the entropies discussed in connection with high energy heavy-ion reactions [8] refer to the one-body density as the relevant degrees while many-body correlations are considered to be irrelevant.

3. Condensation Phenomena in the Fast Expansion of Hot Nuclear Systems

3.1. Introduction

Two nulei with a relative velocity close to the speed of light collide. On a time scale of about 10 fm/c a highly excited and compressed nuclear system is formed at a considerably high entropy. How does such a system expand and finally disassemble when passing through a regime of dynamical instabilities? While earlier disregarded, the interest in this question has grown recently. Quite a variety of models have been developed to discuss this question. Most of them, however,

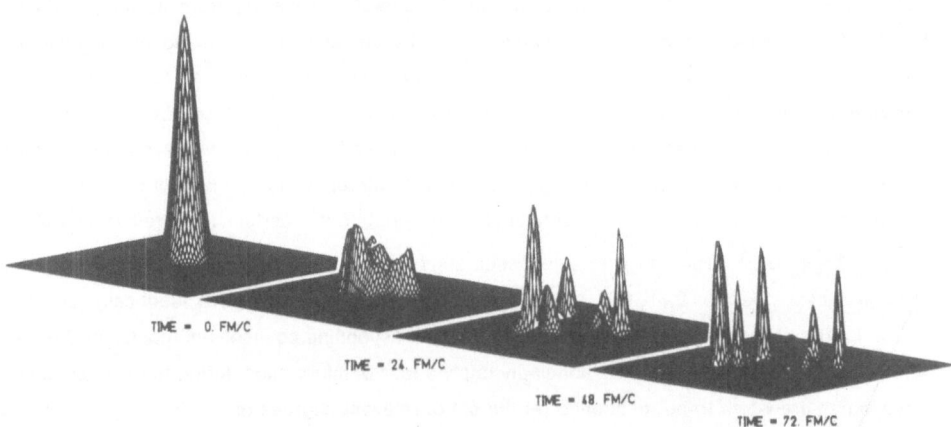

Fig. 1. Disassembly dynamics of a hot and compressed nuclear system from model calculations in a configuration space of two dimensions. The figur displays the time evolution of the density ρ(r,t) of a certain event at four different time steps. The perspective plot shows vertically the density which initially exceeds the saturation value by a factor 2.5, at a temperature of T = 50 MeV.

by-pass the dynamical evolution of the process and discuss the fragment formation in terms of a quasi-static concept: the freeze-out. These models range from the simple picture of coalescence in phase-space [9], percolation models [10], the thermo-chemical chemical equilibrium among nuclei [11] or nuclear droplets (condensation model [12]) up to the discussion of nuclear matter at strained densities discussing the phase coexistence of a nuclear fluid and a vapor phase [13]. Only few attemps have been under taken to study the implications of the dynamics of the fragmentation process. They range from a quantum one-body evolution picture [14], over a description by means of a classical Vlasov equation [15], a Vlasov equation supplemented with collision term [16], up to an entirely classical model in the frame work of a molecular dynamics approach [17]. I like to report upon the first kind of model which was actually the first one discussing these questions.

3.2. One-body Evolution at Finite Entropies

It was important to realize that mean-field approaches in the sense of treating the whole statistical ensemble by one mean field are principly not capable to treat the phenomenon of multifragmentation. Rather, the individual fluctuations of a certain state out of the statistical ensemble together with the instabilities implied by the saturating nature of the nuclear forces are the decisive factor for fragmentation.

As nuclear binding is of genuine wave mechanical nature (the Pauli-pressure determines the nuclear sizes, not the range of the forces) we consider a quasi-quantum mechanical model. It comprises three ingredients: a) the definition of the initial configuration at the onset of the expansion phase, b) the dynamics that cranks this configuration through the instability region, and c) a definition of the final stable fragments:

a: We assume that up to a certain time ($t = 0$) which may be the moment of highest compression the reaction can be described by a macrodynamical approach like the cascade or the hydrodynamical models. This may provide sufficient information as to determine the initial state of the expansion phase, e.g. in form of the one-body density matrix $f^0(r,p)$ at time $t = 0$. In line with the high entropy at this time we represent the system by a statistical ensemble of pure states (product wave functions), such that the ensemble average reproduces f^0.

b: Given a Skyrme type of interaction with proper saturation properties each of these pure states (here to after called event) is then propagated by self consistent one-body dynamics. The one-body field which governs the evolution of the event is deduced from the time-dependent one-body density of this product wave function. In this way the dynamics has the knowledge on its proper one-body density and is capable to form fragments which ultimately remain stable due to the nonlinearity of the dynamical equations.

c: A fragment is defined as the connected area in space where the density exceeds a certain threshold value (at present taken as 0.1 of the saturation value of ρ^0).

So far the study is confined to a $2+1$ dimensional model world (two space and one time dimension), and anti-symmetrization is neglegted. We calculated the evolution of a mass 40 system (40 single particle wave functions per event) on a physical theatre of 60 fm by 60 fm size. No additional symmetry is used. A two parameter Skyrme force with saturation at a Fermi-momentum of $k= 1.2$/fm (with the saturation density $\rho^0 = k^2/\pi$) and binding energy of 16 MeV has been used. We also discuss results with a spin-isospin dependent force.

For our initial study we bypass the compression phase and assume the initial phase-space distribution f^0 to be of a Gaussian form both in coordinate and momentum space with an initial compression of $\kappa = 2.5$. The momentum part is parametrised by a Maxwell distribution of 'temperature' T. The single particle wave functions are taken as Gaussian wave functions with random centroids in position and momenta (boost) in accordance with the distribution f^0. Each event of the statistical ensemble corresponds to one random set of positions and momenta for the 40 single particle wave functions.

A few comments are in order: We recall that anti-symmetrization is neglected so far; the study of genuine Fermion systems is in progress [18]. The problems are at the side of formulating appropriate stochastic initial conditions while the dynamical evolution is the same due to the Skyrme nature of the effective interaction. As compared to ordinary mean field theories at finite temperature which treat the whole statistical ensemble in a single common (i.e. ensemble averaged) mean field, here each individual state evolves in its proper one-body field. Thus we account for fluctuations in the one-body field, which in turn generate the various ways the system can granulate. Our picture implies that the evolution for $t \geq 0$ is isentropic.

3.3. Time Development of a Hot and Compressed Mass 40-System

Let us first consider the developement of one pure state in time as displayed by fig. 1. The sequence shows the density distribution $\rho(r,t)$ at four different time steps $t = 0, 24, 48, 72$ fm/c, for an initial value of $T = 50$ MeV. Due to the internal stress caused by the large momentum spread the events immediately expand and overstress around $t = 24$ fm/c, i.e. the density falls below ρ^0 throughout. One recognizes density depressions (bubbles) in the interior and some areas of higher density at this time initiating the disassembly of the system. At a later stage these density islands are capable to recover back to saturation density sucking in the matter of their local environment. It turns out that all islands which have at least a mass of two nucleons remain stable and therefore form ultimately stable soliton-like fragments. Indeed this performance cannot be regarded as a surface evaporation process. It's a boiling dynamics, with bubble formation in the interior while first fragments fly away from the surface.

3.4. Mass Spectra and Multiplicity Distributions

Let us finally come to the results not of one arbitrarily and randomly chosen pure state but of the full ensembles. Computing a series of events pertaining to a specific ensemble (T) one can compile multiplicity and mass distributions, fig. 2. For the moment we study the dependence on T as well as the dependence on both the initial state correlations and the nature of the binding forces. In detail the following scenarios are discussed:

a: the initial state is spin-isospin correlated, i.e. always the four different spin-isospin wave functions are identical to one another, and the binding force has no spin-isospin dependence.

b: the initial state is not spin-isospin correlated, and the force has no spin-isospin dependence.

c: the initial state is not spin-isospin correlated as in (b), however the force has a spin-isospin dependence, which focusses into the valley of stability ($N = Z$).

The case (c) relative to case (b) clarifies that the nature of the binding force has strong implications on the final mass spectra. An attraction toward equal spin and isospin is incorporated in the force of case (c). Clearly this force focusses nucleons which emerged as free nucleons in case (b) to bound fragments up to mass 4 in case (c). Note that the initial conditions in case (c) and (b) are taken precisely identical from event to event. In case (a) we see that the initial correlations are carried over into the distribution of final fragments, favouring mass 4-type fragments.

Besides the spin-isospin dependence discussed so far, a quite general behaviour of the resulting mass spectra as a function of the excitation energy (comprised in T) can be observed. The tendency of manufactoring smaller droplets with increasing T-parameter is obvious and seems so far neither to be strongly dependent on the number of nucleons of the system nor on the details of the force. At low excitation energies a highly excited compound nucleus remains, or at most,

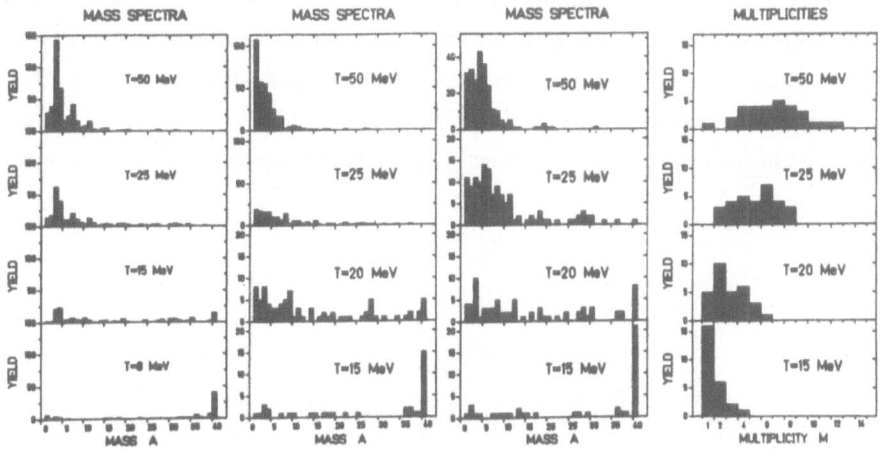

Fig. 2:   Mass spectra at four values of T and initial compression of 2.5 ρ^0 resulting from the ensemble average of thirty runs for each T. The spectra on the r.h.s. are extracted from 60 runs for each T. The figure contains from left to right the mass spectra for the cases (a) to (c) as explained in the text and the multiplicity distribution of case (b).

a binary fragmentation occurs. Interesting is the characteristic change of the mass spectra around $T = 20$ MeV, where the spectrum is relatively flat, indicating strong fluctuations in the resulting mass. In analogy to the theory of condensation, one may ask whether such a flat mass spectra represents a signal of a critical behaviour of the system.

3.5. The Role of Initial Fluctuations

The picture employed is such that given the initial configuration the final outcome of the fragmentation dynamics is determined deterministically by solving the the equations of motion. To this extent the description preserves the entropy, certainly a limitation of the model as it omitts possible collision terms. In turn, however, given this model, one can ask the question which pieces of information contained in the initial state actually determine the fragmentation. In our case the initial state is entirely give by the initial position and momentum (cetroids of the corresponding Wigner distribution) of all the single-particle wave functions at $t = 0$. For this purpose we performed the following analysis: During the time evolution each single particle wave function fractions and goes likewise into any of the fragments or into the back ground which does not belong to any of the fragments with a certain probability. However, it turns out that more than 50% of its probability goes into only one fragment or into the back ground density. We then looked into the distribution of all classical initial positions of the wave functions going essentially into one fragment, and likewise we did for the initial momenta. As a result of this analysis we stated a strong correlation among the distribution of the initial momenta, fig. 3, and the final fragmentation, while such a correlation could not be established for the initial positions. Evidently a coalescence or percolation picture employed in momentum space may be a much better phenomenological approach to describe the dynamics than employed in coordinate space or actually the product space of both. Yet, a definite answer cannot yet be given as we may not have performed the analysis at a time which one likes to call the moment of freeze-out. This still has to be established. In a way it is a work

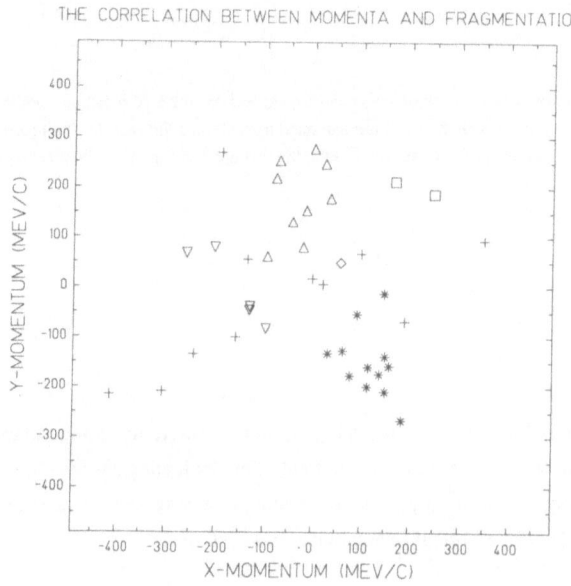

Fig. 3 Distribution of the initial momenta of the single particle wave functions of one event and their assignment to the different final fragments. The + denote those wave functions which essentially go into the back groud, i.e. into no stable fragment. Else the initial momenta of wave functions going into the same fragment are plotted by the same symbol.

similar to that of analysing an experiment, and a lot of experience is still needed to come behind the secrets of multifragmentation.

3.6. Conclusions

Within a field theoretical model we have shown that a series of numerical experiments with random initial conditions corresponding to hot nuclear systems leads to final stable fragments in a quite natural way. Besides, it is expected that this time-dependent break-up description could give a stimulation on future calculations for mass spectra and at least some hints concerning the questions of a quasi-critical behaviour of a finite system.

Certainly there are a lot of open questions relative to the presented approach. One concerns the neglect of the residual interactions in our approach. There are physical arguments in favour that they may not be that important once one considers a statistical ensemble, as we do. Second concerns the definition of the initial state. Here a coupling to an other dynamical theory for the initial entropy generating phase is certainly desirable. The third concerns the generalisation to a genuine $3+1$ dimensional model. Work is in progress for an accurate account of the Pauli principle [18]

In conclusion, we think of having presented a full scale micro-dynamical model which opens a challenging perspective of studying many of the pending questions on micro-dynamical instabilities. Furthermore we expect some insight from the comparison of this model study with quasi-static, i.e. freeze-out consideration to the same dynamical system. Some investigations in this direction are in progress [18].

4. Fermi Motion Versus Co-operative Effects in Subthreshold Pion and Energetic Gamma Production

In some fraction of the events of a collision of two nuclei particles like pions, kaons or energetic photons are produced such that from bare kinematical considerations they cannot stem from a collision of two nucleons at the corresponding relative velocity [19-27]. Such processes are quoted as subthreshold reactions. Although normally not discussed within this context also the emission of nucleons into extreme kinematical regions, as in proton backward scattering [28-31] or the promptly emitted particles PEPs or Fermi jets [32] do belong to this category of processes. They all have in common that the production mechanism has to couple to the only form of energy initially available, the relative kinetic energy of the two colliding nuclei. That is, even if the basic process happens on a microscopic scale, hence by nucleon-nucleon (NN) collisions, one needs the surrounding nuclear medium. These medium effects make such reactions very interesting.

As an example fig. 4 shows the production cross sections of pions at different bombarding energies for the system of ^{12}C on ^{12}C. With a reaction cross section of about 1 barn one realizes the rareness of such reaction channels. This figure gives an impressive account of the present experimental possibilities to follow the cross sections over more than 8 orders of magnitude down to close to the absolute threshold, which is around 20MeV/nucl. in this case. Due to the rareness of these channels one may expect them to contain information about the reaction dynamics other than those resulting from the bulk of the reaction channels. Fig. 5 shows an energy spectrum resulting from such reactions [33]. Despite the very light projectile/target combination this example is selected for two reasons. It shows the tremendous fall of the cross section towards the absolute threshold where ultimately the discrete states of the fused nuclear system are resolved (so called pionic fusion) [34]. These discrete final channels can only be understood by a coherent quantum description. The continuous part of the spectrum certainly involves a lot of unresolved final states,

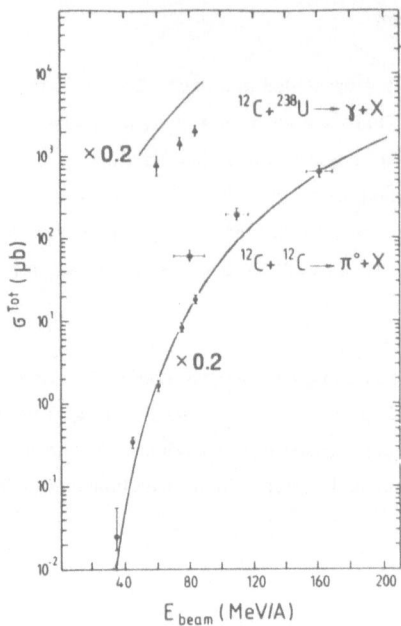
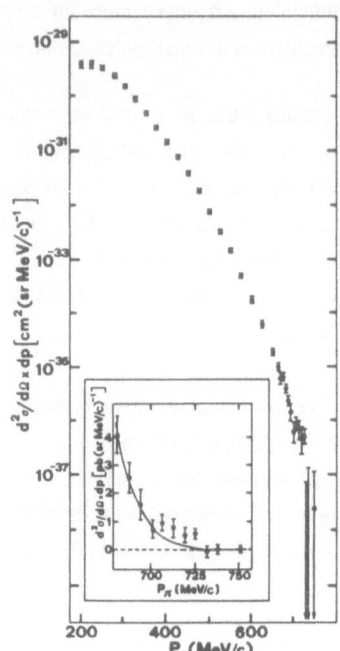

Fig.4. Production cross section of neutral pions and and gammas above 50MeV as a function of beam energy. Calculations from the co-operative model discussed in sect.4.3

Fig.5. Momentum spectrum of positive pions up to the kinematical limit; from ref [33].

so that their description is much more amenable by statistical concepts. This picture also clarifies that the form of the energy spectra is not necessarily exponentially falling. Rather the slopes become steeper the closer one approaches the absolute threshold.

Theoretical attempts to understand the production mechanism range from *simple* collisions among the nucleons, over pictures which employ the *co-operative* action of a couple of nucleons, up to fully *collective* phenomena. The *simple* multiple collision models treat the nucleons as *classical* particles employing a *classical* collision term of Boltzmann or Boltzmann-Uehling-Uhlenbeck type. That is, they conserve energy in the classical sense for each NN collision and respect the Pauli principle classically through the one-body occupation probabilities in phase-space. Since subsequent collisions are less energetic, it suffices to consider only the first collision at subthreshold energies. This defines the NN single collision model (NNSC) [35,36,25,37]. In such a picture the initial Fermi motion is the only possible source to provide the necessary extra boost for the production process. On the *quantum* level, however, the situation can be quite different. Between two subsequent NN collisions the energy can be quite off-shell. Therefore in a quantum scheme multiple collisions lead to a by far more flexible sharing of the energy than allowed by classical scattering. This prompted to consider a *co-operative* model [38], where, due to the mutual interactions, a group of nucleons may share all its available energy. In addition to the above collision dominated pictures the role of the mean field may become more and more important at lower bombarding energies. Entirely *collective* are the models where all the nucleons help simultaneously to produce the observed particle. Into this category falls the Bremsstrahlung picture [39], where the change in the relative motion of the two nuclei due to the reaction causes the radiation. This picture is *collective* on the nuclear level and it is even *coherent* in the sense of adding amplitudes for the radiative field. alternatively also the statistical radiation from the fused compound system [40] has been considered, however it failed to reproduce the mass dependence of the absolute cross sections for asymmetric systems.

It is the purpose of this contribution to debate the different theoretical concepts. After a discussion of the gross properties of subthreshold reactions, we will examine the picture of classical collision approaches. The next part summarizes the results of a co-operative model studied by R. Shyam and myself.

4.1. Gross properties

Some simple quantities of the resulting spectra can be discussed as a function of the beam energy and the combination of projectile and target masses, A and B, respectively. These are the total production cross section σ_λ where λ stands for any of the produced particles, the slopes of the spectra, called E_0, their nonisotropy relative to an optimally chosen frame of reference and the velocity of this frame (source velocity, v_{source}) for asymmetric mass combinations. A convenient reference for the total production cross section is the integrated participant cross section which in a simple straight-line estimate results to [41]

$$\sigma_{part} = A\sigma_B + B\sigma_A = <part> \sigma_{react}.$$ (4.1)

where $\sigma_A \cong \pi r_0^2$ is the reaction cross section on nucleus A, and similar for B. For the Carbon on Carbon case it is about 5 barn, and about 35 barn for $A=B=40$. In a statistical picture one expects a production yield at momentum $\mathbf{p}_\lambda$, c.f. eq(2.17)

$$W(\mathbf{p}_\lambda) = g \frac{V}{(2\pi\hbar)^3} \frac{D(E - E_\lambda)}{D(E)}$$ (4.2)

Here D(E) is the level density of the radiating 'source' at total energy E, and $E_\lambda, \mathbf{p}_\lambda$ are the energy and momentum of the radiated particle λ in the source rest frame. The volume V comes in from level counting of the observed particle, g is its degeneracy. Relating the source size V to the number of participants through a density ρ_0, ($<V> = <part>/\rho_0$), one obtains from impact parameter integration the cross section

$$\frac{d^3\sigma}{dp_\lambda^3} = \frac{g}{(2\pi\hbar)^3\rho_0} \frac{D(E - E_\lambda)}{D(E)} \sigma_{part}$$ (4.3)

provided the ratio of level densities

$$\frac{D(E - E_\lambda)}{D(E)} \cong \exp(-E_\lambda/E_0), \qquad \text{where } 1/E_0 \cong d/dE \; ln \, [\, D(E) \,], \qquad (4.4)$$

does not vary too much with impact parameter. We see from (4.3) and (4.4) a close relation between the slopes of the energy spectra and the expected absolute yields in such a picture. Furtheron, besides kinematical factors one expects about the same cross section for different particles to occur at about the same observed energy (including the rest mass), if all particles were emitted statistically. With $\rho_0 = 0.16$ fm^{-3} one arrives at the following absolute yields for pions of one charge state (employing non relativistic kinematics, $E_0 \ll m_\pi$) and for all photons with energies beyond E_γ

$$\sigma_\pi = 7.2 \; 10^{-3} \; (E_0/20\text{MeV})^{3/2} \; \exp[\; -m_\pi/E_0] \; \sigma_{part} \qquad (4.5a)$$

$$\sigma_\gamma = 1.3 \; 10^{-3} \; (E_0/20\text{MeV})^3 \; (1 + E_\gamma/E_0 + \frac{1}{2}(E_\gamma/E_0)^2) \; \exp[\; -E_\gamma/E_0] \; \sigma_{part.} \qquad (4.5b)$$

With a slope E_0 of about 20MeV extracted from the pion and gamma spectra at a beam energy of 80MeV/nucl. (c.f. E. Grosse these proceedings) one estimates $\sigma_\pi \cong 42$ µb for the ^{12}C on ^{12}C collision which is not far away from the data. Yet, the yield of photons with energies beyond $E_\gamma = 50$MeV estimated to 3mb exceeds the measured result by about one order of magnitude. This meets the conclusion of E. Grosse drawn from the energy spectra at same gamma and pion energy [27].

The predicted mass dependence with σ_{part} is quite resonable, c.f. fig.10 below.

For asymmetric collision systems $A \neq B$ a third quantity is of interest, the so called source velocity v_{source}. It refers to the velocity of that frame in which the observed spectrum looks the most isotropic or at least balanced with respect to the forward/backward hemisphere. It is to be compared to three velocities: the velocity of the NN c.m. frame v_{NN}, the mean velocity of the participant matter, and eventually with that of the fused compound system. In non-relativistic kinematics they are simply

$$v_{NN} = \frac{1}{2} v_{beam}, \qquad (4.6a)$$

$$v_{part} = A\sigma_B/(A\sigma_B + B\sigma_A) \; v_{beam.} \cong A^{1/3}/(A^{1/3} + B^{1/3}) \; v_{beam} \qquad (4.6b)$$

$$v_{comp} = A/(A+B) \; v_{beam}, \qquad (4.6c)$$

$$\text{note} \quad v_{comp} \le v_{part} \le v_{NN}. \qquad (4.6d)$$

The experimental results [26,27] give a source velocity between v_{part} and v_{NN}, which even is seen to change if different energy cuts for the observe particle are taken, c.f. the discussion of E. Grosse for the γ-production. The closeness of v_{source} to v_{NN} has induced the conclusion that first chance NN-collisions are eventually the dominant mechanism. This is not yet conclusive to my opinion. From event to event the number of participants fluctuates, inducing a preference for those events with higher production rates. In the statistical picture these are events with a favourable ratio of level densities, i.e. a large E_0. Not to close to the absolute threshold, E_0 can be identified with the temperature which is maximal for symmetric participant combinations. This shifts v_{source} towards v_{NN}! In the co-oporative picture discussed later this effect is nicely seen.

In conclusion of this qualitative discussion, we see at present no objection to a statistical interpretation. For the pions the level-density parameter E_0 extracted from the slopes of the energy spectra is seen to match the predicted yields with the data within half an order of magnitude. The dependence on mass of the absolute yields, the small change of the slopes with mass and also possibly the source velocities follow the predictions. Only for the photons the absolute yields are overestimated by about an order of magnitude relative to the pion yields. This effect is expected, since the photon couples only pertubatively to the nuclear currents.

In the co-operative model discussed below we will employ a specific model in order to estimate the level densities.

4.2. Quasi-classical Collision Dynamics

All dynamical schemes so far employed to discuss the subthreshold production rates treat the nuclear motion on a one-body level. That is, one describes the evolution of the nuclear one-body density by the influence of a one-body force (mean field) plus a collision term. In the TDHF approch of the Gießen group [42] the nuclear motion is treated entirely within the one-body TDHF-dynamics, while the pion or γ-production proceeds via 2p-2h transitions relative to the time-dependent TDHF state. The approach omits a collision term with the advantage that within the TDHF limit the overall energy conservation is well treated by a time Fourier-transformation. W. Cassing [43] approximates the TDHF evolution by a two-centre shell model and includes a quasi-classical collision term of Uehling-Uhlenbeck form, both for the nuclear relaxation dynamics and in pertubative treatment for the production process. Aichelin [44] stepped further towards the quasi-classical limit, in as much as also the one-body force acts classically (Vlasov part of the equation of motion). In this Monte-Carlo simulation they also include the reabsorption process (important for pions, kaons and lambda-particles).

The Boltzmann-Uehling-Uhlenbeck collision term is of the following form

$$\frac{d}{dt}\ f_\lambda(\mathbf{x},\mathbf{p}_\lambda,t)\ =\ \frac{4}{m^2(2\pi\hbar\ )^3}\ \int \frac{d^3\sigma}{dp_\lambda^3}\ \ f(\mathbf{x},\mathbf{p}_1,t)\ f(\mathbf{x},\mathbf{p}_2,t)\ (1 - f(\mathbf{x},\mathbf{p}_3,t))\ (1 - f(\mathbf{x},\mathbf{p}_4,t)).$$

$$\delta^3(\mathbf{p}_1+\mathbf{p}_2\text{-}\mathbf{p}_3\text{-}\mathbf{p}_4\text{-}\mathbf{p}_\lambda)\ \delta(E_1+E_2\text{-}E_3\text{-}E_4\text{-}E_\lambda)\ dp_1{}^3\ dp_2{}^3\ dp_3{}^3\ dp_4{}^3$$

$$(4.7)$$

It uses the free NN production cross-section and conserves besides the momentum also the energy in each collision in a quasi-classical sense. That is, the energies E_i are the quasi-classical Thomas-Fermi energies of the nucleons and the particle λ. i.e.

$$E_i\ =\ E_{TF}(\mathbf{x},\mathbf{p}_i)\ =\ \sqrt{m_i(\mathbf{x})^2+\mathbf{p}_i{}^2}\ +\ V_i(\mathbf{x}).$$

$$(4.8)$$

Here m_i is the local effective mass and V_i is the potential energy of the i-th particle (so far only rest masses and no potential energy for the produced particle, i.e. $V_\lambda = 0$, have been used in the literature). The distributions $f(\mathbf{x},\mathbf{p},t)$ and $f_\lambda(\mathbf{x},\mathbf{p},t)$ are the classical phase-space occupations of the nucleons and the created particle. The factors $(1 - f)$ care for the Pauli principle in the quasi-classical limit.

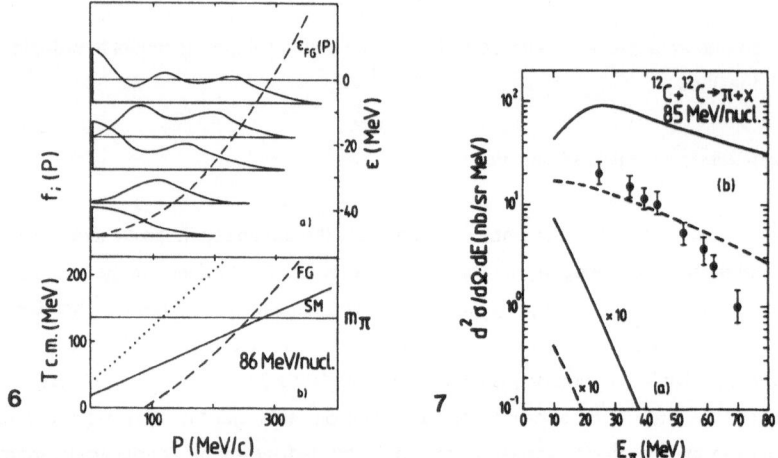

Fig.6. (a) Display of the momentum distributions $f_i(p)$ of five closed harmonic oscillator shell with their corresponding binding energies; the dashed line shows the Thomas-Fermi energy momentum correlation. (b) Total kinetic energy available for a pair of nucleons with intrinsic momenta p and -p, respectively. The beam energy is 85MeV/nucl.; the full line gives the shell model kinematics, the dashed line the Thomas-Fermi kinematics, and the dotted line gives the shell model kinematics, if two on two nucleons share their energy.

Fig.7. Pion production cross section at 90° laboratory angle, data from ref [25]. Calculations with the shell model energies (a), and the inconsistent prescription (b); solid and dashed lines correspond to the NN→dπ and the NN→NNπ channels, respectively.

It has become the habit to use the quantum Wigner functions for the initial nuclear distributions rather than Thomas-Fermi distributions. This is considered to be an improvement. I rather see a great risk of inconsistency! These Wigner functions are not consistent with the energy and Pauli prescription in the collision term. This may not always be a danger, in particular if one never really relies on the classically forbidden momentum components. However I do have my concerns for the issue discussed here. Already the ground state Wigner function with its high momentum components leads in eq. (4.7) to collisions which are neither prevented from energy conservation nor the Pauli factor in this equation. Thus, the ground state radiates pions, a *perpetual motion* of particular kind!

If one wants to comply with the conservation laws, and I think everybody will agree that this is important for the problem, there is an *either/or:*

either: one wants to use a classical collision term then the distribution f(**x**,**p**,t) has to comply with this and be also classical,

or: if one thinks the quantum high momentum components are important, one has to stick to the quantum energy conservation, throughout. The latter is, of course, cumbersome, as the energy conservation only applies to the asymptotic states. Therefore it requests a quantum treatment from the initial state fully through to the final one.

To corroborate this point R. Shyam and myself studied carefully the single collision picture. Due to the assumption that *only* a single collision occurs for the producing pair of nucleons, both, the initial state energy and the final state energies can be precised. We employed the shell model for the initial state. There in the rest frame of each nucleus each nucleon has a given sharp energy irrespective of its momentum, namely the energy of the shell it belongs to. That is, the correct energy-momentum distributions in the initial state take the form

$$f_{s.m.}(\mathbf{x,p},E) = \sum_{i\in F} f_i(\mathbf{x,p})\delta(E-\varepsilon_i) \qquad \text{with } \varepsilon_i = m_0 - B_i \tag{4.9}$$

which for five closed harmonic-oscillator shells is illustrated in fig. 6a. Here the ε_i are the shell model energies relative to the respective nuclear rest frame, and f_i is the Wigner distribution of each occupied shell $i \in F$. Note, that 4.9 gives the precise quantum energies in the shell model picture. For the moving projectile one only has to Lorentz-transform the four-vector (**p**,E) of each nucleon into the lab-frame. This leads to fully consistent formulation where the initial state energy always complies with the shell model precription. In the inconsistent picture used in the literature one still employes the one nucleon Wigner distributions given by the shell model

$$f(\mathbf{x,p}) = \sum_{i\in F} f_i(\mathbf{x,p}), \tag{4.10}$$

however, one replaces the shell model energies by the Thomas-Fermi energies E_{TF}. This leads to the following inconsistent energy- momentum distributions for the initial state

$$f_{incons}(\mathbf{x,p},E) = f(\mathbf{x,p})\delta(E-E_{TF}(\mathbf{x,p})). \tag{4.11}$$

It violates the energy account exactly there where one explores the classically forbidden parts of the quantum Wigner function. Both, the correct QM-treatment of the energy (4.9) and the prescription (4.11) lead to significantly different kinematical situations as illustrated in fig. 6b. Precisely where the momentum components become non-classical, i.e. beyond 250MeV/c, with prescription (4.11) the c.m. energy of a considered NN-collision exceeds the QM-prescription. This has far reaching consequences for the resulting cross sections, fig. 7. The correctly treated case leads to spectra which are orders of magnitude smaller than the inappropriately treated ones with by far too steep slopes! Thus, as a result, the single collision picture in its pure QM form is seen to be insufficient to explain the observed pion yields.

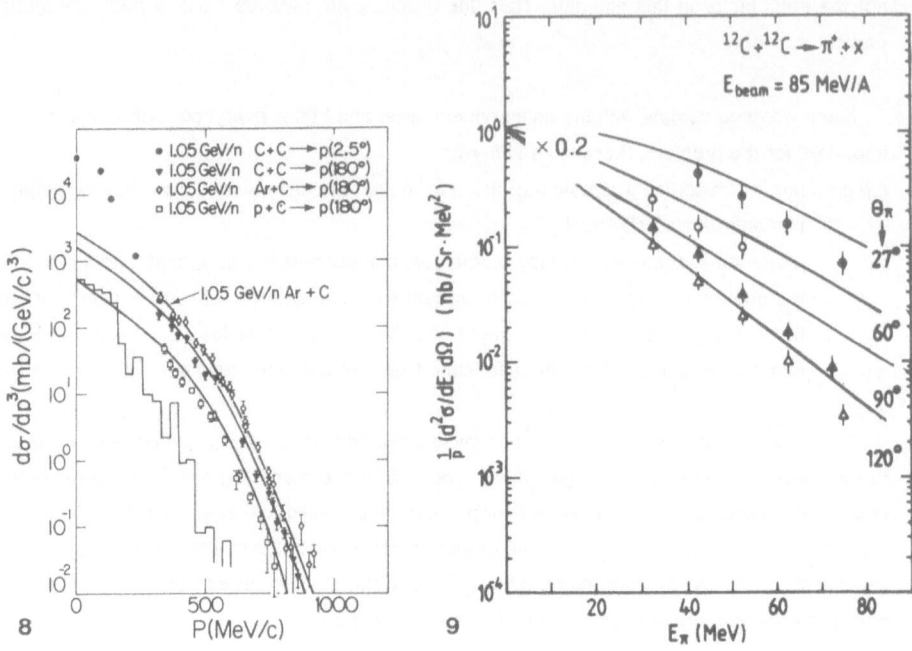

Fig.8. Inclusive proton spectra at forward/backward angles. Data from ref [29], the histogramm gives the cascade calculation for the proton induced reaction; the full lines give the phase-space model results [31], discussed below.

Fig.9. Inclusive charged pion data from ref [25] compared to the results from the co-operative model, full line. Note a normalization by a factor 0.2.

Of course there are many if's. The most essential is that in the discussed calculation the nucleonic states after the collision are taken as free plane waves with their corresponding energies. After the scattering, however, they are in the medium and their wave functions can be quite different. That is just the point to make. If they are in the medium, one needs further interactions: either by the one-body force (e.g. by distorted waves), or subsequent NN collisions. Then, however the classical energy conservation is no longer needed for the first collision, because one can and will go highly off the energy shell. Then an intermediate energy uncertainty results from the time it takes between first and subsequent collisions. Thus, for a precise treatment, one needs information about multi-nucleon space-time correlations. In ref. [43] an interpretation is given how the initially available non-classical momentum components are put on the classical energy (4.8) by the time-dependent mean field, however the author sees this mechanism limited to beam energies above 50 MeV/nucl. and for heavy collision systems.

In conclusion of this discussion, one can say that one needs a careful treatment of multi-nucleon processes in order to be sure about a correct treatment of the energy question. Hints that such multi-nucleon processes are important come already from other interesting data, namely proton-backward scattering [28,31], fig.8. Here it was absolutely impossible to reach the yields at high backward energies with a conventional cascade model, i.e. with a semi-classical collision term, unless one employs momentum distributions for the Fermi-motion which do not comply with electron scattering data. However, if one does the following: Rather than taking the resulting cascade spectra, one reestimates the momentum spectra for those nucleons which were in mutual interaction contact by the available phase-space, one arrives at spectra which fit perfectly the data over several orders of magnitude and for different collision systems. This study shows the importance of multi-nucleon processes which cannot be represented by successive classical collisions for a case which in a way can also be considered as subthreshold (there is no backscattering in free NN collisions).

4.3. The Co-operative Model

Ref [38] gives a detailed description of this model. Therefore we limit ourselves to present briefly its main features.

As just discussed in a quantum mechanical multiple collision picture off-shell collisions allow a by far more flexible sharing of the available energy than permitted by a sequence of on-shell scatterings (as in a cascade model). Such a genuine co-operative mechanism allows the pooling of the energies of all those nucleons which are in mutual interaction contact during the collision. Therefore one assumes that in any event of the collision between two nuclei, through the collision dynamics, all nucleons can be grouped into, what one likes to call *virtual clusters*. Each virtual cluster contains all those target and projectile nucleons which are interacting with each other. As a consequence, one body observables, like the inclusive cross section for observing a certain produced particle λ, can be expressed as an incoherent sum over the contributions arising from all the different virtual clusters

$$E_\lambda \frac{d^3\sigma}{dp_\lambda^3} = \sum_{MN} \sigma_{AB}(M,N) \, F_{MN}^\lambda(\mathbf{p}_\lambda). \tag{4.12}$$

Here the labels M and N denote the numbers of projectile and target nucleons respectively in each contributing virtual cluster. In eq.(4.12) the yield of each virtual cluster is factorized into a formation cross-section $\sigma_{AB}(M,N)$ and a properly normalized probability distribution function F_{MN}^λ. While the formation cross-section specifies the occurrence of a given virtual cluster, the spectra F_{MN}^λ determines the partial flux that goes into the observed channels out of that virtual cluster (M,N), i.e. to observe a particle λ with energy E_λ and momentum $\mathbf{p}_\lambda$. As such, eq. (4.12) is quite general and embodies in itself a variety of multiple collision approaches. Different prescriptions to calculate σ_{AB} and F_{MN}^λ lead to different models.

Let us summarize the main assumptions, merits and drawbacks of the co-operative model of Shyam and myself. The *formation cross sections* $\sigma_{AB}(M,N)$ are analyzed by a three-dimensional cascade study [45]and then approximated by an analytic formula of Glauber type. This may deserve further improvements since the cascade study was only done for light systems, however we see only a minor sensitivity on the one parameter which enters here. The *spectral distributions* are estimated in the spirit of Fermi's statistical picture. That is, the partial cross sections for the different final channels just follow the number of available states as already discussed in sect. 2. The phase-space calculations employed comply with energy and momentum conservation between the initial and the final configuration, and are therefore in line with the quantum mechanics. The initial Fermi motion is included by an appropriate folding over the initial momenta and obey the shell model energy prescription. For the final states it is important that the considered phase-space includes all the relevant channels. Below 400MeV/nucl. these are in particular those channels where the virtual cluster can break up into stable nuclei of finite mass, i.e. besides nucleons into deuterons, tritons and so on. The present investigation includes all possible partitions of the virtual clusters into stable nuclear fragments up to a mass of 12. This is an important improvement

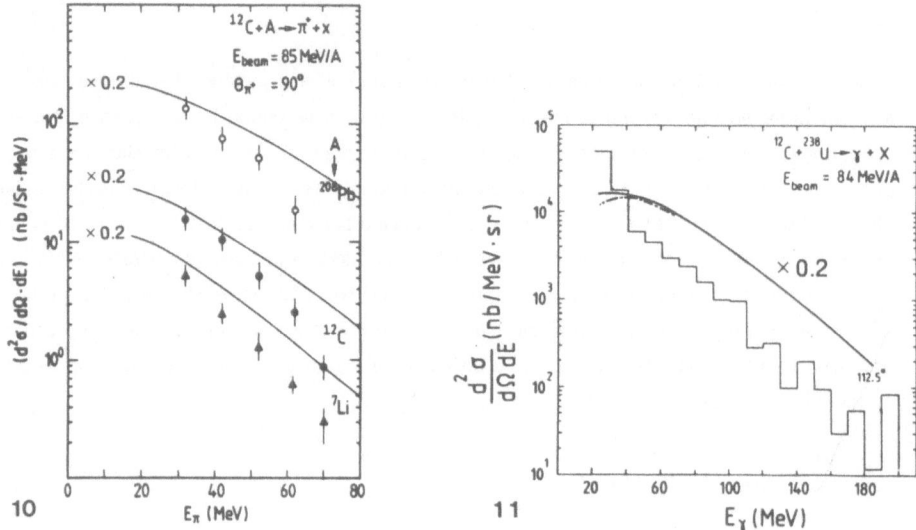

Fig.10. Energy spectra of charged pions at 90° laboratory angle for different collision systems.

Fig.11. Gamma spectra at 84MeV/nucl.; data from [22]; the full line gives the result of the co-operative model including the Bose-Einstein statistical correction; the dashed dotted line is without this correction. Note again the factor 0.2 by which the calculations are normalized.

relative to the earlier model of Bohrmann and Knoll [46]. The now possible composite nuclei formation implies a heating of the system which leads to a change of the fragment composition with bombarding energy. It is clear that only those states have to be counted which are in reach during the interaction. This introduces to specify an interaction volume which by means of a density parameter ρ_0 is related to the number $M+N$ of nucleons participating in that virtual cluster. The results show a sensitivity to this parameter since it influences the occurrence of heavier fragments in the final state. In the calculations $\rho_0 = 0.17\,\mathrm{fm}^{-3}$ has been chosen which is not too unreasonable in value.

At this point we have to insert an *erratum* concerning the calculations so far published [38]. Due to a typing error in one of the program tables the spectra came out to be given by a few cluster configurations rather than by all of them. We also realized that we did not correct the classical phase-space counting in order to comply with the indistinguishability of identical particles. This is not so important for the composite nuclei, since most of the time a particular nucleus occurs at most once in the final state. It is, however, relevant for the nucleons, for which less states are available quantum mechanically. This all has been revised. With the same set of parameters as used in refs. [38] the new cross sections result to about a factor five to six above the old calculations for all mass combinations and also beam energies. Therefore the new results are consistently by about this factor above the data. In view of the simplicity of the model we do not think that this is a serious problem. There are many facts that could modify the quantitative result since the number of states is calculated by an interaction free gas of nuclear compounds.

We rather see some importance in the fact that one has a well defined simple model with physically transparent assumptions which comply with all relevant conservation laws: the nucleon number, the momentum and the energy conservation. Besides the just discussed factor five it follows the dependence of the cross section on beam energy, on mass both for the absolute rates and the differential energy and angle spectra. The following figures illustrate the results of the model. It is important to realize that the model underestimates the cross sections by two orders of magnitude if the final channels include only free nucleons and no composite nuclei, fig.9, dashed line*. Yet, the inclusion of the formation of composite nuclei in the final states causes two changes. It brings the cross sections up, now even above the data, and helps to improve on the slopes of the energy spectra [47]. Be aware that the calculations are multiplied by a factor 0.2 in order to facilitate the qualitative comparison. Figs. 11 to 12 show some further results. For the photon spectra we note a further overestimation of the spectra by this model. As already discussed, this may be expected. The photon couples perturbatively to the nuclear current, so that, although the nuclear interactions may cause all available nuclear states to occur with equal probability, channels which contain a photon are suppressed all by the same factor due to the electromagnetic coupling constant! Except for the γ spectra below 50MeV also the slopes of the energy spectra are reproduced about correctly. A proper account of the Bose-Einstein statistics could enhance the cross sections at lower photon energies. However, this effect turned out to be not as important as to improve the photon energy spectra significantly.

4.4. Conclusion

Various theoretical models to investigate the production of pions and energetic photons under subthreshold conditions have been discussed. We realized that most of the approaches lack a careful treatment of the energy. Many models suffer from the fact either that the energy is only conserved on the mean and not strictly, or that a classical Thomas-Fermi energy is ascribed to diffractive, i.e. classically forbidden components of the QM wave functions, or that one assumes

---

* Note that this result depends neither on the density parameter ρ_0 nor on the appropriate quantum counting.

that the radiated quantum represents only a perturbation relative to the nuclear motion. Under the extreme kinematical conditions as discussed here any of the above deficiencies may easily lead to an inconsistency, since the quantum could result from a form of energy that is spurious due to the approximations involved. Therefore it is time to ask for a careful investigation of the validity conditions of the different approaches proposed.

The co-operative model discussed in more length has the advantage that it treats all relevant conservation laws carefully. The initial energy results from a shell model prescription so that all nucleons remain bound initially irrespective of their momenta. No energy conservation is enforced during the collision dynamics, allowing all nucleons in interaction contact to share their energy in the most optimal way. Only the final states which contain also composite nucei as fragments are again in line with the energy conservation. Besides an overall overestimation by about a factor 5 the model gives a good reproduction of the pion data, while the photon data are overestimated even further by another factor 3 to 5. The analysis with the model asserts the necessity of co-operative phenomena, in particular the formation of composite fragments in the final channel. Still, the results as preliminary as they might be viewed are encouraging further investigations. I also think that from a thorough theoretical point of view there is still a lot to be done before careful conclusions can be drawn about the physics which rules these reactions.

References

1. A. Katz, Priciples of statistical mechanics - the information theory approach (Freeman, San Francisco, 1967)
2. R. Balian and M. Veneroni, Saclay Preprint S.Ph.T/86-110, submitted to Ann.Phys.(NY)
3. E.T. Jaynes,Phys. Rev. 106(1957)620
4. B. Robertson, Phys. Rev. 144(1966)151
5. P. Buck and H. Feldmeier, Phys. Lett. 129B(1983)172
6. R. Hagedorn, Relativistic kinematics (Benjamin, New York, 1963)
7. J. Knoll, Phys. Rev. C20(1979)773
8. P.J. Siemens and J.I. Kapusta, Phys. Rev. Lett. 43(1979)1487
9. H.Gutbrod et al., Phys. Rev. Lett. 37(1976)667
   H.Sato, K.Yazaki, Phys. Lett. 98B(1981)153
10. W. Bauer et al. Phys.Lett.150B(1985)53
    X. Campi, J. Desbois, Proc. 23rd Intn. Winter Meeting on Nucl. Phys., Bormio 1985 p.497
    T.S. Biro, J. Knoll, J. Richert, Nucl.Phys.A459(1986)692
11. A.Z.Mekjian, Phys.Rev.Lett. 38(1977)640, Pys.Rev. C17(1978)1051
    J.Randrup,S.Koonin Nucl. Phys. A356(1981)223
    D.H.E. Gross et al.,Phys.Rev.Lett.56(1986)1544, and earlier refs in there
    H.W. Barz et al., Nucl.Phys.A448(1986)753 and earlier refs in there
12. M.E.Fisher, Physics 3(1967),255
    P.J.Siemens, Natur 305(1983)410
    A.L.Goodman,J.I.Kapusta,and A.Z.Mekjian, Phys.Rev.C in print
13. M.W.Curtin,H.Toki,and D.K.Scott, Phys. Lett.123B(1983)289
    G.Bertsch,P.J.Siemens, Phys.Lett.126B(1983)9
    H.Schulz et al., Phys.Lett. 113B(1983)141
    H.H.Jaqaman et al., Phys. Rev. C27(1983)2782
    H.Stöcker et al., Nucl.Phys.A400(1983)63c
    J.Cugnon, Phys.Lett.135B(1984)374
    J.Knoll et al.Phys.Lett. 112B(1982)13
14. J. Knoll, B. Strack,Phys.Lett.149B(1984)45,
    J. Knoll, in 7-th High Energy Heavy Ion Study, Darmstadt Oct 1984, GSI 85-10
    J. Knoll, in Proc. of the Topical Meeting on Phase Space Approach to Nuclear Dynamics, ed. M. Di Toro, W. Nörenberg, M. Rosina, S. Stringari; Oct. 1985, World Scientific
15. S. Das Gupta et al., to be published in Phys.Rev.C

16. D. Boal, Proc. of the Intern. Conference on Heavy Ion Nucl. Coll. in the Fermi Energy Domain, Journ. de Physique Coll C4(1986)409
    J. Aichelin, H. Stöcker, to be publ. in Phys.Lett.B
17. A. Vincentini et al.,Phys.Rev.C31(1985)1783
18. J. Knoll, J. Wu, to be publ.
19. W. Benenson et al., Phys. Rev. Lett. *43* (1979) 683
20. T. Johansson et al., Phys. Rev. Lett. *48* (1982) 732; V. Bernard et al., Nucl. Phys. *A423* (1984) 511
21. S. Nagamiya et al., Phys. Rev. Lett. *48* (1982) 1780
22. H. Noll et al., Phys. Rev. Lett. *52* (1984) 1284
23. H. Heckwolf et al., Z. Physik *A315* (1984) 243
24. P. Braun-Munzinger et al., Phys. Rev. Lett. *52* (1984) 255
25. B. Jakobsson, Phys. Scri. *T5* (1983) 207
26. H. Nifenecker et al.,XXIV Intern. Winter Meeting on Nucl. Phys., Bormio/Italy, Jan.1986; R. Bertholet et al., Proc. of the Intern. Conference on Heavy Ion Nucl. Coll. in the Fermi Energy Domain, Journ. de Physique Coll C4(1986)201
27. E. Grosse, et al., preprint GSI 86-9, and E. Grosse, Proc. of the Intern. Conference on Heavy Ion Nucl. Coll. in the Fermi Energy Domain, Journ. de Physique Coll C4(1986)197
28. S. Fraenkel et al., Phys. Rev. Lett. *36*(1976)642
29. L.S. Schroeder et al. Phys. Rev. Lett. *43* (1979) 1787
30. V.I. Komarov et al. Nucl. Phys. *A326*(1979)297
31. J. Knoll, Phys. Rev. *C20*(1979)773; A. Blin, S. Bohrmann, J. Knoll, Z. Phys. *A306*(1982)177
32. J.P. Bondorf et al., Nucl. Phys. *A333*(1980)285
33. E. Aslanides et al., Phys. Rev. Lett. *43*(1979)1466
34. K. Klingenbeck, M. Dillig and M.G. Huber, Phys. Rev. Lett. *47* (1981) 1654
35. W.G. McMillan and E. Teller, Phys. Rev. *72* (1947) 1
36. G. Bertsch, Phys. Rev. *C15*(1977)713
37. R. Shyam and J. Knoll, Phys. Lett. *136B* (1984) 221
38. R. Shyam and J. Knoll, Nucl. Phys. *A426* (1984) 606; and Nucl. Phys. *A448* (1985) 322
39. D. Vasak et al., Phys. Scr. *22* (1980) 25; Phys. Lett. *93B* (1980) 243; Nucl. Phys. *A428* (1984) 291c
40. J. Aichelin and G.F. Bertsch, Phys. Lett. *138B* (1984) 350
41. J. Hüfner and J. Knoll, Nucl. Phys. *A290* (1977) 460
42. M. Tohyama, R. Kaps, D. Masak and U. Mosel, Phys. Lett. *136B* (1984) 226; and Nucl. Phys. *A437*(1985)443
43. W. Cassing, Habilitationsschrift TH Darmstadt, GSI Report 86-6
44. J. Aichelin, Phys. Lett. *164B*(1986)
45. J. Cugnon, J. Knoll and J. Randrup, Nucl. Phys. *A360* (1981) 444
46. S. Bohrmann, J. Knoll, Nucl. Phys. *A356*(1981)498
47. see also the thesis of D. Durand, Univ. of Caen

FRAGMENTATION IN MEDIUM ENERGY HEAVY ION COLLISIONS [+]

A. Bonasera   Cyclotron Laboratory
               Michigan State University
               East Lansing, MI 48824 - USA

M. Di Toro   Dipartimento di Fisica
               Università di Catania
               57, Corso Italia. 95129 Catania - Italy

Ch. Grégoire   GANIL, BP 5027 - 14021 Caen - France

ABSTRACT

A modified participant-spectator model is discussed, suitable to study medium energy heavy ion collisions, and applied to the analysis of peripheral fragmentation. Phase space constraints and one-body dissipation terms are shown to imply noticeable modifications in the high energy fireball model. Fragmentation and damped collisions are found with their relative importance. Results and predictions are given for many observables and compared to existing experimental data.

INTRODUCTION

Fragmentation is the dominant reaction mechanism at high energy. For central collisions large particle multiplicity have been observed suggesting multifragmentation processes. For more peripheral collisions the evidence is for a three body exit channel, formed from projectile-like (PLF), target-like (TLF) and fireball fragments, quite well described within a participant-spectator (or abrasion-ablation) model /1-2-3-4/, consistent with a dynamics ruled by two-body collisions. In the medium enrgy region we also have participant-spectator-like processes /5-6/ but with clear signatures of a transition from deep inelastic to projectile fragmentation mechanisms:

+) Presented by M. Di Toro

i)   PLF-TLF mass-mass correlations are compatible with an abrasion pic-
       ture, where geometrical overlaps between collision partners are
       removed from the fragments, but only with some renormalized reduced
       radius r_o, characterizing the interaction size of each nucleus.

ii)   PLF angular distributions are not found peaked at zero degrees. Some
       evidences for orbiting during the collision are found, at variance
       with the high energy case.

iii)   Energy distributions of PLF's can be decomposed into three parts:
       high energy component corresponding to transfer into continuum, a
       low energy tail related to widely damped reactions and an intermedi-
       ate energy fragmentation peak. For fragmentation part the mean velo-
       city is found smaller by 10-15% than the beam velocity, clear indica-
       tion that dissipation is mixed with abrasion processes.

iv)   The momentum widths associated to the fragmentation peak, parallel
       $\sigma_{//}$ and perpendicular $\sigma_{\perp}$, are also directly measured. $\sigma_{//}$ varies
       with the PLF mass according to the Goldhaber scaling law /2/ verified
       at higher energy. On the contrary $\sigma_{\perp}$ is much larger than in the high
       energy case and the scaling law fails completely to reproduce data.

v)   For the fragment isobaric distribution a clear feature is the neutron
       enrichment of PLF's for interactions with neutron rich targets. The
       widths of distributions can be simply related to a zero point motion
       of a Giant Dipole Resonance for the whole system, before fragmentation,
       in agreement with a picture of a sudden separation of the fragments.

MEDIUM ENERGY FIREBALL MODEL: DISSIPATIVE FRAGMENTATION

     A revised participant-spectator reaction mechanism can be introduced
where all the medium energy features are understood just using quite simple
phase-space arguments /7-8/. To consider the interplay between dissipative
processes and the high energy collisional dynamics a two-stage mechanism is
analysed. In the first stage we have a one-body dissipation process, with
neck formation and incoherent nucleon exchange, quite well described within
the window formula picture. Some radial kinetic energy is converted into
intrinsic energy and some orbital rotational energy into intrinsic spins.
During the first step the dinuclear system suffers some deflection with respect to the beam axis. The medium energy case differs from the usual deep
inelastic picture since nucleons from one nucleus into continuum states of
the partner nucleus are not contributing to exchanges through the neck, and
the window formula must be suitably renormalized.

     The second stage is the abrasion mechanism, modified not only by the
previous radial and orbital dissipation, but also because the participant
nucleons have only part of the projectile (target) momentum distribution
available since the two Fermi spheres are partially overlapping at these
energies (Fig.1). In particular, also assuming a zero mean free path behavi-
hour, the number of abraded nucleons forming the fireball is modified by
$A_{Fire} = A_{Geom} (1-R(P_{rel}))$, where A_{Geom} is the geometrical abrasion value and
$R(P_{rel})$ is the ratio between the overlapping volume of the momentum distributions and the volume of a Fermi sphere (dashed area in Fig.1). This is an
essential change of the fireball which strongly affects excitation energies
and decays. The use of a pure fireball model at medium energies can be quite
uncorrect. Of course at higher energy ($\gtrsim$ 150 Mev/u), when $P_{rel} \gtrsim 2P_F$, the

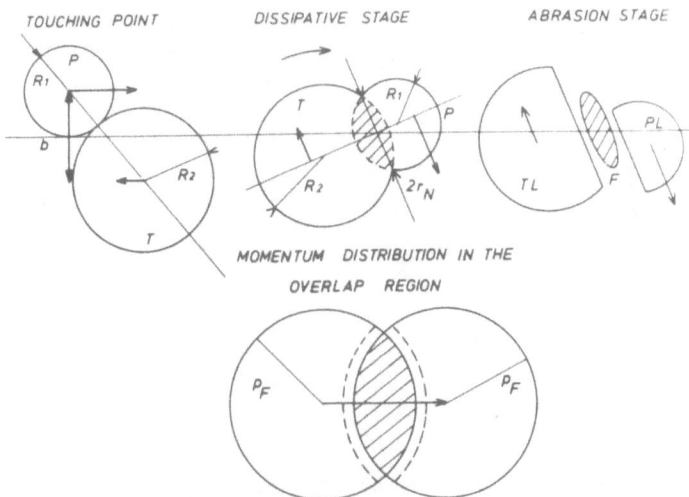

TOUCHING POINT   DISSIPATIVE STAGE   ABRASION STAGE

MOMENTUM DISTRIBUTION IN THE
OVERLAP REGION

Fig. 1. Coordinate and momentum picture of a dissipative-participant-
spectator reaction mechanism.

participant-spectator picture is fully recovered.

Kinematical properties of the three fragments are obtained in the final
state by conservation laws at each step of the reaction. At a given beam
energy we have a one to one correspondence between the impact parameter and
the PLF or TLF mass, kinetic and excitation energy, deflection angle. Finally fragmentation is expected at medium energy only if energetic and momentum
conditions are properply fulfilled: the model automatically contains indications concernùng the transition between deep inelastic or fusion reactions
and fragmentation. Fig. 2. shows the onset of fragmentation for a given PLF
mass in the system ^{40}Ar + ^{27}Al. For the same system, where excellent data
are existing also with some coincidence measurements /9-10/ at beam energy
44 Mev/u, we show several comparisons of the model with data. In Fig.s 3 and
4 we have the production cross section for PL-fragments and the mass-mass
correlations measured in PLF-TLF coincidences.

Fig. 5 shows the correlation between PLF masses and emission angles
(in the lab. system) predicted by the model. It is interesting the quite
flat behaviour of ϑ_{PLF} for masses 23-37, that could be used to choose
the best detection angle to select fragmentation residues.

Fig. 6 is related to the behaviour of the dissipated window energies as
a function of the impact parameter (and then the PLF mass number) for the
reaction ^{40}Ar + ^{58}Ni at 44 Mev/u, with the corresponding angular momentum
L_F transferred to the PLF. L_F turns out to be maximum of few ℏ units. However for quite asymmetric systems (e.g. like Ar + Au) we can easily reach
values of 40-50 ℏ (before evaporation) that can open a new series of interesting coincidence experiments. For the same reaction Fig. 7 contains a

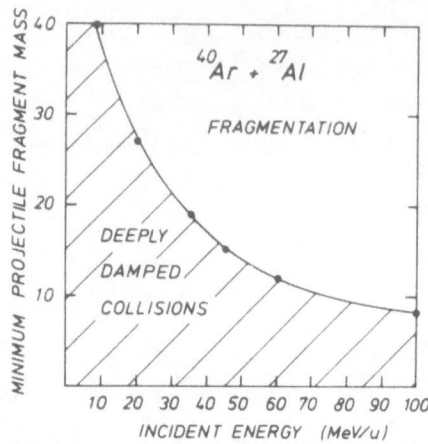

Fig. 2. Damped collisions vs. fragmentation as a function of
initial incident energy.

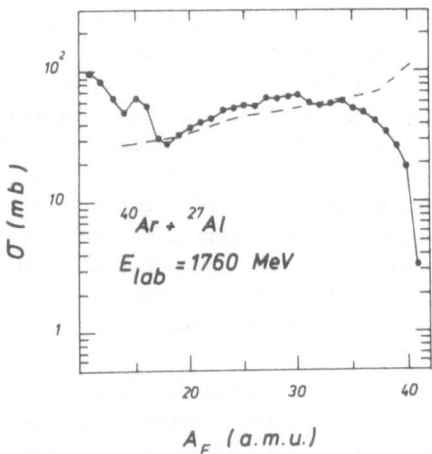

Fig. 3. PLF production cross-section. Data (black dots) are
from ref. 9.

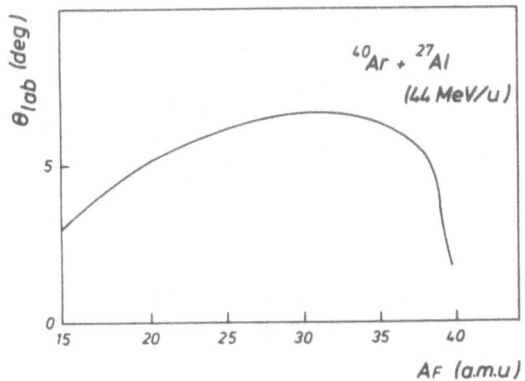

Fig. 4. PLF-TLF mass-mass correlations. Data from ref. 10.

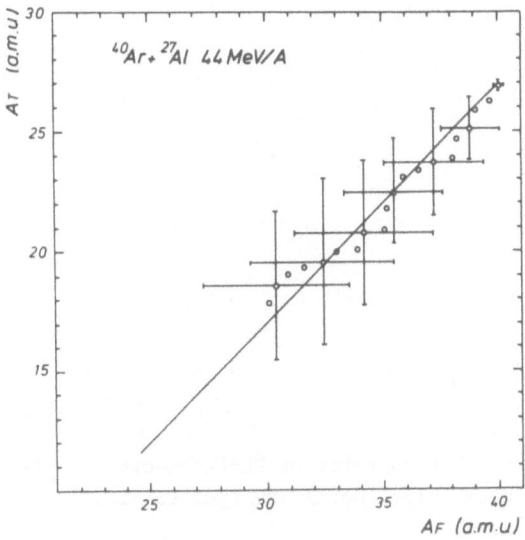

Fig. 5. Orbiting angle vs. PLF masses.

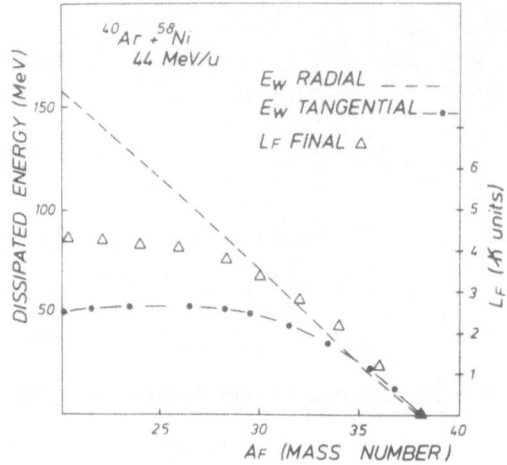

Fig. 6. Dissipation energies and PLF transferred angular
momentum as a function of the PLF mass.

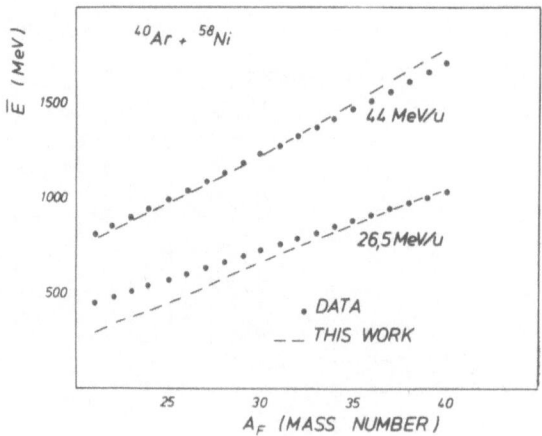

Fig. 7. Kinetic energies of PL-fragments vs. PLF masses.
Data (black dots) are from ref 18.

comparison of the PLF mean kinetic energies with the corresponding experimental positions of the fragmentation peak, at two different beam energies.

A nice feature of the new reaction mechanism can be seen in the analysis of the fragment momentum widths. As already mentioned, now the Pauli blocking prevents the access of participating nucleons to a part of the momentum space. Since the Fermi sphere overlap is depending on the relative momentum at the moment of fragmentation, the final structure of $\sigma_{/\!/}$, $\sigma_{\perp}$ will also depend on the dissipation phase and therefore on the impact parameter. Moreover the orbiting will introduce an explicit tilting in momentum space with a further correction, related to the ϑ_{PLF} angle.

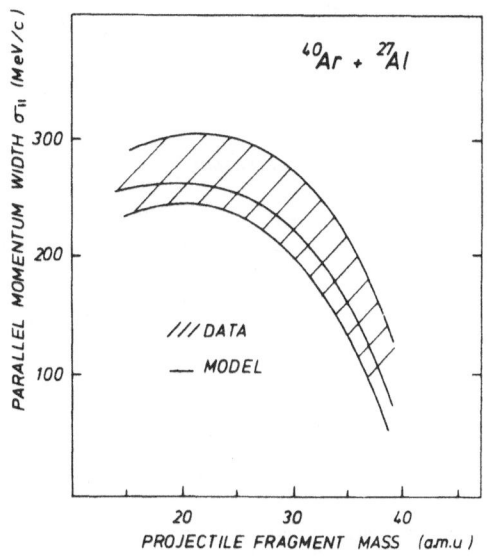

Fig. 8. PLF mass dependence of parallel momentum widths.
Dashed area corresponds to data of ref. 10.

In Fig.s 8 and 9 we show some results for parallel and orthogonal widths. For $\sigma_{\perp}$, also including entrance channel deflection effects /11/ we are still below the data. It seems really essential to consider collective distortions in momentum space that will further enhance the $\sigma_{\perp}$ / $\sigma_{/\!/}$ ratio /7-12-13/. The presence of collective excitations in the approaching phase is quite clear: independent considerations of the fragment energy spectra /13-5/ are leading to the same conclusion. This is also supported from the evidence of a strong enhancement for the excitation probability of giant resonances in grazing collisions at medium energies /14-15-16-17/.

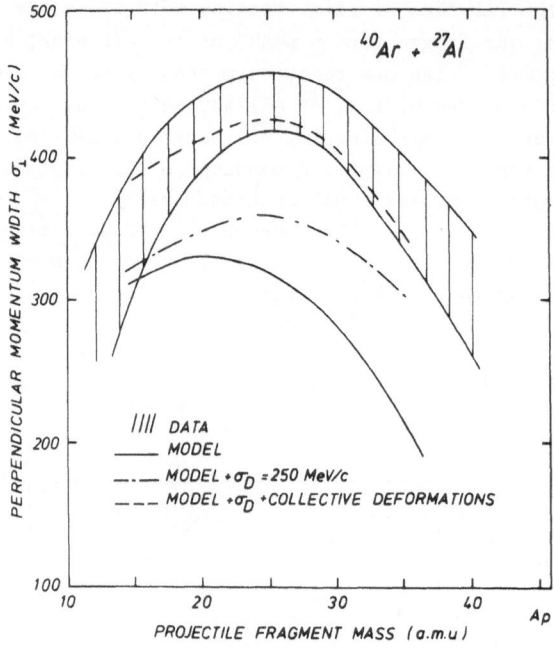

Fig. 9. PLF mass dependence of orthogonal widths.
Dashed area corresponds to data of ref. 10.

CONCLUSION

Using simple phase space arguments and standard one-body dissipation
terms we are able to get a quantitative evaluation of low energy mean field
effects on the high energy participant-spactator model to describe peripheral fragmentation. The mean field is responsible for three fundamental sources of corrections which must be included in any medium energy fireball
model:

i) Pauli blocking effects which reduce the phase space available to the
participant nucleons and affects the fireball dynamics, acting both
on the number of abraded nucleons and on their mean momentum.

ii) One-body dissipation effects. The exchanged nucleon flow leads to a
radial and tangential friction that can be analysed on the basis of
a modified window formula.

iii) Collective excitations, mainly of isoscalar giant quadrupole type,
in the initial stage of the reaction, as collective polarizations
in the two approaching ions. The main dynamical effects are related
to the correspondent collective distortions in the nucleon momentum
distributions.

We can evaluate a quite impressive amount of observables with nice experimental comparisons and interesting predictions.

Due to the interplay between one-body and collision terms a microscopic dynamical descriptin of medium enrgy heavy ion collisions is very difficult and in any case not much transparent. Our approach shows that it is possible to develop reasonable models, combining mean field effects to a short mean free path behaviour, which correctly include the main physics of medium energy heavy ion reactions.

REFERENCES

1.   H.Feshbach and K.Huang, Phys.Lett. 47B (1973) 300
2.   A.S.Goldhaber, Phys.Lett. 53B (1974) 306
3.   G.D.Westfall et al. Phys.Rev.Lett. 37 (1976) 1202
4.   J.Gosset et al. Phys. Rev. C16 (1977) 629
5.   D.Guerreau, Nucl. Phys. A447 (1985) 37c
6.   C.Grégoire and B.Tamain, Ann. Phys. Fr. 11 (1986)
7.   M.Di Toro and C.Grégoire in "Phase Space Approach to Nuclear Dynamics" Ed.s M.Di Toro, W.Nörenberg, M.Rosina and S.Stringari, World Sci.Publ. Singapore 1986, p.256
8.   A.Bonasera, M.Di Toro and C.Grégoire, "Dissipative effects in projectile fragmentation" Preprint GANIL 86-19, Nucl.Phys A, in press.
9.   R.Coniglione et al. Nucl. Phys. A447 (1986) 95c
10. R.Dayras et al., preprint SACLAY DPh N2329 (1986), Nucl.Phys.A, in press
11. K.Van Bibbier et al. Phys.Rev.Lett. 43 (1979) 840
12. M.Di Toro in "Fundamental Nuclear Physics", Ed.s K.Dietrich, M.Di Toro and H.J.Mang, World Sci.Publ., Singapore 1985, p.451.
13. R.Coniglione et al., XXIII Int. Winter Meeting on Nuclear Physics, Bormio 1985, p.417.
14. N.Frascaria, XXIV Int. Meeting on Nuclear Physics, Bormio 1986
15. V.Bellini et al., Bormio 1985 as in ref.13, p.58.
16. D.Vautherin and Ph.Chomaz, Phys. Lett. 139B (1984) 244
     Ph.Chomaz, N.Van Giai and D.Vautherin "Collective excitations of closed shell nuclei in heavy ion graziong collisions" Orsay preprint IPNO/TH 86/32, April 1986.
17. F.Catara and U.Lombardo, Nucl.Phys. A455 (1986) 158
18. V.Borrel et al., Z.Phys. A324 (1986) 205.

SOME PLANNED EXPERIMENTS WITH RELATIVISTIC HEAVY IONS

Peter Braun-Munzinger

Department of Physics
State University of New York
Stony Brook, N.Y. 11794

ABSTRACT

We discuss briefly selected aspects of some experiments planned with relativistic heavy ions as projectiles. Special emphasis is put on very peripheral collisions and the possibility to produce fragile exotica by multiple Coulomb excitations. Also discussed are the prospects for thermalization and transverse energy production in central nucleus-nucleus collisions.

INTRODUCTION

The imminent availability of ions accelerated to relativistic energies ($E_{lab}/A \approx 15$ GeV at Brookhaven, $E_{lab}/A \approx 200$ GeV at Cern) has produced a high level of activity both in the preparation of experiments and in theoretical investigations. A recent survey can be found in Ref. 1. The central goals of this exciting new field are to study new forms of matter which conceivably can be produced in collisions between two heavy nuclei at relativistic energies. These prospects are reviewed with emphasis on quark-gluon plasma formation in the lectures of G. Baym (these proceedings). Here we focus on two physics questions which are specific to relativistic heavy ions. One of these, transverse energy production, is related to the question of nuclear stopping and thermalization of the initial center of mass energy. Its understanding is very important for the design of future experiments and for estimating the energy densities produced in such reactions. The other topic is the study of very peripheral collisions with emphasis on the strong Coulomb excitation expected if $\gamma = E_{lab}/M \gg 1$. Such

Coulomb excitation processes can lead to cold production of exotic neutron rich objects. Both areas will be addressed soon in an experiment presently under construction at Brookhaven and we will present some of the experimental aspects insofar as they relate to the physics topics discussed above.

PERIPHERAL COLLISIONS

We focus first on collisions where the impact parameter exceeds the sum of the nuclear radii R_T and R_P of target and projectile. In such a collision, the projectile sees a very strong electromagnetic pulse as it passes through the strongly Lorentz contracted electric field of the target (we consider the situation from the rest frame of the projectile). In the limit $\gamma \gg 1$ the electromagnetic field looks (spatially) like a plane wave of photons impinging on the projectile. Since the projectile moves very nearly with the speed of light, the duration of the impulse is $\Delta T \approx R_T/\gamma \cdot c$. The wave numbers of the photon wave can, therefore, be very large, leading to very strong excitation of the projectile.

More quantitatively, this can be nicely described by the Weizsacker-Williams method[2]. This approach allows one to calculate the number of virtual photons of energy E_γ as seen by the projectile in the collision with impact parameter b:

$$N(E_\gamma,b) = 1/E_\gamma \cdot Z_T^2 \, \alpha /(\, \pi^2\beta^2 b^2) \cdot \xi^2 \, k_1^{\,2}(\xi) \qquad (1)$$

Here, Z_T is the charge number of the target, $\alpha = e^2/hc \approx 1/137$ βc is the speed of the projectile in the lab system and K_1 is the modified spherical Bessel function. The adiabaticity parameter ξ determines the cut-off at high energies of the virtual photon spectrum. Numerically, $\xi = E_\gamma(MeV) \cdot b(fm)/197 \cdot \gamma$. As a typical example at $\gamma=14.5$ (relevant for the Brookhaven AGS) and $b \approx 10$ fm as appropriate for $^{16}O+^{238}U$ collisions, $\xi=0.15$ for $E_\gamma =50$ MeV and $\xi^2 K_1^{\,2}(\xi)$ is very close to one. Since the giant resonance strength for nuclear excitations is concentrated below $E_\gamma =50$ MeV such collisions should lead to very strong Coulomb excitation of the projectile. Because of the nearly plane wave character of the virtual photon spectrum one might expect equal excitations of all multipolarities in the projectile. However, the electric dipole resonance is so dominant that one is essentially restricted to dipole excitations.

We can make simple estimates of the excitation probability assuming that the impact parameter range $[b_{min}, b_{max}]$ contributes to the reaction. Since $N(E_\gamma) \approx 2/\pi \, Z_T^2 \, \alpha / E_\gamma \cdot \ln b_{max}/b_{min}$, the precise choice of the impact parameter range is very not important and we assume $b_{min} \approx 10$ fm $\approx R_p + R_T$ for $^{16}O + ^{238}U$ and $b_{max} = 20$ fm. In the Weizsäcker-Williams approach, one can then quickly calculate the mean number of photons absorbed in the projectile per collision as

$$m = \int N(E_\gamma) \sigma(E_\gamma) dE_\gamma / \pi (b_{max}^2 - b_{min}^2) \qquad (2)$$

where $\sigma(E_\gamma)$ is the cross section for (real) photon absorption at energy E_γ on the projectile nucleus. Using a simple parameterization for $\sigma(E_\gamma)$ one obtains approximate values of $m_G \approx 0.045$ for absorption into the giant dipole resonance of ^{16}O and $m_T \approx 0.085$ for absorption into the energy range $10 < E_\gamma \leq 140$ MeV. In approximately 5% of the collisions into the range $[b_{min}, b_{max}]$ a virtual photon is absorbed into the giant dipole resonance of ^{16}O, indicating that cross sections in the hundreds of mb range are expected in such collisions. If the probability to absorb a single photon is so large one also expects considerable multi-photon excitation. In keeping with our simple approach we estimate the multi-photon absorption probability P_{n_γ} via a simple Poisson distribution

$$P_{n_\gamma} \approx m^{n_\gamma} e^{-m} / n_\gamma ! \qquad (3)$$

This probability is displayed, as a function of the number of photons n_γ absorbed in the projectile, in Fig. 1 for various reactions and choices of m. From this figure we estimate $P_4(^{16}O + ^{238}U) \approx 2 \cdot 10^{-7}$ and $\sigma_4 = P_4 \cdot \pi (b_{max}^2 - b_{min}^2)$ is of the order of μb, i.e., easily accessible experimentally. Also note that since $P_{n_\gamma} \sim 1/b^{2 \cdot n_\gamma}$, the cross section will be very strongly concentrated near b_{min}.

The above simple estimate was first discussed in a proposal[3] to the AGS program committee for a heavy ion experiment at Brookhaven. The numbers have since been put on very solid ground by the quantum mechanical calculations of Baur and collaborators[4] where very similar results are obtained in a full coupled channels approach. In the rest of this chapter we want to very briefly discuss what possible signatures one might find for these multi-giant resonance excitations and how one can go about detecting them experimentally.

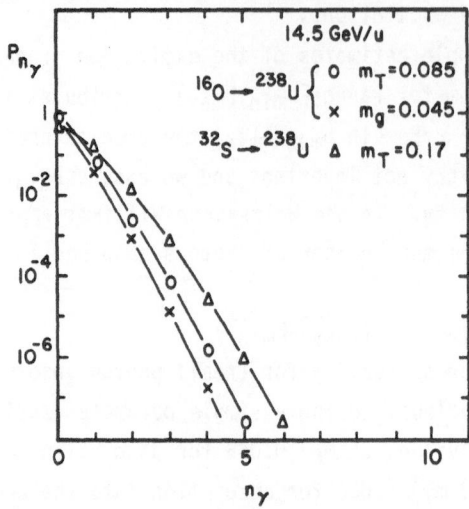

Fig. 1. Probability P_{n_γ} to absorb n_γ photons per collision into the projectile for reactions of 14.5 GeV/u ^{16}O, ^{32}S on ^{238}U. The average number of photons absorbed per collisions is $m_{T,G}$ (see text).

Multiple excitation of the giant dipole resonance in a nucleus might lead to states with very unusual structure. In light nuclei the giant dipole resonance consists mainly of an oscillation of all the neutrons vs. all the protons (Goldhaber-Teller mode). In a two-fluid model the maximum separation between protons and neutrons can be estimated as $d_{max} \approx 0.8$ fm for the nucleus ^{16}O oscillating after absorption of a giant resonance photon. After absorbing n protons, this distance increases (in the harmonic approximation) as $d_{max}^{(n)} = \sqrt{n} \cdot d_{max}$ (see Fig.2). For n=4 which seems possible (see above) this leads to separations of the order of one half of the nuclear radius. Coupled with the fact that the total excitation energy for n=4 is close to the binding energy of ^{16}O we expect decay of such states to lead to very neutron-and proton rich fragments. Of course, this scenario neglects the damping or thermalization of these initially completely collective states. This will be an important effect and needs further investigation. Here we only note that the state produced after n-fold giant dipole excitation will dominantly be of isospin $T^{(n)} = n \cdot T^{(1)} = n$. The decay and spreading widths of such high isospin states might not be too much different from that of the lowest giant dipole resonance. Finally the time scale $\Delta t \approx 1$ fm/c of these reactions is very short compared to typical nuclear vibration or damping periods.

How would one detect the fragments produced in such a peripheral reactions? The problem is mainly two-fold: (i) selecting pure projectile fragmentation by requiring that little or no energy has been transferred to the target during the reaction and (ii) measuring in a high resolution spectrometer the strongly forward focussed projectile fragments with enough resolution to reconstruct the projectile excitation energy to about 10-20% accuracy.

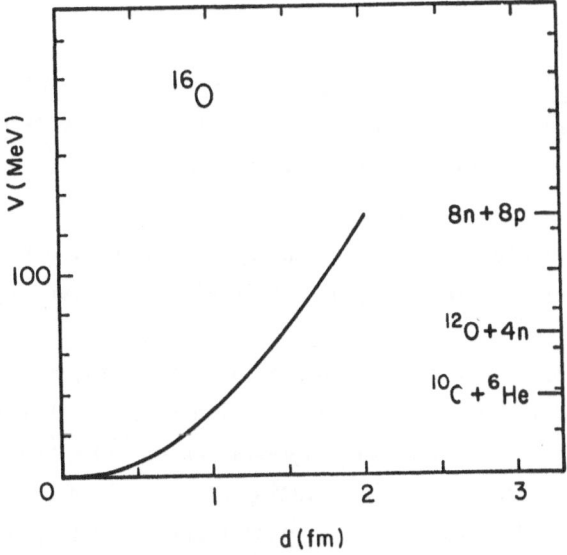

Fig. 2. Potential energy in the Goldhaber-Teller giant dipole mode[5] versus proton neutron separation d in the center of mass. Also shown are Q-values of selected projectile break-up channels.

The implementation of such a scheme is presently under construction as experiment E814 at Brookhaven National Laboratory. Fig. 3 shows the layout of the experiment. The beam enters from the left and hits the target inside the target calorimeter. Nearly 4π calorimetry (target calorimeter, participant calorimeter) allows the detection of any energy flow produced in the reaction outside a forward cone with opening angle $\Delta\theta \approx 1^\circ$. Due to the strong kinematic focussing all decay products following pure projectile excitation will end up in this forward cone and, after passing through a hole in the participant calorimeter, will be analyzed in the very long (40 m) forward spectrometer. The spectrometer consists of two large dipole magnets along with appropriate tracking chambers (DC1,DC2,DC3) for trajectory reconstruction and of scintillator hodoscopes and more hadronic

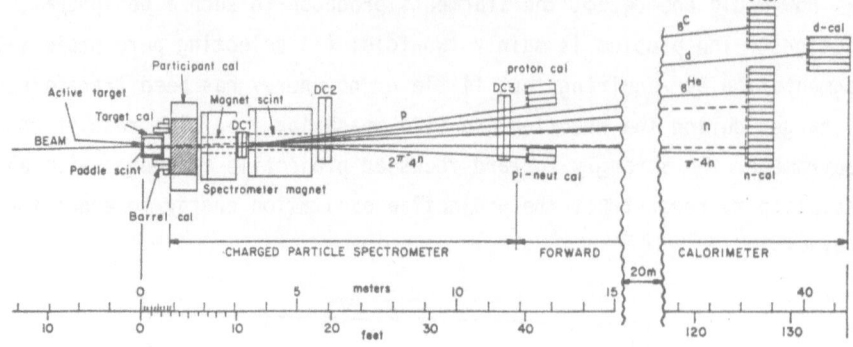

Fig. 3. Schematic drawing of the experimental set-up under construction at Brookhaven National laboratory (Experiment E814). For further explanations see text.

sampling calorimeters to measure the charge and total energy of the high energy fragments. Details of the attainable resolutions are discussed in Ref. 3.

Here we note that this setup should allow a completely exclusive measurement of the projectile fragmentation region with full particle identification. For a typical fragmentation of $^{16}O \to {}^{10}C + {}^{6}He$ we present, in Fig. 4, the expected Q-value resolution as determined by a Monte Carlo simulation including experimental resolutions and multiple scattering in the target and the tracking chambers. The resulting Q-value resolution in the decaying ^{16}O-projectile is of the order of 10%, may be not quite of nuclear physics standards but sufficient to narrow the range of possible projectile

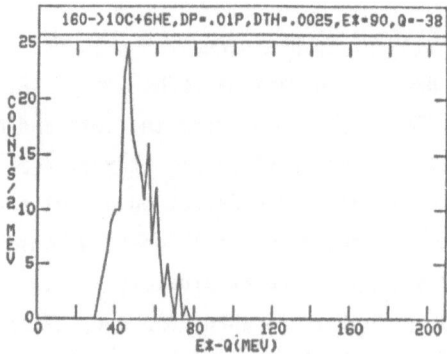

Fig. 4. Q-value resolution from invariant mass reconstruction of the projectile. The calculation uses a Monte Carlo simulation of $^{16}O \to {}^{10}C + {}^{6}He$ decays detected in the set-up described in Fig. 3.

excitation energies considerably. Note also that the spectrometer will spatially separate neutron-rich and proton-rich fragments and even a single event of unusual charge to mass ratio (see Fig. 3) will be very striking.

CENTRAL COLLISIONS

The apparatus described in Fig. 3 is, alternatively, very useful in studying flow of matter and energy in central collisions. A key question to be addressed in the first round of experiments with relativistic heavy ions is the amount of kinetic energy a projectile loses as it traverses the center of a nucleus. A second important question is how much of this lost energy is actually deposited in the target and usable to heat and compress nuclear matter

In the following we will outline some of the basic facts of these processes and discuss simple interpretations of recent p-nucleus scattering. Then possible consequences for nucleus-nucleus scattering are mentioned. Finally we will briefly describe how some of these topics are addressed in experiment 814.

We begin by recalling some of the kinematical relations necessary to understand the data. Particles are classified by 4-vectors $P_\mu = (E, \vec{p})$. Along a certain axis z (usually the beam direction) the momentum $\vec{p}$ is then split in $\vec{p} = (p_\parallel, p_\perp)$ in obvious notation. Rather than by velocities particles are classified by their rapidities $y = 1/2 \ln (E+p_\parallel)/(E-p_\parallel)$. The y-variable has the simple property that under Lorentz-boost in the z-direction $y \rightarrow y - \Delta$ with $\Delta = 1/2 \ln (1+\beta_{boost})/(1-\beta_{boost})$ so that rapidities can simply be added under Lorentz transformations and rapidity shifts leave cross sections shapes invariant. Also, since $E^2 = P_\parallel^2 + P_\perp^2 + m^2 = P_\parallel^2 + m_\perp^2$ the energy and parallel momentum component of a particle can be expressed in terms of its rapidity and perpendicular mass as $E = m_\perp \cdot \cosh y$ and $P_\parallel = m_\perp \cdot \sinh y$. Numerically, for a beam of 14.5 (200) GeV/nucleon the corresponding rapidity is $Y_B = 3.5(6.0)$.

Imagine now a proton at 200 GeV incident energy impinging on a Pb-nucleus. Due to collisions with the nucleons in Pb, it will lose rapidity while traversing the nucleus. If the initial rapidity (before collision) is y_0 and the final rapidity is y then the energy lost in the collision can approximately be related to the rapidity shift via $E_f = E_0 \cdot \exp(-(y_0-y))$. A rapidity shift of two units, e.g., reduces the initial energy by a factor of $e^{-2} = 0.14$.

The question which rapidity shift a proton undergoes when it hits a Pb-nucleus head on has recently been a subject of many discussion (see Ref. 1). Experimental information is still rather sketchy. However, there is a growing consensus that the mean rapidity shift in such a reaction is rather large. To illustrate this we show, in Fig. 5, recent bubble chamber data from the E565/510 experiment at Fermilab[6] In this experiment protons at 200 GeV are used to bombard Au targets. Central collisions are selected by requiring a rather large associated multiplicity. The leading (i.e., fastest) particle is not identified except by charge and is assumed to be a proton.

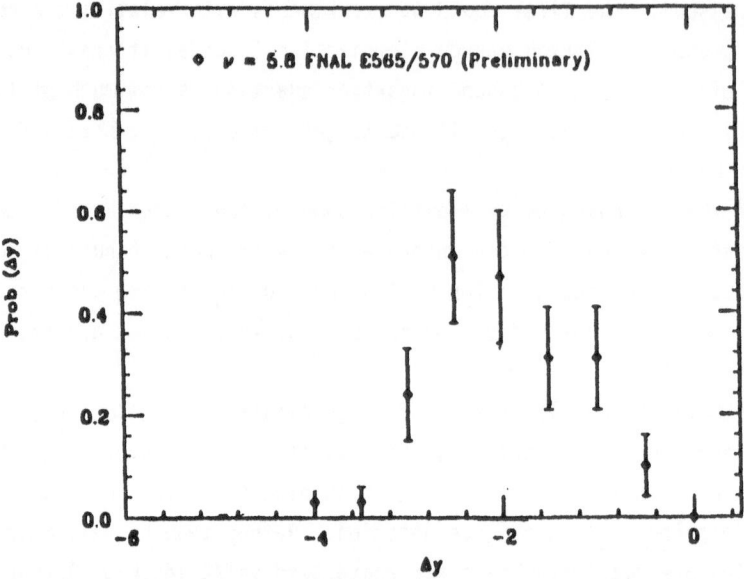

Fig. 5. Probability distribution of rapidity shifts of leading particles in 200 GeV p+Au collisons (Ref. 6).

From such data the probability for a rapidity shift $\Delta y = y_0 - y$ has been deduced and is plotted versus Δy in Fig. 5. The uncertainties from the lacking particle identification notwithstanding these data indicate that large rapidity shifts up to about three units are not uncommon and that the mean rapidity shift is $\Delta y = 2.1$.

As discussed above this implies very large energy losses in such a reaction. The question crucial for the field of relativistic heavy ion physics is how much of this lost energy is thermalized, i.e., used to heat up and compress nuclear matter in such reactions. One way to address this problem is to study experimentally the probability for transverse energy production in nucleon-nucleus and nucleus-nucleus reactions. The total transverse energy produced in a reaction is simply defined as a sum over the energies of all particles in the exit channel, multiplied by the sine of their respective laboratory production angle, i.e. $E_T = \sum_i E_i \cdot \sin \theta_i$. Note that if all particles are assumed to be produced isotropically in the laboratory the mean transverse energy is simply given by $\langle E_T \rangle = \pi/4 \cdot E$. This (somewhat unrealistic) estimate provides an upper limit on transverse energy production. Due to the weighting with $\sin\theta$ it is intuitively clear that large transverse energy production goes along with a very large energy deposit in the reaction although the details of this relation are by no means clear. In the following we will briefly discuss some of the recent experimental results on transverse energy production in p-nucleus collisions and mention some of the implications for nucleus-nucleus collisions.

Perhaps the most surprising and instructive aspect of recent data on transverse energy production in proton-nucleus collisions is the apparent strong dependence of the measured transverse energy spectra on the angular interval covered experimentally. In discussions of such data the scattering angle of a particle is usually related to its pseudorapidity $\eta = -\ln \tan \theta/2$. The quantity η can also be obtained if one replaces in the expression for rapidity the total energy E by P , indicating the close relationship between η and y. The dependence of transverse energy distributions on the pseudorapidity interval covered experimentally is shown in Fig. 6 where we compare data from p-Pb collisions from Fermilab experiment E557 at 800 GeV with those from the Helios collaboration at 200 GeV. Despite the lower incident energy the Helios data exhibit a much shallower slope and extend to considerably larger E_T values than the Fermilab data. To compare these data we note that the Helios experiment covers the pseudorapidity range $0.8 \leq \eta_{lab} \leq 2.4$ while in E557, the coverage is $2.84 \leq \eta_{lab} \leq 4.6$. Since the (pseudo) rapidity of the nucleon-nucleon system is 3.02 (Helios) and 3.7 (E557) we see that Helios covers mostly `backward' angles, i.e., the so-called target fragmentation region while E557's detectors span the region around the nucleon-nucleon center-of-mass, i.e., the so-called central region. The kinematical limit for E_T production from 200 GeV fixed target pp scattering is obtained when both protons scatter into 90^0 (in the nucleon-nucleon center-of-mass) and amounts to $E^{max}_T(pp) = \sqrt{2M_N E_{lab}} = 19.4$ (38.8) GeV at $E_{lab} = 200$ GeV (800 GeV). From Fig. 6 we see that the Helios data exceed well beyond the kinematical limit from pp scattering.

A possible interpretation of this difference has recently been worked out[9]. The main idea behind this analysis is connected to the question of how much transverse energy can be produced by the rescattering in the target matter of secondary particles produced in the sequence of p-nucleon collisions which takes place if a proton traverses the center of a Pb nucleus. If this rescattering is substantial then much of energy will appear at comparatively large angles, i.e., in the target fragmentation region, thereby strongly enhancing the probability for production of large E_T values as has apparently been observed in the Helios experiment.

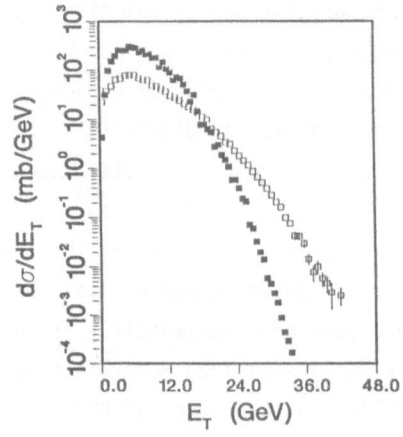

Fig. 6. Transverse energy distributions from p-nucleus
■ ■ 800 GeV p+Pb   Ref. 7   □ □ 200 GeV p+Pb   Ref. 8.

The key to understanding which secondaries can rescatter has to do with the hadronization time of these produced particles. Imagine a Lorentz frame in which the produced particle has only transverse motion. From the uncertainty principle such a particle will appear on the mass shell after a formation time $\Delta t \approx 1/m_\perp$. Hadrons of laboratory energy E will therefore be formed at distances $L_c = E/m \cdot \Delta t = E/m_\perp^2 = \cosh y_c/m_\perp$. If L_c exceeds nuclear dimensions there will be no rescattering, implying a critical rapidity $y_c \approx \cosh^{-1} m_\perp R_T$. For a Pb nucleus y_c is limited to values less than about 3, taking $m_\perp \gtrsim 0.4$ GeV. This limitation implies that only a fraction of the energy created within a nucleus is available for rescattering and showering. In fact this available energy depends only little on beam energy as long as the beam rapidity substantially exceeds y_c. Despite this limitation the total energy produced inside a nucleus in each primary collision is still quite substantial (of the order of 10 GeV for 200 GeV p+Pb). If only a fraction of this energy reappears via rescattering in the target fragmentation region that might be enough to explain the differences observed in Fig. 6. In Ref. 9 some schematic calculations towards this end are discussed.

Similar effects should also appear in nucleus nucleus collisions at high energy. In fact, first results from very recent experiments at Cern at 200 GeV/nucleon [16]O+Pb indicate again[10,11] very large transverse energy production (exceeding 200 GeV) and substantial enhancements if the target fragmentation region is covered. For more detailed discussions see Refs. 9-11.

The simple remarks discussed above and the simple multiple scattering model of Ref. 9 shed some light on the mechanism of stopping and transverse energy production. However, critical pieces of experimental information are yet missing. We need to understand, e.g., the correlation between the rapidity distribution of leading particles and transverse energy production. The dependence of energy deposition on impact parameter needs to be studied. The leading baryons (including neutrons) need to be identified and separated from pions to get a clearer picture of stopping.

Some of these topics can be very well addressed within experiment E814. With the experimental set-up shown in Fig. 3 one can measure distributions in transverse energy in the central region (via the participant calorimeter) and in the target fragmentation region (via the target calorimeter). Both instruments have sufficient segmentation to allow a detailed measurement of energy flow in p-nucleus and nucleus-nucleus collisions. Furthermore, this information can be correlated with the corresponding distributions of leading baryons and/or projectile fragments detected and identified in the forward spectrometer (see Fig. 3). We consider these studies as important steps toward an understanding of the energy densities attainable in nuclear collisions at high energy.

CONCLUSION

These brief remarks should indicate the richness of this new field of high energy nucleus nucleus collisions. If I have been able to convey some of the enthusiasm which brought us to work in this enterprise, I consider these lectures a success.

REFERENCES

*Supported in part by the National Science Foundation.

1. Proc. 5th International Conference on Ultra-Relativistic Nucleus-Nucleus Collisions, Nucl. Phys. A461 (1987) 1,2.

2. E. Fermi, Z. Physik 29 (1924) 315. C.F. Weizsäcker, Z. Physik 88 (1934) 612. E.J. Williams, Phys. Rev. 45 (1934) 729.

3. Study of Extreme Peripheral Collisions and of the Transition from Peripheral to Central Collisions in Reactions Induced by Relativistic Heavy Ions, AGS Experiment 814, Brookhaven National Laboratory, Oct. 1985.

4. G. Baur and C.A. Bertulani, Phys. Letters 174B (1986) 23.

5. W.D. Myers et al., Phys. Rev. C15 (1977) 2002.

6. R.J. LeDoux, M. Bloomer and H.Z. Huang, Proc. 2nd Int. Workshop on Local Equilibrium in Strong Int. Physics, P. Carruthers and D. Strottman, eds., World Scientific, Singapore 1986, p. 24.

7. R. Gomez et al, preprint, Fermilab-Conf-86/85-E; E-557 Collaboration, also C. Halliwell in Ref. 6, p. 239.

8. T. Akesson et al., Nucl. Phys. A447 (1987) 475c and Contr. to 23rd Int. Conf. on High Energy Physics, Berkeley, July 1986.

9. G. Baym, P. Braun-Munzinger and V. Ruuskanen, submitted to Phys. Letters B.

10. Helios $^{16}O+^{208}Pb$ data, Cern Bulletin, No. 50/86.

11. A. Bamberger et al. Phys Letters B (in print).

ELECTROMAGNETIC EXCITATION PROCESSES IN RELATIVISTIC HEAVY ION COLLISIONS

G. Baur and C.A. Bertulani

Institut für Kernphysik
Kernforschungsanlage Jülich
D-5170 Jülich, FRG

1. INTRODUCTION

Certainly the most important purpose of relativistic heavy ion machines is the study of nuclear matter under extreme conditions (see e.g. the lectures of G. Baym at this school). In central nucleus-nucleus collisions one hopes to observe new forms of nuclear matter, like the quark-gluon-plasma. On the other hand, very strong electromagnetic fields for a very short time are present in distant collisions with no nuclear contact. This is essentially due to the Lorentz-contraction. Such fields can also lead to interesting effects, which we want to discuss here (see e.g. also the lecture of P. Braun-Munzinger at this school, where also experimental problems are discussed).

A theoretical framework to study electromagnetic effects in relativistic heavy ion collisions (RHIC) is provided by the so-called equivalent photon method. The original idea was laid on by Fermi, Weizsäcker and Williams[1]. The time-varying electromagnetic field is replaced by a spectrum of equivalent photons which will interact with the ions. This spectrum can be calculated classically. Later on, Winther and Alder[2] have shown a semiclassical solution to the problem and calculated the full composition of multipole interactions between the ions. Bertulani and Baur[3,4] performed a quantum calculation of the same process and showed the connection between all different approaches. Specially, it was shown how the equivalent photon numbers for all multipolarities can be deduced and that the Fermi-Weizsäcker-Williams method accounts properly only for the E1 multipolarity of the interaction (which is the most important one).

In section 2 we briefly describe the equivalent photon method, we discuss the semiclassical[2] as well as the quantummechanical approach[3,4] and compare them with each other. Then we mention the application of this method to various electromagnetic processes which occur in RHIC, like excitation of discrete nuclear states, Bremsstrahlung, e^+e^- production, π-production or atomic ionization.

In section 3 the possibility of multiphoton processes is studied in more detail. Of special interest is the possibility of exciting new multiphonon giant dipole resonance (GDR) states. RHIC with their strong electric fields seem to be a good probe for exciting such collective vibrations of all protons against all neutrons (see Fig. 1).

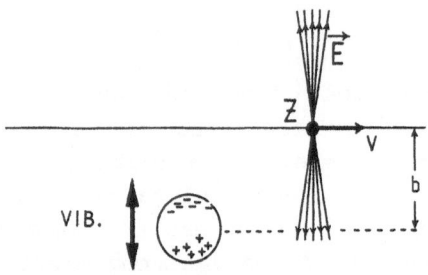

Fig. 1. Electromagnetic excitation of a target by a relativistic projectile with impact parameter b and velocity v. The collision time is given by $\tau_{coll} \approx \frac{1}{\gamma}\frac{b}{c}$, where $\gamma = (1-v^2/c^2)^{-1/2}$, the maximum electric field by $E = Z\frac{e}{b^2}\gamma$. It tends to separate neutrons from protons in a sudden impact.

We study especially a rather transparent harmonic oscillator model for the GDR, which gives a reliable estimate of the cross section. The cross sections for exciting multiphonon GDR states turn out to be quite large, which will make interesting experiments possible[5]. In this context it is especially important to take the width of the states into consideration. This applies to the influence of the width on the excitation process itself and also on the question of the decay properties (statistical and direct) of the hitherto unexplored multiphonon states. Due to the large separation of neutrons and protons, exotic effects could be expected (see also ref. 5). Conclusions and an outlook are given in chapter 4.

2. THE EQUIVALENT PHOTON METHOD AND ITS APPLICATION TO RHIC

The equivalent photon method consists of replacing the effect of the time varying electromagnetic field of the colliding heavy ions by a spectrum of real photons. In this method, the excitation cross section of one nucleus by an energy amount $\hbar\omega$ due to the electromagnetic interaction with the other nucleus is given by[3,4]

$$\sigma_c = \sum_{\pi\ell} \int \frac{d\omega}{\omega} \, n_{\pi\ell}(\omega) \sigma_\gamma^{\pi\ell} \tag{1}$$

where $n_{\pi\ell}(\omega)$ is the equivalent number of photons with the frequency ω and multipolarity $\pi\ell$ (π = E or M, ℓ = 1,2,...). The photonuclear absorption cross sections for the excitation energy $\hbar\omega$ and multipolarity $\pi\ell$ are denoted by $\sigma_\gamma^{\pi\ell}(\omega)$. Using the semiclassical method with straight line trajectories, appropriate for the high energy scattering, the excitation amplitudes of the ions for a given impact parameter b were derived in ref. 2. In ref. 3 these amplitudes were related to the equivalent photon numbers. A quantal method, which takes the nuclear absorption into account in a black sphere model, was developed in ref. 3. The strongly forward peaked angular distributions are characterized by a diffraction pattern[3,4]. The relative importance of quantal diffraction and Coulomb repulsion is governed by the product of the charges of the two colliding particles; the ratio of the diffraction angle θ_d to the classically expected Coulomb deflection angle θ_c is given by

$$\frac{\theta_c}{\theta_d} \approx \frac{2e^2}{\hbar c} Z_P Z_T = \frac{2}{137} Z_P Z_T , \tag{2}$$

where $\frac{e^2}{\hbar c} = \alpha = 1/137$. This shows that only for small projectile and/or target charge the quantal diffraction effects will be comparable to the Coulomb deflection. It was shown in ref. 3 that the excitation cross section integrated over the scattering angle is equal to that obtained in the semiclassical method[2], integrated over impact parameters (starting from a minimum impact parameter R, which corresponds to the radius of the black sphere).

The equivalent photon numbers (integrated over impact parameter or scattering angle) for the lowest multipolarities are given explicitly in ref. 3. For $1 < \gamma < 2$, corresponding to the intermediate energy regime of some hundreds of MeV/A, we have $n_{E2} \gg n_{E1} \gg n_{M1}$. For $\gamma \gg 1$, corresponding to the ultrarelativistic regime, all equivalent photon numbers will become equal to

$$n_{\pi\ell} = Z^2\alpha \, \frac{1}{\pi} \, \ell n\left[\left(\frac{\delta}{\xi}\right)^2 + 1\right] \tag{3}$$

with $\delta = 0.68 \ldots$ and $\xi = \dfrac{\omega R}{\gamma v}$. In the limit of large excitation energies, $\omega \gg \gamma \dfrac{v}{R}$, an adiabatic cutoff sets in and
instead of eq. (3) we have

$$n_{\pi \ell} \propto e^{-2\xi} . \tag{4}$$

This means that a useful approximation for some practical
purposes is to use the relation (3) for $\xi < 1$ and $n_{\pi \ell}(\omega) = 0$
for $\xi > 1$. In other words, the spectrum contains equivalent
photons with energies up to

$$(E_\gamma)_{max} \simeq \gamma \dfrac{\hbar v}{R} . \tag{5}$$

For a typical value of R=10 fm one obtains $(E_\gamma)_{max} \simeq 20 \gamma$ MeV.

In ref. 6 the equivalent photon method, as characterized
by eq. (1), is applied to the excitation of highly excited
collective nuclear states, Bremsstrahlung, K-shell ionization, π- and e^+e^- pair production. In ref. 7, the electromagnetic production of heavy lepton pairs ($\mu^+\mu^-$ and $\tau^+\tau^-$) is
studied. Essentially due to the large mass of the heavy ions
Bremsstrahlung is a relatively unimportant process, however,
we found it is interesting to see how the cancellation of E1
Bremsstrahlung for equal charge to mass ratio projectiles in
nonrelativistic scattering disappears for relativistic energies due to retardation effects. Especially large values of
the cross section were found for e^+e^- pair production. For γ-
values of the order of 50 (see Fig. 3 of ref. 6) the cross
section for electromagnetic π-production in $^{238}U-^{238}U$ will
exceed 1 barn. However, many more pions will be produced in
the violent nuclear collisions, due to the high multiplicity.
A very important process is the excitation of high lying
nuclear states which will contribute to the fragmentation
process since those states decay mainly by particle emission.
Let us consider E1, E2 and M1 excitations.

The position of the E1 and E2 giant resonances is
approximately given by (see e.g. ref. 8)

$$E_{GDR} = 80 \ A^{-1/3} \ MeV \tag{6a}$$
and
$$E_{GQR} = 62 \ A^{-1/3} \ MeV . \tag{6b}$$

We assume that, as is characteristic of strongly collective
states, the strength is given by an appropriate sum rule value. For E1 we use the Thomas-Reiche-Kuhn (TRK) sum rule

$$\int \sigma_\gamma^{E1}(E\gamma)dE\gamma \simeq 60 \ \dfrac{NZ}{A} \ MeV \ mb \tag{7}$$

which is exhausted essentially by the GDR. As an example we show in Fig. 2 the cross section for the excitation of E1, E2 and M1 ($E_{1+} = 10.3$ MeV, $B(M1) = 1$ μ_N^2) states in ^{40}Ca projectiles hitting ^{238}U targets as a function of the beam energy. One observes that the E2 fragmentation mode is quite important at intermediate energies; the excitation of M1 states is of less importance. For $v \sim c$ we have $n_{M1} \approx (\frac{v}{c})^2 n_{E1} \approx n_{E1}$, i.e. equal number of M1 and E1 photons, however, for the corresponding γ absorption cross sections we have typically $\sigma_\gamma^{M1}/\sigma_\gamma^{E1} \approx (\frac{\mu_N}{eR})^2 = (\frac{\hbar}{2m_N cR})^2 \ll 1$.

Another very interesting application[4] of the equivalent photon method is the so-called Primakoff-effect. A fast Λ particle will be excited to the Σ^0 in the Coulomb field of a nuclear target. From the absolute value of this cross section the lifetime of the Σ^0 particle can be determined. Due to the short lifetime of $\tau = (5.8 \pm 1.3) \cdot 10^{-20}$ sec there seems no other method for the lifetime measurement. This number is quite important for our understanding of the structure of hadrons with strangeness.

Another application of the equivalent photon method is considered in ref. 9: the Coulomb dissociation of (nonrelativistic or relativistic) projectiles determines the photodissociation cross section. This cross section is related by detailed balance to the corresponding radiative capture cross section. It is hoped that with this method one can determine astrophysically interesting radiative capture processes, like $\alpha(d,\gamma)^6$Li, $\alpha(^3$He,$\gamma)^7$Be or ^{12}C$(\alpha,\gamma)^{16}$O.

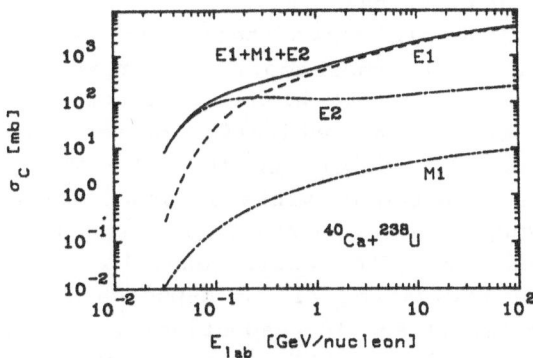

Fig. 2. Electromagnetic excitation cross section of ^{40}Ca projectiles incident on ^{238}U targets. The lower curve corresponds to the contribution of the M1 excitation, the dashed-dotted curve to the E2 excitation and the dashed curve to the E1 excitation. The solid curve is the sum of all these contributions.

3. EXCITATION OF MULTIPHONON GIANT DIPOLE RESONANCE STATES BY MEANS OF MULTIPLE ELECTROMAGNETIC EXCITATION

a) Harmonic Oscillator Model without Damping

The total momentum transfer in a sudden Coulomb collision of two charges Z_P and Z_T is given by (see e.g. ref. 10, p. 619)

$$\Delta p = 2 \frac{Z_P Z_T e^2}{bv} . \tag{8}$$

With this result we obtain for the energy transfer to the whole nucleus, treated as a rigid body with mass Am_N,

$$\Delta E_A = \frac{(\Delta p)^2}{2Am_N} . \tag{9a}$$

The energy transfer to all protons, treated as free particles in a very sudden collision is given by

$$\Delta E_Z = \frac{(\Delta p)^2}{2Zm_N} . \tag{9b}$$

As an example, we obtain for a relativistic ($v \approx c$) $^{238}U-^{238}U$ collision with impact parameter $b = 15$ fm the values $\Delta E_A = 5$ MeV and $\Delta E_Z = 15$ MeV. The difference $\Delta E_{int} = \Delta E_Z - \Delta E_A$ will go into internal excitation energy of the nucleus. This energy corresponds to about the GDR energy in ^{238}U. Since this state is a collective movement of all protons against all neutrons, one expects a very strong excitation of this state already from this simple classical consideration.

A more concise calculation[6,11], based on first order perturbation theory, shows that for heavy systems like $^{238}U+^{238}U$ the first order amplitudes approach unity, i.e. it becomes necessary to take higher order effects into account. In refs. 11 and 12 the excitation process was studied in a harmonic oscillator model for the GDR. The problem of the excitation of a harmonic vibrator under the influence of a time dependent external perturbation can be solved exactly (see e.g. ref. 13). The result leads to a Poisson distribution for the excitation of an N-phonon state. In Fig. 3, taken from refs. 11 and 12 respectively, the total cross sections σ_N for N-phonon excitation are shown for the systems $^{238}U+^{238}U$ and $^{16}O+^{208}Pb$.

Multiphonon excitations have not been identified experimentally up to now. It is worth mentioning that the quite extensive results on Coulomb fragmentation in RHIC, mainly at the BEVALAC, are in good agreement with the theoretical expectations (see ref. 12).

A simple and reliable estimate of the amplitude of the collective neutron-proton dipole vibration can be obtained from the TRK sum rule and the experimentally observed energy position of the GDR states. One obtains for the matrix element of the collective coordinate $\rho \equiv R_p - R_n = \frac{A}{NZ} D$, where D is the dipole operator

$$\rho \equiv |\langle GDR|\rho|0\rangle| = \frac{0.51 \text{ fm}}{\sqrt{NZ}} A^{2/3} .\tag{10}$$

It decreases like $A^{-1/3}$ with A. Thus neutrons and protons are more effectively separated in low mass nuclei. However, the excitation cross sections are smaller.

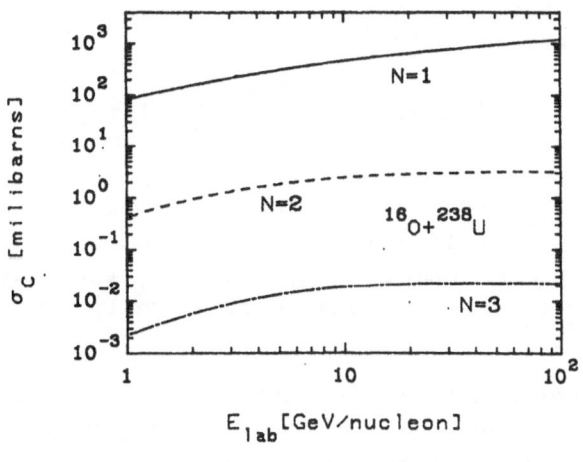

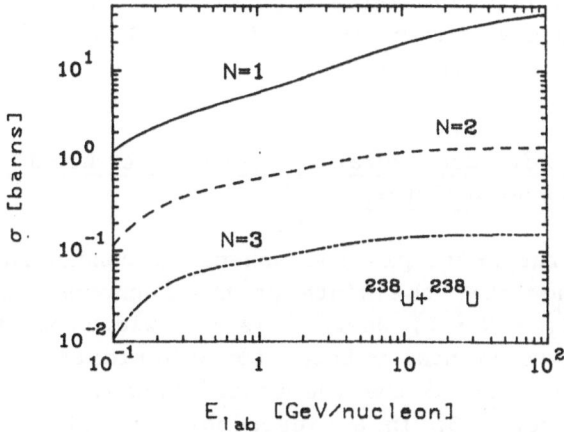

Fig. 3. The total cross sections σ_N for the excitation of N-photon GDR states in $^{238}U-^{238}U$ and $^{16}O+^{238}U$ (excitation in the ^{16}O projectile) collisions as a function of E_{lab}.

b) The Influence of Damping: a Dissipative Quantum Vibrator under External Forces

The giant dipole state is a very short-lived state. It couples strongly to other more complicated states and decays, especially in heavy nuclei, statistically by particle emission. A typical width of $\Gamma \simeq 5$ MeV corresponds to a lifetime of $\tau_{decay} \simeq 10^{-22}$ sec. In a situation where the lifetime of a state is comparable to or even smaller than the collision time, an essential modification of the usual description of Coulomb excitation has to be introduced. This was accomplished by Weidenmüller and Winther[14]. In a phenomenological way we can take the width of the states into account by introducing a term $- \frac{i}{2} \Gamma_N c_N(t)$ in the usual coupled equations for the excitation amplitudes, this takes the loss of flux by the decay into account[12]:

$$i\hbar \, \dot{c}_N(t) = \sum_M \langle N|V(t)|M\rangle e^{i/\hbar(E_N - E_M)t} c_M(t) - \frac{i}{2}\Gamma_N c_N(t) \ . \quad (11)$$

In a RHIC the interaction time is very short, of the order of $\tau_{coll} \simeq \frac{1}{\gamma}\frac{b}{c}$. This means, that the loss term in eq. (11) becomes only effective after the electromagnetic interaction has taken place. One has $\tau_{decay} > \tau_{coll}$ even for moderately high values of γ and multiple excitation is possible in principle[11,12].

Of special interest is the width Γ_N of the multiphonon states ($N \geqslant 2$). Nothing is known experimentally about such high-lying continuum states. In particular, one would also like to know into which channels these states will decay eventually. From the large separation of neutrons and protons one can expect rather exotic final nuclei with unusual N/Z ratios. We feel that only the experiment can tell (see ref. 5).

c) Microscopic Consideration about the Width (Damping) of Multiphonon GDR States

From rather simple and schematic considerations we want to argue now that the width of the N-phonon state is of the order of $\Gamma_N \simeq N \cdot \Gamma$, where Γ is the width of the usual GDR, which is experimentally known for practically all nuclei. For a general review of the theory of damping of nuclear excitations see ref. 15. In a microscopic approach, the GDR state is described by a coherent superposition of one particle-one hole states. One of the many such states is pushed up by the residual interaction to the experimentally observed position of the GDR. This state carries practically all the E1 strength. This situation is most simply realized in a model with a separable residual interaction (Brown-Bolsterli model). We write the GDR state as (one phonon with angular momentum 1M)

$$|1,1M\rangle = A^+_{1M}|0\rangle \qquad (12)$$

where A^+_{1M} is a proper superposition of particle-hole creation operators. If we can apply the quasiboson approximation we can use the usual boson commutation relations and construct the multiphonon states (see e.g. ref. 8). Thus an N-phonon state will be essentially a coherent superposition of N particle-N hole states. The width of the GDR is essentially due to the spreading width, i.e. to the coupling to more complicated quasibound configurations. The escape width (Landau damping) plays only a minor role. We are not interested in a detailed microscopic description of these states here. We use a simple model of the strength function (see e.g. ref. 16). We couple a state $|a\rangle$ (i.e. the GDR state) by some mechanism to more complicated states $|\alpha\rangle$, for simplicity we assume a constant coupling matrix element $V_{a\alpha} = \langle a|V|\alpha\rangle = \langle\alpha|V|a\rangle = v$. With an equal spacing of D of the levels $|\alpha\rangle$ one obtains a width

$$\Gamma = 2\pi \frac{v^2}{D} \qquad (13)$$

for the state $|a\rangle$. We assume the same mechanism to be responsible for the width of the N-phonon state: one of the N independent phonons decays into the more complicated states $|\alpha\rangle$ while the other (N-1)-phonons remain spectators. We write the coupling interaction in terms of creation (destruction) operators $c^+_\alpha (c_\alpha)$ of the complicated states $|\alpha\rangle$ as

$$V = v(A^+_{1M}c_\alpha + A_{1M}c^+_\alpha) . \qquad (14a)$$

For the coupling matrix element v_N, which connects an N-phonon state $|N\rangle$ to the state $|N-1,\alpha\rangle$ (N-1 spectator phonons) one obtains

$$v_N = \langle N-1,\alpha|V|N\rangle = v\langle N-1|A_{1M}|N\rangle = v \cdot \sqrt{N} \qquad (14b)$$

i.e. one obtains for the width Γ_N of the N-phonon state

$$\Gamma_N = 2\pi \ N \ \frac{v^2}{D} = N\Gamma \qquad (15)$$

where Γ is given by eq. (13).

351

4. CONCLUSION AND OUTLOOK

We have seen that the strong electromagnetic field present in distant RHIC causes many processes of nuclear and atomic origin. They are interesting in themselves; with their typically large cross sections, they will have to be considered also as background effects in RHI experiments as well as in the design of future RHI machines. In this paper we have especially studied the excitation of high-lying nuclear states, this is possible because of the high Fourier components present in these very short collisions. Coulomb fragmentation cross sections will already at moderately high γ-values exceed the geometrical cross section. We devoted particular attention to the possibility of exciting new nuclear states, the multiphonon GDR excitations. While the cross sections to excite such states in a RHIC are rather large, it remains to be seen whether one can really observe these states experimentally. The main problem is the question of the width (spreading as well as decay width to specific final fragment channels) of these states.

We acknowledge with pleasure many stimulating discussions during the Erice Summer School, especially with Profs. A. Winther, G. Bertsch and S. Levit.

REFERENCES

1. E. Fermi, Z. Phys. 29:315 (1924);
   C.F. Weizsäcker, Z. Phys. 88:612 (1934);
   E.J. Williams, Phys. Rev. 45:729 (1934);
   see also ref. 10.
2. A. Winther and K. Alder, Nucl. Phys. A319:518 (1979).
3. C.A. Bertulani and G. Baur, Nucl. Phys. A442:739 (1985).
4. C.A. Bertulani and G. Baur, Phys. Rev. C33:910 (1986).
5. P. Braun-Munzinger et al., Proposal 814 submitted to the AGS Program Committee, SUNY at Stony Brook, accepted October 1985.
6. C.A. Bertulani and G. Baur, Nucl. Phys. A458:725 (1986).
7. G. Baur, C.A. Bertulani, Phys. Rev. C, in press.
8. A. Bohr and B. Mottelson, "Nuclear Structure", Vol. II, Benjamin, Reading, MA (1975).
9. G. Baur, C.A. Bertulani and H. Rebel, Nucl. Phys. A458:188 (1986).
10. J.D. Jackson, "Classical Electrodynamics", 2nd ed., Wiley, New York (1975).
11. G. Baur and C.A. Bertulani, Phys. Lett. B174:23 (1986).
12. G. Baur and C.A. Bertulani, Phys. Rev. C, November 1986.
13. K. Alder and A. Winther, "Electromagnetic Excitation", North-Holland, Amsterdam (1975).
14. H.A. Weidenmüller and A. Winther, Ann. Phys. (New York) 66:218 (1971).
15. G.F. Bertsch, P.F. Bortignon and R.A. Broglia, Rev. Mod. Phys. 55:287 (1983).
16. A. Bohr and B. Mottelson, "Nuclear Structure", Vol. I, Benjamin, Reading, MA (1969).

ULTRARELATIVISTIC HEAVY-ION COLLISIONS AND THE

PROPERTIES OF NUCLEAR MATTER UNDER EXTREME CONDITIONS *

Gordon Baym

Loomis Laboratory of Physics, University of Illinois

1110 W. Green St., Urbana, Illinois 61801

1. INTRODUCTION

The two main issues I will discuss in these lectures are the nature of extended nuclear matter at extremely high energy densities, and how, by means of ultrarelativistic heavy-ion collisions in the laboratory, one can create matter at high densities, and thus have the opportunity to learn about its properties experimentally. To set the scale of nuclear energy densities let us note that the total energy in a nucleus in its ground state is essentially the rest mass density of the nucleons. Since nuclear matter has a density ρ_{nm} of order 0.16 $nucleons/fm^3$ and the nuclear mass is of order 940 MeV, the rest mass density is of order 0.15 Gev/fm^3. This energy density is large compared with the scale of low-energy spectroscopy, involving energy densities of order Mev/fm³. The question we are interested in is how we expect nuclear matter to act when we raise its energy density to the range of $1 - 10$ Gev/fm^3 say. What, for example, are its principal degrees of freedom, its thermodynamic properties, and its quantum chromodynamic properties?

Studying matter at such energy densities will require colliding heavy nuclei together at energies well above 1 Gev per nucleon in the center-of-mass frame. A program of fixed-target experiments with lighter nuclear projectiles is currently underway at the CERN SPS and the Brookhaven AGS and is yielding exciting new data on high energy nuclear collisions. At present the CERN experiments are being carried out at lab energies of 60 and 200 GeV per nucleon with ^{16}O beams, and soon with ^{32}S (and possibly ^{40}Ca) beams. The Brookhaven program is at lab energies of $12 - 14$ GeV per nucleon, with ^{16}O and ^{28}Si beams. In addition, Brookhaven is adding a booster

* Supported in part by NSF Grant PHY84-15064.

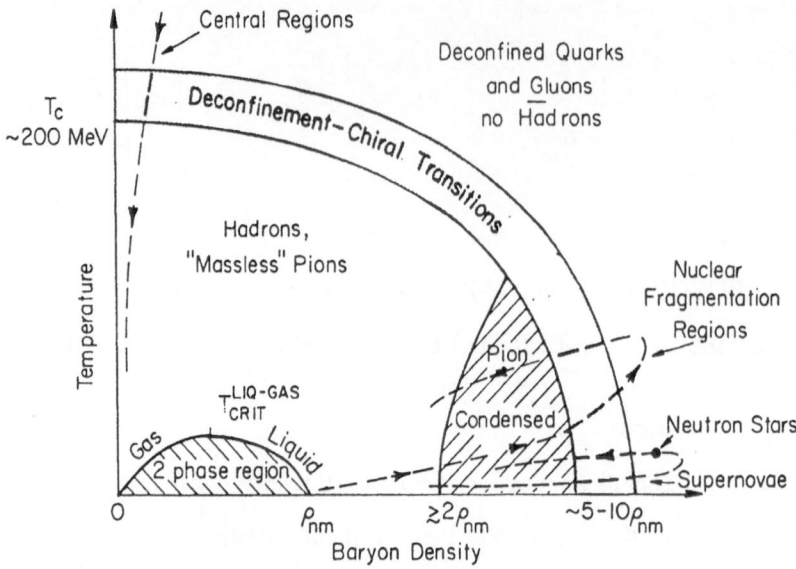

Fig. 1. Phase diagram of nuclear matter in the baryon density, temperature plane showing regions of hadronic and deconfined matter. Normal nuclear matter density ρ_{nm} is 0.16 fm^{-3}.

ring to the AGS which will enable these experiments to be carried out with heavy beams, e.g., ^{197}Au. Farther in the future, the Brookhaven Relativistic Heavy Ion Collider (RHIC), on which construction is expected to begin in 1989, will provide the capability of colliding colliding nuclei as heavy as Au on Au at 100 GeV per nucleon c.m. (equivalent to 20 TeV per nucleon lab).*

One of the most interesting possibilities is that at extremely high energy density, matter will form a new state, the *quark-gluon plasma*, where the quarks and gluons become deconfined. The regions where this new state is expected to occur can be seen in the phase diagram of nuclear matter in the temperature-baryon density plane, fig. 1. In the low temperature - low baryon density region the basic degrees of freedom are hadronic, those of nucleons, mesons and isobars, while in the high temperature or high baryon density regions the basic degrees of freedom should become those of quarks and gluons. Between these regions there may or may not be a sharp phase transition. Later I will come back to the question of how the phase diagram is explored in various physical situations, but first let us consider the elementary properties of quark matter, and the transition between hadronic and quark matter.

---

* Useful general references on relativistic nucleus-nucleus collisions are the proceedings of the ongoing conferences on quark matter, refs. 1-3; see also ref. 4.

2. QUARK-GLUON PLASMA

As matter is heated or compressed its degrees of freedom change from composite to more fundamental. For example, by heating or compressing a gas of atoms, one eventually forms a plasma in which the nuclei become stripped of the electrons, which go into continuum states forming an electron gas. Similarly, when nuclei are squeezed, as happens in the formation of neutron stars in supernovae where the matter is compressed by gravitational collapse, the matter merges into a continuous fluid of neutrons and protons. Since nucleons themselves are made of quarks, one further expects that a gas of nucleons, when squeezed or heated, turns into a gas of uniform quark matter, composed of quarks, and at a finite temperature, antiquarks and gluons as well, which are no longer confined in individual hadrons but are free to roam over the entire volume of the deconfined region.

Let us recall a few features of qcd which are relevant for understanding the elementary physics of the quark-gluon plasma. In quantum electrodynamics, photon exchange produces the basic force between charges, which between static point charges is simply the Coulomb interaction, $\sim e^2/r$. The force between charges of opposite sign is opposite to that between like charges; thus, in qed one can form electrically neutral systems, such as positronium or hydrogen atoms, which do not give rise to long-range Coulomb fields. Qcd has a similar structure, in that the forces between quarks arise from exchange of gluons, and the color degree of freedom functions as a three-valued charge, rather than simply $\pm$, as in qed. Again one can form color singlet or neutral systems which do not give rise to long-range color Coulomb fields. Such a charge scheme requires eight gluons, rather than a single photon, themselves having color and hence coupling directly to themselves, producing the rich non-linear structure of qcd. Because qcd allows color neutrality, quark matter in equilibrium in its state of lowest energy, or free energy at finite temperature, will on average have no long-range color Coulomb fields, as in an ordinary electrically neutral plasma.

Qcd is also asymptotically free. In qed an electron gathers around it a polarization cloud of electron-positron pairs in the vacuum which decreases the net charge seen at large distances; the effective charge of an electron at large distances is given by $e^2/\hbar c = 1/137$. At short distances, inside the polarization cloud, the effective charge on the electron grows, diverging at zero distance – one of the troublesome divergences of qed. In qcd a rather different behavior occurs. Because the gluons themselves carry color, they also screen the bare charges, but their net effect is opposite that of quark-antiquark pairs; the result is that close to a quark, the effective charge does not become infinite, but rather goes to zero – the property of asymptotic freedom. At short distances, corresponding to large momentum scales, interactions become arbitrarily weak, with the effective coupling

$$\alpha(p) = \frac{g^2}{4\pi} = \frac{6\pi}{(33 - 2N_f)ln(p/\Lambda)}, \tag{1}$$

where N_f is the number of quark flavors that are relevant, Λ, the qcd scale parameter is of order $100 - 200 \, MeV$, and p is the momentum scale. At large distances however quite the opposite happens; as colored particles are separated, the forces between them become larger and larger, giving rise to confinement.

Imagine then highly compressing or heating nuclear matter, so that many quark-gluon degrees of freedom are excited. To a first approximation, one can at very high densities treat the system as a non-interacting gas of relativistic quarks, antiquarks and gluons. The reason is that, as in an ordinary plasma, any small region of the matter will, at high densities, be on average color neutral and not produce long-range (color) Coulomb fields, while the residual short distance forces in the region become weak as the interparticle separation becomes small, due to asymptotic freedom. [On closer examination asymptotic freedom does not guarantee absence of all effects of interactions; see refs. 5 and 6 for a careful discussion of interaction effects on long wavelength scales in the high density limit.]

Although for the temperatures and baryon densities of plasmas realistically expected in laboratory collisions, interactions are in fact important, the non-interacting limit provides a useful first handle on the quark-gluon plasma. While ordinary nuclear matter has 4 helicity states, 2 for spin times 2 for isospin, a quark-gluon plasma has many more internal degrees-of-freedom; the quarks have from 24 to 36 helicity states, composed of 2 spin, 3 color, 2 particle-antiparticle, and 2 to 3 flavor degrees-of-freedom, depending on whether strange quarks are also present in addition to the light up and down quarks; the massless gluons have in addition 16 helicity states (2 spin and 8 color). From a thermodynamic point of view a quark-gluon plasma at a given energy density has a high entropy. Hot free quark matter is similar to ordinary black-body radiation with energy density $E \sim T^4$, where T is the temperature; in a system with equal number of u, $\bar{u}$, d, and $\bar{d}$ quarks, as well as gluons,

$$T \simeq 160 \, MeV E^{1/4} \tag{2}$$

and the total density of excitations per fm^3 is

$$n_{exc} \simeq 2.25 E^{3/4}, \tag{3}$$

with E measured in GeV/fm^3. Since the qcd phase transition to a quark-gluon plasma is believed to occur at T of order $200 \, MeV$, we see from (2) that the scale of energy densities that must be deposited in collisions to excite a plasma is of order several GeV/fm³. Because of the slow dependence of T on E, it will not be easy to heat a plasma in a nuclear collision much beyond hundreds of MeV.

Since the temperatures produced in collisions are expected to be at most on the order of a few times Λ, one must in fact take interactions fully into account. They are clearly always important near the deconfinement transition. At first one is tempted to use perturbation theory; the Feynman diagrams for the thermodynamic free energy to order $\alpha^2 \ln \alpha$ are straightforward to evaluate. Taking into account terms of order α^2 is considerably harder. The eventual result, an asymptotic expansion of the free energy in α, is, however, absolutely useless at the coupling strengths of interest. For example, the first correction to the entropy density, s, calculated by differentiating the free energy with respect to T, is given by

$$s = s_0[1 - (54/19\pi)\alpha], \tag{4}$$

where s_0 is the non-interacting entropy density. This first order expression for the entropy turns negative as α increases, with decreasing density, beyond ~ 1. This is impossible – entropies must be positive. What we see is a signal that perturbation theory breaks down early, and is not a satisfactory way to calculate.

The only useful approach so far to calculating effects of interactions is Monte Carlo lattice gauge theory, where by putting the theory of qcd on a space-time lattice, one becomes capable of dealing with all strengths of interaction. (See ref. 7 for a general review.) The calculations are usually done on a lattice with equal number of sites N in each spatial direction and, to describe finite temperature, a different number N_t in the time direction, related to the temperature by $T = N_t/a$, where a is the lattice constant. The calculations require very large computing capabilities, but with supercomputers, one is now able to compute with good statistics on large lattices. Lattice gauge theory is rapidly approaching the point where it will be able to give quantitatively good information on the properties of quark matter over large ranges of temperature and also baryon density. Calculations with finite baryon density are just in their beginning, and I shall give a brief overview of only the finite temperature, zero baryon density results.

Figure 2 shows early Monte Carlo calculations of the energy density, plotted in units of the ideal non-interacting system (Stefan-Boltzmann) energy, $\sim T^4$, in a system with just gluons, and no quarks whatsoever – pure Yang-Mills theory – in a) for[8] $SU(2)$, with just 2 colors, and in b) for[9] $SU(3)$. The temperature is measured in units of a lattice qcd scale parameter Λ_L. In $SU(2)$, the energy density exhibits a clean second order phase transition. At low temperature the system behaves as a gas of massive glueballs, and is confined, while at the transition temperature, T_c, the system turns smoothly into a gas of deconfined gluons, rather rapidly approaching the high temperature T^4 limit. In $SU(3)$, by contrast, one sees a sharp first transition with a large latent heat, of order a few GeV/fm^3. Indeed the Monte Carlo calculations in this case exhibit considerable hysteresis effects, which are a good signal of a first order transition.

In a pure gluon theory one can measure whether the system is confined or not by adding a massive quark-antiquark pair of test particles to the system, and asking how much energy is required to separate the pair to infinity. If the system is confining, it is impossible to separate them and the energy of separation $\epsilon(R)$ diverges. Gluons, which are in a color octet, cannot screen the force between quarks in color triplet states. Thus the "Wilson line," defined by

$$W = exp[-\epsilon(R \to \infty)/T],$$

(5)

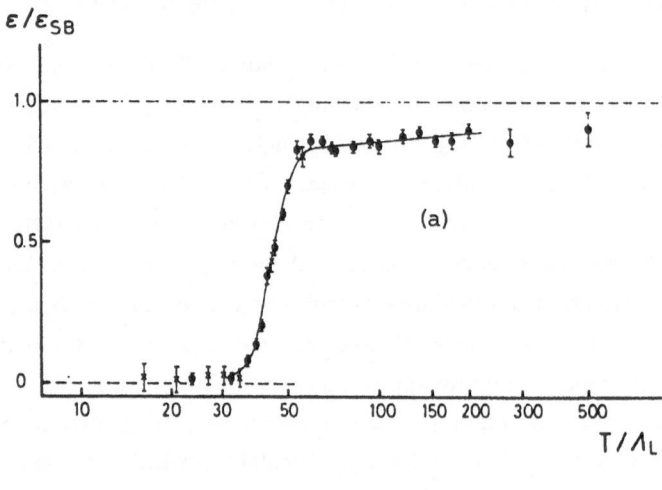

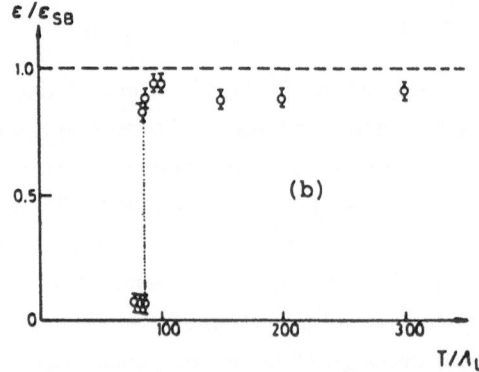

Fig. 2. Energy density of pure gauge theory in units of the ideal gas energy: (a) $SU(2)$, (b) $SU(3)$.

vanishes in the confined state as the separation R goes to infinity, while in the deconfined state it should be non-zero. The Wilson line functions in pure gluon theory as a useful order parameter to distinguish the confined from the deconfined phase. Calculations of W for the pure gluon theory are shown in fig. 3; in a) for $SU(2)$, W begins to rise above the transition point, corresponding to the onset of deconfinement, while in b) it shows a discontinuous jump, consistent with the behavior of the energy density E.

When one begins to include finite mass quark degrees of freedom, q and $\bar{q}$, the simple test of adding a pair of heavy test quarks $Q\bar{Q}$ runs into trouble, since at sufficient separation, it becomes energetically favorable to create a $q\bar{q}$ pair in the system, which screens out the interaction between the test pair; the q binds to the $\bar{Q}$, and the $\bar{q}$ to the Q, creating effectively a pair of mesons which can be separated to infinity with finite energy. The point is that once light quarks are in the system there no longer exists a good measure of whether the system is in a confined or deconfined state, and there need not be a sharp transition between the confined and deconfined phases. The transition between the two phases can be smooth, as occurs for example in ionization of a gas as it is heated, where one goes gradually from gas molecules to electrons and nuclei; the two states are qualitatively different and there is a reasonably rapid onset of ionization, but it is not sharp. Alternatively, the transition may be first order, as in the boiling of water.

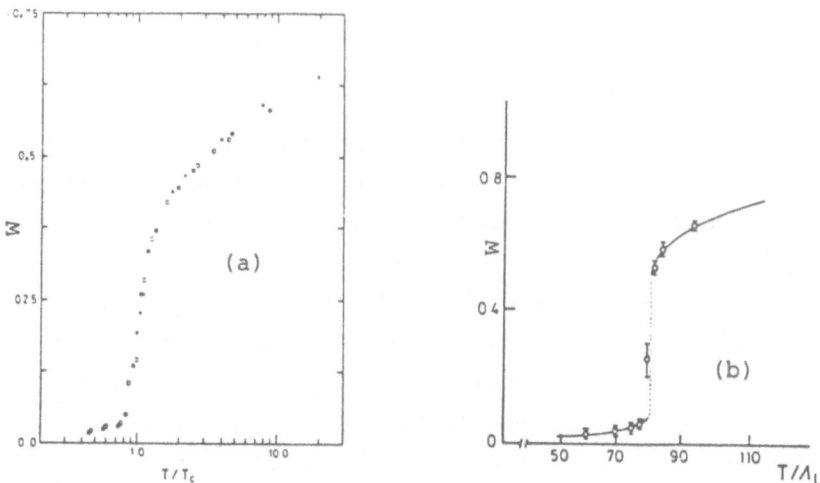

Fig. 3. The Wilson line for (a) SU(2) and (b) SU(3) pure gauge theory.

Qcd matter with light quarks turns out to be analogous technically to a ferromagnet studied in the presence of an external magnetic field, H, in which the transition in zero external field is first order. The quark mass plays a similar role to H^{-1}. For such a ferromagnet the transition becomes weaker and weaker with increasing H, and finally beyond a critical H_c, the transition becomes smoothed out; in enormous H, the spins are all aligned and there is no transition. In the $T - H$ plane, one finds a line of first order transition points, terminating in a critical point at H_c. Similarly a quark-gluon plasma, at large quark mass, has a well-defined transition, which weakens as the mass decreases; the question is whether for realistic quark masses, analogous to large H, there is a sharp transition or not. To take light mass quarks into account accurately in lattice gauge theory requires quite extensive computing, and only very recently has one begun to achieve a reasonable understanding of the nature of the transition (at zero net baryon density).

One further ingredient which must be taken into account is chiral symmetry, the nearly exact $SU(2) \otimes SU(2)$ symmetry of the strong interactions in low energy nuclear physics, generated by the conserved vector current,

$$\vec{V} = \int d^3 r \bar{\psi}(r) \gamma^o \vec{\tau} \psi(r) \tag{6}$$

together with the partially conserved axial vector current (PCAC),

$$\vec{A} = \int d^3 r \bar{\psi}(r) \gamma^o \gamma_5 \vec{\tau} \psi(r). \tag{7}$$

Because the axial current is not precisely conserved, chiral symmetry is not exact; the level of violation is measured by the smallness of the pion mass: $(m_\pi/m_n)^2 \sim 1/50$. From the point of view of the underlying quark structure, chiral symmetry is exact only for zero mass u and d quarks; the violation of chiral symmetry is a reflection of the fact that the light quarks have masses that are not precisely zero, but are on the order of 10 MeV.

Symmetries can be realized in a physical situation in two ways: the Wigner mode in which the states can classified according to the representations of the rotation group, as in atoms and nuclei, and the Goldstone mode, in which the equilibrium state picks out a given direction in the group space, analogous to the situation in a ferromagnet, where the magnetized state chooses a special spatial direction for the magnetization breaking the overall rotational symmetry of the state. (Since the ground state of a ferromagnet in fact belongs to a representation of the rotation group the more precise analogy is to an anti-ferromagnet.) In nuclear physics chiral symmetry is spontaneously broken. The long wavelength small oscillations of the spins in an aligned ferromagnet are low-lying modes, the spin waves; in the case of broken chiral symmetry, the analogous low-lying excitations, or Goldstone bosons – the oscillations of the spontaneously selected direction – are the physical pions.

The implications for qcd are the following. At low temperatures and baryon densities chiral symmetry is spontaneously broken. On the other hand, if at the high momentum scales of very high temperatures or densities the system becomes asymptotically free, then one expects chiral symmetry to be fully restored. Between these two limits a chiral symmetry restoring phase transition should occur. Associated with this transition is a well-defined order parameter, $\langle \bar{\psi}\psi \rangle$ (where ψ is the quark field) which is non-zero in the spontaneously broken phase and zero where chiral symmetry is fully restored. Finite quark mass m makes chiral symmetry only approximate and tends to wash out the chiral transition. But for sufficiently small m the transition associated with the chiral symmetry should be sharp.

Thus with decreasing quark mass, m, the transition between the confined and deconfined phases has the following behavior. For very large m, the transition is sharp and, as the Monte Carlo calculations indicate, first order. With decreasing m the transition becomes washed out. But with further decrease of m, chiral symmetry becomes more and more exact, and eventually the transition sharpens again, driven by restoration of chiral symmetry.

This structure is illustrated in a series of recent calculations of the Illinois Monte Carlo group (refs. 10-12 and earlier references therein) in SU(3) with finite mass quarks, on a $10 \times 10 \times 10 \times (N_t = 6)$ lattice,[10,11] and on $8 \times 8 \times 8 \times (N_t = 4)$ and $4 \times 4 \times 4 \times 4$ lattices.[12] In figs. 4-6, the horizontal scale is the effective coupling strength $\beta = 6/g^2$, which increases monotonically with temperature. The quark mass is given in terms of the transition temperature T_c (generally of order twice the qcd scale parameter Λ) by the values in the figures (there in units of the inverse lattice spacing a^{-1}) times N_t.

Figure 4a, the Wilson line for a system with relatively heavy quarks – over an order of magnitude more massive than the u and d quarks – nicely illustrates how matter with heavy quarks has a first order phase transition; here $N_t = 6$. The Wilson line shows a fairly sharp onset of deconfinement; hysteresis in the transition provides good evidence that it is actually first order. The smoothing out of the transition to the deconfined phase caused by finite quark mass is evident in fig. 4b, for intermediate mass $ma = 0.05$, or $m/T = 0.3$. The behavior is more like a second order or smoothed-out phase transition. Also plotted is the chiral order parameter $\langle \bar{\psi}\psi \rangle$, which goes rapidly from a finite to a small value (not exactly to zero, since the quark mass is finite in these calculations) as deconfinement sets in. Another measure of the sharpness of the transition is the specific heat, which indicates the temperature range over which energy must be put into the system to go from one phase to the other; fig. 5 shows, for the intermediate mass cases $ma = 0.05$ and 0.07, and $N_t = 6$, the specific heat versus β (or roughly the temperature) in the neighborhood of the transition.[11] We again see not a sharp transition, but rather a large but smooth bump;

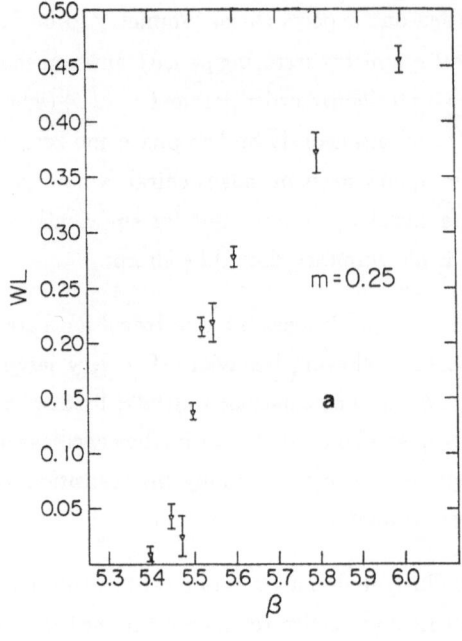

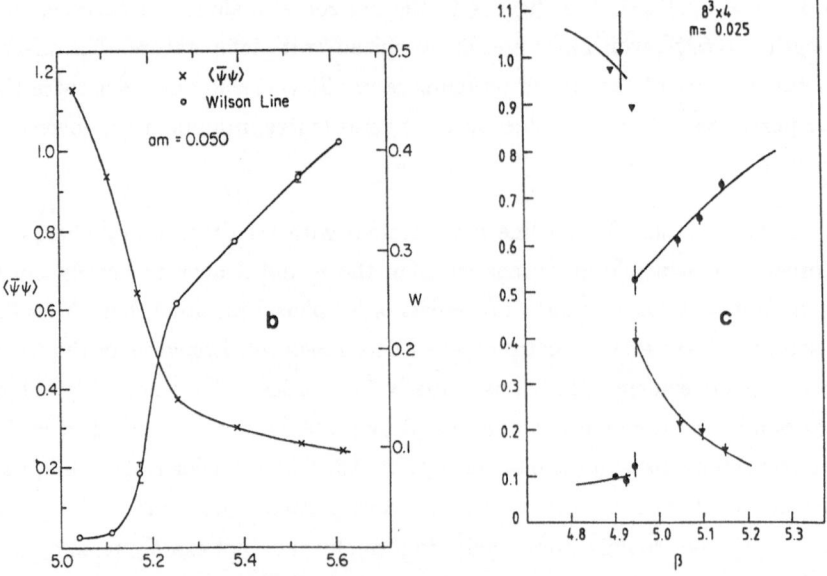

Fig. 4. Wilson line in SU(3) with (a) relatively heavy quarks, (b) intermediate mass
quarks, (c) light mass quarks (filled circles); (b) and (c) (inverted triangles)
show the chiral behavior as well.

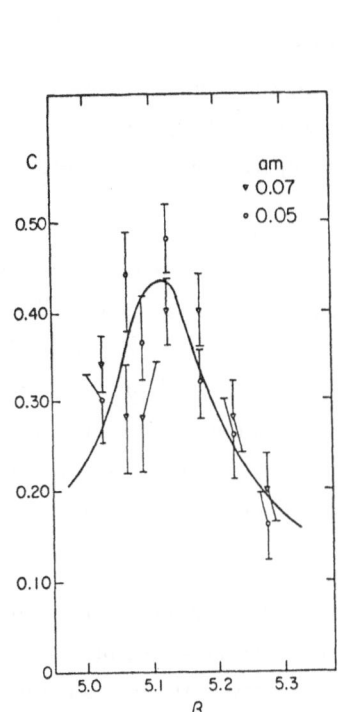

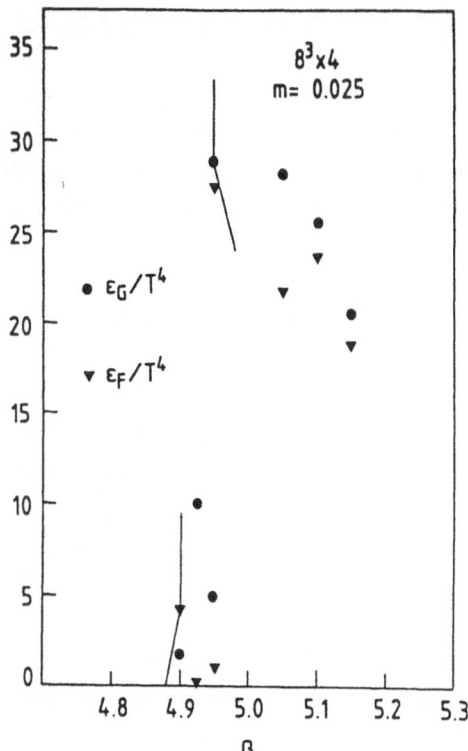

Fig. 5. Specific heat with intermediate mass quarks.

Fig. 6. Gluon and quark contributions to
the energy density, in units of T^4,
for light quarks, $m/T_c = 0.10$.

the total energy under the curve is of order a few GeV/fm³.

However for even lower mass, $ma = 0.025$, on an $8 \times 8 \times 8 \times 4$ lattice, and thus $m/T_c = 0.10$, closer to the realistic case, we see in fig. 4c clear evidence that the restoration of chiral symmetry, and deconfinement, is now via a sharp transition.[12] Figure 6 shows the corresponding contributions of the gluons and fermions to the energy density (divided by T^4) for this light mass case. Above the deconfinement transition, the energy density initially overshoots and then relaxes to the Stefan-Boltzmann limit.

Present lattice gauge theory calculations with finite mass quarks thus indicate that when chiral symmetry is accurately taken into account the transition to the deconfined phase, for realistic mass quarks, is sharp, and apparently first order, with a latent heat of order a few GeV/fm³ (but not yet precisely determined).

3. EXPLORING THE PHASE DIAGRAM

Nuclear matter at high temperature and density plays an important role in astrophysical situations; understanding its properties in these situations is an important motivation for studying it in laboratory collisions of nuclei. Let us spend a moment briefly discussing matter under extreme conditions in supernovae, neutron stars and the early universe.

In a (Type II) supernova explosion, a massive star which has burnt out its fuel at the end of its evolution can no longer support itself against gravitational collapse, and begins to implode.[13] In the infall the matter is crushed to very high densities, several times that of normal nuclear matter. The core of the star bounces back, as shown by the trajectory in the phase diagram; the matter in the core may or may not cross into the deconfined region. The energy output in a supernovae, which depends on the strength of this bounce, may in fact provide a handle on the nature of the matter in the interior reached in the collapse.

In neutron stars the properties of matter under extreme conditions play a particularly crucial role. For example, our present lack of knowledge of the properties of matter at densities beyond $\sim 2\rho_{nm}$ is reflected in uncertainty in the maximum mass of neutron stars, an important quantity in trying to identify black holes unambiguously. Measured masses of neutron stars are ~ 1.4 solar masses, with radii calculated to be $\sim 10~km$. Typical temperatures are very low, less than one MeV. The central conditions in a neutron star are indicated on the phase diagram; the matter in the interior may possibly be deconfined.

One can in fact study the interiors of neutron stars by observing their cooling. In their early years ($\leq 10^5 y$), cooling is governed primarily by neutrino emission. For phase space reasons, cooling via the nucleonic $URCA$ process, $n \rightarrow p + e + \bar{\nu}$, and $e + p \rightarrow n + \nu$, is considerably slower than it would be via the corresponding process with light mass up (u) and down (d) deconfined quarks, $d \rightarrow u + e + \bar{\nu}$, and $u + e \rightarrow d + \nu$.[14] Combining knowledge of the ages of astrophysical objects containing neutron stars with measurements of neutron star surface temperatures, taken with x-ray telescopes, gives one a measure of how rapidly neutron stars cool. Particularly rapid cooling would provide evidence for unusual states of matter in the interior. Present observations, which generally provide an upper bound on surface temperatures, are so far consistent with the interiors being normal nuclear matter, although future satellite observations should sharpen these bounds and provide a more definitive answer to the nature of the matter in the interiors of neutron stars. The recent supernova, Shelton 1987a, has given us the particularly exciting opportunity (unless it failed to leave a neutron star remnant) of measuring, over the next several years, the detailed cooling of a neutron star from its birth.

In the first microseconds of the early universe, the temperature falls as

$$T \simeq \frac{0.5 \; MeV}{\sqrt{t_{seconds}}}, \tag{8}$$

so that prior to $\sim 5 - 10$ microseconds after the big bang, when the temperatures are hundreds of MeV, matter is in the form of a quark-gluon plasma. The matter of the early universe has a relatively small net baryon density, of order 1 part in 10^9 (as inferred from the present photon/baryon ratio). As the universe expands it cools and matter hadronizes, following a downward trajectory practically along the vertical axis of the phase diagram. Matter emerging from the transition is primarily in the form of pions, with a slight baryon excess. Possible astrophysical consequences of the transition from deconfined plasma to hadrons are reviewed in refs. 15 and 16.

4. ULTRARELATIVISTIC HEAVY-ION COLLISIONS

Let us turn now to the question of studying matter at high density by means of very energetic collisions of heavy nuclei in the laboratory. Imagine colliding two heavy nuclei together at energies from 1 to 100 GeV per nucleon in the center-of-mass. At low energies, the regime that will be studied particularly in the AGS experiments, and to a certain extent in the higher energy SPS experiments, one can picture the two Lorentz-contracted colliding nuclei as nearly stopping each other, with reasonable probability of forming, to a crude first approximation, a fireball. [In reality, parts of the nuclei will generally pass through the collision rather than remain in a fireball. One should also note that with light projectiles, as are being initially employed in the AGS and SPS experiments, the collision volume may not achieve the thermal equilibration necessary for a fireball description.] Such heavy ion collisions may reach energy densities of order a few Gev/fm^3 and baryon densities several times ρ_{nm}; the matter may indeed cross into the deconfined region, as shown in fig. 1 in the curve labelled "fragmentation regions," and then expand out. The high density matter produced in such collisions will be relatively baryon rich.

As the beam energy is increased, the nuclei pass through each other, become highly excited internally, and at the same time, leave the vacuum between them in a highly excited state, containing quarks, antiquarks and gluons, as illustrated in fig. 7. Such "nuclear transparency" is very important in the ultrarelativistic regime, above ~ 10 GeV per nucleon c.m. (the energy in the 200 GeV CERN fixed target experiments). The nuclear fragmentation regions, which recede from each other at the speed of light, contain essentially all the baryons of the original nuclei; the central region, to a first approximation, has no baryon excess, and resembles the hot vacuum of the early universe. [In fact, the baryons of the colliding nuclei will spread somewhat into the central region; predicting how much, the "nuclear stopping power," is an important problem on which the present CERN and Brookhaven experiments will shed light. However, the energy per baryon should in general be very high.]

Before entering into the details of relativistic collisions, let us recall the concept of rapidity, which is a particularly convenient variable for describing relativistic particles. The rapidity y of a particle (either in the initial beams, or a product of the collision) is defined by

$$y = \frac{1}{2} \ln \left[\frac{E + p_z}{E - p_z} \right],$$
(9)

where E is the particle energy, and p_z its momentum along the beam axis (taken to be the z direction). For the special case of motion purely along the z axis, the velocity is given by $v/c = \tanh y$. Rapidities, unlike ordinary relativistic velocities, have the nice property of being additive; under a Lorentz transformation along z by velocity $u = c \tanh y_u$, rapidities transform by $y \to y + y_u$. In a collision one generally finds a range of rapidities of the final particles extending roughly from that of the target to that of the projectiles, a total rapidity spread given by

$$\Delta y = 2 \ln(2E/m)$$
(10)

where E is the beam energy per nucleon in the center-of-mass.

In ultrahigh energy collisions, the final rapidities of the detected particles emerging from the collision should be correlated with the spatial structure of the collision region. To see how this correlation works, think (in the center of mass frame) of the fragmentation regions after the collision as receding at the speed of light from each other, with the central region being uniformly stretched out in between, so that its velocity increases linearly with distance, from -c at the left fragmentation region to +c at the right fragmentation region (see fig. 7). Thus, particles observed at large positive rapidities come primarily from the right fragmentation region, those at large negative rapidities from the left fragmentation region, while the intermediate rapidity particles arise from the central regions. The intrinsic motion of the particles with respect to the local average motion will, of course, blur this correspondence somewhat, over ~ one unit of rapidity.

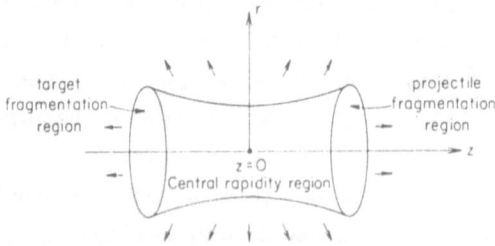

Fig. 7. Nuclear fragmentation and central rapidity regions in an ultrarelativistic central heavy ion collision.

Ultrarelativistic collisions provide sufficient total rapidity spread – for example, as we see from eq. (10), 10.6 units at 100 GeV on 100 GeV – that this correlation enables one to sort out the different collision regions from the rapidities of their final state products; this ability to distinguish different regions experimentally is one of the principal reasons for going to high energies. The baryons will appear, in the center-of-mass frame, predominantly at large absolute rapidities (those of the initial beams), while in the central rapidity region, which has little excess of baryons over anti-baryons, one will see primarily mesons. Indeed, such a structure emerges in proton-proton scattering, as carried out at the CERN ISR, and in $\bar{p}p$ collisions at the CERN S$\bar{p}$pS collider. The charged particle multiplicity versus rapidity for 30 GeV on 30 GeV pp collisions is shown schematically in fig. 8; the meson spectrum (unshaded) is spread out over the central region, while the net baryon density (shaded) is peaked near the rapidities of the two incident colliding beams, with a width of order 2 units in rapidity.

The scale of energy densities expected in ultrarelativistic collisions can be estimated from the observation that pp or $\bar{p}p$ collisions in this energy range produce $\sim 2-3$ charged particles, predominantly pions, per unit of rapidity, with typical transverse energy ~ 400 MeV. Adding in neutral pions as well, we find an energy density of $\sim 1.5 GeV$ per unit of rapidity. A very conservative extrapolation to a nucleus-nucleus collision is to multiply this produced energy by a factor $\sim (1 - 2)A$, i.e., assume that each nucleon of one of the nuclei makes effectively only one or two collisions going through the other nucleus. The resulting estimate of the energy per unit rapidity in an AA collision is $\geq (2 - 4)A$ GeV, and the corresponding energy density is

$$E \sim \frac{0.4 A^{1/3}}{t} GeV/fm^3 \qquad (11)$$

where t is the local time in the collision volume. At a time of 1 fm/c the energy

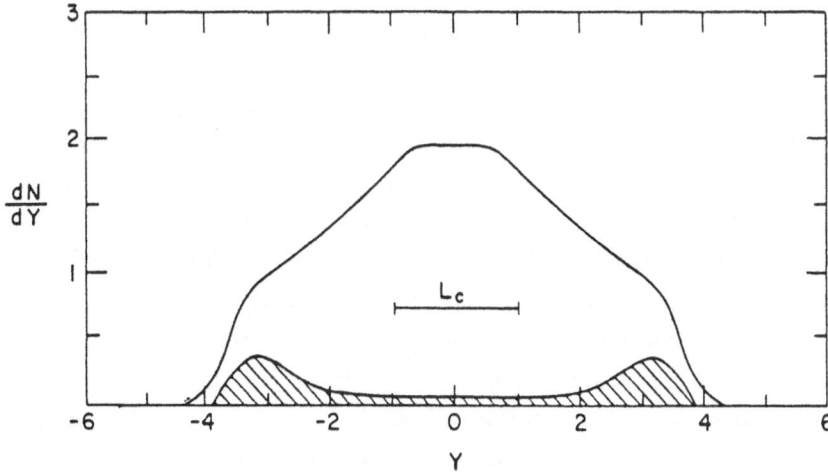

Fig. 8. Charged particle multiplicity in pp collisions at 30 GeV on 30 GeV.

density is at least of order 2.5 GeV/fm^3 in the average central collision.

The actual energy density can be much larger, as can be seen from a similar argument based on extrapolating from final state particle multiplicities rather than energies.[17] If we assume that entropy, essentially proportional to dN/dy, the final multiplicity density in rapidity, is conserved in the evolution of the collision, then at early times,

$$tT^3 \sim (dN/dy)/\pi R^2 \qquad (12)$$

, where T is the temperature, R is the nuclear radius. Clearly the earlier the time, the greater the temperature and hence energy density. If we apply the uncertainty principle to the initial formation time, τ_o, when one can first begin to describe the system in terms of well-defined interacting excitations (quarks, antiquarks and gluons), and the initial temperature, T_o, at that time, then $T_o\tau_o \geq 1$. Let us assume (conservatively) that the mean multiplicity in an AA collision is of order A times that in a pp collision. Combining with the uncertainty principle relation we deduce that the formation time can be as early as $A^{-1/6} fm/c$, that the initial temperature can be $\sim 200A^{1/6} MeV$, and most importantly, the initial energy density one expects to achieve, proportional to T_o^4, can be $E \sim A^{2/3}$ in GeV/fm^3, which is of order 30 GeV/fm^3 for Au or U collision partners. Furthermore, rare events, on which one can certainly trigger, can yield even larger energy depositions. Thus one has good reason to believe that the energy densities in collisions will be sufficiently high in many events to form interesting states of matter.

A further feature of observed pp and $\bar{p}p$ collisions at CERN collider energies is that the multiplicity distributions in the central regions are roughly flat, as can be seen in fig. 8. Since changing rapidity is equivalent to making a Lorentz transformation, the lack of change under a shift of the horizontal rapidity scale implies that conditions in the central region are approximately Lorentz invariant. The assumption of Lorentz invariance in the central region provides a very simple first picture of the evolution of the central region, as is illustrated in the space-time diagram, fig. 9, which shows a slice of the collision along the central axis, z, versus time t. [Here we neglect the longitudinal thickness of the Lorentz contracted nuclei, and finite size effects of the transverse dimension.] The projectile and target approach each other along the light cone at negative times, and collide. The first event in the collision is production of excitations, after a finite formation time, which stream out from the collision point, the origin. In heavy nucleus-nucleus collisions, unlike in pp collisions, sufficiently large numbers of excitations are made, and the system sizes are correspondingly large, that after a further finite time the excitations come into local thermodynamic equilibrium; this means that the system enters a regime where it can be described by local hydrodynamics. Two interesting phenomena occur now. One is that the matter passes through the hadronization transition, as shown in the phase diagram by the curve labelled "central regions." The matter emerges from the transition in the form of hadrons. Eventually the system expands sufficiently that the interactions among the hadrons cease – "freezeout" – and the system becomes a collection of freely streaming particles, which are eventually detected. The lines separating the regions in the space-time diagram are essentially hyperbolae, since as a consequence of the approximate Lorentz invariance in the central region, one expects the same conditions in all central regions at the same proper time, $\tau = \sqrt{t^2 - z^2}$ (corresponding to local velocity z/t along the central axis).

One very important point that this diagram illustrates is that the matter under-goes considerable processing from the initial quark-gluon plasma phase to the finally observed hadronic products. In the collision, the two fragmentation regions move away from each other longitudinally, and the central region undergoes a longitudinal stretching and cooling, as illustrated in fig. 9. In addition, the system begins to un-dergo transverse expansion, which initially occurs hydrodynamically with a rarefaction wave propagating inward from the outer edge at the speed of sound.

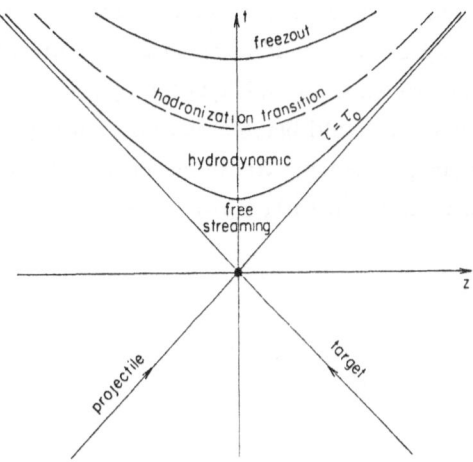

Fig. 9. Space-time picture of the evolution of a central collision.

Let me conclude by briefly reviewing possible probes of collisions and signals of plasma formation. Tracing back from the final state products to the as yet unex-plored states of matter reached in an ultra-relativistic heavy ion collision is clearly a challenging question. In particular, being able to see clearly that deconfinement has occurred requires a fuller understanding of the physics of strongly interacting quark-gluon plasmas, which we may possibly not have before experiments are actually carried out. Many different experimental approaches will be used.[1-3] The first class involves studying the global parameters of events: the spectra of the emitted particles, as functions of beam energy and mass number, A, as indicators of possible phase transitions, to determine entropy density in rapidity of the final states, accelerations as reflected in transverse momentum distributions, etc.; interferometry with final particles, such as pions, to determine source sizes which, when combined with the global data, can yield temperatures and densities of the source regions; and multi-particle correlations in rapidity, indicators of energy flow, hydrodynamic behavior, and collective modes. A second class is direct probes of the interior of the collision through detection of the direct photon and lepton pair production, including coherent, plasma, and Drell-Yan emission processes.

Hadron production processes also give information about the unusual conditions expected in an ultra-relativistic heavy ion collision. For example, the ratio of K mesons to pions measures strange quark production in the initial phases.[18-19] Hadronic jets produced in the collision, in which only one of the two jets passes through the collision volume, are a powerful potential probe of the conditions in the initially produced matter.

Finally, it is tempting to speculate on the rather unusual objects that may possibly be made as a quark-gluon plasma expands and hadronizes, such as multiquark states, hadrons involving heavy quarks, extended metastable quark droplets of large strangeness, fractionally charged particles, and multi-baryon states of unusual chiral topology produced in the chiral symmetry breaking transition, in the way that monopoles can be created in phase transitions in the early universe.

References

1. T. W. Ludlam and H. E. Wegner, eds., "Quark Matter '83, Proc. 3rd Int. Conf. on Ultra-relativistic Nucleus-Nucleus Collisions," Nucl. Phys. A418 (1984).

2. K. Kajantie, ed., Quark Matter '84, "Proc. 4th Int. Conf. on Ultra- relativistic Nucleus-Nucleus Collisions," Lect. Notes in Phys." 221, Springer, Berlin (1985).

3. L. S. Schroeder and M. Gyulassy, eds., "Quark Matter '86, Proc. 5th Int. Conf. on Ultra-relativistic Nucleus-Nucleus Collisions," Nucl. Phys. A461 (1987).

4. G. Baym and L. McLerran, "Ultrarelativistic heavy ion collisions," W. A. Benjamin, Menlo Park (to be published).

5. B. Svetitsky, Nucl. Phys. A461: 71c (1987).

6. J. Polonyi, Nucl. Phys. A461: 279c (1987).

7. H. Satz, Ann. Rev. Nucl. Part. Sci. 35:245 (1985).

8. J. Engels, F. Karsch, I. Montvay and H. Satz, Phys. Lett. 101B:89 (1981).

9. T. Çelik, J. Engels and H. Satz, Phys. Lett. 129B:323 (1983).

10. J. Kogut and D. Sinclair, preprint ILL-(TH)-86-46, Nucl. Phys. B (1987).

11. J. Kogut, Phys. Rev. Lett. 56:2557 (1986).

12. J. Kogut, H. W. Wyld, F. Karsch and D. K. Sinclair, preprint ILL-(TH)-87-6.

13. G. E. Brown, H. A. Bethe and G. Baym, Nucl. Phys. A375:481 (1982).

14. N. Iwamoto, Phys. Rev. Lett. 44:1637 (1980).

15. G. Baym, Nucl. Phys. A447:463c (1986).

16. J. H. Applegate and C. J. Hogan, Phys. Rev. D31:3037 (1985).

17. H. Von Gersdorff, L. McLerran, M. Kataja and P. V. Ruuskanen, Phys. Rev. D34:794 (1986).

18. T. Matsui, B. Svetitsky and L. McLerran, Phys. Rev. D34:783, 2047 (1986).

19. K. Kajantie, M. Kataja and P. V. Ruuskanen, Phys. Lett. B179:153 (1986).

SIS/ESR:

A HEAVY ION SYNCHROTRON AND COOLER FACILITY AT GSI

Paul Kienle

Gesellschaft für Schwerionenforschung mbH

D-6100 Darmstadt, West-Germany

1. INTRODUCTION

The eighteenth century philosopher Lichtenberg said: "Man muß etwas Neues tun, um etwas Neues zu sehen", or translated: "You have to do something new in order to see something new". In honour of Darmstadt's famous son, Lichtenberg was born in a little village Oberramstadt near Darmstadt, the concept of our new GSI project is guided rather closly by Lichtenberg's words.

Fig. 1 shows a view of the future GSI accelerator facilities for heavy ions[1], the construction of which has been started in 1985 and should be completed during 1989. They consist of an upgraded UNILAC used as a 19 MeV/u injector into a medium energy (1-2 GeV/u) heavy ion synchrotron SIS 18 which is connected with a storage cooler ring ESR of half the circumference of SIS 18. The combination of these two rings should allow to produce completely stripped heavy ion beams up to U^{92+} with the highest possible phase space densities achievable by various beam cooling techniques. In addition SIS/ESR will provide beams of radioactive nuclei in the energy range from several MeV/u up to 1-2 GeV/u again cooled to the highest possible phase space densities. The beams in the ESR may be used either circulating with high currents or extracted with a great variety of time structures and intensities. They may be also reinjected into SIS for further acceleration or deceleration. There will be a large experimental area with several experiments set up on beams from both SIS and ESR. Further experimental areas are located directly behind SIS, between SIS and ESR and around the ESR. For the future one can dream of injecting the high phase space density completely stripped beams in superconducting collider rings with small apertures, modest size and prize to achieve very high c.m. energies (> 20 GeV/u) at as high as possible luminosities.

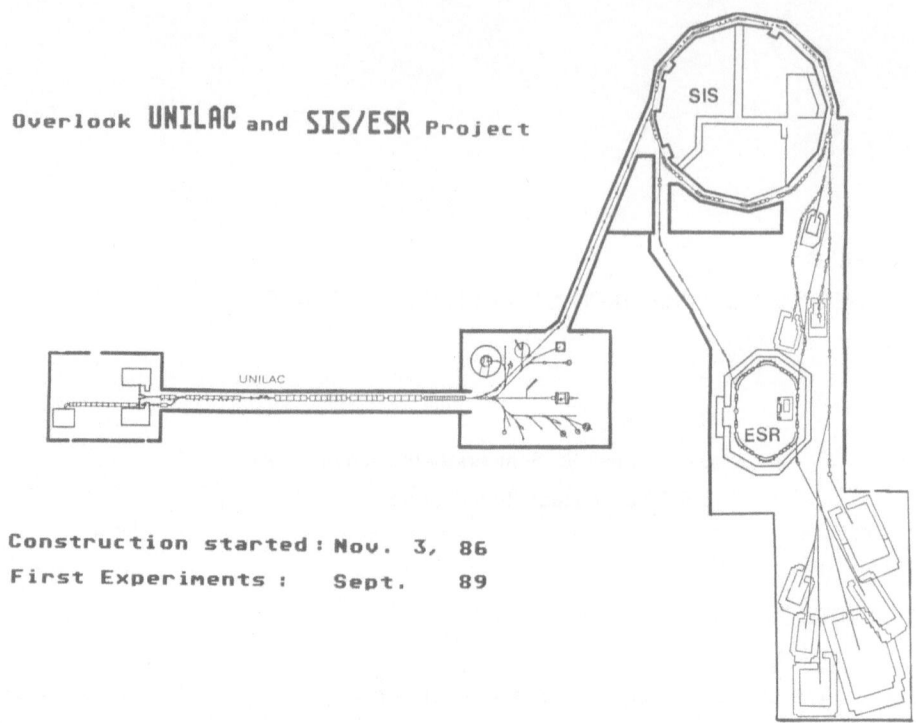

Construction started: Nov. 3, 86
First Experiments :   Sept.   89

Fig. 1   Layout of the upgraded UNILAC, SIS and ESR.

1. THE UPGRADED UNILAC, THE HEAVY ION SYNCHROTRON SIS 18
AND THE EXPERIMENTAL STORAGE RING ESR

Fig. 2 shows a layout plan of the heavy ion synchrotron SIS 18[2,3] to be built at GSI for acceleration of heavy ions up to an energy of 2 GeV/u connected with the experimental storage ring ESR,[4] which can be used for accumulation, storage and phase space density increase of heavy ions up to uranium with energies between 834 MeV/u (Ne^{10+}) and 556 MeV/u (U^{92+}), and a hall for experiments.

The synchrotron (Fig. 3) with a circumference of 216.72 m and a maximum bending power $B\rho = 18$ Tm accelerates heavy ions up to uranium with a cycling rate of 3 Hz (up to 1.2 T) and 1 Hz to the highest energies (1.8 T).

A heavy ion beam accelerated in the UNILAC up to 19 MeV/u, and stripped to an adequate high charge state for the desired energy and intensity, is injected into SIS 18 during 10 to 30 turns and accelerated to maximum energies, depending on the charge states of the ions as shown in Fig. 4.

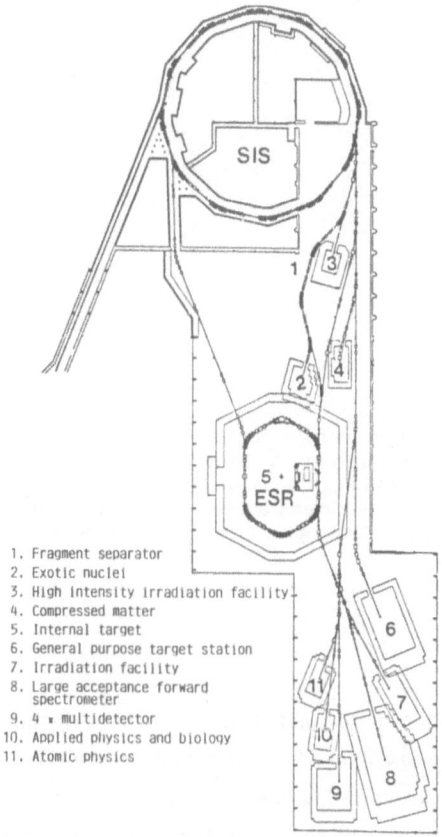

1. Fragment separator
2. Exotic nuclei
3. High intensity irradiation facility
4. Compressed matter
5. Internal target
6. General purpose target station
7. Irradiation facility
8. Large acceptance forward spectrometer
9. 4 × multidetector
10. Applied physics and biology
11. Atomic physics

Fig. 2   Layout plan of the heavy ion synchrotron SIS 18 and the experimental storage ring ESR and the experimental hall.

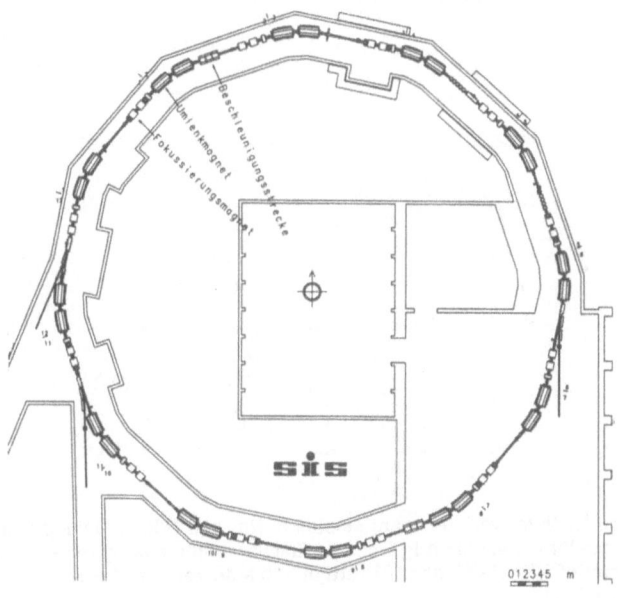

Fig. 3   Layout of the heavy ion synchrotron SIS 18

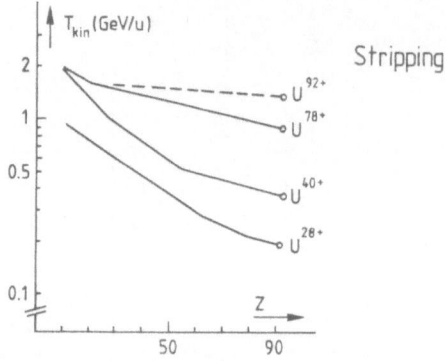

Fig. 4   Maximum achieveable energies at SIS 18 as a function of nuclear charge, the energies are given for a gas- or a foil-stripper at an energy of 1.4 MeV/u, resulting in relatively low degrees of ionization. If a second stripper at 11.4 MeV/u is added or if completely ionized particles from the experimental storage ring ESR are reinjected into the synchrotron higher energies can be achieved.

For uranium ions with a charge state of $q = 78$, after stripping behind the UNILAC with a foil target, 1 GeV/u is achieved as maximum energy. The maximum beam intensities from SIS 18 are shown in Fig. 5 for Ne- and U-ions of various ionic charges, depending on the stripping procedure, as function of their specific energies. The decrease of the intensities towards higher energies is caused by the change of the synchrotron repetition rate; the drop for 1 GeV/u Ne and 500 MeV/u U is due to a change of the repetition rate from 3 to 1 Hz.

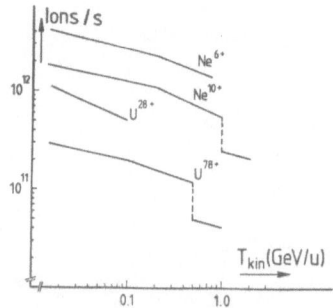

Fig. 5   Beam currents for various charge states of Ne- and U-ions, gained with the stripping procedures described in Fig. 2 as function of the energy. The intensity drops by a factor of 3 for Ne^{10+} and U^{78+} are due to a decrease of the repetition rate from 3 Hz to 1 Hz.

In order to achieve the beam intensities shown in Fig. 5 a new injector for the UNILAC is under construction(Fig. 6). It is based on recently developed high intensity ion sources[5] for low charge states (U^{2+}), which are accelerated in RFQ-structures to energies of 130 keV/u[6] After stripping to higher charge states (U^{9+}), they will be injected into the second section of the Wideröe-linac. Based on this scheme, 8 mA of Kr^{1+} ions have been accelerated in 3 RFQ-linacs to energies of 10 keV/u[7], and recently 5 mA of Ar^{1+} in 5 RFQ-linacs to 45 keV/u. For SIS injection up to the space charge limit, the necessary pulse currents of 2×10^{15} p/s Ne^{10+} and 2×10^{14} p/s U^{78+} can be achieved using this scheme.

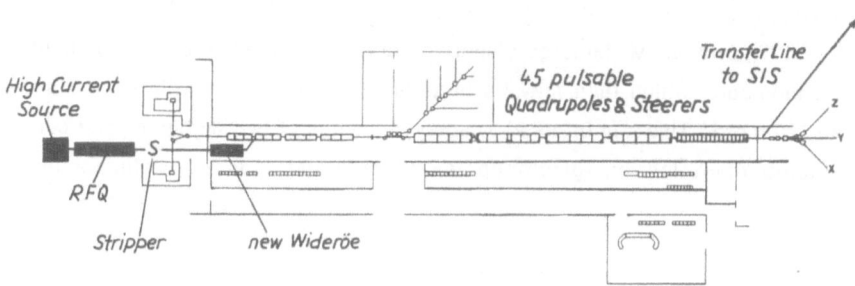

Fig. 6   Layout for the modification of the UNILAC with a RFQ high current injector.

The Unilac beam will thus be used for injection into SIS but during the synchrotron acceleration phase also for an experimental program at variable energies in the present experimental hall. For this the transverse beam optics in the linacs must be switched between the excitation level for the injection into SIS at 19 MeV/u and a selectable value between 3.5 and 20 MeV/u for a low energy program.

Between SIS 18 and ESR the beam may be stripped once more to the highest desired charge state. The ESR with a bending power of $B\rho = 10$ Tm allows to store ions up to U^{92+} with the following maximum energies: Ne^{10+} (834 MeV/u), Ar^{18+} (709 MeV/u), Kr^{36+} (656 MeV/u), Xe^{54+} (609 MeV /u) and U^{92+} (556 MeV/u). The uranium ions can be fully stripped at this energy with an efficiency of 60 % in a Cu-target of several g/cm^2 thickness.[8] The stripping yield increases strongly with decreasing nuclear charge charge, thus one expects a yield of 70 % for Pb^{82+}-ions (574 MeV/u) and already 100 % for Xe^{54+}-ions (609 MeV/u). Alternatively one can install a reaction target for projectile fragmentation. The favourable kinematic focussing of the products around the beam direction and velocity allows effective mass-separation in a special mass-separator between SIS and ESR, followed by accumulation of radioactive beams with the ESR, which accepts beams with $\delta p/p = \pm 0.5$ % and transverse emittances of 20 π mm mrad.

Fig. 7 shows the layout plan of the ESR, with two 9.5 m long straight experimental sections, in one of which an electron cooling device will be installed. The other 4 straight sections will be used for the installation of rf cavities, slow and fast extraction elements. The rf cavities are used for acceleration, deceleration and especially also for bunching of the beam together with the electron cooling for reduction of the occupied longitudinal phase space volume. With the fast extraction system of the ESR one can transfer a highly ionized and cooled beam back to SIS 18 for further acceleration or specially also deceleration. The optics of the ring allows three modes of operation, one with moderate dispersion along the ring specially suited for accumulation of beams with large momentum spread ($\Delta p/p = 1$ %) and emittance ($E_{hv} = 20\ \pi$ mm mrad), one with zero dispersion in the straight sections, which allows multi-charge operation (U^{89+} - U^{92+}) and one w ith large dispersion to accomodate two beams of slightly different momenta, which then may be brought to merge with a well defined angle of about 100 mrad.[9] This can be used to study collisions of *two* highly ionized beams at fixed target equivalent energies of up to 7.2 MeV/u and an energy definition of better than 10 %.

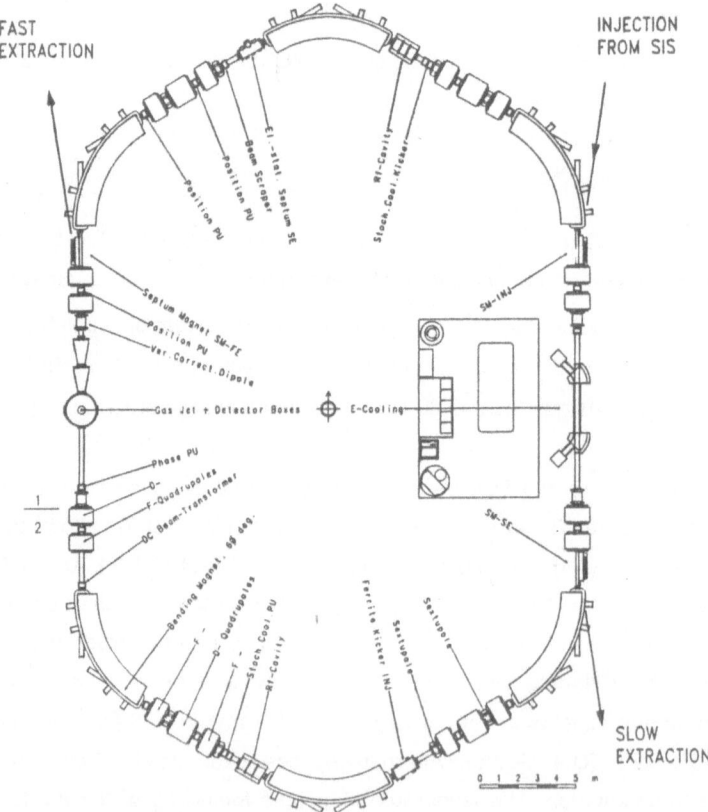

Fig. 7   Layout plan of the experimental storage ring ESR with a circumference of 108.36 meters and 10 Tm bending power. In two straight sections each with a length of 9.5 m experiments and the facility for cooling can be mounted. The rf cavities are for acceleration, deceleration and bunching.

The most important facilities of the ESR are various cooling devices which can be applied complementary. For low phase space density secondary beams stochastic pre-cooling may be used. For cooling to very high phase space density, electron cooling of completely stripped heavy ions is foreseen in an interaction zone of 2 m length. A "cool" electron beam of 5-10 A is focussed within an area of 5 cm diameter collinearly along the ion beam at the corresponding average velocity. For cooling of beams between 30 MeV/u and 560 MeV/u, electron energies in the range of 16.5 keV and 310 keV are required. With an electron beam current density of up to 1 A/cm² and ion beams of initially $\delta p/p = 0.1\ \%$ and $2\ \pi$ mm mrad the following cooling times (in ms) for U^{92+} and Ne^{10+} are expected:

Energy	30 MeV/u	500 MeV/u
U^{92+}	30	90
Ne^{10+}	125	630

Note that the cooling time scales with A/Z^2 and thus reaches shorter values for heavier ions. The loss of ions due to capture of electrons and associated change of charge states - note that this assumption overestimates the loss time of circulating beams in rings which can transport several charge states - can be reliably calculated

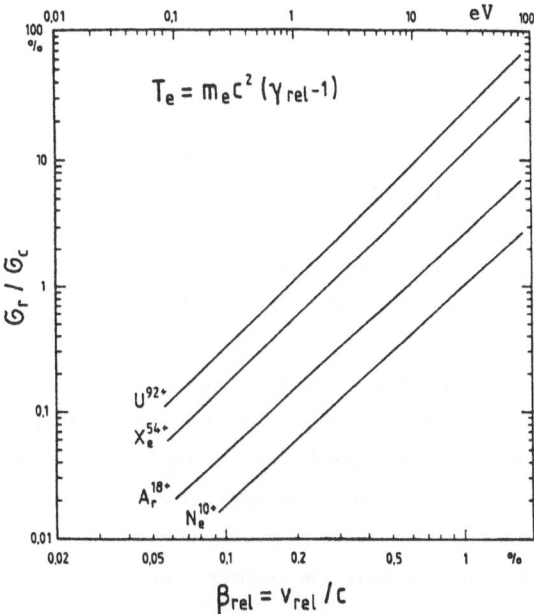

Fig. 8 The quotient of the cross-sections for recombination and for cooling as a function of the relative velocity between ions and electrons. The average values of the equivalent electron energies T_e at the beginning of the cooling vary from several eV (SIS 18 beam) to more than 10 eV (beam of fragments). The temperature of the electron beam is around 0.5 eV.

for radiative electron capture, the only recombination mechanism active for completely stripped ions.

Fig. 8 shows the ratio of the recombination to the cooling cross-section (σ_R/σ_C) as function of the relative velocity between cooling electrons and ions and the mean value of the equivalent electron energy in eV respectively. This ratio amounts to several percent, depending on the ion charge, at the beginning of the cooling process (T_e is several eV for a SIS beam and more than 10 eV for a fragment beam) to decrease proportionally to v_{rel}^2 during the cooling process, which ends at the "temperature" of the electron beam (0.3 eV). One can expect that this method leads to cooled heavy ion beams with emittances as small as 0.1 π mm mrad and momentum spreads of 10^{-5}. Space charge effects may limit the number of ions to be cooled in a circulating beam.[10]

While the cooled beam circulates in the ring, it may be used in the second straight section for the study of collision processes with internal targets, which may be atomic or electron beams (unpolarized or even polarized), gas jets or fibres. For all experiments which need thin targets a high gain in luminosity may be achieved compared with a single pass experiment due to the increase of the circulating beam current ($\sim 2 \times 10^6$). Also the interaction of collinear laser and electron beams with the circulating ions of high intensity and small momentum spread may be favourably studied.

3. RADIOACTIVE BEAMS

Following the decision in April 1985, to extend the GSI experimental facilities[1] with a heavy ion synchrotron SIS and a storage ring ESR an intensive programmatic discussion for a search of the first generation experiments started. The initial ideas proposed are documented to some extent in the proceedings of workshops held in Heidelberg[11], Rauischholzhausen[12] and Darmstadt[13].

My review tries to sketch the present status of the discussion without a claim of being complete. It is just intended to stimulate the appetite of as many scientists as possible to make the best use of our facility. We address our invitation especially to our friends abroad to make efficient use of our new accelerators.

It has been shown previously that electromagnetic dissociation and fragmentation of medium energy projectiles may be favourably used to produce radioactive beams[14]. For electromagnetic dissociation high Z-targets for generating a high electromagnetic field are preferred whereas for fragmentation by nuclear interaction light targets serve equally well. What makes these reactions favourable for the production of radioactive beams, is the fact that at bombarding energies well above 100

MeV/u only a small momentum is transfered during the reaction [15,16]. This means that the fragments have a small momentum spread ($\Delta p/p < 1$ %). They are focused in the direction of the beam ($\Theta_{1/2} < 0.5^0$-1^0) and retain the beam velocity. Thus in principle they can be separated from the beam with high efficiency using their different magnetic rigidities. Because of the high collection efficiencies, the production rates become competitive with low energy isotope separators and do not depend on chemical properties. With a beam of 10^{11} projectiles/s one expects fragments in a rate of $\dot{N} = 3 \times 10^7$/mbs. Using this method several new isotopes have been discovered at the Bevalac[17] and at lower beam energies but higher intensities at GANIL[18]. Recently neutron rich beams of He and Li were produced by using 800 MeV/u ^{11}B and ^{20}Ne incident on Be and applied to measure their total reaction cross sections from which the interaction nuclear radii of all particle stable He and Li nuclei relative to ^{4}He were deduced[14].

Fig. 9 shows the principle of a projectile fragmentation mass separator following a scheme used for the first time at GANIL[19]. A high energy beam from SIS hits a target, in which projectile fragments are produced with similar velocities as the beam. A

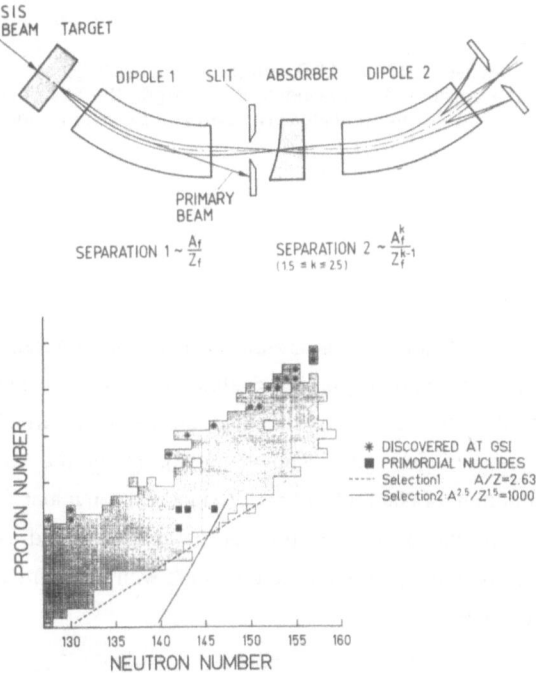

Fig. 9   Principle of projectile fragmentation mass separation using an absorber between two magnetic deflection systems (upper part). The lower part shows a chart of nuclides with examples of separation lines of dipole 1 (dashed curve) and dipole 2 following the absorber (solid line).

magnetic field (dipole 1) separates the primary beam and the fragments, if A/Z is sufficiently different. The bottom part of Fig. 9 shows a chart of nuclides with a certain selection of A/Z = 2.63 (dashed line). Note that all nuclei with this selected A/Z ratio will be contained in the beam after the slit. For a complete separation of nuclei with identical A/Z, they are passed through an absorber in which they loose energy proportional to Z^2. Thus nuclei with higher Z leave the absorber with lower velocities. A second magnetic field (dipole 2) can finally separate essentially one nucleus (crossing of dashed and solid selection line). By shaping the absorber chromaticity corrections may be introduced.

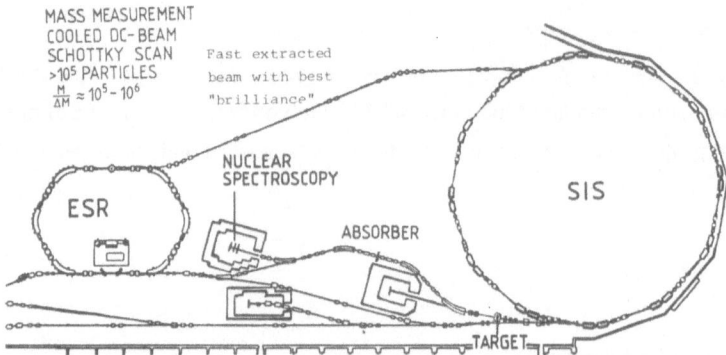

Fig. 10　Schematic layout of SIS and ESR with the projectile fragmentation separator connecting both facilities. The positions of the target, the absorber, and the nuclear spectroscopy facility are indicated. The ESR can be used for mass measurements.

Fig. 10 shows the optics of the separator[20], which links the synchrotron with the ESR. To achieve the necessary dispersion S-shaped achromatic magnet arrays are placed before and after the absorber. The kind of resolution which one hopes to achieve with such an arrangement is shown in Fig. 11 for the selection of ^{234}Ac produced by fragmentation of 600 MeV/u ^{238}U beams. On a nuclear chart, the relative intensities of various fragments leaving the separator are plotted. Note that for this most difficult situation only the neighbouring elements with the same Z/A as ^{234}Ac are present in the percent level.

As shown in Fig. 10 the radioactive nuclei are available directly behind the separator for nuclear spectroscopy experiments and as low phase space density secondary beams.

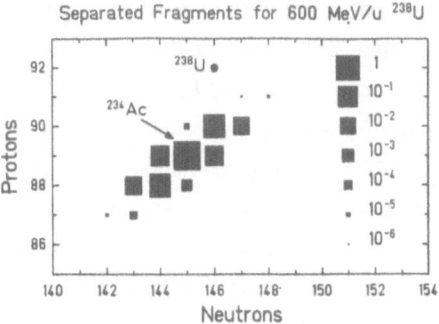

Separated Fragments for 600 MeV/u ^{238}U

Fig. 11   Presentation of the separation power of the projectile/fragmentation mass separator for the example of ^{234}Ac using a 600 MeV/u ^{238}U beam.

First experiments will be focused on the measurements of projectile fragmentation cross sections for heavy nuclei. The good resolution, the high efficiency of the separation and the high beam intensity will allow to detect nuclei with very small yields.

There are various spectroscopy experiments envisaged with neutron rich and neutron deficient nuclei far off the stability line, such as the investigation of the structure of N = 126 isotopes of neutron rich elements. A large field of research is anticipated to evolve which is very interesting for nuclear structure and astrophysics.

As pointed out before, the separated radioactive beams may be deflected into the ESR, accumulated and cooled by stochastic precooling and electron fine cooling. Their energy may be adjusted in a large range by acceleration or deceleration in the ESR. These high phase space density radioactive beams may be used either in the ESR as high current circulating beams or they may be slowly extracted from the ESR and transported into the experimental hall. There are first thoughts[21] of using the ESR as a high resolution mass spectrometer by measuring the orbiting frequency of cooled dc beams containing more than 10^5 particles with the help of Schottky scans. A mass resolution A/ΔA ~ 10^5 - 10^6 may be attainable.

Another idea[22] suggests to use cooled circulating radioactive beams of 200 - 400 MeV/u energy for high resolution nuclear reaction spectroscopy on nuclei far off stability. All standard quasi elastic reactions, like inelastic scattering and transfer reactions may be investigated by bombarding atomic beam targets of H, D, T, ^{3}He, ^{4}He, ^{6}Li,^{7}Li etc. with circulating cooled, radioactive beams and detecting the light recoils at angles θ_R around $90°$, which corresponds to forward angles in the c.m.-system. For large mass ratios of projectile and target nuclei and Q-values smaller than $E/A_p \cos^2\theta_R$ the energy of the recoil is given by $E_R = A_R(E/A_p) \bullet \cos^2\theta_R + 2Q$. Thus by measuring the recoil energy E_R and θ_R, the reaction Q value may be determined with a resolution of the order of 70 keV. The resolution is critically dependent on the emittance of the cooled beam, which determines essentially the accuracy of the recoil angle measurement. As targets one can also use polarized atomic beams, thus gaining the full power of high resolution reaction spectroscopy for radioactive nuclei at bombarding energies where reaction models are most reliable to extract transition densities.

4. FIRST EXPERIMENTS IN THE ESR

There was an extensive discussion of some experiments, which might be carried out in an early phase of the ESR operation[23]. One class which does not need electron cooling nor an internal target is concerned with the β-decay of a completely stripped nucleus to its isobar with the decay electron bound in the 1s state. This process which is interesting for the nuclear synthesis[24] and neutrino physics has not been observed before. Because the final state is energetically favoured relative to the initial one by the binding energy of the 1s electron, nuclei which are stable as atoms may decay if the following condition is fulfilled:

$$Q = [m(Z) - m(Z+1)] \bullet c^2 + [B(Z) - B(Z+1)] + |B(1s)|_{Z+1} > 0.$$

In this expression $[m(Z) - m(Z+1)]c^2$ is equal to the mass difference of the neutral atoms, $[B(Z) - B(Z+1)]$ denotes the total binding energy difference of the electrons in the atom Z and $Z+1$, and $|B(1s)|$ $Z+1$ the binding energy of the electron captured in the 1s-state of the nucleus with atomic number $Z+1$.

The simplest example which has been discussed as a test case, is the detection of the $^{163}Dy^{66+}$ ($5/2^-$) decay into $^{163}Ho^{66+}$ ($7/2^-$) with one electron in the 1s-shell. Neutral ^{163}Dy is stable with an abundance of 24.9 %, while neutral ^{163}Ho decays via electron capture decay to ^{163}Dy. For the allowed bound state β-decay of $^{163}Dy^{66+}$ one expects a Q value of about 50 keV, associated with a lifetime of 0.1 y. Thus one expects for 10^{10} stored $^{163}Dy^{66+}$-ions about 400 $^{163}Ho^{66+}$ nuclei be created per second. They stay in the ESR because they have the same magnetic rigidity. Various ^{163}Ho-detection schemes are under discussion. The most simple one would be stripping of $^{163}Ho^{66+}$ to $^{163}Ho^{67+}$ by a foil and magnetic deflection. This method works only if the ratio of ^{163}Dy and ^{163}Ho nuclei which is equal to λdt (λ = decay constant, dt = storage time) becomes smaller than the ratio of the 1s electron stripping cross section to the nuclear charge exchange cross section, because ^{163}Ho can also be produced by nuclear charge exchange when ^{163}Dy passes through a stripper foil. From a rough estimate one expects σ (1s-stripping)/σ (charge exchange) ~ 10^7 which would permit this method to be applied to the case of ^{163}Dy.

Two interesting applications come in mind. One is using the abundance of ^{187}Re as a nuclear thermometer[24,25] the other uses ^{205}Tl as a low energy time integrating neutrino detector[26,27]. As neutral atom ^{187}Re(5/2$^+$) has a half-life of 4 x 10^{10}y because of a second unique forbidden β$^-$-decay to the groundstate of ^{187}Os(1/2$^-$). Completely stripped ^{187}Re^{75+} can decay with a half life of about 10 y with a first forbidden β$^-$-decay to the 3/2$^-$, 10 keV excited state of ^{187}Os^{75+}. In a high temperatur plasma of a star, this fast decay may take place[24,25]. Thus from the relative abundance of ^{187}Re and ^{187}Os one can learn about the temperature of nucleosynthesis processes. As an important input one has to know the lifetime of ^{187}Re^{75+}, which might become measureable with the ESR.

Another interesting example is the study of the β$^-$-decay of completely stripped ^{205}Tl to ^{205}Pb. The decay rate gives a weak interaction matrix-element which is not measurable in any other fashion. This transition probability is basic for the interpretation of experiments using ^{205}Tl as low energy neutrino detector[26,27]. ^{205}Pb(5/2$^-$) decays with a unique forbidden EC transition to ^{205}Tl(1/2$^+$) groundstate. But this transition is not important for ν-induced reactions on ^{205}Tl, because of its weakness. The main transition is expected to a 1/2$^-$, 2 keV excited state of ^{205}Pb. For the interpretation of results with ^{205}Tl as neutrino detector, one needs this transition matrix element. It can be gained by studying the decay of completely stripped ^{205}Tl^{81+}, which is expected to decay[28] by a bound state β-transition to the 1/2$^-$, 2 keV excited state of ^{205}Pb^{81+}.

For the use of the ESR as a cooler ring it is most important to study the effectiveness of electron cooling. One method, which is also of general interest is the measurement of radiative capture (REC) of the cooling electrons by the ions[29]. For completely stripped ions REC is the only relevant recombination, if one ignores cooperative effects of the cooling electrons in the potential of the ions. The energy of the REC-photons for transitions to boundstates with energies $E_{n,l,j}$ is given by

$$= m_o c^2 \{\gamma_e\text{-}1 + E_{n,l,j}/m_o c^2\}$$
$$\text{with } \gamma_e = E_e/m_o c^2$$

The cross section to a state with quantum number n is given by the relation ·

$$\sigma_n = 2 \times 10^{-22} \frac{(Z^2 \cdot Ry)^2}{n[Z^2 \, Ry + n^2 \, Te] \, Te} \; cm^2$$

with Te denoting the electron energy in the cm system. A measurement of the REC photon energy distribution as well as σ_n can be used as diagnostic tool for the cooling device. Of fundamental importance is the possibility to measure E_n, the binding energy of an electron with good precision. It is also noteworthy that the REC of the cooling device represents an intense photon source with variable energy up to about 400 keV.

An interesting problem is the limit of electron cooling. Related to this is an expected phase transition of the statistical particle distribution into an ordered state of a Wigner-Coulomb-lattice[30]. Monte Carlo-calculations[31] in a large system predict such a transition, if the order parameter $\Gamma = (Z^2 e^2 / a)/kT$ exceeds 160. Recent model calculations[32] for a straight beam held together by harmonic focusing forces showed for $\Gamma = 110$ the ordering given in Fig. 12.

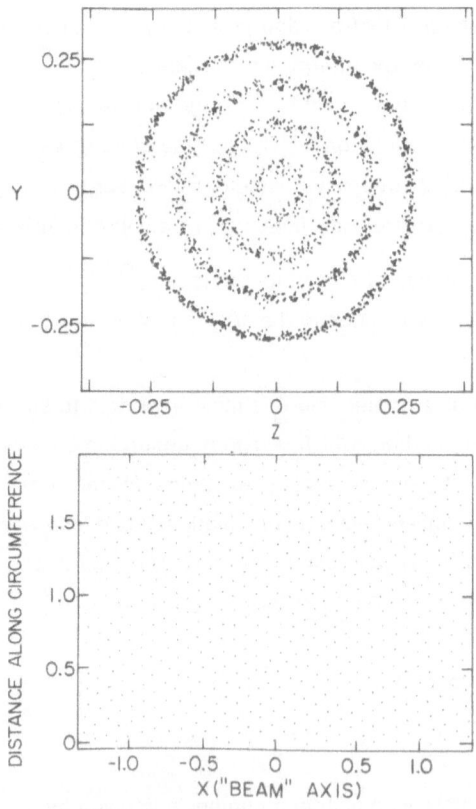

Fig. 12 Upper part: Projection of 2000 particles in a molecular-dynamics calculation onto the plane perpendicular to the beam (x-axis) for $\Gamma = 180$. Lower part: distribution of particles in the outer shell with the shell unfolded into a plane. All shells but the innermost show a similar pattern.

There is an interesting proposal for a precision measurement of the 1 $s_{1/2}$ - hyperfine splitting in hydrogen like heavy ions by laser induced M1 transitions[33]. Note that a large range of nuclei can be covered by making use of the large Doppler shift tuning possibility given by the relation $\lambda = \lambda_0(1 \pm \beta)/\sqrt{1-\beta^2}$, where $\beta_{max} \sim 0.7$.

Such measurements may also be used to determine the anomalous part of the magnetic moment of a strongly bound electron. A more direct way to measure the anomalous magnetic moment of an electron in a strong Coulomb field would be to induce magnetic transitions between the states of a bound electron exposed to an external magnetic field. Polarized atoms or circularly polarized microwaves are needed for such experiments.

Precision X-ray spectroscopy[34] on one, two and three electron systems up to uranium, will profit from the availability of cooled and slowed down beams of heavy few electron atoms. Special high resolution X-ray diffraction spectrometers with arrangements to minimize the Doppler shift are needed. Furthermore methods like measuring the shifts using different beam velocities may help to eliminate systematic errors. Of special interest for studying QED corrections, would be the investigation of (2s - 1s), M1-transitions in heavy systems for reduction of the natural line width, which otherwise limits the resolution.

At the end we just would like to mention experiments to study resonant dielectronic recombination for determination of all binding energies in ions with one or more electrons left. An electron target with variable energy is needed for such experiments. With polarized electron targets, polarized excited atomic states may be prepared.

5. MEDIUM ENERGY NUCLEUS-NUCLEUS-COLLISIONS

Following the exploring work on the properties of heated, compressed and excited nuclear fireballs produced by medium energy nucleus-nucleus collisions at the Bevalac[35], second generation experiments are in discussion with the aim to study also rare processes like the production of leptons, γ-rays, strange hadrons and antiparticles.

Fig. 2 shows the layout of the beam transport system, designed to serve the area between the two rings and the new target hall. For the latter possible locations of six larger experimental facilities (6-11) are indicated. Five of them can be served with beams from SIS and cooled beams from the ESR (7-11). Position 6 has SIS-beam option only. Position 4 located close to SIS will be provided with high intensity beams bunched by the synchrotron for the creation and study of high density plasma. Beams of radioactive nuclei will be available after separation (1) in the area 2. The ESR will be equipped with a gas jet target located in the experimental area 5.

The study of central collisions will be persued at SIS to learn more about the properties of nuclear fireballs produced by nucleus-nucleus collisions with laboratory energies up to 1-2 GeV/u. Of particular interest is the question of nuclear compression in context with the study of the equation of state of nuclear matter and the possibility . to produce Δ-rich nuclear matter[35]. For a relevant experimental program the construction of a second generation electronic 4π-detector is under consideration[36]. It will be designed to cope with high multiplicity events and to detect high energy γ-rays. In one of the detector versions under discussion (Fig. 13), the 4π-detector is divided into two parts, a forward cone with an opening angle of 45°, relative to the beam axis and an inner sector radius of 2 m, completed by a ball in the backward direction with 25 cm inner radius. The forward sector is composed of 3645 detector modules. Each detector is made out of a sandwich of 5 mm thick fast plastic (NE 102), ~ 50 mm slow plastic (NE 115) and 350 mm lead glass mounted on a common photomultiplier tube. The plastic ball is used for charged particle identification by time of flight and energy loss measurement; it stops particles up to an energy of 80 MeV/u. The lead glass scintillators are used for π^0, η, and high energy γ-spectroscopy. Their thickness corresponds to about 14 radiation length and their entrance area (~ 3.5 x 3.5 cm^2) to the Moliere radius of the shower. The backward ball consists of about 590 detector elements with 10 cm^2 entrance surface and 7° opening angle of the acceptance cone. As detector materials CsI or the fast but expensive BaF$_2$ are in discussion.

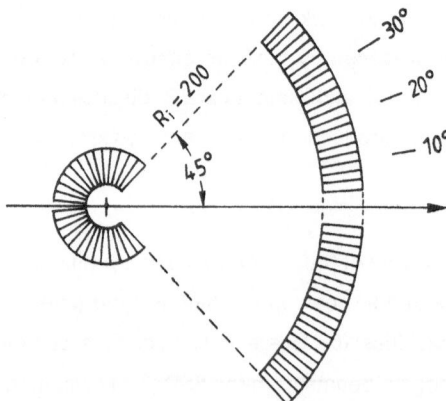

Fig. 13 Concept of a new 4π-detector consisting of 3645 fast/slow plastic and Pb-glass modules in the forward cone and a ball of 590 CsI or BaF$_2$ detectors covering the remaining solid angle.

The basic concept of the new 4π-detector is its ability to study neutral decay products of the fireball, like γ-rays, π⁰ and η-decays. The latter allows the investigation of strangeness production in excited hadronic matter. High energy γ-spectroscopy permits to study the slowing down process in nucleus-nucleus collisions. It may serve to determine the energy density and temperature achieved in the fireball and also for the study of the γ-decay of baryonic resonances. One has to solve the experimental problem to separate out the γ-background from π⁰ decay by kinematic reconstruction.

With the highest SIS energies it will be feasible to study strangeness production and possibly creation of antiparticles. Especially subthreshold production of K^+ mesons seems to be an interesting process to study systematically in order to learn more about the property of the fireballs. For such experiments we discuss the construction of a superconducting orange spectrograph[37] as sketched in Fig. 14. Such an instrument should provide for example high efficiency in detection of K^+ mesons from

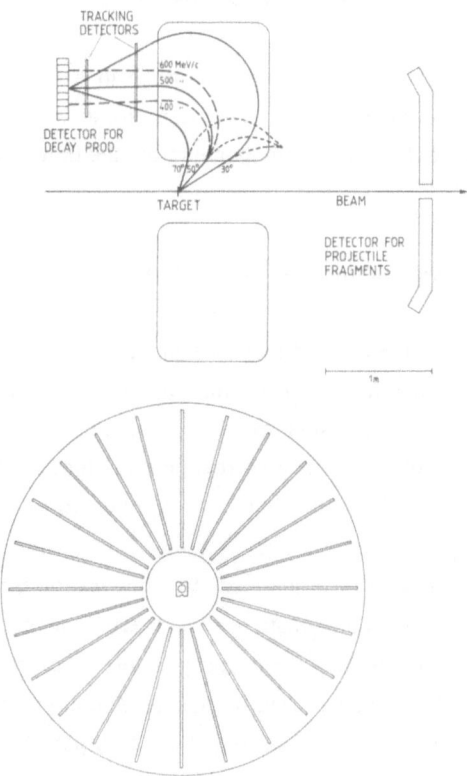

Fig. 14   Concept of a superconducting orange type spectrometer for $K^\pm$-spectroscopy.

subthreshold nucleus-nucleus collisions. The K^+ may be identified by momentum, time of flight and energy loss measurement followed by the detection of the fast decay products in Cerenkov-detectors.

For more peripheral collisions non equilibrium processes which lead to fragmentation[38] or fast fission[39] may be favourably studied using inverse reaction kinematics, i.e. by bombarding lighter targets with heavier projectiles. Such experiments are needed for a better understanding of various break up mechanisms in fast nucleus-nucleus collisions (multifragmentation). There are interesting suggestions[40] to make use of the extremely large transverse electric fields of relativistic heavy ions for multiple excitation of the giant resonances in nuclei. Such a process may possibly lead to a cold separation of neutrons and protons with the result that fragments of unusual N/Z ratios may be produced and studied. Discussions are in progress to construct a large solid angle magnetic spectrometer for investigation of these fragmentation processes.

Experiments have been suggested[41] to study Δ-excitation in nuclei using "recoil free" quasielastic transfer reactions. Such experiments, which call for a high resolution spectrometer, seem to be possible by a modification of the design of the fragmentation mass spectrometer placed between SIS and ESR.

Experiments with slowed down highly stripped ions for studying atomic processes and radioactive beams for nuclear structure physics may be performed on position 11 (Fig. 2) using cooled low energy beams from the ESR or SIS.

6. EXPERIMENTS FOR PRODUCTION OF HIGH POWER DENSITY PLASMAS

Computer simulations showed[42] that by bombarding cylindrical Au-targets with well bunched, high intensity heavy ion beams from SIS, one can produce cylindrical shock waves (sidewise) and conical shock compression (in the front). In the latter one expects pressures of 2×10^8 bars which correspond to 4 times solid densities. Thus it seems to be possible to produce and study matter under extreme conditions.

Heavy ion beams may be favorably used to pump lasers. Since heavy ion pumping of infrared lasers has been demonstrated[43] with 100 W quasi continuous heavy ion beams, lasers as short as 100 nm should be possible to be pumped with a power density of about 1 MW/cm^2 using UNILAC beams. A crude extrapolation (pumping power $\sim \lambda^{-3}$) indicates that soft x-ray lasers (20 nm) may possibly be pumped with SIS beams[44].

ACKNOWLEDGEMENT

I would like to thank all the many discussion partners which helped us to work out our plans in a very short time. We anticipate that the new facilities will be useable for experiments in 1989. I should like to use this opportunity to invite the international nuclear physics community to participate in the use of SIS/ESR beams for experiments.

REFERENCES

1.   Die Ausbaupläne der GSI, March 1984
2.   SIS - Ein Beschleuniger für schwere Ionen hoher Energie, GSI-Bericht 82-2
3.   K. Blasche, D. Böhne, B. Franzke, H. Prange, 1985 Particle Acc. Conf. 1985, Vancouver, IEEE Trans. NS32, (1985)
4.   B. Franzke et al., Zwischenbericht zur Planung des Experimentier-Speicherrings (ESR) der GSI, GSI-SIS-INT/84-5 August 1984
5.   R. Keller et al., Proc. Int. Ion Engineering Congress, Kyoto (1983)
6.   R.W. Müller, V. Kopf, J. Bolle, S. Arai, P. Spädtke, Proc. of the 1984 Linear Accelerator Conf., Seeheim (1984) p. 77
7.   R.W. Müller, GSI Scientific Report 1985, p. 375
8.   H. Gould et al., Phys. Rev. Lett. 52, 180 (1984)
9.   B. Franzke, Ch. Schmelzer, GSI-Scientific Report 1984, p.341
10. I. Hofmann, GSI Scientific Report, 1985, p. 387
11. Workshop on the Physics with Heavy Ion Cooler Rings, Heidelberg, May 1984.
12. Arbeitstreffen über Experimente am geplanten Experimentier- Speicherring ESR der GSI, Schloß Rauischholzhausen 1984.
13. Workshop on Experiments at External Beam Lines of SIS and ESR, Darmstadt, July 1985.
14. I. Tanihata et al. Phys. Lett., B. 160, 380 (1985)
15. W. Greiner et al. Phys.Rev., Lett. 35, 152 (1975).
16. A. S. Goldhaber and H. H. Heckman, Am. Rev. Nucl. Part. Sci. 28, 161 (1978)
17. G. D. Westfall et al. Phys. Rev. Lett. 43, 1859 (1979)
18. M. Langevin et al. Phys. Lett. 150B, 71 (1985)
19. J. P. Dufour, 7th High Energy Heavy Ion Study GSI, Darmstadt (1984)
     C. Mueller, Proc. of the Int. Conf. on Atomic Masses and Fundamental Constants, AMCO-7, Seeheim (1984), 696
20. H. Geissel, G. Münzenberg et al.: ref. 12
21. B. Franzke, Proceedings: "10 Years of Uranium Beam".
22. W. Wagner in ref. 12, p. 227
23. see F. Bosch, Proc. Int. Nuclear Physics Conference, Harrogate 1986
24. K. Takahashi and K. Yokoi, Nucl. Phys. A404 578 (1983)
25. J.M. Luck et al., Nature 283, 256 (1980)
26. M.S. Friedmann et al., Science 193, 1117 (1976)
27. see: Workshop on the Feasibility of the Solar Neutrino Detection with ^{205}Pb by Geochemical and Accelerator Mass Spectroscopical Measurements. Ed. E. Nolte, GSI 86-0 Report, April 1986
28. M.S. Friedmann has estimated the decay constant of ^{205}Tl^{81+} to 5.67×10^{-5} per h assuming a decay energy of 40 keV.
29. D. Liesen in Ref. 12, p. 20
30. J.P. Schiffer and P. Kienle, Z. f. Physik, A321, 181 (1985)
31. J.P. Hansen, Phys. Rev. A11, 1025 (1975)
32. A. Rahman, J.P. Schiffer, Phys. Rev. Letters 57, 1133, (1986)
33. G. Huber and T. Kühl, ref. 12, p. 149
34. R.D.Deslattes, see ref. 11
35. R. Stock, et al., Phys. Rev. Lett. 49, 1236 (1982)
     S. Nagamiya and M. Gyulassy, "High energy nuclear collisions", in Advances in Nuclear Physics 13, (Plenum Press, New York, 1984).
36. see A. Gobbi et al., ref. 13, p. 5
37. see E. Grosse, ref. 13, p. 135
38. W.F.J. Müller, ref. 13, p. 57
39. P. David, ref. 13, p. 32
40. H. Emling, ref. 13, p. 83
41. W. v. Oertzen, ref. 13, p. 96
42. R. Arnold et al., ref. 13, p. 200
43. A. Ulrich et al., Appl Phys. Lett. 42, 782 (1983)
44. A. Ulrich, ref. 13, p. 217.

THE NUCLEAR STRUCTURE FACILITY AT DARESBURY

P.J. Twin

SERC Daresbury Laboratory
Daresbury
Warrington WA4 4AD, U.K.

ABSTRACT

The present facilities for nuclear physics at Daresbury are described
and the current development plans are outlined.

INTRODUCTION

The Daresbury Laboratory houses the major research facility for
nuclear structure physics research in the United Kingdom. It consists of
a 20 MV electrostatic tandem accelerator and an excellent range of experimental equipment. The success of the facility since it became operational
three and a half years ago has been in a large part due to the simultan-
eous provision of both a reliable large tandem and the investment in
experimental instrumentation. Work has commenced on extending the
accelerator facilities by the addition of superconducting linac modules
and this will be followed by further capital investment on experimental
equipment.

THE ACCELERATOR

The Tandem Van de Graaff

The tandem Van de Graaff accelerator was designed and built in the
United Kingdom. It incorporates a hard vacuum accelerator tube, a ladder-
tron for the charging system, and focusing elements inside the pressure
vessel. These include a charge state separator following the first
stripper (gas or foil) in the centre terminal, thus ensuring that only the
one selected charge state is transmitted through the second part of the
accelerator. The tandem is currently operating up to 20 MV and it has
already accelerated fifty-seven different beam species for experiments.

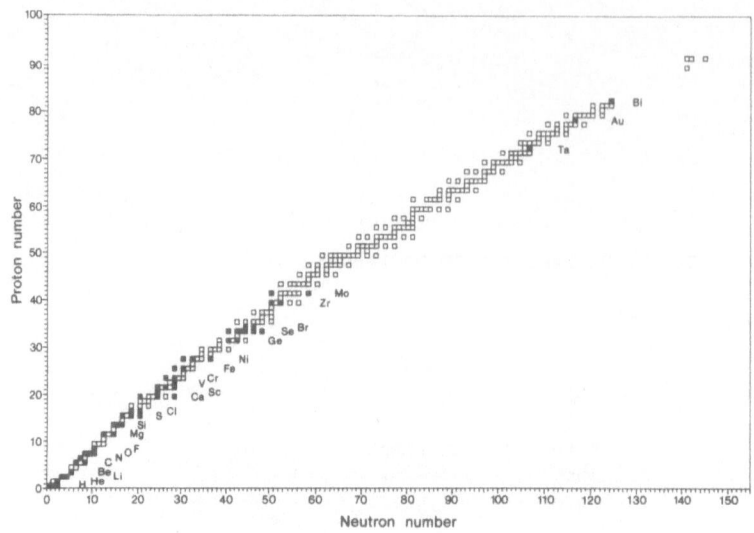

Fig. 1.   The table of isotopes. Dark squares indicate isotopes
          accelerated by the tandem at Daresbury.

These are illustrated in Fig. 1 and include the low natural abundance
isotopes ^{36}S, ^{48}Ca and ^{74}Se. Their provision is made possible by the
sputter source which efficiently operates with small pellets of separated
isotopes. A second sputter source is dedicated to radioactive beams and
it has been used for tritium and ^{14}C. A beam buncher is available which
compresses 60% of the D.C. beam into 1-2 ns pulses. Commissioning is
under way of the associated beam chopper which will remove the remaining
40% of the beam lying between the bunches. A polarised ion-source for
lithium and sodium ions is nearing completion.

The Addition of Linac Modules

        The present range of beam energies as a function of mass is given in
Fig. 2. Coulomb barrier phenomena represent the lower limit of 'interacting' energies and these effects commence at about 4 MeV per nucleon.
Thus the useful beams are now limited to masses up to A = 80 or 90. An
efficient and effective means of increasing this range is to add some
linear accelerating modules. Superconducting split ring cavities had been
purchased for the 10 MV tandem at Oxford University and following the
financial decision to withdraw support for this facility the modules are
being transferred to Daresbury. They consist of a sub-nanosecond super-
buncher and three modules with a total of nine resonators equivalent to an

energy gain of about 6 MeV per charge. A new 300 W liquid helium refrigerator system is being installed (a major part of the cost of installing the modules) and this will have sufficient capacity to handle a further twelve resonators which it is planned to install at a later date and thus provide a total energy gain of 14 MeV per charge.

A number of laboratories are installing linac modules and there is a great variation amongst the voltages of tandems and the number of modules. The six-module system at Daresbury is small compared with other facilities, especially as it is planned to obtain beams of mass 200 at 5 MeV per

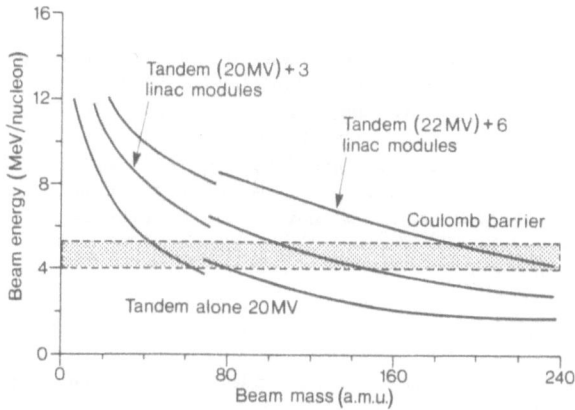

Fig. 2.   A graph of the maximum beam energy, as a function of mass, currently available from the tandem operating at 20 MV.   Curves are also shown of the energy gain with the three ex-Oxford linac modules (6 MeV per charge) and with a further three modules (14 MeV per charge).

nucleon.   The different modes of operation of a tandem and linac are illustrated in Fig. 3.   The charge states and energy gains are given for a beam of around mass 150.   In Fig. 3(a) a foil stripper is assumed to be used in the tandem.   However, this is unrealistic for masses as high at 150 due to the lifetime of the foil in a beam of 100 nA being of the order of minutes.   The foil lifetimes shorten as the accelerator voltage is reduced and useful foil lifetimes restrict Daresbury to masses below 80-90 and lower voltage tandems to smaller masses.   The other problem with this mode of operation is that the velocity of the very heavy ions becomes too

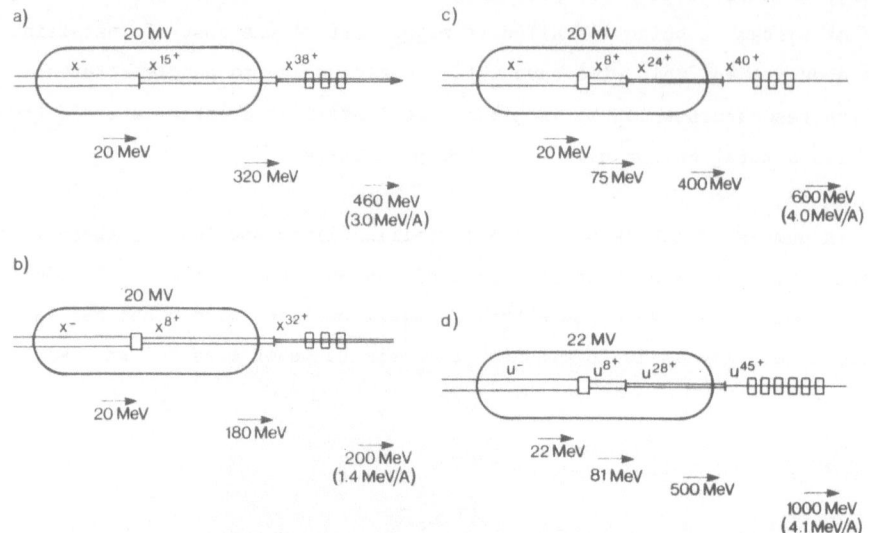

Fig. 3.   The energy gain of a tandem and β = 0.1 linac modules for beams
            of around mass 150 with various modes of stripper operation.
            (a) Foil stripper in the centre terminal. (b) Gas stripper in
            the centre terminal. (c) Gas stripper in the centre terminal
            plus a second foil stripper at two-thirds of the voltage in the
            high energy beam line. (d) Same stripper conditions as (c) with
            additional linac modules and a beam of uranium.

low for efficient acceleration in a standard β = 0.1 resonator. Thus less
efficient low-β resonators have to be included at the beginning of the
linac.

    The foil lifetime problem can be overcome by using a gas stripper in
the tandem (Fig. 3(b)). However, the charge state is lowered considerably, reducing the voltage gain in both the tandem and the linac. The
velocities for very heavy ions are lowered further and low-β modules are a
necessity. Such a mode of operation is making inefficient use of a large
tandem.

    A third mode of operation is to use a gas stripper in the centre
terminal of the tandem and introduce a second foil stripper part ·way down
the high energy accelerating section. At Daresbury this stripper is at
the intershield potential which is two-thirds of the full voltage. The
increased voltage gain in the tandem ensures that the ion velocity is well

matched to β = 0.1 modules and also that the optimum charge state produced at the entrance foil of the linac is over 40$^+$, thus greatly increasing the voltage gain in the linac. However, there is loss of beam intensity due to the addition of a third stripping stage. This should not be a major problem due to the new high current sputter sources and the ability at Daresbury to select only one charge state at the centre terminal thus reducing the loading on the high energy accelerator tube.

One possible scenario for Daresbury is to operate with slightly increased terminal voltage (22 MV) and a 14 MV gain in linac modules. Figure 3(d) shows that this produces 5 MeV per nucleon beams up to mass 200.

The layout of the modules is shown in Fig. 4. They will be situated in part of the present building which was used as a construction area for the original tandem. The linac installation will have minimal impact on the beam provision for the current experimental areas. Several new beam stations for charged particle spectroscopy and γ-ray arrays will be

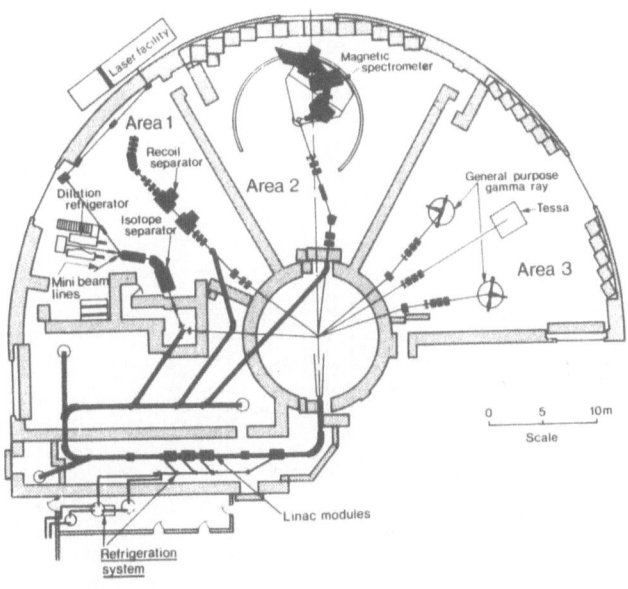

Fig. 4.   The proposed layout of the experimental areas and the linac modules.

constructed. The beam will also be transported to the isotope separator, recoil separator and magnetic spectrometer. The first stage of the development, the installation of the three Oxford modules, should be operational by the end of 1988.

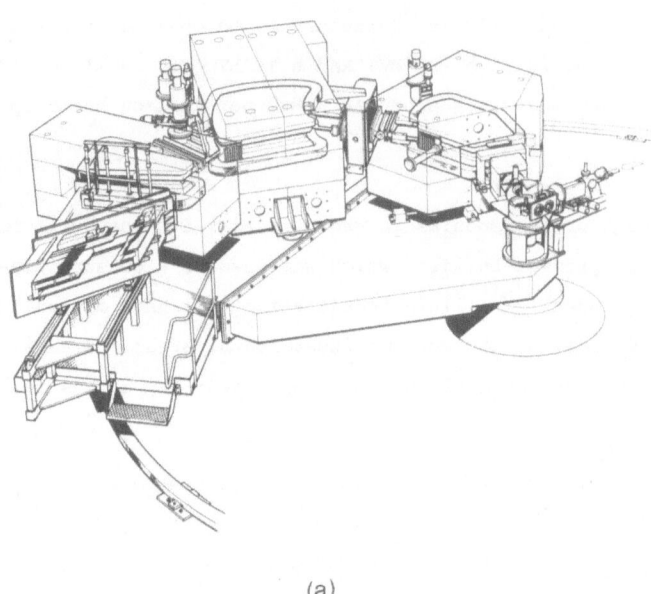

(a)

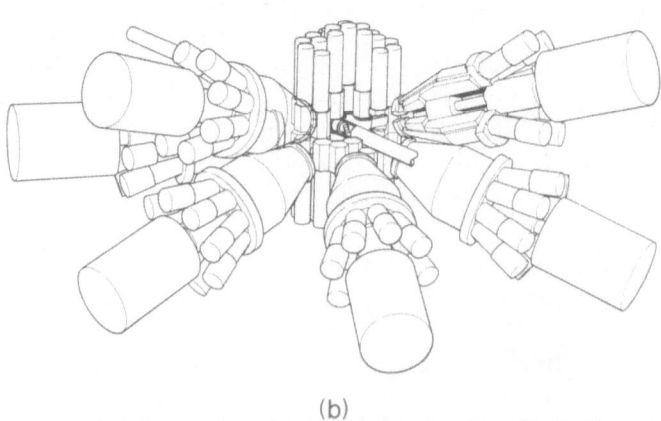

(b)

Fig. 5. Experimental equipment on the tandem at Daresbury, (a) magnetic spectrometer, and (b) TESSA 3 gamma-ray array.

EXPERIMENTAL EQUIPMENT

There is an extensive repertoire of advanced instrumentation for experimental exploitation of the Daresbury accelerator. Some of these are illustrated in Figs. 5 and 6.

Charged Particle Instrumentation

A Magnetic Spectrometer (QMG2) (Fig. 5(a)) has been operational for a number of years. It has been used, in particular, for high resolution transfer reaction work with masses up to 40. The present emphasis is on developing large charged particle arrays with multi-detector telescopes. These will be used for light-ion transfer studies, break-up phenomena, resonant particle spectroscopy, etc.

Multi-detector Gamma Arrays

High resolution germanium detector arrays with Compton-suppressed spectrometers have been pioneered at Daresbury. The TESSA 2 (Total Energy Suppression Spectrometer Array) configuration was operational from 1982-1985. It consisted of six suppressed germanium detectors and a 50-element inner ball (or calorimeter). At the end of 1985 TESSA 3 (Fig. 5(b)) came on-stream with its twelve suppressed germanium detectors and the same inner ball of bismuth germanate to measure the general

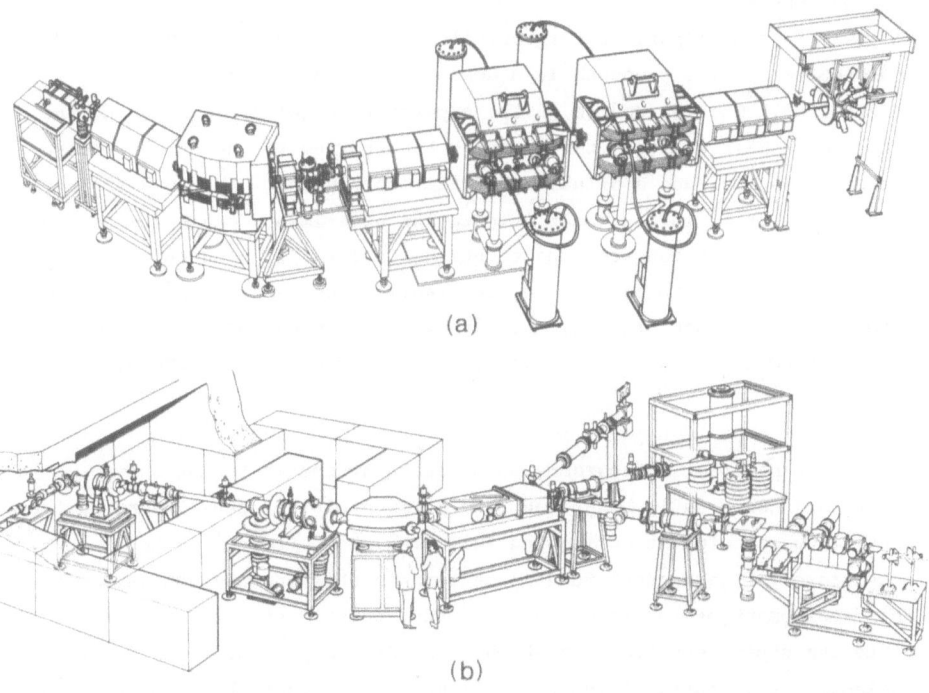

(a)

(b)

Fig. 6. Experimental equipment on the tandem at Daresbury, (a) recoil mass separator, and (b) isotope separator.

properties of the γ-ray shower. TESSA 3 can now operate with sixteen germanium spectrometers and these spectrometers will be reconfigured in early 1987 together with other spectrometers from collaborating laboratories within Europe to set up a 30-detector array based on a 32-faced "soccer" ball structure. The major research topic of the TESSA family of arrays has been high spin states in nuclei.

Recoil Mass Separator

In 1985 the recoil mass separator became operational (Fig. 6(a)). It consists essentially of two velocity filter elements (crossed magnetic and electric fields), a dipole magnet for mass (or A/q) separation, and various focusing quadrupole elements. It was designed for coincidence measurements between the recoiling ions and γ-rays emitted at the target position. Initially the γ-ray array was only six suppressed germanium detectors but the full 20-spectrometer POLYTESSA array is now operational. In 1987 the TESSA 3 array with the BGO ball and 10 suppressed germanium detectors will be available. Initially the major research topic on the recoil mass separator has been nuclei far from stability with some work on high spin states, sub-barrier fusion and heavy-ion transfer below the Coulomb barrier.

Isotope Separator

The isotope separator (Fig. 6(b)) is an instrument which separates radioactive species produced in a nuclear reaction by the tandem accelerator. The ion source of the separator is selective in Z and the ion beam is mass analysed before being transported to one of the instruments. These include a dilution refrigerator in which the ions are cooled down to 15 mK so that nuclear orientation experiments can be performed. Another instrument concerns the technique of laser fluorescence in a collinear geometry to measure nuclear radii. The latest experiments have involved coincidence measurements with the scattered ions or atoms improving the sensitivity of the method by several orders of magnitude.

SUMMARY

The experimental programme on nuclear physics at the Daresbury Laboratory has greatly benefited from a large capital expenditure on the experimental equipment during and following the building of the accelerator. The present funding pattern is shifting the emphasis back to the development of the accelerator by the addition of linac modules to raise the upper mass limit at which beams can be provided at 5 MeV per nucleon. It is anticipated that this must be followed by a recapitalisation of instrumentation for the experimental programme.

THE NEW UPPSALA ACCELERATOR FACILITIES AND THEIR EXPERIMENTAL PROGRAMS

Lars Westerberg

The Svedberg Laboratory
Box 533
S-751 21 Uppsala, Sweden

INTRODUCTION

A new laboratory for the three accelerators in Uppsala, the EN tandem, the Gustaf Werner synchrocyclotron and the CELSIUS cooler ring, has just been formed by merging the Gustaf Werner Institute and the Tandem Accelerator Laboratory. The laboratory is called The Svedberg Laboratory after The Svedberg, founder and first director of the Gustaf Werner Institute and Nobel prize winner of chemistry in 1926. The laboratory operates the three accelerators and has also some research positions. At the same time, the new Department of Radiation Sciences was established for teaching and research in elementary particle physics, nuclear physics, ion physics and physical biology. In the laboratory overview of fig. 1, the Gustaf Werner synchrocyclotron and CELSIUS can be seen together with their experimental areas. The synchrocyclotron will serve as injector machine of CELSIUS, but is has also a wide experimental program of its own.

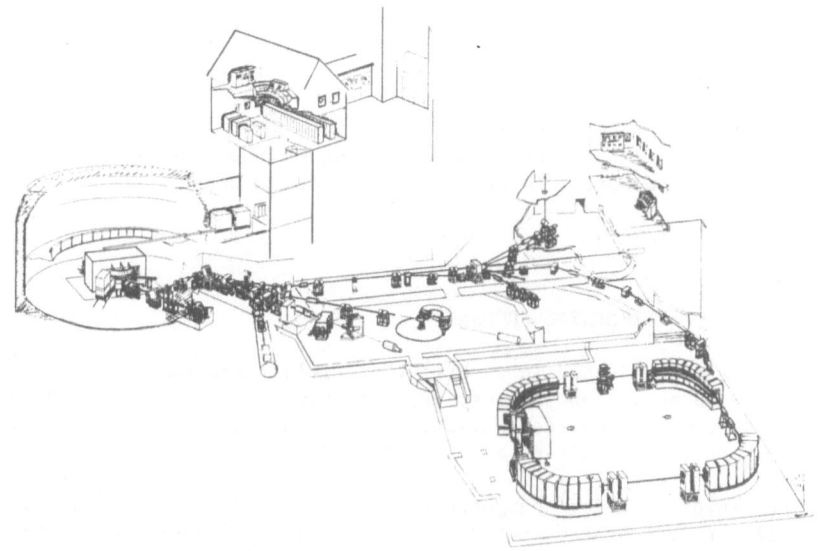

Fig. 1. Overview of the synchrocyclotron and CELSIUS accelerator complex.

THE GUSTAF WERNER SYNCHROCYCLOTRON

The Gustaf Werner synchrocyclotron first operated from 1951 until 1977, when it was shut down to be rebuilt from flat-pole geometry into an iso-chronous sector-focusing synchrocyclotron having 3 sectors. From the original machine now only the 600 ton iron yoke remains. The pole diameter is 2.8 m, and the ions are extracted at a radius of 1.2 m by a combination of an elec-trostatic deflector and an electromagnetic channel. The maximum average field is 1.8 T. Frequency modulation is used for protons above 110 MeV and ^{3}He above 250 MeV.

The maximum energy/nucleon is given by $200 \cdot (Q/A)^2$, where Q and A are the charge state and mass of the ion. The ion source now used is of the standard internal PIG type. It will give the following maximum energies: 200 MeV p, 50 MeV/A α-particles and 35 MeV/A ^{12}C. An external ECR heavy-ion source is under construction and is planned to be ready in 1988. It will increase the maximum energies for light heavy ions to the maximal value of 50 MeV/nucleon as can be seen in fig. 2. An external polarized light-ion source has now also been funded. The minimum energy is, during the first operation, 10 MeV/nucleon and later, when also using the 4th harmonic of the rf frequency, it will be possible to go down to 2.7 MeV/nucleon. The first internal beam, 45 MeV α-particles, was achieved on Nov. 6 1986 and the first external beams for physics experiments are scheduled for spring 1987.

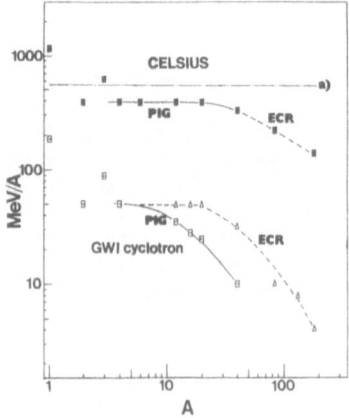

Fig. 2. Maximum energies per nucleon as a function of mass number for the synchrocyclotron and CELSIUS ring using PIG and ECR ion sources. The line a) denotes the upper limit of electron cooling.

SYNCHROCYCLOTRON EXPERIMENTAL PROGRAM

Experiments are planned in a variety of fields: neutron-induced reac-tions, light- and heavy-ion physics, physical biology and medicine. We will here concentrate on the nuclear physics program and give a short description of some experimental facilities and the planned research.

The new (n,p) facility[1] has been built up by groups from Uppsala and Lund. It will produce neutron beams in the energy range 50-200 MeV with 1 MeV energy resolution using (p,n) reactions on a ^{7}Li target. In order to get a clean neutron beam, the charged particles from the production target are deflected by 45° in a system of magnets and is dumped in a well shielded beam dump far away from the (n,p) target. The research program involves studies of isovector monopole and Gamov-Teller giant resonances.

An Uppsala-Osaka collaboration has developed a new type of pair spectro-meter, PACMAN[2], for studies of radiative capture of protons on nuclei. This reaction is of interest for understanding the mechanism for pion production in proton-nucleus interactions leading to discrete states in the final nuc-leus. The PACMAN spectrometer has a solid angle of 70 msr, a momentum accep-tance $\Delta p/p$ of 50 % and an energy resolution for a 200 MeV pair event of 0.6 MeV.

A group from AFI, Stockholm is planning g-factor measurements of high-spin states in shell-model nuclei using TDPAD technique. Special equipment for downscaling of the beam pulse frequency by deflection of beam pulses will be installed in the long corridor before entering the gamma cave, in which the experiments will be done. Also in the gamma cave experiments in nuclear spectroscopy, such as Coulomb excitation are prepared. The NORDBALL detector system[3], a 32-fold geometry gamma-, particle- and neutron detector system owned by research groups in the Nordic countries, will be used in these and other heavy-ion experiments.

Heavy-ion reaction mechanism studies are planned by a Gothenburg-Stockholm-Uppsala collaboration specially emphasizing fast emission of neu-trons and light ions in 10 to 35 MeV/A heavy-ion reactions on Sn nuclei[4]. A collaboration Bergen-Copenhagen-Gothenburg-Lund-Uppsala is setting up for inclusive and large-angle correlation experiments on light-particle produc-tion in heavy-ion reactions and tests of detectors for heavy-ion experiments in CELSIUS.

THE CELSIUS STORAGE AND COOLER RING

CELSIUS stands for Cooling with ELectrons and Storing of Ions from the Uppsala Synchrocyclotron. The project was funded in 1983. Forty magnets from the ICE (Initial Cooling Experiment)[5] were bought from CERN and trans-ported to Uppsala. A new underground 28 m x 36 m experimental hall was built for the ring. The project is a collaboration between The Svedberg Laboratory, Uppsala, Royal Institute of Technology, Stockholm (development of the elec-tron cooler), NFL, Studsvik (internal target developments) and CERN (specialist technical support).

Lattice

A more detailed layout of the CELSIUS ring is given in fig. 3. The 40 bending magnets, each weighing 12 tons, are arranged in four quadrants with 7.00 m bending radius. The four straight sections will mainly be used for: injection (extraction), diagnostics, electron cooling plus rf, and internal-target experiments, respectively. There are quadrupole and sextupole magnets in the diagnostic and target straight sections.

The magnetic field in the ring can be varied between 0.05 and 0.89 T. This gives a maximum rigidity of 6.25 Tm (at 2500 A) and a maximum momentum/Z of 1.875 GeV/c. The working point is Q_X=1.68 (just above the resonance Q_X= 1.667 for a future third integer resonance extraction) and Q_Z=1.9.

Injection, acceleration and possible extraction

There are two ways for injection of ions from the synchrocyclotron into CELSIUS, multiturn and stripping injection. The first mode involves an elec-tromagnetic septum placed in the middle of the injection straight section, an electrostatic septum 2 m downstream and the two bumper magnets in the diagnostic and target straight sections. Typically 10 turns (revolutions in CELSIUS) can be injected in this mode. In the stripping injection mode a beryllium stripper foil is placed in the first bending magnet after the injection straight section. The electrostatic septum is withdrawn and not used. It is possible to inject the order of 100 turns in this mode.

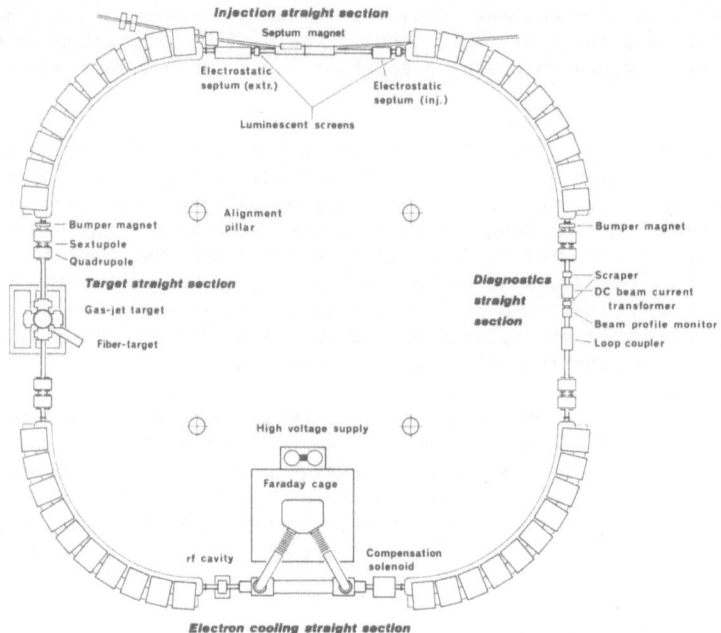

Fig. 3. Lattice layout of the CELSIUS ring

Place has been reserved in the injection straight section for a future extraction by adding a second electrostatic septum between the end of quadrant 4 and the electromagnetic septum, which can also be used for extraction. As can be seen in the layout drawing of the laboratory (fig. 1), there is an area outside the ring in the bottom right corner for an extracted beam experiment. Extraction could become interesting for heavier ions which are far from completely stripped and therefore would get too short a life-time in the ring due to charge exchange in the internal target. The maximum energies in CELSIUS are given in fig. 2. The bending magnets are not lami-nated. Therefore a special scheme has been developed to compensate for Eddy-current effects. This gives an acceleration time to full energy of 10 s, but since storage times around a minute are expected the duty factor will be high.

Electron cooler

Electron cooling was first developed by Budker[6] in Novosibirsk in 1966. It is now of current interest in many laboratories over the world since it is a promising way to shrink the phase space of storage ring beams. Our electron cooler has an interaction length of 2.5 m, See fig. 4. It will pro-duce an electron beam with a maximum current of 3 A, a diameter of 2 cm and and energy variable between 10 and 300 kV. This is equivalent to a maximum energy of cooled ions of 550 MeV/nucleon. It is possible for the future to add second acceleration tubes in the electron gun and collector to increase the cooler voltage to 600 kV in order to be able to cool also the highest energy p and ^{3}He beams.

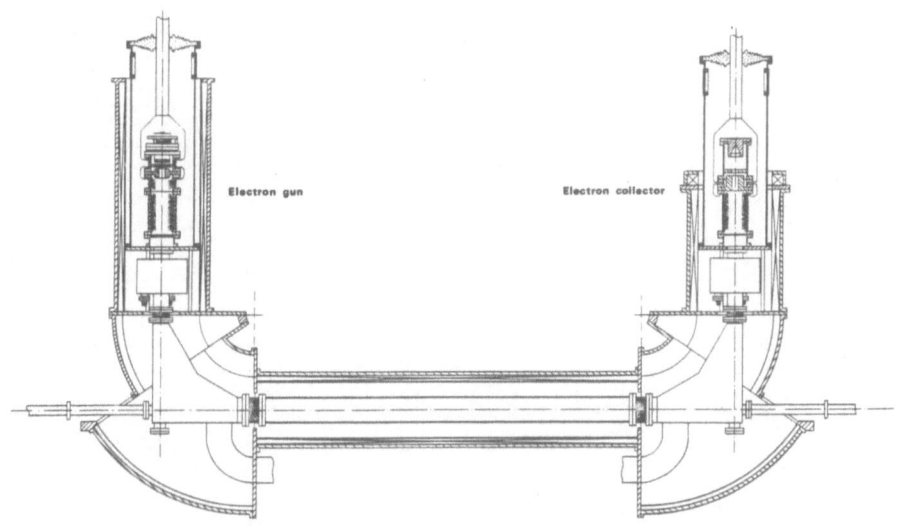

Fig. 4. Cross-sectional view of the electron cooler

An example of a calculation of the time evolution of horisontal and vertical positions in the electron cooler of a sample of 100 fully stripped Ar ions, out of a total of 10^{10} ions in the ring, using the SPEC program of A. Wolf[7], is illustrated in fig. 5.

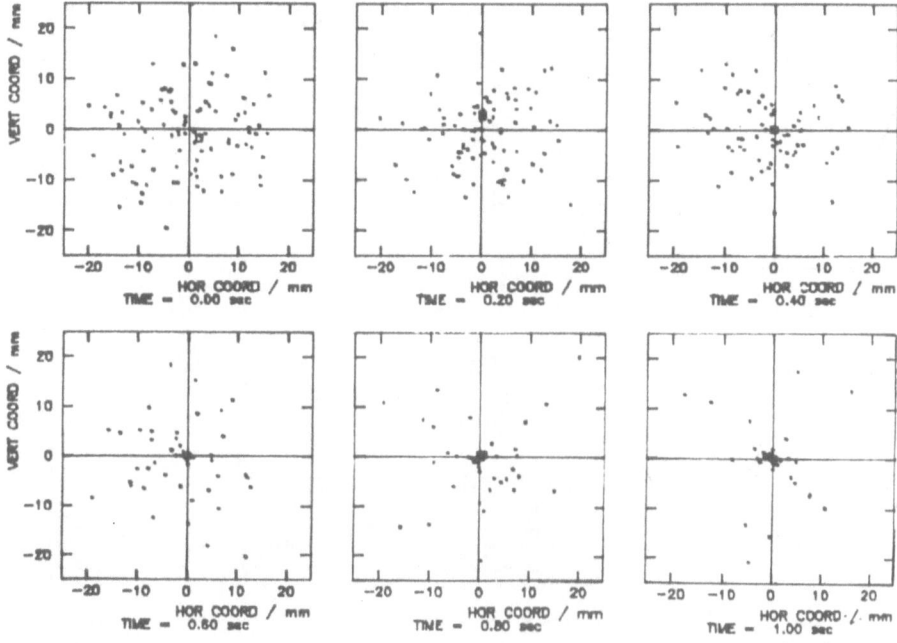

Fig. 5. Calculation of electron cooling of Ar^{18+} ions at 46.8 MeV/A in CELSIUS.

Internal targets

A supersonic gas-jet cluster-beam target has been built for CELSIUS[8] and is now being tested in Studsvik. A side view of the target is given in

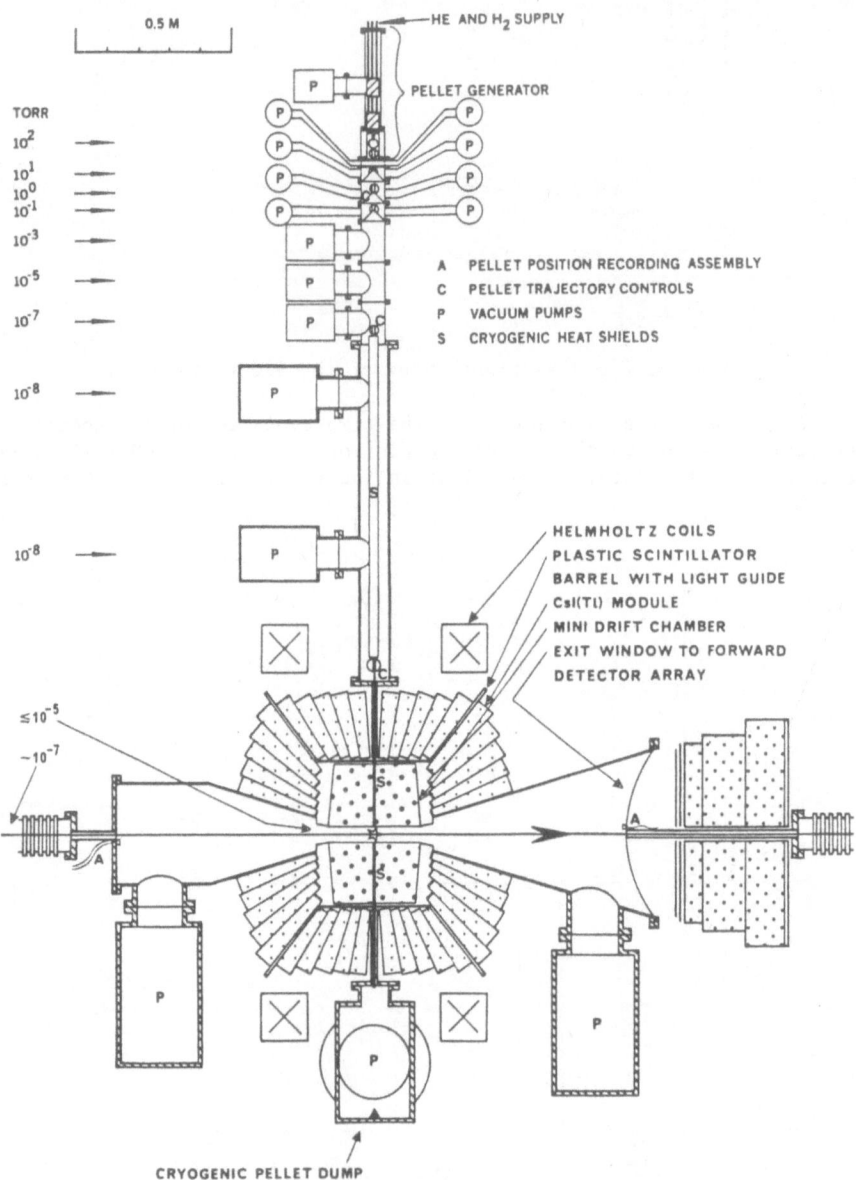

0.5 M

HE AND H₂ SUPPLY

P

PELLET GENERATOR

TORR
10²
10¹
10⁰
10⁻¹

10⁻³

10⁻⁵

10⁻⁷

10⁻⁸

10⁻⁸

A   PELLET POSITION RECORDING ASSEMBLY
C   PELLET TRAJECTORY CONTROLS
P   VACUUM PUMPS
S   CRYOGENIC HEAT SHIELDS

HELMHOLTZ COILS
PLASTIC SCINTILLATOR
BARREL WITH LIGHT GUIDE
CsI(Tl) MODULE
MINI DRIFT CHAMBER
EXIT WINDOW TO FORWARD
DETECTOR ARRAY

≤10⁻⁵
~10⁻⁷

CRYOGENIC PELLET DUMP

Fig. 8. The proposed WASA detector system

fig. 6. The design is based on the SPS gas-jet target at CERN. For hydrogen the design intensity is 10^{14} atoms/cm^2, equivalent to a luminosity of 2×10^{30} cm^{-2}s^{-1} for 10^{10} protons at 200 MeV. For heavier targets the target thickness is reduced by a factor of 10. In the ion beam crossing area the gas-jet is 7 mm wide and 10 mm long. Using a cooled beam the intersection volume will reduce since the ion-beam diameter then is around 2 mm.

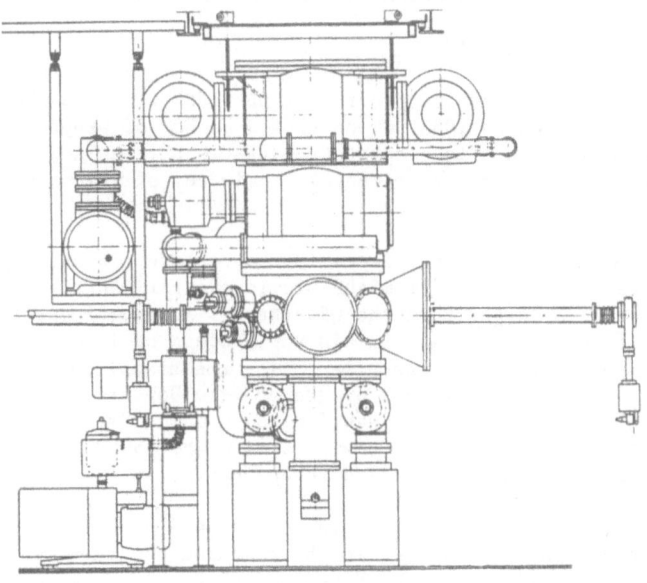

Fig. 6. Cross-sectional view of the gas-jet target. The ion beam will enter from the left

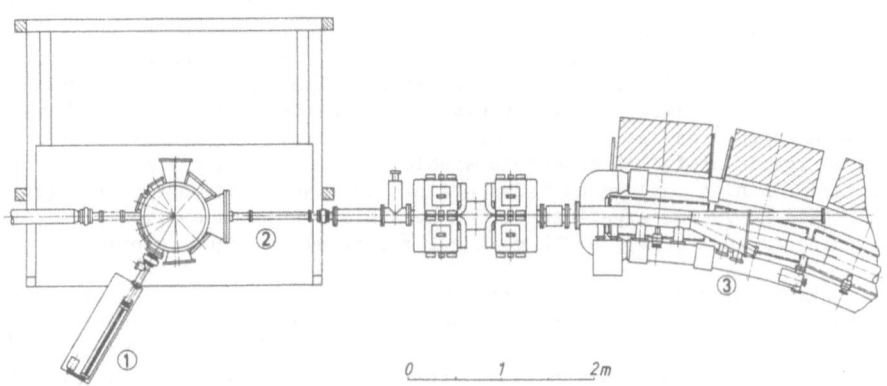

Fig. 7. Fiber target arrangement, 1, layout of the area around the target, 2, and detector ports for small angle scattering, 3.

407

Using a 5 MeV p beam, our target development group has performed test experiments with thin fibers of C, Mo and W, having thickness 7, 13 and $4\mu m$, respectively. The result is, after scaling up the energy, that the fibers can withstand up to 1.3, 0.35 and 0.9×10^9 200 MeV protons in the ring. Luminosities of the order of $10^{32} cm^{-2} s^{-1}$ are obtained with the fiber targets. A fiber target holder and vacuum lock has been built. It can be mounted in the gas-jet experimental chamber (see fig. 7), or in a smaller experiment chamber of its own designed for the first test experiments. The group is also performing a design study of a frozen pellet target for $10^{32} cm^{-2} s^{-1}$ luminosity hydrogen and deuterium target experiments, (see fig. 8).

CELSIUS EXPERIMENTAL PROGRAM

A workshop on the physics program of CELSIUS[9] was held shortly after the project was funded in 1983. Since then a number of proposals and letters of intent for experiments in elementary particle physics and nuclear physics has been submitted to the experimental committee, of which some will be mentioned here.

The rare $\pi^0 \to e^+ + e^-$ decay is estimated to have a branching ratio of only 10^{-7}. The 65(22) events from the previous experiments give a branching ratio a factor of 3 larger than the QED theory. An experiment is proposed to find out if this theory can explain the process or if more exotic decay modes such as a pseudoscalar boson mediating the transition from annihilation to creation. In the WASA proposal by groups from Uppsala, Stockholm, Warsaw and Osaka a 4π -detector system (see fig. 8) combined with a pellet target is proposed for the reaction $p + p \to p + p + \pi^0$ at 550 MeV. The group is also proposing a search for dibaryon resonances.

Hyperon production near threshold is proposed by a collaboration Bonn-Freiburg-Jülich-Uppsala e.g. in the reaction $p + d \to \Lambda + K^+ + d$, which has a threshold of 1.127 GeV. The Uppsala-Osaka collaboration is proposing a study of two-pion production to study the ABC effect.

The CHIC collaboration (CELSIUS Heavy Ion Collaboration - Bergen-Copenhagen-Gothenburg-Lund-Stockholm-Uppsala) is proposing a series of heavy-ion reaction studies of subthreshold π^+ production and heavy fragment - light particle correlations. The inclusive π^+-production cross section for $^{12}C + ^{12}C$ varies 8 orders of magnitude between 25 and 400 MeV/A. This whole energy range can be covered in a CELSIUS experiment. The experiments involve detection of pions, neutrons, protons and light particles, projectile fragments and specially target fragments (recoils) down to very low energies.

Of the special features a cooler and storage ring like CELSIUS offers, the proposed rare π^0-decay experiment would utilize the windowless target to prevent spurious background and the high luminosity of a pellet target. In the search for dibaryons it is planned to use the possibility of sweeping the beam energy between preset limits to hunt for resonances. The well-defined beam energy is useful in threshold studies. The broad energy range available and the high maximum energies are of importance for most of the proposed light- and heavy-ion experiments. The heavy-ion projects specially utilize the thin gas-jet target, which does not retard slow target recoils.

Taking full advantage of the high momentum resolution is closely linked to the development and funding of a suitable high-resolution magnetic spectrometer. Some experiments of this kind can, however, be performed in angles close to 0° using the quadrupole and the first two magnets in the bend after the experimental section as a spectrometer. See fig. 7.

CONCLUSIONS

The synchrocyclotron has now internal beam and will soon be ready for the first experiments. According to the time schedule the first injection test in CELSIUS will start in Sept.-Oct. 1987 followed by a running-in period where some fiber-target experiments could be performed. The gas-jet target will be installed early 1988 and the electron cooler is expected to be ready towards the end of 1988.

Since the time needed for injection into CELSIUS is of the order of a second it will, at least in principle, be possible to run experiments simultaneously both in the cyclotron experimental area and in CELSIUS.

A more detailed description of the two accelerators is given in a recent article by S. Holm et al.[10]. Forthcoming information on the Uppsala accelerators is available in our newsletter Uppsala Accelerator News[11].

REFERENCES

1. H. Condé, S. Crona, A. Håkansson, O. Jonsson, A. Lindholm, L. Nilsson, P.-U. Renberg, G. Tibell, I. Bergqvist and P. Ekström, The Uppsala(n-p) facility, Workshop on Isovector Excitations in Nuclei, Vancouver Oct. 7-8 1986, to be published.

2. B. Höistad et al., Gustaf Werner Biennial Report 1984/85.

3. B. Herskind, The NORDBALL - a Multidetector System for the Study of Nuclear Structure, Nuclear Physics A447:395c (1985).

4. S. E. Arnell, S. Mattsson, H. A. Roth, M. Rydehell, Ö. Skeppstedt, A. Johnson, J. Nyberg and L. Westerberg, Preequilibrium Reactions with 118 MeV ^{12}C on ^{118}Sn, Physica Scripta 34:484 (1986).

5. M. Bell, J. Chaney, H. Herr, F. Krienen, P. Møller-Petersen, and G. Petrucci, Nucl. Instr. and Meth. 190:237 (1981).

6. G. I. Budker, The 1966 Proc. Int. Symp. Electron and Positron Storage Rings, Saclay, Atomnaya Energya 22:346 (1966).

7. A. Wolf, Elektronenkühlung für niederenergetische Antiprotonen, Kernforschungszentrum Karlsruhe, KFK 4023 (1986).

8. C. Ekström, B. Holmquist and H. Sterner, Target developments for CELSIUS, AIP conference Proceedeings 128:106 (1984).

9. Workshop on the physics program at CELSIUS, vol. 1 and 2, Uppsala Nov. 7-9 1983, Ed. B. R. Karlsson and G. Tibell.

10. S. Holm, A. Johansson, S. Kullander and D. Reistad, New Accelerators in Uppsala, Physica Scripta 34:513 (1986).

11. Uppsala Accelerator News, G. Tibell, ed. (Available upon request from the The Svedberg Laboratory).

INDEX